Particle Acceleration and Detection

Particle Acceleration and Detection

springer.com

The series *Particle Acceleration and Detection* is devoted tomonograph texts dealing with all aspects of particle acceleration and detection research and advanced teaching. The scope also includes topics such as beam physics and instrumentation as well as applications. Presentations should strongly emphasize the underlying physical and engineering sciences. Of particular interest are

- contributions which relate fundamental research to new applications beyond the immediate realm of the original field of research
- contributions which connect fundamental research in the aforementioned fields to fundamental research in related physical or engineering sciences
- concise accounts of newly emerging important topics that are embedded in a broader framework in order to provide quick but readable access of very new material to a larger audience

The books forming this collection will be of importance for graduate students and active researchers alike.

For further volumes:
http://www.springer.com/series/5267

Valery Lebedev • Vladimir Shiltsev
Editors

Accelerator Physics at the Tevatron Collider

Springer

Editors
Valery Lebedev
Fermi National Accelerator Laboratory
Accelerator Division
Batavia, IL
USA

Vladimir Shiltsev
Fermi National Accelerator Laboratory
Accelerator Physics Center
Batavia, IL
USA

ISSN 1611-1052
ISBN 978-1-4939-0884-4 ISBN 978-1-4939-0885-1 (eBook)
DOI 10.1007/978-1-4939-0885-1
Springer New York Heidelberg Dordrecht London

Library of Congress Control Number: 2014939030

Printed on acid-free paper

Springer is part of Springer Science+Business Media (www.springer.com)

We dedicate this book to Fermilab staff, who made possible the Tevatron—the particle accelerator which shaped the world high energy physics for more than quarter of century.

Preface

The intent of this book is to present major advances in accelerator physics and technology implemented at the Tevatron proton–antiproton collider at the Fermi National Accelerator Laboratory in Batavia, IL, USA, during its quarter of century long quest for better and better performance. The collider was arguably one of the most complex research instruments ever to reach the operation stage and is widely recognized for many technological breakthroughs and numerous physics discoveries. In this book we have tried to coherently describe the contributions to the physics of colliding beams made at the Tevatron. Both theoretical and experimental works are presented in uniform fashion. Throughout the text, we use the same symbol definitions and provide references which are readily available for the reader. For example, all the references to the proceedings of the International, European and IEEE Particle Accelerator Conference series (PACs) can be found at the JACOW website http://accelconf.web.cern.ch/accelconf/. All cited Fermilab technical publications are available at *inSPIRES* http://inspirehep.net/.

In Chap. 1 we outline the basics of the colliding beams technique and brief history of the Tevatron, describe the Fermilab accelerator complex, and overview the collider luminosity progress. Other chapters are devoted to special topics, such as beam optics methods used in the Tevatron accelerators (Chap. 2), accelerator magnets and magnetic field effects on beam dynamics (Chap. 3), novel longitudinal beam manipulation methods widely used at the Tevatron (Chap. 4), high intensity beam issues and instabilities (Chap. 5), beam emittance growth and halo collimation (Chap. 6), production and cooling of the antiprotons (Chap. 7), the beam–beam effects (Chap. 8), and beam instrumentation (Chap. 9).

Batavia, IL

Valery A. Lebedev
Vladimir D. Shiltsev

Preface

Acknowledgements

We are most grateful to many of our colleagues who have gratuitously contributed to the chapters presented in this book: Jerry Annala, Chandra Bhat, Alexei Burov, Richard Carrigan, David Harding, Stephen Holmes, Andreas Jansson, Nikolai Mokhov, Ronald Moore, Sergei Nagaitsev, Ralph Pasquinelli, John Peoples, Lionel Prost, Kiyomi Seiya, Alexander Shemyakin, Giulio Stancari, James Steimel, Dean Still, Michael Syphers, Cheng-Yan Tan, Alexander Valishev, James Volk.

We would especially like to thank Frank Zimmermann for his suggestion to write this book and encouragement to publish it in the Springer "Particle Acceleration and Detection" series, and our editorial team from Springer Verlag for their patience and valuable help.

Symbols

Symbol(s)	Meaning	Units
L	Luminosity per IP	$cm^{-2}\ s^{-1}$
$I_L = \int L\ dt$	Integrated luminosity	pb^{-1}/week, fb^{-1}
$E_{p,a}$	Proton(antiproton) beam energy	GeV
$C, R = C/2\pi$	Ring circumference, radius	m
$c = 2.9979 \times 10^8$ m/s, v $\beta = v/c;\ \gamma_{p,a} = (1-\beta^2_{p,a})^{-1/2}$	Speed of light, velocity, relativistic factors (protons, antiprotons)	
$f_0 = C/v, f_{RF}, h = f_{RF}/f_0$	Revolution frequency, RF frequency, harmonics number	MHz
x, y, z, s	Horizontal, vertical, and longitudinal displacements, longitudinal coordinate	
$Q_{x,y,s}$	Horizontal, vertical, synchrotron tune	
N_b	Number of bunches	
t_b	Bunch spacing	ns
$N_{p,a}$	Protons (antiprotons)/bunch	10^9
$\varepsilon_{(p,a)\ (x,y,L)}$	RMS normalized emittance (proton, antiproton), (horizontal, vertical, longitudinal)	π μm, eV s
$\sigma_{(p,a)(x,y,z)}$	RMS beam size (proton, antiproton), (horizontal, vertical, longitudinal)	μm, m
$\sigma_{E,\delta}$	RMS energy spread, relative energy spread	
$\beta_{x,y}, \alpha_{x,y}, D_{x,y}, \beta^*_{x,y}$	Beta- and alpha-beam optics functions (horizontal, vertical), dispersion, beta-function at IP	m, cm
$x,y,z = (2J_{x,y,z}\beta_{x,y,z})^{1/2} cos(\psi_{x,y,z})$	Coordinates, actions, variables, and phases	
V_{RF}	RF voltage amplitude	MV

(continued)

Symbol(s)	Meaning	Units
$Z_{\parallel}, Z_{\perp}, Z_0$	Longitudinal, transverse impedance, free space impedance	Ohm, Ohm/m, 377 Ω
$H(\sigma_z, \beta^*, \ldots)$	Hour-glass factor	
$e = 2.71828 \ldots$		
$e = 1.602 \times 10^{-19}$ C	Electron charge	
$m_{\mathrm{p}} = 938.27$ MeV/c^2	Proton mass	
$m_{\mathrm{e}} = 511$ keV/c^2	Electron mass	
$r_p = e^2/m_pc^2 = 1.535 \times 10^{-18}$ m	Proton classical radius	
$r_e = e^2/m_ec^2 = 2.818 \times 10^{-15}$ m	Electron classical radius	

Abbreviations

AA	Antiproton Accumulator at Fermilab
ANL	Argonne National Laboratory, USA
BNL	Brookhaven National Laboratory, USA
BPM	Beam position monitor
CERN	European Organization for Nuclear Research, Switzerland
DESY	Deutsches Elektronen-Synchrotron Laboratory, Germany
FNAL	Fermi National Accelerator Laboratory, USA
Fermilab	Fermi National Accelerator Laboratory, USA
HERA	Hadron-Elektron Ring-Anlage at DESY
IP	Interaction point
ISR	Intersection Storage Ring at CERN
LHC	Large Hadron Collider at CERN
Linac	Linear accelerator
MI	Main Injector synchrotron at FNAL
Quad	Quadrupole magnet
RR	Recycler Ring at FNAL
Sp(p)S	Super Proton (antiproton) Synchrotron at CERN
SSC	Superconducting Super Collider, USA
Tevatron	TeV proton antiproton collider at Fermilab

Parameters of Fermilab Accelerators

		BS	MI	RR	DB	AA	TeV	
Particles		*p*	*p*, *pbar*	*pbar*	*pbar*	*pbar*	*p*, *pbar*	
Circumference	C	474	3,319	3,319	505	474	6,283	m
Injection energy (kinetic)	E_{inj}	0.4	8	8	8	8	150	GeV
Peak energy (kinetic)	E	8	150	8	8	8	980	GeV
Cycle time		1/15	2.2	-	2.2	-	ramp 84	s
Harmonic number	h	84	588	-	90	84	1,113	
Transition gamma	γ_t	5.5	21.6	20.7	7.7	6.2	18.6	
Maximum RF voltage	V_{RF}	0.75	4.0	0.002	5.1	0.04	1.4	MV
β_{max} in cells		34/20 (h/v)	57	55	16	52/40	100	m
β^* at collision points		–	–	–	–	–	0.28	m
Maximum dispersion	D_x	3.2	2.2	2	2.1	9	8	m
Tune (approx.)	$Q_{x,y}$	6.7	22.42	25.45/ 24.46	9.76/ 9.78	6.68/ 8.68	20.59	
Bend magnet length		2.9	6.1/4.1	4.3/2.8	1.6	1.5/3/4.6	6.1	m
Half-cell length		19.76	17.3	17.3	4.4		29.7	m
Bend magnets/ cell		4	4	4	2		8	
Bend magnets total		96	300	344	66	24	774	
Phase advance per cell		96	90	79/87 (v/h)	60		68	°
Cell type		FOFDOOD	FODO	FODO	FODO		FODO	

BS booster, *MI* main injector, *DB* debuncher, *AA* antiproton accelerator, *RR* recycler, *TeV* tevatron, *p* protons, *pbar* antiprotons

Parameters of Fermilab Accelerators

Contents

Chapter 1
Introduction

S. Holmes, R. Moore, J. Peoples, and V. Shiltsev

1.1 Colliding Beams and the Tevatron Collider

Particle accelerators have been widely used for research in physics since the early twentieth century and greatly progressed both scientifically and technologically since then. To get an insight into the physics of elementary particles, one accelerates them to very high kinetic energy, let them impact on other particles, and detect products of the reactions that transform the particles into other particles. The center of mass energy E_{cm} for a head-on collision of two particles with masses m_1 and m_2 and energies E_1 and E_2 is

$$E_{\mathrm{cm}} = \left[2E_1E_2 + \left(m_1^2 + m_2^2\right)c^4 + 2\sqrt{E_1^2 - m_1^2c^4}\sqrt{E_2^2 - m_2^2c^4}\right]^{1/2}. \qquad (1.1)$$

For many decades, the only arrangement of accelerator experiments was a fixed target setup when a beam of particles accelerated with a particle accelerator hit a stationary target set onto the path of the beam. In this case, as follows from Eq. (1.1), for high energy accelerators $E \gg mc^2$, the CM energy is $E_{\mathrm{cm}} \approx (2Emc^2)^{1/2}$. For example, 1,000 GeV protons hitting stationary protons can produce reactions with about 43 GeV energy. A more effective colliding beam setup in which two beams of particles are accelerated and directed against each other has much higher center of mass energy of $E_{\mathrm{cm}} \approx 2(E_1E_2)^{1/2}$. In the case of two equal mass of particles (e.g., protons and protons, or protons and antiprotons) colliding with the same energy E of 1,000 GeV, one gets $E_{\mathrm{cm}} = 2E$ or 2,000 GeV. Such an obvious advantage secured the

S. Holmes • R. Moore • J. Peoples
Accelerator Physics Center, Fermi National Accelerator Laboratory, Batavia, IL 60510, USA

V. Shiltsev (✉)
Accelerator Physics Center, Fermi National Accelerator Laboratory, MS221, PO Box 500, Batavia, IL 60510, USA
e-mail: shiltsev@fnal.gov

V. Lebedev and V. Shiltsev (eds.), *Accelerator Physics at the Tevatron Collider*, Particle Acceleration and Detection, DOI 10.1007/978-1-4939-0885-1_1,

place of the colliders as frontier high energy physics machines since 1960s [1]. The first electron–positron colliders were built in the early 1960s almost simultaneously three laboratories: AdA collider at the Frascati laboratory near Rome in Italy, the VEP-1 collider in the Novosibirsk Institute of Nuclear Physics (USSR), and the Princeton–Stanford Colliding Beam Experiment at Stanford (USA). Their center of mass energies were 1 GeV or less. Construction of the first hadron (proton–proton) collider the Intersecting Storage Rings began at CERN (Switzerland) in 1966, and in 1971 this collider was operational and eventually reached $E_{\mathrm{cm}} = 63$ GeV. In the case of electrons and positrons the synchrotron radiation results in fast damping of betatron and synchrotron oscillations and creates an effective way to accumulate large currents. The synchrotron radiation has negligible effect on damping of particle oscillations in proton (hadron) colliders and, therefore, a construction of proton–antiproton colliders required damping based on a mechanism other than the synchrotron radiation. The invention of stochastic cooling technique in 1969 led to the construction of the first proton–antiproton collider Sp$\bar{\mathrm{p}}$S in CERN in 1982. Many electron–positron and hadron colliders were built over the past half a century—see, e.g., [2]. The Tevatron proton–antiproton collider [3] was the world's highest energy collider for almost 25 years since it began its operation in December 1985 until it has been overtaken by the LHC in 2009.

The aim of the Tevatron collider was to explore the elementary particle physics phenomena with center of mass collision energies of up to 1.96 TeV. The number of events per second generated in the Tevatron collisions is given by:

$$dN_{\mathrm{event}}/dt = L \cdot \sigma_{\mathrm{event}}, \tag{1.2}$$

where σ_{event} is the cross section for the event under study, and L is the machine luminosity. The machine luminosity depends only on the beam parameters and for a Gaussian beam distribution is:

$$L = \gamma f_0 \frac{N_{\mathrm{b}} N_{\mathrm{a}} N_{\mathrm{p}}}{4\pi\beta^* \varepsilon} H(\sigma_s/\beta^*), \tag{1.3}$$

where $N_{\mathrm{p,a}}$ is the number of particles (protons or antiprotons) per bunch, N_{b} the number of bunches per beam, ε is the average rms normalized emittances of two round beams $(\varepsilon_{\mathrm{a}} + \varepsilon_{\mathrm{p}})/2$, $H(x)$ is the geometric luminosity reduction factor ("hourglass factor") which depends on the ratio of the rms bunch length σ_s and beta-function at the collision point β^*, γ is the relativistic factor, and f_0 is the revolution frequency. The exploration of rare events in the Tevatron collisions therefore requires not only high beam energy but also a high beam brightness. The initial design luminosity of the Tevatron was $10^{30}\ \mathrm{cm}^{-2}\ \mathrm{s}^{-1}$, however due to many upgrades, the accelerator has been able to deliver luminosities up to $L = 4.3 \times 10^{32}\ \mathrm{cm}^{-2}\ \mathrm{s}^{-1}$ to each of two high luminosity experiments, CDF [4] and D0 [5]. It is configured as a particle–antiparticle collider in which both circulating beams share a common vacuum and magnet system. In its latest operation with 36 bunches in each beam and a nominal bunch spacing of 396 ns, the common beam pipe implies 72 collision points per turn for each beam.

Table 1.1 Basic parameters of the Tevatron collider

Colliding particles	Protons and antiprotons
Particle energy	980 GeV
Circumference	2,000 π m ≈ 6,283 m
Magnetic lattice	Alternating gradient focusing, separated function
Focusing order	FODO
Number of main bending magnets	774[a]
Bending magnetic field	0.664 T at injection, 4.33 T at the maximum energy
Bending magnet	Nb–Ti conductor at 4.3 K, cold-bore, iron at ambient temperature
Quadrupole field gradient	11.4 T/m at injection, 74.5 T/m at maximum energy
Betatron tunes	20.583 (*h*), 20.585 (*v*)
Energy ramp time	85 s (in operation, 15 s minimum)
Long straight sections	Number—6, length—50 m (incl. two low-beta insertions)
RF system (tunable)	Eight cavities (4 per beam), 53.1 MHz, 1.2 MeV/turn total maximum voltage for each beam, independent voltage control for each beam
Vacuum chamber	Stainless steel, rounded square 70 mm full aperture (*h*, *v*) in dipoles

[a]This number of dipoles implies that all magnets would have the same field and length. Actually two Tevatron dipoles are replaced by special purpose magnets. That leaves 772 standard 6.12 m long dipoles used for beam bending

The high voltage electrostatic separators excite helical orbits of opposite signs for protons and antiprotons making the beams separated at the undesirable collision points. At injection the beams are separated everywhere. The total relative separation, in the units of the rms beam horizontal and vertical sizes σ_x and σ_y, $\sqrt{x^2/{\sigma_x}^2 + y^2/{\sigma_y}^2}$, is limited by available aperture and stays in the range of 4–12 σ (5–16 mm in the arcs). The beam sizes decrease with beam acceleration and after ~500 GeV the beam separation becomes limited by the strength of the electrostatic separators. It results in that at the end of acceleration the relative beam separation is approximately the same as at the injection in spite of larger ratio between the available aperture and the beam size. At collisions the beams are separated by 5–15 σ (3.5–8.5 mm) everywhere except the two experimental interaction points where beams collide head-on. For computation of the above beam separation the reference rms beam emittance of 3.2 mm-mrad and the reference rms momentum spread of 5.2×10^{-4} at the injection and 1.2×10^{-4} at the top energy were used. The aperture of the vacuum pipe in Tevatron quadrupole and dipole magnets is about 70 mm. The major aperture limitation comes from the electrostatic separators which limit it to 50 mm.

The maximum beam energy that can be reached in the Tevatron is limited by the peak dipole field in the superconducting magnets. The nominal field is 4.33 T, corresponding to the energy of 980 GeV. The maximum field attainable in the ring is some 2 % higher, and such an operational margin is needed to tolerate for the heat load and temperature raises due to the beam losses. The operation at the top energy is extremely sensitive to the beam losses and they have to be minimized. The basic Tevatron parameters are given in Table 1.1.

There are several important performance limitations of the collider. Production of antiprotons is the most challenging one, as it requires operation of sophisticated antiproton complex consisting of the antiproton target area and three 8 GeV antiproton synchrotrons: Accumulator, Debuncher, and Recycler in which stochastic and electron cooling systems are utilized for collection, beam size reduction, and accumulation. The complex has set all the world's records in the antimatter production, including the most antiprotons accumulated in 1 hour: 28.5×10^{10}, in 1 week: 4.1×10^{13}, and the total number of antiprotons produced—some 10^{16}, or 17 ng (that is more than 95 % of the world's total number of antiprotons ever produced).

The maximum particle density per bunch is limited by the nonlinear beam–beam interaction that each particle experiences when bunches of both beams collide with each other. The strength of beam–beam interaction is measured by the linear tune shift, which for round beams is given by:

$$\Delta Q_{\mathrm{p,a}} = N_{\mathrm{IP}}\xi_{\mathrm{p,a}} = N_{\mathrm{IP}}\frac{N_{\mathrm{a,p}}}{4\pi\varepsilon_{\mathrm{a,p}}}, \tag{1.4}$$

where $r_{\mathrm{p}} = e^2/(m_{\mathrm{p}}c^2)$ is the classical proton radius. The Tevatron experience is that the total linear tune shift summed over its two main interaction points (IPs) should not exceed 0.025–0.03 for operation with acceptable performance loss. This implies that the linear beam–beam tune shift for each IP should satisfy $\xi < 0.012$–0.015. The collisions in other interaction points also strongly affect the strength of beam–beam effects (see details in Chap. 8).

Magnetic field quality errors compromise the particle stability in the storage ring, and hence a loss-free operation requires a high field quality of few 10^{-4} of the main field at a reference radius of 1 in. The Tevatron was the first large accelerator operating with superconducting magnets and the first to deal with decay of persistent currents and their "snap back" at the beginning of the ramp resulting in some 5 G field errors. Achieving small beam losses therefore required a tight control of the magnetic field errors during magnet production and careful compensation during machine operation.

At nominal operation a total beam current of two beams is about 0.10 A. It corresponds to a stored beam energy of about 2.1 MJ. In addition to the energy stored in the circulating beams, the Tevatron magnet system has a stored electromagnetic energy of approximately 400 MJ. Both the beam and magnetic stored energies must be absorbed safely at the end of each run or in the case of a malfunction or an emergency. That has been taken care of with the beam dumping system and the magnet quench-protection and energy dump systems.

The luminosity in the Tevatron is not constant over a physics run (also called "HEP store"), but decays due to the degradation of intensities and emittances of the circulating beams. The main causes of the luminosity decay during nominal Tevatron operation are beam size (emittance) increase due to intra-beam scattering (IBS), the beam loss from collisions and beam–beam effects, reduction of the hourglass factor due to bunch lengthening, in turn caused by the intra-beam scattering. The emittance blow-up can also be caused by the scattering of particles on residual gas, the nonlinear force of the beam–beam interaction, RF noise, and other external noises. The initial decay time of the Tevatron luminosity was 10–15 h in the early years of relatively low-luminosity operation, while at the ultimate high luminosities in excess of 3×10^{32} cm^{-2} s^{-1} the initial decay time is about 5–6 h.

Filling the Tevatron requires special magnet preparation ramps and multiple injections of protons and antiprotons. Together with ramping up and down time and several minutes needed for low-beta squeeze and beam halo scraping, the average turnaround time for the collider is of the order of 70–90 min, though it took many years to optimize the whole collider accelerator complex operation to achieve that.

The collider has been typically operated for about 40 weeks/year with some 100–120 h of colliding store time a week. At the end the Tevatron Run II it has been delivering about 2.4 fb^{-1} of integrated luminosity a year with record performance of 0.073 fb^{-1}/week and 0.273 fb^{-1}/month. A peak luminosity over 4×10^{32} cm^{-2} s^{-1}—that is 400 times its original design value—has been reached by continuous upgrade of all the systems, introducing advanced accelerator technologies and streamlining the operational procedures. The Tevatron has delivered an integrated luminosity of about 12 fb^{-1} from its start to end of 2011.

Highlights from the Tevatron collider high energy physics research can be found in [6] and include the discovery of the top quark in 1995, the observation of B-mesons oscillations in 2006, and the indications in 2010 that B-meson decays produce slightly more muons than antimuons, which may help explain the dominance of matter over antimatter, and, in 2011, a strong evidence for a never-before-seen particle with some 140 GeV mass that does not fit into any of our established theories. In addition, a large range of mass was excluded for the elusive Higgs boson.

1.2 History of Fermilab and Its Accelerators

Fermi National Accelerator Laboratory (Fermilab), located just outside Batavia near Chicago, Illinois, is a US Department of Energy (DOE) national laboratory specializing in high energy particle physics. The moment of creation of Fermilab was December 16, 1966, when the US Atomic Energy Commission (AEC) announced that it had selected the 6,800 acre Weston, Illinois site for the 200 GeV project. Very soon, thereafter, "members of Universities Research Association, Inc. (URA) and the AEC met to discuss arrangements for the new laboratory" and in January 1967, first, the AEC awarded URA a temporary contract to prepare for the construction of the 200 GeV project, and then, URA selected Robert R. Wilson as the Director of the new laboratory (Fig. 1.1).

When Wilson accepted the Directorship he promised the AEC and the Joint Committee on Atomic Energy (JCAE) to reach the highest energy that the project funding of about $250 million would enable. He also insisted that URA name the Laboratory the National Accelerator Laboratory (NAL) so that it would be clear that it would have the national character that the physics community wanted. It was renamed in honor of Enrico Fermi in 1974.

Fig. 1.1 Robert Rathbun Wilson (courtesy of Fermilab Video Media Services)

1.2.1 Building the 400 GeV Main Ring

Starting with the well-developed Berkeley design the NAL team explored changes which would reduce the cost and time to completion while creating the ability to significantly exceed 200 GeV. The essential elements of the design were a 750 kV Cockcroft–Walton Pre-Accelerator, which delivered protons to a 200 MeV proton Linac, which in turn delivered protons to a 15 Hz, 8 GeV rapid cycling strong focusing synchrotron, and the 8 GeV protons were injected into the Main Ring, the multi-hundred GeV proton synchrotron, and accelerated to the peak energy and then slowly extracted and transported to the external experimental areas. The radius of the Main Ring was set to 1 km and the length of each external beam transport line from the Main Ring to the experimental areas was made long enough to increase the energy at a later date. The team selected the separated function dipoles and quadrupoles for the magnet lattice as opposed to the more traditional alternating gradient dipoles. All of these choices opened up the possibility of reaching 400 GeV or even 500 GeV. While Wilson and his team briefly considered the use of superconducting magnets as a way of reaching an even higher energy, they concluded that superconducting magnet technology was not sufficiently advanced at the time to use it in the project. Nevertheless, space was left for such an accelerator above and below the Main Ring.

Ground breaking for the Linac Gallery was held in December 1968. The Linac group purchased a complete Cockcroft Walton from Switzerland, adopted the Brookhaven 200 MeV Linac design, which it was building, capitalized on their work to build the first tank of the drift tube Linac, borrowed parts from other laboratories and from the pieces assembled a system in temporary building in the village. The first 10 MeV beam from the ion source was realized in June 1969 in the village, although the whole system had to be taken apart and reinstalled in the Linac Gallery when it was finished less than 6 months later. The traditional sequence of designing prototyping components and then building production components was dropped in favor of building the Linac and putting it into service as the injector for the next machine in the cascade. The faster pace and the lack of a prototype program saved money. The 8 GeV Booster followed a similar hectic pace. As soon as the Booster enclosures were available installation of components began and by January 1971 all of the magnets were in place and Booster commissioning was underway. The construction of the Main Ring followed the same template and the last Main Ring magnet was installed in April 1971. Once it was possible to accelerate beams in one of the machines it was done. Commissioning the "air cooled" Main Ring with protons from the Booster began in the summer of 1971 before the low conductivity water system was complete. Main Ring magnets began to fail at an alarming rate and for a while threatened the continuation of the project; ultimately 350 magnets out of 1,014 magnets failed and were repaired. These problems were sufficiently under control to allow protons to be stored and then accelerated in the Main Ring at the beginning of 1972. Protons were accelerated to 200 GeV on March 1, 1972, just 40 months after ground breaking. There was still enormous amount of work remaining to complete accelerator complex, including completing the Main Ring low conductivity water system, and the conventional construction of the experimental areas had barely begun. Nevertheless, by July 1972, 200 GeV protons were transported through the neutrino area to the 30 in. bubble chamber, which had been acquired from neighbor Argonne National Laboratory, and the first collisions with 200 GeV protons at NAL were photographed.

A year later the Meson, Neutrino, and Proton experimental areas were just complete enough to start the initial experiments, albeit under very primitive conditions. Preliminary results of these experiments were presented to the 1973 summer conferences. By May 1974, less than 6 years after ground breaking, the entire complex was in full operation and it was dedicated as the Fermi National Accelerator Laboratory. Moreover, the Main Ring had accelerated protons to 400 GeV and delivered them to the Neutrino Area and the Proton Area for experiments. However the Main Ring magnets continued to develop ground faults at the rate of several per month as the energy was increased thereby reducing the available beam time for experiments.

1.2.2 The 400 GeV Experimental Program

By 1975 the Main Ring and all the experimental areas were routinely operating at 400 GeV. The effort to reach the highest energy that the project budget could afford, paid off handsomely in 1977 when Leon Lederman's group discovered the Upsilon the Υ(1s) and its first excited state the Υ(2s) and had strong evidence for a second excited state Υ(3s). These results clinched the evidence for the *b* quark and the third generation of quark states. It is doubtful that this discovery could have been made if the proton energy was only 200 GeV.

In 1976 Carlo Rubbia, David Cline, and Peter McIntyre proposed that the *W* and *Z* vector bosons could be discovered by operating the CERN SPS or the Fermilab Main Ring as a proton–antiproton collider if a luminosity of 10^{29} cm^{-2} s^{-1} could be achieved. They further proposed implementation schemes to produce proton–antiproton (often referred as p–pbar) collisions at a center of mass energy of about 500 GeV both at CERN and at Fermilab [7]. Each scheme required a new facility for the accumulation and cooling of antiprotons that could produce and cool approximately 10^{11} antiprotons per day. The CERN Council approved the antiproton accumulation facility and the conversion of the SPS to a pbar–p collider in late 1977. The commissioning of the CERN AA (Antiproton Accumulator) Ring with protons from the PS and antiprotons produced by the PS was underway by mid-1980. Collider operations commenced in 1981 and the discovery of the *W* and *Z* bosons was announced in early 1983. Initially the SPS supported collisions at 540 GeV (center-of-mass) and then later at 630 GeV. A quick path to the discovery of the *W* and *Z* with the Fermilab Main Ring was far more risky because the Main Ring had to first accelerate and deliver 120 GeV protons to the antiproton production target for collection, and then once collection was done send them back to the Main Ring so that they could collide with protons after accelerating protons and antiprotons to the peak energy that the Main Ring could sustain. The Main Ring magnets could not operate as a storage ring much above 150 GeV and even in 1976 they were still failing at a rate of one or more per month. If the Doubler was used as the collider ring and the Main Ring was used to produce pbars, those risks would be removed but at the cost of ceding the race for the *W* to CERN. In 1978 Fermilab finally decided once and for all that proton–antiproton collisions would be supported in the Tevatron, at roughly 2,000 GeV (center-of-mass).

1.2.3 The Energy Doubler

The Energy Doubler concept conceived as a way of doubling the energy of the Fermilab complex from 500 to 1,000 GeV. At first it was an informal part time effort focused on the development of superconducting magnets and the related cryogenic technology for a synchrotron in the Main Ring enclosure. Prior to 1972 the Laboratory had carried out some studies of superconducting magnets and cryogenic technology, but these efforts were officially set aside when the Main Ring magnet crisis erupted in 1971 with the intention of resuming the work once the Main Ring experimental program was underway. Between 1972 and 1974 a prototype magnet program was launched and this gradually evolved into a full time effort for the people who worked on it.

The original name, the "Energy Doubler," reflected the goal of doubling the energy and dated back to at least 1970. The name was changed by DOE to the Energy Saver/Doubler in 1977 because it highlighted the benefit of reduced power utilization through the use of superconducting magnets [3]. The name Tevatron had been coined in 1975 with the intention of renaming the project "The Tevatron" once it was finished.

Even in 1974 a great deal of development was needed because superconducting magnet technology was still in its infancy. The Doubler Group approached this challenge by rapidly trying many ideas with one foot model magnets, while changing only feature at a time. Ideas were only incorporated into larger magnets built in the prototype magnet production facility when their success was demonstrated in the one foot program. By 1978, in parallel with building accelerator quality prototype magnets, Fermilab began to develop plans to use the Doubler for both fixed target physics in the existing experimental areas and pbar–p collider physics at 2,000 TeV in the pbar–p center of mass system.

While the Doubler Group was making substantial progress on magnets and cryogenic technology, subpanels of HEPAP (High Energy Physics Advisory Panel—the critical advisory committee for the DOE) did not give the Doubler a high priority for new construction. That convinced Wilson that the Doubler would never be finished unless the level of funding for the Doubler increased. When his direct appeal to the Secretary of Energy did not improve the prospects of funding the Energy Doubler Wilson offered his resignation. In February 1978 the URA Board of Trustees accepted his resignation. In August 1978 Leon Lederman was offered the Directorship and he accepted in October 1978. Lederman consolidated the confusing multiplicity of collider proposals into a single scheme and set building and commissioning the Energy Saver/Doubler as the Laboratory's highest priority. The Tevatron I proposal was submitted in 1979 and subsequently funded. Tevatron I included an antiproton source, the accelerator components that were needed to make the Energy Saver/Doubler a pbar–p collider, such as a low-beta magnet insertion for tighter focusing at the B0 location of the Ring (CDF experiment) and Main Ring overpasses at B0 and D0, and other improvements to the accelerator complex that would enable colliding beams in the Tevatron.

1.2.4 Tevatron I: The Fermilab Collider

The antiproton source concept that Fermilab submitted in 1979 to DOE for funding consisted of a target station, beam transport lines, and two storage rings that collected and cooled the antiprotons. This concept used stochastic cooling in the first ring and electron cooling in the second ring. It was based on the experience gained with 200 MeV electron cooling ring that had been started in 1977 and successfully cooled 200 MeV protons from the Linac. This design was reconsidered in 1981 once it was realized that it had a very limited potential for future improvements, and this effort lead to the current scheme of two 8 GeV antiproton storage rings which only employ stochastic cooling. The Tevatron Collider I scheme as established in 1982 (see Fig. 1.2) included: the high energy storage ring—Tevatron at 900 GeV per beam; the antiproton producer—120 GeV beam from the Main Ring (MR) sent to a target; the antiproton source—two 8 GeV rings the Debuncher and

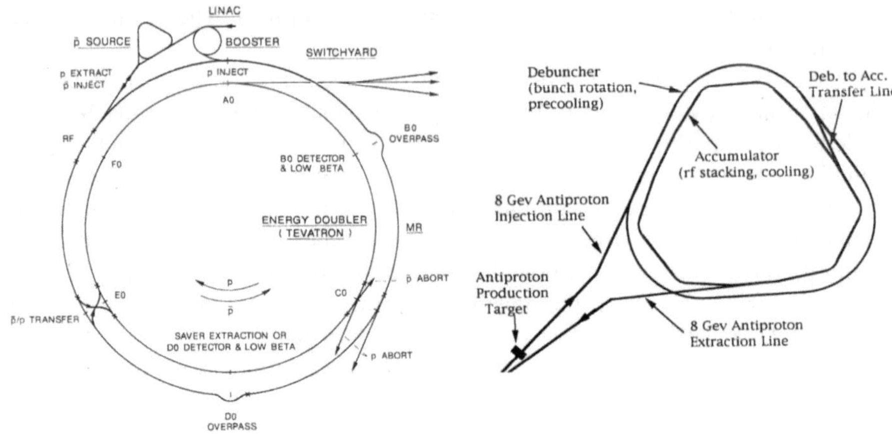

Fig. 1.2 Schematics of the Tevatron I accelerator complex (*left*) and its antiproton source (from [19] and [18])

the Accumulator both employing a stochastic cooling method. The MR would also accelerate antiprotons and protons for injection into the Tevatron. The MR was equipped with two overpasses to provide space for the large detectors located in two crossing points of the Tevatron. The construction of the antiproton source took place in the period 1982–1985. The conventional construction for the Antiproton Source began in mid-1983 and by early 1985 the installation of components in the Antiproton Source enclosures and the target station was underway. Commissioning with protons began that spring and by late summer antiproton cooling and accumulation had been demonstrated, albeit at a rate well below the design goals. The accumulation rate was sufficient to justify a dedicated month long run to testing colliding beams in the Tevatron beginning in September 1985. The first collisions were observed at 1,600 GeV center of mass energy on October 13, 1985, with a partially built CDF detector. Within days accelerator operations in the Main Ring/ Tevatron enclosure were suspended for the next 9 months to build the conventional facilities for the CDF overpass at the B0 location and the collisions hall at D0 location. During this shutdown the accelerator systems needed for colliding beams were added to Main Ring and the Tevatron and the antiproton source was completed. After a recommissioning run that started in late 1986, colliding beams for physics with an essentially complete CDF detector began in January 1987 and continued to the end of May 1987. Copious production of intermediate vector bosons was observed at 1,800 GeV and the results verified that expected large increase in the cross section with center of mass energy relative to production at 640 GeV in the SPS. Energy mattered.

After a technical stop to repair the superconducting magnet current leads and a Tevatron fixed target run in 1988 colliding beams operations for physics were resumed. Between the two collider runs a large number of small modifications were made to improve the efficiency of the beam transfers between machines and to increase the antiproton accumulation rate. The benefits of these improvements were quickly reaped and the original luminosity goal of 10^{30} cm^{-2} s^{-1} was met and exceeded. The luminosity was limited by the beam tune shift created by the six bunches of protons and antiprotons which collided at 12 locations. The R&D that would in time lead very significant increases in the collider luminosity, including reducing the number of beam crossing from 12 to 2, was also set in motion.

1.2.5 The Path to High Luminosity in the Tevatron

From 1988 to 1998, the lab was directed by John Peoples. Major accelerator undertakings during this period included installation of HV separators needed for a helical orbit beam separation scheme and new low-beta insertions at D0 and B0 interaction regions in 1990–1992, the 1992–1993 Collider Run Ia at 1.8 TeV c.o.m. with two simultaneously working detectors CDF and D0, construction and commissioning of a new 400 MeV Linac (1992–1993), the Collider Run Ib (1994–1996) culminating in the top quark discovery, and finally, initiation, construction, and commissioning of a new injector synchrotron, the Main Injector, to increase the average proton beam power on the antiproton target and to eliminate the perturbation of the experiments by the MR overpasses (1993–1999). The Main Injector is a synchrotron of 3.32 km circumference and a minimum repetition time of 1.4 s. It is located in a new tunnel and provides the protons at 120 GeV for antiproton production. It also accelerates the antiprotons and the protons from 8 to 150 GeV for injection into the Tevatron (see following section for more details).

From 1999 until 2005, the lab was run by Michael S. Witherell succeeded by Piermaria Oddone. Since 2007, Fermilab is operated by the Fermi Research Alliance, a joint venture of the University of Chicago and the Universities Research Association (URA). Fermilab's highest priority over that period of time was to explore the energy frontier at the Tevatron final run (called Run II). The initial goal of the collider was set to deliver over several years an integrated luminosity of at least 15 fb^{-1} to each CDF and D0 detector that would be enough for a five-standard-deviation signal of Higgs boson particle heavier than 115 GeV. That goal was factor of 100 higher than what was achieved in the preceding Run I and posed significant challenges. At first, the shutdown between the two Runs lasted 2 years longer than anticipated. And when the collider did finally start running again in March 2001, the Tevatron's performance over the first few years was not fully up to expectations. A number of internal and external reviews appreciated that the Tevatron collider is, by far, the most complex accelerator ever to reach the operation stage and recognized that at the frontier of several technologies, achievement of required beam parameters turned out to be tougher than anticipated. The laboratory responded by setting in motion reorganizations, reanalysis, and revitalized efforts to maximize the total output of the Collider Run II. In 2003, a DOE committee approved the course of actions and noted that "...(the Collider) Run II is not a construction project. Run II is a complex campaign of operations, maintenance, upgrades, R&D and studies."

All that paid off in the end, and the ultimate performance of Run II exceeded expectations by roughly 70 %. With the startup of the higher energy Large Hadron Collider at CERN, the Tevatron was scheduled to be shut off on September 30, 2011. Excellent performance of the machine and detectors, and the LHC startup delays raised hope among particle physicists to extend the life of the Tevatron through 2014—as recommended by the HEPAP—but the budgetary situation did not allow the required additional funding. Additional details on the history of Fermilab accelerators can be found in [8–10].

1.3 Overview of the Fermilab Accelerator Complex

The accelerator complex at Fermilab supports a broad physics program including Tevatron Collider Run II [11], neutrino experiments using 8 and 120 GeV proton beams, as well as a test beam facility and other fixed target experiments using 120 GeV primary proton beams. This section provides a brief description of each of the accelerators in the chain as they operate at the end of the Collider Run II (2011) and an outline of the Collider shot-setup process (cycle of injection, acceleration, low-beta squeeze, and collisions) (Fig. 1.3).

1.3.1 Accelerators

The Proton Source consists of the Pre-Accelerator (Pre-Acc), Linac, and Booster. For operational redundancy, there are two independent 750 kV Pre-Acc systems which provide H^- ions for acceleration through the Linac. Each Pre-Acc is a Cockcroft–Walton accelerator having its own magnetron-based H^- source running at a 15 Hz repetition rate, a 750 kV Haefely voltage multiplier to generate the 750 kV accelerating voltage, and chopper to set the beam pulse length going into the Linac. The typical H^- source output current is 50–60 mA.

The Linac accelerates H^- ions from 750 keV to 400 MeV. Originally, the Linac was a 200 MeV machine made entirely of Alvarez-style drift tube tanks [12], but a 1991 upgrade replaced some of the drift tubes with side-coupled cavities to allow acceleration up to 400 MeV [13]. Today, the low energy section (up to 116 MeV) is made of drift tube tanks operating with 201 MHz RF fed from tetrode-based 5 MW power amplifier tubes. The high energy section (116–400 MeV) consists of seven side-coupled cavities powered by 805 MHz 12 MW klystrons providing a gradient of $\approx$7 MV/m. A transition section between the two Linac sections provides the optics matching and rebunching into the higher frequency RF system. The nominal beam current out of the Linac is 34 mA.

The Booster is a 474 m circumference, rapid cycling synchrotron ramping from 400 MeV to 8 GeV at a 15 Hz repetition rate. (Note that while the magnets ramp at 15 Hz, beam is not present on every cycle.) Multi-turn injection is achieved by passing the incoming H^- ions through 1.5 μm thick (300 μg/cm^2) carbon stripping

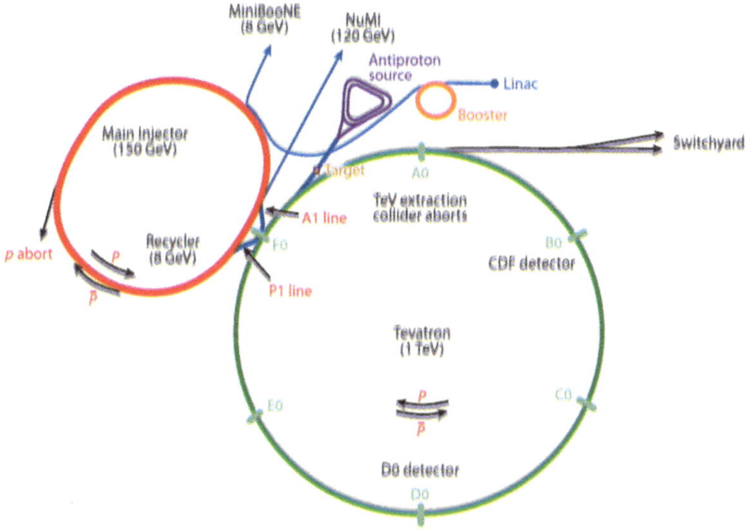

Fig. 1.3 Layout of the Fermilab accelerator complex (from [26])

foils as they merge with the circulating proton beam on a common orbit. The 96 10-ft long combined-function Booster gradient magnets are grouped into 24 identical periods in a FOFDOOD lattice [14]. The Booster RF system (harmonic number = 84) consists of 19 cavities (18 operational + 1 spare) that must sweep from 37.9 to 52.8 MHz as the beam velocity increases during acceleration. The ferrite tuners and power amplifiers are mounted on the cavities in the tunnel. The cavities provide a sum of ≈750 kV for acceleration. The Booster transition energy is 4.2 GeV which occurs at 17 ms in the cycle. Figure 1.4 illustrates Booster efficiency for various beam intensities during an acceleration cycle; the efficiency is 85–90 % for typical beam intensities of $4.5\text{–}5.0 \times 10^{12}$ protons per pulse. Collimators are used to localize as much of the lost beam as possible to reduce the radiation dose absorbed by technicians during maintenance periods. Figure 1.5 shows how Booster throughput has increased remarkably over the Booster operational history. A majority of the proton flux through Booster is delivered to the 8 and 120 GeV neutrino production targets.

The Main Injector [15] (MI) is 3,319.4 m in circumference and can accelerate beam from 8 GeV up to 150 GeV. It has a FODO lattice using conventional, separated function dipole and quadrupole magnets. There are also trim dipole and quadrupoles, skew quadrupole, sextupole, and octupole magnets in the lattice. Since the Main Injector circumference is seven times the Booster circumference, beam from multiple consecutive Booster cycles, called batches, can be injected around the Main Injector. In addition, even higher beam intensity can be accelerated by injecting more than seven Booster batches through the process of slip-stacking: capturing one set of injected proton batches with one RF system, decelerating them slightly, then capturing another set of proton injections with another independent RF system, and merging them prior to acceleration (see Fig. 1.6). There are 18 53 MHz RF cavities grouped

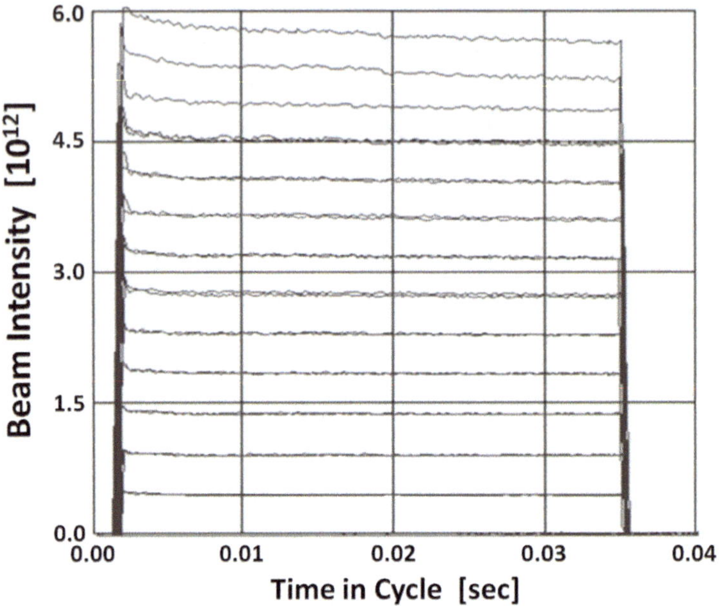

Fig. 1.4 Booster proton intensity for various beam intensities in an acceleration cycle. The injected beam intensity varies from 1 to 13 turns of Linac pulses. The average efficiency is ~90 %

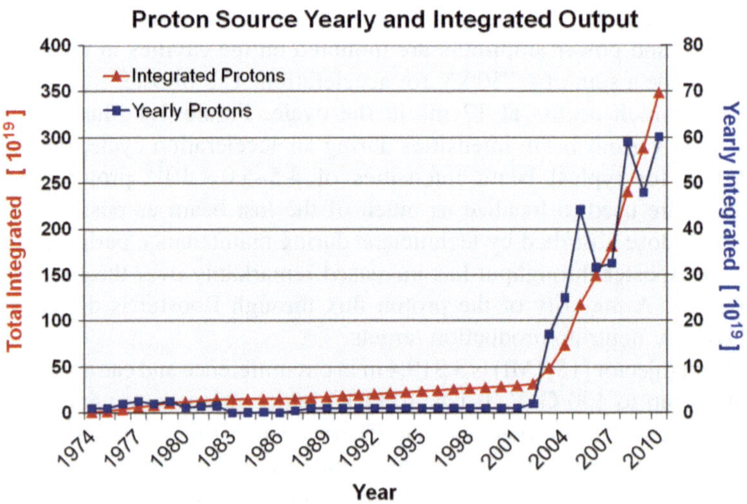

Fig. 1.5 Yearly and integrated proton flux from the Fermilab Booster. The sharp increase in 2003 corresponds to the initiation of the 8 GeV neutrino program

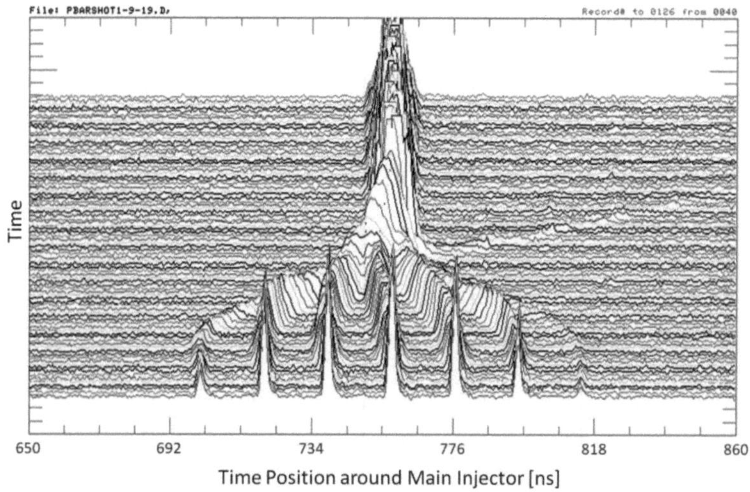

Fig. 1.6 Illustration of the bunch coalescing process in Main Injector

into two independently controlled systems to allow slip-stacking and flexibility when an RF station is switched off for maintenance. Beam-loading compensation and active damping systems have been implemented to help maintain beam stability. For beam injections into the Tevatron, coalescing of several 53 MHz bunches of protons and antiprotons into single, high intensity bunches also requires 2.5 MHz system for bunch rotations and a 106 MHz cavity to flatten the potential when recapturing beam into the single 53 MHz bunch. Like the Booster, a set of collimators was installed in the Main Injector to help localize beam losses to reduce widespread activation of components that technicians need to maintain (Fig. 1.7).

The Main Injector supports various operational modes for delivering beam across the complex. For antiproton and neutrino production, 11 proton batches from Booster are injected and slip-stacked prior to acceleration. After reaching 120 GeV, two batches are extracted to the antiproton production target while the remaining nine batches are extracted to the NuMI neutrino production target—see Fig. 1.5. At peak performance, the Main Injector can sustain 400 kW delivery of 120 GeV proton beam power for 2.2 s cycle times. The Main Injector also provides 120 GeV protons in a 4 s long slow-spill extracted to the Switchyard as primary beam or for production of secondary and tertiary beams for the Meson Test Beam Facility and other fixed target experiments. In addition, the Main Injector serves as an effective transport line for 8 GeV antiprotons being transferred from the Accumulator to the Recycler for later use in the Tevatron. Protons from Booster and antiprotons from Recycler are accelerated up to 150 GeV in the Main Injector and coalesced into higher intensity bunches for injection into the Tevatron for a colliding beam store.

The Recycler [16] is a permanent magnet 8 GeV/c storage ring whose components are hung from the ceiling above the Main Injector. The Recycler is used as an intermediate storage ring for accumulating significantly larger stashes of antiprotons that can be accommodated in the Antiproton Accumulator. The main

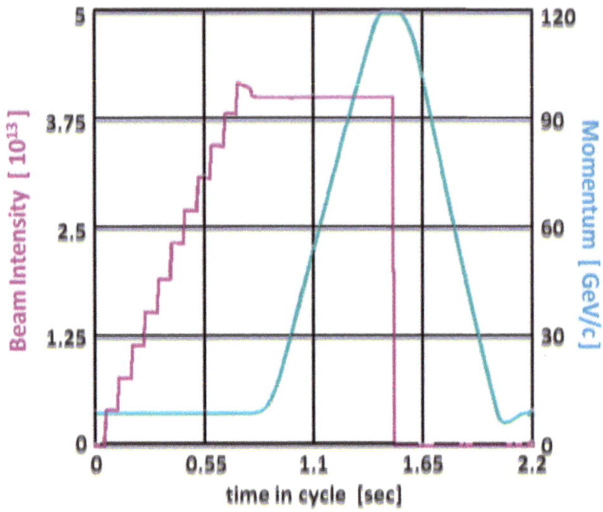

Fig. 1.7 A 120 GeV Main Injector cycle illustrating 11 batch proton injection and acceleration

Recycler magnets are combined-function strontium ferrite permanent magnets arranged in a FODO lattice. Powered trim magnets are used to make orbit and lattice corrections. An important feature of the Recycler is an electron cooling section to augment its stochastic cooling of the antiprotons. A Pelletron (electrostatic accelerator of the Van-der-Graaf type) provides a 4.3 MeV electron beam (up to 500 mA) that overlaps the 8 GeV antiprotons in a 20 m long section and cools the antiprotons both transversely and longitudinally. After becoming operational in September 2005, electron cooling in the Recycler [17] allowed significant improvements in Tevatron luminosity by providing higher intensity antiprotons with smaller emittances. With electron cooling, the Recycler has been able to store up to 600×10^{10} antiprotons. In routine operation, the Recycler accumulates 350–450×10^{10} antiprotons with ~200 h lifetime for injection to the Tevatron.

The Antiproton Source [18] has three main parts: the Target Station, the Debuncher, and the Accumulator (AA). Each of these is described briefly below while outlining the steps of an antiproton production cycle. In the Target Station batches of 120 GeV protons (8×10^{12} per batch) transported from the Main Injector strike one of the Inconel (a nickel–iron alloy) layers of the target every 2.2 s. The beam spot on the target can be controlled by a set of quadrupole magnets. The target is rotated between beam pulses to spread depletion and damage uniformly around the circumference. The shower of secondary particles emanating from the target is focused both horizontally and vertically by a pulsed, high current lithium lens that can provide up 1,000 T/m gradient. Downstream of the Li lens is a pulsed dipole magnet that steers negatively charged particles with 8 GeV/c momentum into the AP2 transport line toward the Debuncher. A collimator between the lens and pulsed magnet was installed to help protect the pulsed magnet from radiation damage as the incoming primary proton beam intensity increased with proton slip-stacking in Main Injector.

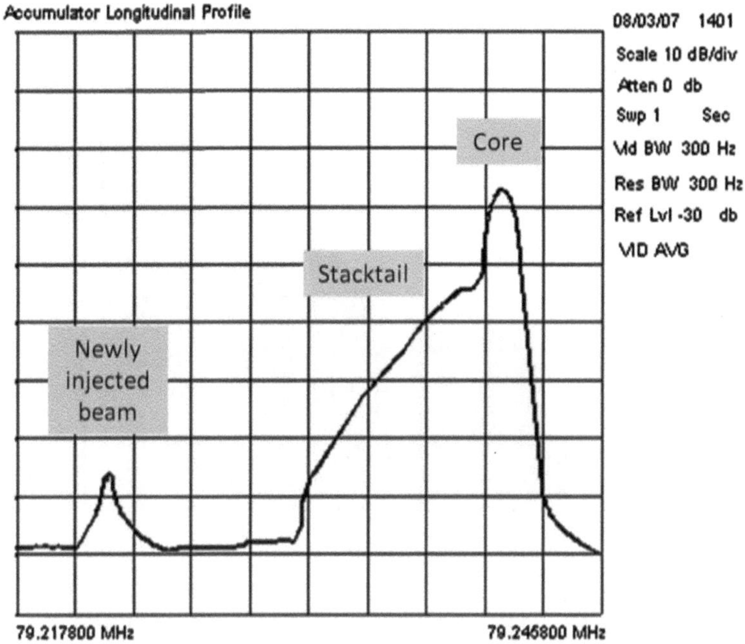

Fig. 1.8 Frequency (energy) distribution of antiprotons in the Accumulator highlighting incoming antiprotons (*left*), the stacktail beam (*middle*) being cooled and decelerated toward the core (*right*). Higher energy is on the *left*, lower energy is on the *right*

The Debuncher and Accumulator are both triangular-shaped rings of conventional magnets sharing the same tunnel. While the Debuncher has a FODO lattice, the Accumulator lattice has particular features needed for cooling and accumulating antiprotons with stochastic cooling systems. The ~2×10^8 antiprotons entering the Debuncher from the AP2 line retain the 53 MHz bunch structure from the primary protons that struck the production target. A 53 MHz RF system (harmonic number = 90) is used for bunch rotation and debunching the antiprotons into a continuous beam with low momentum spread. An independent 2.4 MHz RF system provides a barrier bucket to allow a gap for extraction to the Accumulator. Stochastic cooling systems reduce the transverse emittance from 30 to 3 π mm-mrad (rms, normalized) and momentum spread from 0.30 to <0.14 % prior to injection into the Accumulator.

In the Accumulator, antiprotons are momentum-stacked and cooled by a series of RF manipulations and stochastic cooling. The incoming antiprotons are captured and decelerated 60 MeV by a 53 MHz RF system (harmonic number = 84) to the central orbit where the beam is adiabatically debunched. Before the next pulse of antiprotons arrives (every 2.2 s), the so-called stacktail momentum stochastic cooling system decelerates the antiprotons another 150 MeV toward the core orbit where another set of independent betatron and momentum stochastic cooling systems provides additional cooling while building a "stack" of antiprotons. Figure 1.8 illustrates the frequency (energy) distribution of antiprotons in the Accumulator.

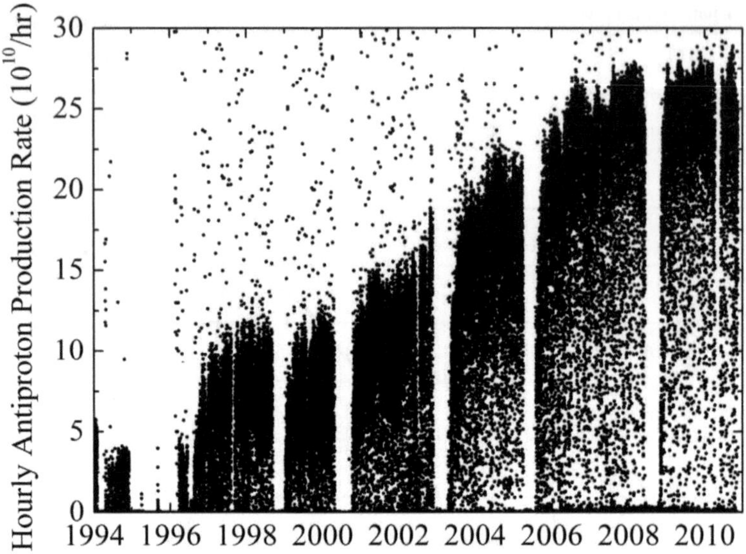

Fig. 1.9 Average antiproton accumulation rate since 1994 and during all of Collider Run II (including production in the Antiproton Source and storage in the Recycler) [20]

Figure 1.9 shows the average antiproton accumulation rates since 1994; typical values for recent Run II operation are in the range 24–26 $\times$ 10^{10} per hour.

The Tevatron proton–antiproton collider is a 1-km radius superconducting magnet synchrotron [19] that accelerates beam from 150 to 980 GeV and provides head-on collisions at two interaction points for the CDF and D0 experimental detectors. The Tevatron magnets, arranged in a FODO lattice, are single-bore, warm-iron wound with Nb–Ti superconductor and cooled to 4.5 K with liquid helium. A common bus powers the main dipoles and quadrupoles. Other magnetic elements, such as dipole and quad correctors, skew quads, normal and skew sextupoles, and octupoles, are located in so-called spool-packages adjacent to the lattice quadrupoles. The interaction points are centered within straight sections with dedicated low-beta quadrupole triplets that can squeeze the beams to a β^* of 28 cm. Since both protons and antiprotons circulate in a single beampipe, electrostatic separators are used to kick the beams onto separate helical orbits. Both beams have 36 bunches—3 trains of 12 bunches with 396 ns spacing (corresponding to 21 buckets of the 53 MHz RF system.) The eight RF cavities in the Tevatron are phased to provide independent control of the protons and antiproton beams. A two-stage collimation system (tungsten primary scatterers and stainless steel secondary absorbers) are used to reduce backgrounds from beam halo in the experiments.

1.3.2 Antiproton Flow and HEP Shot-Setup

As mentioned previously, stacks of freshly produced antiprotons are stored temporarily in the Accumulator. Since the Accumulator's stochastic cooling power is limited, antiproton stacks are periodically transferred to the Recycler where electron cooling allows a much larger antiproton intensity to be accumulated with smaller emittances. Typically 22–25 $\times$ 10^{10} antiprotons are transferred to the Recycler every 60–70 min. Prior to electron cooling in the Recycler, antiprotons destined for the Tevatron were extracted from the Accumulator only or from both the Accumulator and Recycler. Since late 2005, all Tevatron antiprotons were extracted from the Recycler only. Figure 1.10 depicts the flow of antiprotons between the Accumulator, Recycler, and Tevatron over a 1-week period.

A typical collider fill cycle is shown in Fig. 1.11. First, proton bunches are injected one at a time on the central orbit. Then, electrostatic separators are powered to put the protons onto a helical orbit. Antiproton bunches are injected four bunches at a time into gaps between the three proton bunch trains. After each group of three antiproton transfers, the gaps are cleared for the subsequent set of transfers by "cogging" the antiprotons—changing the antiproton RF cavity frequency to let them slip longitudinally relative to the protons. Once the beam loading is complete, the beams are accelerated to the top energy (86 s) and the machine optics is changed to the collision configuration in 25 steps over 125 s (low-beta squeeze). The last two stages include initiating collisions at the two collision points and removing halo by moving in the collimators. The experiments then commence data acquisition for the duration of the high energy physics (HEP) store.

1.4 Collider Runs I and II: Goals, Strategy, and Luminosity Performance

Funding for Tevatron I was initiated in 1981 and the Tevatron was completed, as a fixed target accelerator, in the summer of 1983. The Antiproton Source was completed in 1985 and first collisions were observed using operational elements of the CDF detector (then under construction) in October 1985. First operations of the collider for data taking took place over the period June 1988 to June 1989. Over that period a total of 5 pb^{-1} were delivered to CDF at 1,800 GeV (center-of-mass) and the first western hemisphere W's and Z's were observed. The initial operational goal of 1 $\times$ 10^{30} cm^{-2} s^{-1} luminosity was exceeded during this run. Table 1.2 summarizes the goals established during the Tevatron I design phase and the actual performance achieved.

1.4.1 Performance Limitations

The luminosity achievable in a proton–antiproton collider can be written as:

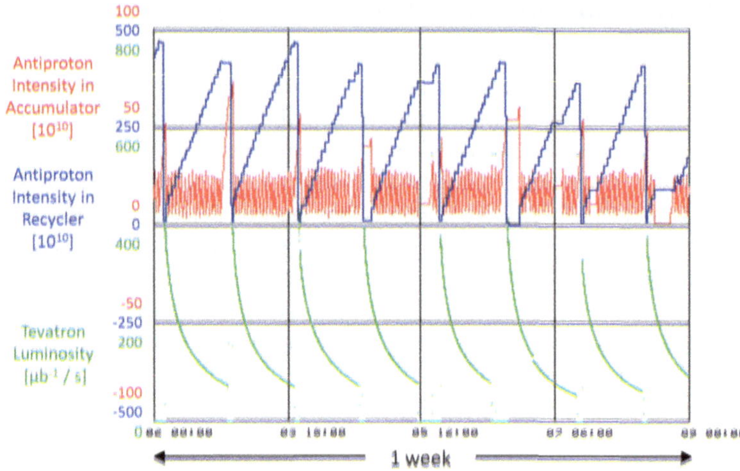

Fig. 1.10 Production and transfers of antiprotons between the Accumulator and Recycler over 1 week of operation. While the Tevatron has a colliding beam store, small stacks of antiprotons are produced and stored in the Accumulator, and then periodically transferred to the Recycler in preparation for the subsequent Tevatron fill

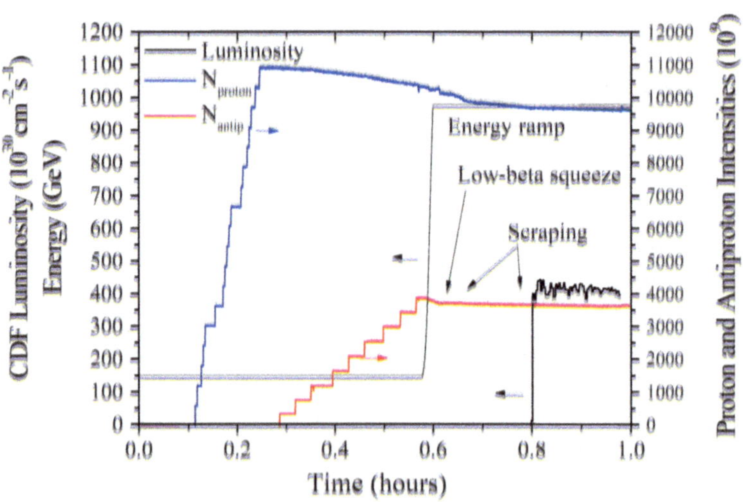

Fig. 1.11 Collider fill cycle for store 8709 (May 2011)

$$L = \frac{f_0 N_b N_p N_a}{2\pi\left(\sigma_p^2 + \sigma_a^2\right)} H\left(\frac{\sigma_z}{\beta^*}\right) = \frac{\gamma f_0 \left(N_p/\varepsilon_{pn}\right)\left(N_b N_a\right)}{2\pi\beta^*\left(1 + \varepsilon_{an}/\varepsilon_{pn}\right)} H\left(\frac{\sigma_z}{\beta^*}\right). \tag{1.5}$$

The luminosity formula is written in this form to highlight the limitations inherent in operations of the Tevatron:

Table 1.2 Design and actually achieved performance parameters for the 1988–1989 Collider Run (typical luminosity at the beginning of a store)

	Tevatron I design	1988–1989 actual
Energy (center-of-mass) (GeV)	1,800	1,800
Protons/bunch N_p	6×10^{10}	7.0×10^{10}
Antiprotons/bunch N_a	6×10^{10}	2.9×10^{10}
Number of bunches N_b	3	6
Total antiprotons $N_b N_a$	18×10^{10}	17×10^{10}
Proton emittance (rms, normalized) ε_{pn} (π mm-mrad)	3.3	4.2
Antiproton emittance (rms, normalized) ε_{an} (π mm-mrad)	3.3	3
IP beta-function β^* (cm)	100	55
Luminosity (cm^{-2} s^{-1})	1×10^{30}	1.6×10^{30}

- N_p/ε_{pn} is the number of protons per bunch divided by the beam emittance. This quantity is directly proportional to the beam–beam tune shift $\xi_a = r_p N_p / 4\pi\varepsilon_{pn}$ experienced by the antiprotons for each head-on encounter with the protons. With six bunch operations there are a total of 12 such encounters per revolution. During the 1988–1989 Run it was observed that the total available tune shift that could be tolerated was determined by the tune space available between resonances below about tenth order. The Tevatron was operated between the 2/5 and 3/7 resonances, allowing a total tune shift (summed over the 12 encounters) of about 0.028. It was observed that the Tevatron could operate in this tune region with no deleterious impacts from the 12th order resonance lying in between the fifth and seventh resonances.
- $N_b N_a$ is the total number of antiprotons in the collider. Two limitations existed: the antiproton accumulation rate, and the ability to cool and store a suitably large number of antiprotons for delivery to the Tevatron. The antiproton accumulation rate is dictated by the rate at which protons can be delivered to the antiproton production target, and by the aperture and stochastic cooling capabilities of the Antiproton Source. The accumulation rate was roughly 2×10^{10} antiprotons per hour during the 1988–1989 Run based on 2×10^{12} protons delivered to the antiproton production target every 2.6 s. The total number of antiprotons that could be delivered to the Tevatron was determined by an interplay between the available Main Ring aperture and the correlation between emittance and antiproton stack size imposed by the stochastic cooling systems in the Antiproton Source. It was determined that for the emittance that could fit through the Main Ring aperture the maximum antiproton stack size was about 6×10^{11}. At this level the Main Ring transmission efficiency for antiprotons was about 60 %.
- $\gamma = E_p/m_p c^2$ is a measure of the beam energy. For all other parameters fixed, the luminosity is proportional to the beam energy, so any increase in beam energy will increase the luminosity. In addition beam energy increases have the added benefit of increasing the cross sections for the production of high mass states.
- The luminosity is inversely proportional to the beta-function at the interaction point, so any decrease in the β^* improves the luminosity. The impact of this is

ultimately limited by the form factor, also called hourglass factor, $H(\sigma_z/\beta^*)$ when the ratio of the rms bunch length to the beta-function σ_z/β^* becomes larger than 1.

1.4.2 Strategy

Based on the above considerations a long-term strategy was developed in the late 1980s for Tevatron upgrades to a luminosity of 5×10^{31} cm^{-2} s^{-1}, a factor of 50 beyond the original goal [21]. The primary elements of the strategy were:

- Electrostatic separators: A total of 20, 3 m long, electrostatic separators operating at up to ±150 kV over a 5 cm gap were installed into the Tevatron. This installation allowed operations with protons and antiprotons traveling on separated helical orbits in the Tevatron. This installation was aimed at mitigating the beam–beam limitations (the head-on collisions would now happen only in the two interaction points located inside the CDF and D0 particle physics detectors) and allowed an increase in the number of bunches (thus keeping the interactions/crossing seen in the detectors under control) as the luminosity increased. As a result of operations with separated orbits the beam–beam effect ceased to be a limitation even as the proton intensity and number of bunches were increased.
- Low-beta systems: The 1988–1989 Run did not have a matched insertion for the interaction region at B0 (where CDF was situated). Two sets of high performance quadrupoles were developed and installed at B0 and D0 (which came online for Run I). These systems ultimately allowed operations with β^* less than 30 cm later at the end of the collider Run II.
- Cryogenic cold compressors: Cryogenic cold compressors were introduced into the Tevatron to lower the operating temperature by about 0.5 K, thereby allowing the beam energy to be increased to 1,000 GeV, in theory. In operational practice 980 GeV was achieved.
- 400 MeV Linac Upgrade: The 200 MeV Linac was upgraded to 400 MeV in order to increase the beam brightness from the 8 GeV Booster through the mitigation of space-charge effects at the injection energy. The total intensity delivered from the Booster increased from roughly 3×10^{12} per pulse to about 5×10^{12}. This resulted in more protons being transmitted to the antiproton production target and, ultimately, more protons in collision in the Tevatron.
- Antiproton Source Improvements: A number of improvements were made to the stochastic cooling systems in the Antiproton Source in order to accommodate the increase antiproton flux generated by continuously increasing numbers of protons on the antiproton production target. Improvements included the introduction of transverse stochastic cooling into the Debuncher and upgrades to the bandwidth of the core cooling system. These improvement supported an accumulate rate of 7×10^{10} antiprotons per hour in concert with the above listed improvements.

The above items constituted the improvements associated with Collider Run I. The luminosity goal of Run I was 1×10^{31} cm^{-2} s^{-1}, a factor of 10 beyond the original Tevatron design goal.

- Main Injector: The Main Injector was designed to significantly improve antiproton performance by replacing the Main Ring with a larger aperture, faster cycling machine [15]. The goal was a factor of 3 increase in the antiproton accumulation rate (to 2×10^{11} per hour), accompanied by the ability to obtain 80 % transmission from the Antiproton Source to the Tevatron from stacks containing as many as 2×10^{12} antiprotons. An antiproton accumulation rate in excess of 2×10^{11} per hour was eventually achieved in the later years of the Collider Run II, and transmission efficiencies beyond 80 % at large stacks were routine.
- Recycler: The Recycler was added to the Main Injector Project midway through the project (utilizing funds generated from an anticipated project under run). As conceived the Recycler would provide storage for very large numbers of antiprotons (up to 6×10^{12}) and would increase the effective stacking rate by recapturing unused antiprotons at the end of collider stores [16]. The Recycler was designed with stochastic cooling systems, but R&D on electron cooling was initiated in anticipation of providing improved performance. Antiproton stacks above 5×10^{12} were ultimately achieved although routine operation was eventually optimized around 4×10^{12} antiprotons. Recycling of antiprotons was never implemented for reasons described below.

The Main Injector and Recycler constituted the improvements associated with Collider Run II [11]. The formally established luminosity goal of Run II was $8 \times 10^{31}\ \mathrm{cm}^{-2}\,\mathrm{s}^{-1}$, a factor of 5 beyond Run I. However, incorporation of the Recycler into the Main Injector Project was projected to provide up to an additional factor of 2.5.

1.4.3 Run I

The Run I improvements were all implemented in the early to mid-1990s and supported operations of Collider Run I over the period from August 1992 through February 1996. Run I consisted of two distinct phases, Run Ia which ended in May 1993, and Run Ib which was initiated in December 1993. The 400 MeV Linac upgrade was implemented between Run Ia and Run Ib. Run I ultimately delivered a total integrated luminosity of 180 pb^{-1} to each of two detectors: CDF and D0. By the end of the run the typical luminosity at the beginning of a store was about $1.6 \times 10^{31}\ \mathrm{cm}^{-2}\,\mathrm{s}^{-1}$, a 60 % increase over the Run I goal.

1.4.4 Run II

The Main Injector and Recycler were completed in the spring of 1999 with the Main Injector initially utilized in the last Tevatron fixed target run. Run II was initiated in March 2001 and continued through 2011. A number of difficulties were experienced in the initial years of operations. They were in large due to underestimate of the amount of work to be done to bring the accelerator complex up after numerous

Table 1.3 Design and actually achieved performance parameters for Collider Runs I and II (typical luminosity at the beginning of a store)

	Run Ia	Run Ib	Run II
Energy (center-of-mass) (GeV)	1,800	1,800	1,960
Protons/bunch	1.2×10^{11}	2.3×10^{11}	2.9×10^{11}
Antiprotons/bunch	3.1×10^{10}	5.5×10^{10}	8.1×10^{10}
Bunches	6	6	36
Total antiprotons	19×10^{10}	33×10^{10}	290×10^{10}
Proton emittance (rms, normalized) (π mm-mrad)	3.3	3.8	3.0
Antiproton emittance (rms, normalized) (π mm-mrad)	2	2.1	1.5
β^* (cm)	35	35	28
Luminosity (cm^{-2} s^{-1})	5.4×10^{30}	16×10^{30}	340×10^{30}
Luminosity (design goal) (cm^{-2} s^{-1})	5×10^{30}	10×10^{30}	200×10^{30}

upgrades and switch to six times higher number of bunches. These were ultimately overcome through a lot of experience accumulated in the course of operation and the organization and execution of a "Run II Upgrade Plan." A critical element in the evolution of Run II was the successful introduction of electron cooling into the Recycler in the summer of 2005. Prior to electron cooling luminosities had approached, but not exceeded, 1×10^{32} cm^{-2} s^{-1}, as had been anticipated with the design of the Main Injector. The success of electron cooling supported typical luminosities well in excess of 3×10^{32} cm^{-2} s^{-1}, with two stores exceeding 4×10^{32} cm^{-2} s^{-1}. At the end of Run II (September 2011) nearly 12 fb^{-1} had been delivered to each detector (CDF and D0).

While the ultimate performance of Run II exceeded expectations by roughly 70 %, the means of achieving this performance differed from the initial plan. In particular, antiproton recycling (the recovery of unspent antiprotons at the end of stores) was never implemented. Difficulties in the removal of protons at 980 GeV in the Tevatron prior to antiproton deceleration proved problematic, and the stunning success of electron cooling and an optimization of store duration time removed the imperative for antiproton recycling.

Also, a plan to implement 132 ns bunch spacing [22], allowing operations with roughly 100 bunches was eventually abandoned. The primary motivation for 132 ns was to limit the number of interactions per crossing in the detectors to roughly two to three as the luminosity increased. However, the utilization of 132 ns would have required the introduction of a crossing angle in the Tevatron and a corresponding reduction in luminosity of roughly a factor of 2. More decisively, the CDF and D0 experiments developed methods for dealing with the higher number of interactions per crossing (up to about eight interactions/crossing) without compromising their performance.

Table 1.3 summarizes the performance achievements of Run I and Run II. Figure 1.12 displays the history of luminosity performance for Runs I and II. Performance in Run II ultimately exceeded the original Tevatron Collider goal by a factor of more than 300.

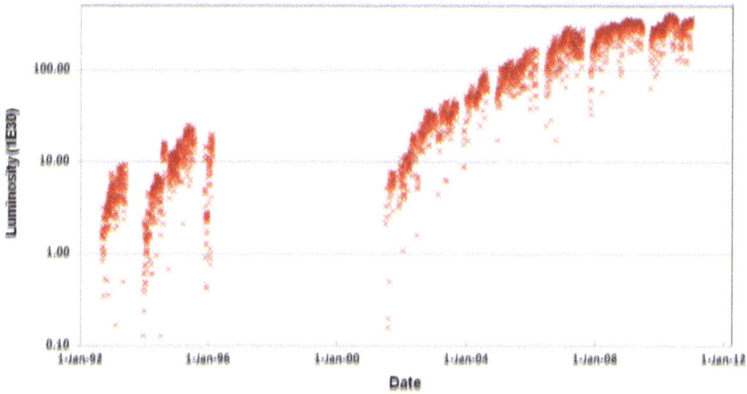

Fig. 1.12 Initial Luminosity for all stores in Collider Runs I and II [26]

1.5 Tevatron Luminosity Evolution Analysis, Collider Complexity

Analysis of the evolution and prediction of high energy colliders' luminosity progress is of great importance for many: it tells machine physicists whether their scientific and technical decisions taken years ago were correct; for the experimental high energy physicists, it is the basis for their schedules and upgrade plans; for the management and funding agencies, it is an important input on the future facilities and projects. The Tevatron luminosity history—see Fig. 1.13—gives several important lessons in that regard. The luminosity increases occurred after numerous improvements, some of which were implemented during operation, and others introduced during regular shutdown periods.

The major improvements, listed in the Table 1.4, took place in all the accelerators of the Collider complex, they were addressing all the parameters affecting the luminosity—proton and antiproton intensities, emittances, optics functions, bunch length, losses, reliability and availability, etc.—and led to fractional increase varying from few % to some 40 % with respect to previously achieved level.

As the result of some 32 improvements in 2001–2011, the peak luminosity has grown by a factor of about 54 from $L_i \approx 8 \times 10^{30}\,\mathrm{cm}^{-2}\,\mathrm{s}^{-1}$ to $L_f \approx 430 \times 10^{30}\,\mathrm{cm}^{-2}\,\mathrm{s}^{-1}$, or about 13 % per step on average. In principle, such complex percentages—"N% gain per step, step after step, with regular periodicity"—should result in the *exponential* growth of the luminosity $L(t+T)/L(t) = \exp(T/C)$. Nevertheless, the pace of the luminosity progress was not always constant. As one can see from Fig. 1.13, the Collider Run II luminosity progress was quite fast with $C \approx 0.7$ year in the period from 2001 to mid-2002 when previous Run I luminosity level was (re)achieved; stayed on a steady exponential increase path with $C \approx 2.0$ year from 2002 till 2007, and significantly slowed down afterward, $C \approx 8.6$. It was shown that most of the other colliders had very similar features of the performance evolution, which can be summarized as (so-called CPT theorem for accelerators [24]):

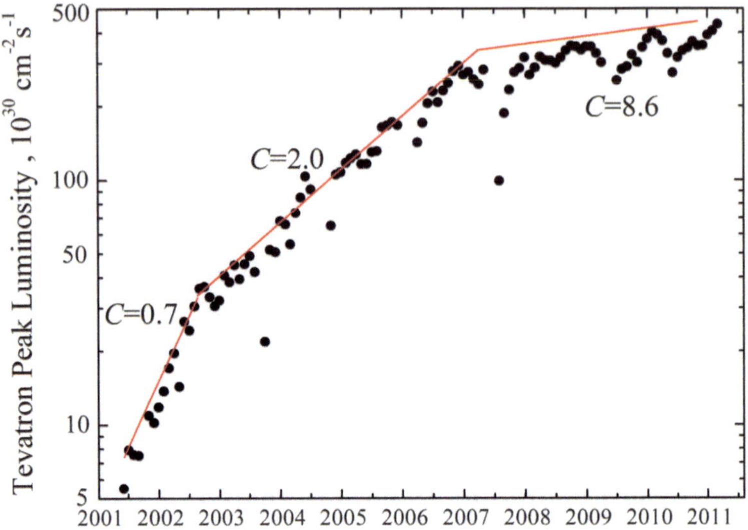

Fig. 1.13 Tevatron peak luminosity progress during Run II [20]

$$C \cdot P = T, \tag{1.6}$$

where the factor $P = \ln(\text{luminosity})$ is the "performance" gain over time interval T, and C is a machine-dependent coefficient equal to average time needed to increase the luminosity by $e = 2.71\ldots$ times, or boost the "performance" P by 1 unit. Both T and C have dimension of time, and the coefficient C was called "complexity" of the machine, as it directly indicates how hard or how easy was it to push the performance of individual machine.

Table 1.5 compares calculated complexity coefficients C for several collider facilities [25]. One can see that compared with SLC, LEP, SppS, ISR, CESR, HERA, and RHIC, the Tevatron is on the lower end of the range, i.e., demonstrated faster progress. Differences in the machine complexity factors C may be due to various reasons: (a) first of all, beam physics issues are quite different not only between classes of machines (hadrons vs $e+e-$) but often between colliders from the same class—all that affects how fast and what kind of improvements can be implemented; (b) the complexity depends on how well understood is the physics and technology of the machine [26]; (c) accelerator reliability may affect the luminosity progress, especially for larger machines with greater number of potentially not-reliable elements; (d) another factor is capability of the team running the machine to cope with challenges, generate ideas for improvements, and implement them; (e) and, of course, the latter depends on resources available for operation of the facility.

It is to be noted that the exponential growth of advancement is characteristic to other areas of science and technology [25]. For example, the maximum beam energy achieved in particle accelerators grew by factor of 10 approximately every 12 years over many decades. It is often presented in semi-log "Livingston plot" [27]

Table 1.4 Tevatron Collider Run II major luminosity improvements history, from [23]

Improvement		Luminosity increase
Pbar injection line AA → MI optics	12/2001	25 %
Tevatron quenches on abort stopped by TEL-1	02/2002	0 %, reliability
Pbar loss at Tevatron squeeze step 13 fixed	04/2002	40 %
New Tevatron injection helix	05/2002	15 %
New AA lattice reduces IBS, emittances	07/2002	40 %
Tevatron injection lines tuned up (BLT)	09/2002	10 %
Pbar coalescing improved in MI	10/2002	5 %
Tevatron C0 Lambertson magnets removed	02/2003	15 %
Tevatron sextupoles tuned/SEMs taken out of pbar lines	06/2003	10 %
New Tevatron helix on ramp, losses reduced	08/2003	2 %
Tevatron magnets reshimming and realignment	12/2003	10 %
MI dampers operations/store length increased	02/2004	30 %
2.5 MHz AA → MI transfer improved/cool shots	04/2004	8 %
Reduction of β^* to 35 cm	05/2004	20 %
Antiprotons shots from both RR and AA	07/2004	8 %
RR e-cooling operational	01–07/2005	~25 %
Slip-stacking in MI	03/2005	~20 %
Tevatron octupoles optimized at 150 GeV	04/2005	~5 %
Reduction of β^* to 28 cm	09/2005	~10 %
"Pbar production task force"	02/2006	~10 %
Tevatron 150 GeV helix improved, more protons	06/2006	~10 %
Tev collision helix improved, better lifetime	07/2006	~15 %
New RR WP, smaller pbar emittances	07/2006	~25 %
Fast transfers AA → RR (60 → 1 min)	12/2006	~15 %
New Pbar target/higher gradient	01/2007	~10 %
Tevatron sextupoles for new WP	2007	~10 %
Tevatron zero second order chromaticity	2008	~5 %
Shot-setup time reduction/multi-bunch proton injection	2008–2009	~5 %
Scraping protons in MI	2008	~5 %
Pbar size dilution at collisions/B0 aperture increased	2008	~5 %
Booster proton emittances reduced/P, A1 lines tuneup	04/2010	~10 %
Tevatron collimators used during low-beta squeeze	04/2011	~8 %

Table 1.5 "Complexities" of colliding beam facilities

	C (years)	Interval
SLC *e + e−*	*1.6 ± 0.1*	1989–1997
Tevatron Run II *p–pbar*	*2.0 ± 0.2*	2002–2007
RHIC *p–p*	*2.2 ± 0.3*	2000–2004
HERA *p–e*	*2.8 ± 0.4*	1992–2000–2005
SPPS *p–pbar*	*3.3 ± 0.2*	1982–1990
LEP *e + e−*	*3.3 ± 0.3*	1989–1995
ISR *p–p*	*3.7 ± 0.3*	1972–1982
CESR *e + e−*	*4.4 ± 0.4*	1984–1997

and corresponds to $C = 5.2$. The most cited example of the exponential growth of performance of a complex system is the "Moore's Law" [28] that describes about half a century trend in the history of computing hardware, namely, that the number of transistors that can be placed on an integrated circuit (IC) has doubled approximately every 2 years, yielding $C = 2/\ln(2) = 2.9$ years.

References

1. D.W. Kerst et al., Attainment of very high energy by means of intersecting beams of particles. Phys. Rev. **102**(2), 590–591 (1956)
2. M. Tigner, A. Chao (eds.), *Handbook of Accelerator Physics and Engineering* (World Scientific, Singapore, 1999)
3. R.R. Wilson, The Tevatron. Phys. Today **30**, 23–30 (1977)
4. CDF-II Collaboration (R. Blair et al.), *The CDF-II detector: Technical design report*, FERMILAB-Pub-96/390-E (1996)
5. D0 Collaboration (B. Pifer et al.), *The D0 experiment at the Fermilab antiproton-proton collider: design report*, FERMILAB-PUB-83-111-E, PRINT-84-0306 (FERMILAB), FERMILAB-DESIGN-1983-04 (1983)
6. C. Quigg, CERN Courier. **51**(8), 20 (2011)
7. C. Rubbia, P. McIntyre, D. Cline, in *Proceedings of the International Neutrino Conference Aachen 1976*, eds. by H. Faissner, H. Reithler, P. Zerwas (Vieweg, Braunschweig, 1977), p. 683
8. L. Hoddeson, A. Kolb, C. Westfall, *Fermilab: Physics, the Frontier, and Megascience* (University of Chicago Press, Chicago, 2008)
9. R. Dixon, CERN Courier. 51(**9**), 28 (2011)
10. S. Holmes, arXiv:1109.2937 (2011)
11. S.D. Holmes, *Tevatron Collider Run II Handbook*. Fermilab Preprint TM-2484 (1998)
12. C.D. Curtis et al., Linac H – beam operations and uses at Fermilab. IEEE Trans. Nucl. Sci. **26**(3), 3760–3762 (1979)
13. C.W. Schmidt, The Fermilab 400-MeV Linac Upgrade, in *Proceedings of Particle Accelerator Conference*, Washington, DC, 1993, pp. 1655–1659
14. E. Hubbard et al., *Booster synchrotron*. Fermilab Technical Memo TM-405 (1973)
15. S.D. Holmes, Design criteria and performance goals for the Fermilab main injector. Part. Accel. **58**, 39 (1997)
16. G. Jackson (ed.), *Fermilab Recycler Ring Technical Design Report*, Fermilab Preprint TM-1991 (1997)
17. S. Nagaitsev et al., Experimental demonstration of relativistic electron cooling. Phys. Rev. Lett. **96**, 044801 (2006)
18. M. Church, J. Marriner, The antiproton sources: design and operation. Annu. Rev. Nucl. Part. Sci. **43**, 253–295 (1993)
19. H. Edwards, The Tevatron energy doubler: a superconducting accelerator. Annu. Rev. Nucl. Part. Sci. **35**, 605–660 (1985)
20. V. Shiltsev, Mod. Phys. Lett. A **27**(1), 1230001 (2012)
21. S.D. Holmes, in *Proceedings of the Meeting of the Division of Particles and Fields of the American Physical Society* (Storrs, 1989), p. 317
22. P. Bagley et al., FERMILAB-Conf-96/392. Presented at Snowmass'96 Workshop (1996)
23. S. Holmes, R. Moore, V. Shiltsev, JINST **6**, T08001 (2011)
24. V. Shiltsev, Fermilab Preprint FNAL-Conf-04-126 (2004)
25. V. Shiltsev, Mod. Phys. Lett. A **26**(11), 761–772 (2011)
26. S.D. Holmes, V.D. Shiltsev, The legacy of the Tevatron in the area of accelerator science. Annu. Rev. Nucl. Part. Sci. **63**, 435–465 (2013)
27. M.S. Livingston, *High Energy Accelerators* (Interscience Publishers, Inc., New York, 1954)
28. G. Moore, Electronics, April 19 (1965), see also in *Proceedings of ISSCC, 2003*

Chapter 2
Beam Optics and Orbits: Methods Used at the Tevatron Accelerators

V. Lebedev, V. Shiltsev, and A. Valishev

The success of the Tevatron Run II would not be possible without detailed work on the linear and nonlinear beam optics. The scope of optics work included all major stages: the optics design, optics measurements, and optics correction. Optics of all transport lines and rings was measured and corrected. This work resulted in a significant reduction of the emittance growth for beam transfers and increased the acceptances of the rings and transfer lines. The most spectacular improvements are related to the improvements of antiproton beam transport from the Accumulator to the Main Injector (MI) and optics improvements in Tevatron, Debuncher, and Accumulator. The electron cooler beam transport presented significant challenge for both the optics design and its commissioning.

2.1 Linear Optics with Coupling Between Degrees of Freedom

The major part of optics work has been focused on the linear optics problems. In this section we consider the fundamentals of betatron motion with coupled degrees of freedom. The significant fraction of Run II optics work has been based on this formalism and otherwise would hardly be possible. In particular the beam transport in the electron cooler (see Chap. 7) is completely *x*–*y* coupled, and the Tevatron lattice has significant coupling terms and the tune working point close to the difference resonance that it cannot be accurately described using the perturbation theory. The most of material is related to the *x*–*y* coupled motion. However, it can be directly applied to coupling of any two degrees of freedom. An extension to three degrees of freedom is straightforward and is not presented to keep text and equations compact.

V. Lebedev (✉) • V. Shiltsev • A. Valishev
Fermi National Accelerator Laboratory, PO Box 500, Batavia, IL 60510, USA
e-mail: val@fnal.gov; shiltsev@fnal.gov

V. Lebedev and V. Shiltsev (eds.), *Accelerator Physics at the Tevatron Collider*, Particle Acceleration and Detection, DOI 10.1007/978-1-4939-0885-1_2,

First, we describe the equations of motion and notations. Second, we consider the relationship between eigenvectors, emittances, and the particle 4D ellipsoid in the phase space. Third, we consider the parameterization of particle motion based on an extension of the Mais–Ripken parameterization [1, 2] presented in [3]. Finally, we consider a perturbation theory for the case where the unperturbed motion is Hamiltonian.

2.1.1 Equations of Motion and Conditions of Symplecticity

In the absence of dissipative processes, the particle motion is Hamiltonian. For the linear motion the Hamiltonian is a second-order form of particle coordinates and momenta. For two-dimensional motion, it can be presented in the following matrix form:

$$H(x, p_x, y, p_y, s) = \mathbf{x}^{\mathrm{T}}\mathbf{H}\mathbf{x}, \tag{2.1}$$

where $\mathbf{x} = [x, p_x, y, p_y]^{\mathrm{T}}$, x and y are the particle coordinates, $p_x = x' - R\ y/2$ and $p_y = y' + R\ y/2$ are its canonical momenta,[1] $x' = dx/ds$ and $y' = dy/ds$ are the particle angles, s is the longitudinal coordinate (time coordinate), $R = eB_s/Pc$, B_s is the longitudinal magnetic field, and P is the total momentum of the reference particle. Following the standard procedure for obtaining the equations of motion [4],

$$\frac{dp_i}{dt} = -\frac{\partial H}{dx_i}, \quad \frac{dx_i}{dt} = \frac{\partial H}{dp_i}, \tag{2.2}$$

one comes to

$$\frac{d\mathbf{x}}{ds} = \mathbf{U}\mathbf{H}\mathbf{x}, \tag{2.3}$$

where

$$\mathbf{U} = \begin{bmatrix} 0 & 1 & 0 & 0 \\ -1 & 0 & 0 & 0 \\ 0 & 0 & 0 & 1 \\ 0 & 0 & -1 & 0 \end{bmatrix} \tag{2.4}$$

is the unit symplectic matrix. In the case of flat orbit, the Hamiltonian and the corresponding equations of motion are [2]

[1] Note that in practical optics calculations the difference between particle angles and their canonical momenta does not usually exist because most optics codes compute transfer matrices between points where the longitudinal magnetic fields are equal to zero.

$$\mathbf{H} = \begin{bmatrix} K_x{}^2 + k + \frac{R^2}{4} & 0 & N & -\frac{R}{2} \\ 0 & 1 & \frac{R}{2} & 0 \\ N & \frac{R}{2} & -k + \frac{R^2}{4} & 0 \\ -\frac{R}{2} & 0 & 0 & 1 \end{bmatrix}, \tag{2.5}$$

$$\begin{aligned} &x'' + \left(K_x{}^2 + k\right)x + \left(N - \frac{1}{2}R'\right)y - Ry' = 0 \quad , \\ &y'' - ky + \left(N + \frac{1}{2}R'\right)x + Rx' = 0 \quad . \end{aligned} \tag{2.6}$$

Here $K_x = eB_y/Pc$, $k = eG/Pc$, $N = eG_s/Pc$, and B_y is the vertical component of the magnetic field; G and G_s are the normal and skew components of the magnetic field gradient (the skew component is obtained by +45° rotation around the s axis in the right-handed coordinate frame).

For any two solutions of Eq. (2.3), $\mathbf{x}_1(s)$ and $\mathbf{x}_2(s)$, one can write that

$$\frac{d}{ds}\left(\mathbf{x}_1{}^{\mathrm{T}}\mathbf{U}\mathbf{x}_2\right) = \frac{d\mathbf{x}_1{}^{\mathrm{T}}}{ds}\ \mathbf{U}\mathbf{x}_2 + \mathbf{x}_1{}^{\mathrm{T}}\ \mathbf{U}\frac{d\mathbf{x}_2}{ds} = \mathbf{x}_1{}^{\mathrm{T}}\mathbf{H}^{\mathrm{T}}\mathbf{U}^{\mathrm{T}}\mathbf{U}\mathbf{x}_2 + \mathbf{x}_1{}^{\mathrm{T}}\mathbf{U}\mathbf{U}\mathbf{H}\mathbf{x}_2 = 0,$$

and, consequently,

$$\mathbf{x}_1{}^{\mathrm{T}}\mathbf{U}\mathbf{x}_2 = \mathrm{const}, \tag{2.7}$$

where the following properties of the unit symplectic matrix were employed: $\mathbf{U}^{\mathrm{T}}\mathbf{U} = \mathbf{I}$ and $\mathbf{U}\mathbf{U} = -\mathbf{I}$; $\mathbf{I}$ is the identity matrix. The integral of motion in Eq. (2.7) is called the Lagrange invariant.

Let us introduce the transfer matrix from coordinate 0 to coordinate s, $\mathbf{x} = \mathbf{M}(0, s)\mathbf{x}_0$. Taking into account that the invariant of Eq. (2.7) does not change during motion, we can write that

$$\mathbf{x}_1{}^{\mathrm{T}}\mathbf{U}\mathbf{x}_2 = \mathbf{x}_1{}^{\mathrm{T}}\mathbf{M}(0,s)^{\mathrm{T}}\mathbf{U}\mathbf{M}(0,s)\mathbf{x}_2 = \mathrm{const}.$$

As the above equation is satisfied for any $\mathbf{x}_1$ and $\mathbf{x}_2$, it yields

$$\mathbf{M}(0,s)^{\mathrm{T}}\mathbf{U}\mathbf{M}(0,s) = \mathbf{U}. \tag{2.8}$$

Equation (2.8) expresses the symplecticity condition for particle motion. It is equivalent to $n^2 = 16$ scalar equations, but taking into account that the matrix

$\mathbf{M}^{\mathrm{T}}\mathbf{U}\mathbf{M}$ is antisymmetric, only six ($(n^2 - n)/2 = 6$) of these equations are independent [5]. Consequently, only 10 of 16 elements of the transfer matrix are independent. Thus, the symplecticity condition imposes more severe limitations than the Liouville's theorem [4], which imposes only one condition, $\det(\mathbf{M}) = 1$, and leaves 15 independent parameters.

Multiplying both sides of Eq. (2.8) by $\mathbf{U}$ on the left and by $\mathbf{M}^{-1}$ on the right, we obtain that the inverse of matrix $\mathbf{M}$ is

$$\mathbf{M}(0,s)^{-1} = -\mathbf{U}\mathbf{M}(0,s)^{\mathrm{T}}\mathbf{U}. \tag{2.9}$$

Then, multiplying Eq. (2.9) by $\mathbf{M}$ on the left and by $\mathbf{U}$ on the right, we obtain an alternative expression of symplecticity condition:

$$\mathbf{M}(0,s)\mathbf{U}\mathbf{M}(0,s)^{\mathrm{T}} = \mathbf{U}. \tag{2.10}$$

Note that Eqs. (2.9) and (2.10) are not related by matrix transposition.

2.1.2 *Eigenvalues, Eigenvectors, and Condition of Motion Stability*

Consider a circular accelerator with the one-turn transfer matrix $\mathbf{M}$. The transfer matrix has four eigenvalues, λ_i, and four corresponding eigenvectors, $\mathbf{v}_i$ ($i = 1, 2, 3, 4$):

$$\mathbf{M}\mathbf{v}_i = \lambda_i \mathbf{v}_i. \tag{2.11}$$

Then, the turn-by-turn particle motion can be presented in the following form:

$$\mathbf{x}_n = \sum_{i=1}^{4} \lambda_i^n c_i \mathbf{v}_i, \tag{2.12}$$

where c_i are the coefficients determined by particle initial coordinates.

Comparing the two equations below

$$\begin{aligned}
&\det(\mathbf{M} - \lambda\mathbf{I}) = \lambda^4 + \cdots + \det(\mathbf{M}) = \lambda^4 + \cdots + 1, \\
&\det(\mathbf{M} - \lambda\mathbf{I}) = (\lambda - \lambda_1)(\lambda - \lambda_2)(\lambda - \lambda_3)(\lambda - \lambda_4) = \lambda^4 + \cdots + \lambda_1\lambda_2\lambda_3\lambda_4,
\end{aligned}$$

one obtains that the product of all eigenvalues is equal to 1:

$$\lambda_1\lambda_2\lambda_3\lambda_4 = 1. \tag{2.13}$$

Matrix $\mathbf{M}$ is a real matrix. Therefore, the complex conjugate of an eigenvalue and corresponding eigenvector are also an eigenvalue and eigenvector. As follows from Eq. (2.12) the motion stability requires $|\lambda_i| \le 1$. Combining that with the requirement of Eq. (2.13), one obtains that the four eigenvalues split into two complex conjugate pairs confined to a unit circle, $|\lambda_i| = 1$. We denote them as λ_1, λ_1^*, λ_2, and λ_2^* and the corresponding eigenvectors as $\mathbf{v}_1$, $\mathbf{v}_1{}^*$, $\mathbf{v}_2$, and $\mathbf{v}_2{}^*$, where * denotes the complex conjugate value. Note that if any eigenvalue is equal to ± 1, its complex conjugate partner has the same value; consequently, the solution is degenerate and an infinitesimally small perturbation makes the system unstable.

For any two eigenvectors, the symplecticity condition of Eq. (2.8) yields the identity:

$$0 = \lambda_j \mathbf{v}_j{}^{\mathrm{T}} \mathbf{U}(\mathbf{M}\mathbf{v}_i - \lambda_i \mathbf{v}_i) = (\mathbf{M}\mathbf{v}_j)^{\mathrm{T}} \mathbf{U}\mathbf{M}\mathbf{v}_i - \lambda_j \mathbf{v}_j{}^{\mathrm{T}} \mathbf{U}\lambda_i \mathbf{v}_i = (1 - \lambda_j \lambda_i) \mathbf{v}_j{}^{\mathrm{T}} \mathbf{U}\mathbf{v}_i,$$

which results in that the product $\mathbf{v}_j{}^{\mathrm{T}}\mathbf{U}\mathbf{v}_i$ can be different from zero only if $\mathbf{v}_i$ and $\mathbf{v}_j$ represent a complex conjugate pair. The product $\mathbf{v}_j{}^{+}\mathbf{U}\mathbf{v}_i$ is purely imaginary, indeed:

$$(\mathbf{v}^{+}\mathbf{U}\mathbf{v})^{*} = (\mathbf{v}^{+}\mathbf{U}\mathbf{v})^{+} = \mathbf{v}^{+}\mathbf{U}^{+}\mathbf{v} = -\mathbf{v}^{+}\mathbf{U}\mathbf{v},$$

where $\mathbf{v}^{+} = \mathbf{v}^{*\mathrm{T}}$ denotes the Hermitian conjugate, and we took into account that the transpose of a scalar is equal to itself. That allows us to introduce the symplectic orthogonality conditions:

$$\begin{aligned} \mathbf{v}_1{}^{+}\mathbf{U}\mathbf{v}_1 &= -2i, & \mathbf{v}_2{}^{+}\mathbf{U}\mathbf{v}_2 &= -2i, \\ \mathbf{v}_1{}^{T}\mathbf{U}\mathbf{v}_1 &= 0, & \mathbf{v}_2{}^{T}\mathbf{U}\mathbf{v}_2 &= 0, \\ \mathbf{v}_2{}^{T}\mathbf{U}\mathbf{v}_1 &= 0, & \mathbf{v}_2{}^{+}\mathbf{U}\mathbf{v}_1 &= 0. \end{aligned} \tag{2.14}$$

Other combinations can be obtained by applying the transposition and/or the complex conjugation to Eq. (2.14). Note that the sign choice in the two top equations determines which of two vectors in each complex conjugate pair is the primary vector (see Sect. 2.1.5). The normalization value is chosen to make the matrix $\mathbf{V}$ introduced in the next section symplectic. Similarly as for the transfer matrix elements, there are only six independent real scalar equations among Eq. (2.14). Note that the two equations in the second line are identities because $\mathbf{a}^{+}\,\mathbf{U}\,\mathbf{a} = 0$ for any $\mathbf{a}$.

2.1.3 *Mode Emittances and Emittance Ellipsoid in 4D Phase Space*

The turn-by-turn particle positions and angles can be represented as a linear combination of four independent solutions,

$$\begin{aligned}\mathbf{x} &= \mathrm{Re}\left(A_1 e^{-i\psi_1}\mathbf{v}_1 + A_2 e^{-i\psi_2}\mathbf{v}_2\right)\\ &= A_1(\mathbf{v}_1{}'\cos\psi_1 + \mathbf{v}_1{}''\sin\psi_1) + A_2(\mathbf{v}_2{}'\cos\psi_2 + \mathbf{v}_2{}''\sin\psi_2),\end{aligned} \tag{2.15}$$

where four real parameters, A_1, A_2, ψ_1, and ψ_2, represent the betatron amplitudes and phases. The amplitudes remain constant in the course of betatron motion, while the phases are incremented after each turn.

Let us introduce the following real matrix:

$$\mathbf{V} = [\mathbf{v}_1{}', -\mathbf{v}_1{}'', \mathbf{v}_2{}', -\mathbf{v}_2{}''] . \tag{2.16}$$

This allows one to rewrite Eq. (2.15) in the compact form

$$\mathbf{x} = \mathbf{V}\mathbf{A}\boldsymbol{\xi}_A, \tag{2.17}$$

where the amplitude matrix $\mathbf{A}$ is

$$\mathbf{A} = \begin{bmatrix} A_1 & 0 & 0 & 0 \\ 0 & A_1 & 0 & 0 \\ 0 & 0 & A_2 & 0 \\ 0 & 0 & 0 & A_2 \end{bmatrix} \equiv \mathrm{diag}(A_1, A_1, A_2, A_2), \tag{2.18}$$

and

$$\boldsymbol{\xi}_A = \begin{bmatrix} \cos\psi_1 \\ -\sin\psi_1 \\ \cos\psi_2 \\ -\sin\psi_2 \end{bmatrix} . \tag{2.19}$$

Applying the orthogonality conditions given by Eq. (2.14), one can prove that matrix $\mathbf{V}$ is a symplectic matrix. It can be seen explicitly as follows:

$$\mathbf{V}^{\mathrm{T}}\mathbf{U}\mathbf{V} = \left[\frac{\mathbf{v}_1 + \mathbf{v}_1{}^*}{2}, -\frac{\mathbf{v}_1 - \mathbf{v}_1{}^*}{2i}, \frac{\mathbf{v}_2 + \mathbf{v}_2{}^*}{2}, -\frac{\mathbf{v}_2 - \mathbf{v}_2{}^*}{2i}\right]^{\mathrm{T}}$$

$$\mathbf{U}\left[\frac{\mathbf{v}_1 + \mathbf{v}_1{}^*}{2}, -\frac{\mathbf{v}_1 - \mathbf{v}_1{}^*}{2i}, \frac{\mathbf{v}_2 + \mathbf{v}_2{}^*}{2}, -\frac{\mathbf{v}_2 - \mathbf{v}_2{}^*}{2i}\right] = \mathbf{U}.$$

Here we took into account that every matrix element in matrix $\hat{\mathbf{V}}^{\mathrm{T}}\mathbf{U}\hat{\mathbf{V}}$ can be calculated using vector multiplication of Eq. (2.14).

Let us consider an ensemble of particles, whose motion at the beginning of the lattice (or any other point of a ring) is contained in a 4D ellipsoid. A 3D surface of this ellipsoid is determined by particles with extreme betatron amplitudes. For any of these particles, Eq. (2.17) describes the 2D subspace of single-particle motion, which is a subspace of the 3D surface of the ellipsoid, described by the bilinear form

$$\mathbf{x}^{\mathrm{T}}\boldsymbol{\Xi}\mathbf{x} = 1. \tag{2.20}$$

This ellipsoid confines the motion of all particles. To describe a 3D surface, in addition to parameters ψ_1 and ψ_2 of Eq. (2.19), we introduce the third parameter ψ_3 so that the vector $\boldsymbol{\xi}$ would belong to a 3D sphere with a unit radius, according to the equation

$$(\boldsymbol{\xi}, \boldsymbol{\xi}) = 1, \tag{2.21}$$

where

$$\boldsymbol{\xi} = \begin{bmatrix} \cos\psi_1 \cos\psi_3 \\ -\sin\psi_1 \cos\psi_3 \\ \cos\psi_2 \sin\psi_3 \\ -\sin\psi_2 \sin\psi_3 \end{bmatrix}. \tag{2.22}$$

Then, we can rewrite Eq. (2.17) in the following form:

$$\mathbf{x} = \mathbf{V}\mathbf{A}\boldsymbol{\xi} \tag{2.23}$$

which describes a 3D subspace confining all particles of the beam (water-bag particle distribution). In other words we can consider that the amplitudes of the boundary particles are parameterized by ψ_3 ($A_1 \to A_1\cos\psi_3, A_2 \to A_2\sin\psi_3$), so that we would obtain a 4D ellipsoid.

Expressing $\boldsymbol{\xi}$ from Eq. (2.23) and substituting it into Eq. (2.21), one obtains the quadratic form describing a 4D ellipsoid containing all particles:

$$\mathbf{x}^{\mathrm{T}}\left((\mathbf{V}\mathbf{A})^{-1}\right)^{\mathrm{T}}(\mathbf{V}\mathbf{A})^{-1}\mathbf{x} = 1. \tag{2.24}$$

Comparing Eqs. (2.20) and (2.24) and using Eq. (2.9) for matrix inversion, one can express the bilinear form, $\boldsymbol{\Xi}$, as follows:

$$\boldsymbol{\Xi} = \mathbf{U}\mathbf{V}\hat{\boldsymbol{\Xi}}\mathbf{V}^{\mathrm{T}}\mathbf{U}^{\mathrm{T}}, \tag{2.25}$$

where $\hat{\boldsymbol{\Xi}} = \mathbf{A}^{-1}\mathbf{A}^{-1} = \mathrm{diag}\left(A_1{}^{-2}, A_1{}^{-2}, A_2{}^{-2}, A_2{}^{-2}\right)$ is a diagonal matrix depending on two amplitudes A_1 and A_2, and we took into account that matrices $\mathbf{A}^{-1}$ and $\mathbf{U}$ commute. Inversion of Eq. (2.25) yields

$$\hat{\boldsymbol{\Xi}} = \mathbf{V}^{\mathrm{T}}\boldsymbol{\Xi}\mathbf{V}, \tag{2.26}$$

i.e., a symplectic transform $\mathbf{V}$ reduces matrix $\boldsymbol{\Xi}$ to its diagonal form.

To determine the beam emittance (volume of the occupied 4D phase space) described by Eq. (2.20), we note that due to symplecticity $\det(\mathbf{V}) = 1$. Consequently, the coordinate transform $\mathbf{x} = \mathbf{V}\mathbf{x}'$ corresponding to Eq. (2.26) does not change the ellipsoid volume. Then, in the new coordinate frame, the 3D ellipsoid enclosing the total 4D phase space of the beam is described by the following equation:

$$\hat{\Xi}_{11}x'^2 + \hat{\Xi}_{22}p'^2 + \hat{\Xi}_{33}y'^2 + \hat{\Xi}_{44}y'^2 = 1.$$

It is natural to define the beam emittance as a product of the ellipsoid semiaxes (omitting the factor $\pi^2/2$ correcting for the real 4D volume of the ellipsoid) so that

$$\varepsilon_{4D} = \frac{1}{\sqrt{\hat{\Xi}_{11}\hat{\Xi}_{22}\hat{\Xi}_{33}\hat{\Xi}_{44}}} = \frac{1}{\sqrt{\det(\hat{\Xi})}} = A_1{}^2A_2{}^2. \tag{2.27}$$

Thus, the squares of amplitudes A_1 and A_2 can be considered as 2D emittances ε_1 and ε_2 corresponding to the eigenvectors $\mathbf{v}_1$ and $\mathbf{v}_2$. Their product is equal to the total 4D emittance: $\varepsilon_1\varepsilon_2 = \varepsilon_{4D}$. We will call them the mode emittances. Consequently, one can write matrix $\hat{\Xi}$ as

$$\hat{\Xi} = \mathrm{diag}\left(\varepsilon_1{}^{-1}, \varepsilon_1{}^{-1}, \varepsilon_2{}^{-1}, \varepsilon_2{}^{-1}\right). \tag{2.28}$$

2.1.4 Eigenvectors and Particle Phase Space Ellipsoid

Similarly to the one-dimensional case, the particle ellipsoid shape, described by matrix Ξ, determines the mode emittances ε_1 and ε_2 and the eigenvectors $\mathbf{v}_1$ and $\mathbf{v}_2$. In this case the mode emittances are reciprocal to the roots of the following characteristic equation:

$$\det(\Xi - i\lambda\mathbf{U}) = 0. \tag{2.29}$$

One can prove the above using Eq. (2.25) as follows:

$$\begin{aligned}\det(\Xi - i\lambda\mathbf{U}) &= \det\left(\mathbf{U}\mathbf{V}\hat{\Xi}\mathbf{V}^{\mathrm{T}}\mathbf{U}^{\mathrm{T}} - i\lambda\mathbf{U}\right) = \det\left(\hat{\Xi} - i\lambda\mathbf{U}^{\mathrm{T}}\mathbf{V}^{\mathrm{T}}\mathbf{U}\mathbf{V}\mathbf{U}\right)\\ &= \det\left(\hat{\Xi} - i\lambda\mathbf{U}\right) = \left(\frac{1}{\varepsilon_1{}^2} - \lambda^2\right)\left(\frac{1}{\varepsilon_2{}^2} - \lambda^2\right) = 0.\end{aligned} \tag{2.30}$$

Knowing the mode emittances and, consequently, $\hat{\Xi}$, one can obtain from Eq. (2.25) a system of linear equations for matrix $\mathbf{V}$:

$$\boldsymbol{\Xi}\mathbf{V}\mathbf{U} = \mathbf{U}\mathbf{V}\hat{\boldsymbol{\Xi}}. \tag{2.31}$$

Multiplying the above equation by u_l, one obtains two equations for the eigenvectors:

$$\left(\boldsymbol{\Xi} - \frac{i}{\varepsilon_l}\mathbf{U}\right)\mathbf{v}_l = 0, \tag{2.32}$$

where $l = 1, 2$, and

$$\mathbf{u}_1 = \begin{bmatrix} 1 \\ -i \\ 0 \\ 0 \end{bmatrix}, \quad \mathbf{u}_2 = \begin{bmatrix} 0 \\ 0 \\ 1 \\ -i \end{bmatrix}. \tag{2.33}$$

We also took into account that $\mathbf{V}\mathbf{u}_l = \mathbf{v}_l$, $\mathbf{U}\mathbf{u}_l = -i\mathbf{u}_l$, and $\boldsymbol{\Xi}'\mathbf{u}_l = \frac{1}{\varepsilon_l}\mathbf{u}_l$.

Taking into account Eq. (2.20) a Gaussian distribution function for coupled beam motion can be written in the following form:

$$f(\mathbf{x}) = \frac{1}{4\pi^2\varepsilon_1\varepsilon_2}\exp\left(-\frac{1}{2}\mathbf{x}^{\mathrm{T}}\boldsymbol{\Xi}\mathbf{x}\right). \tag{2.34}$$

Then, the second-order moments of the distribution function are

$$\Sigma_{ij} \equiv \overline{x_i x_j} = \int x_i x_j f(\mathbf{x}) dx^4 = \frac{1}{4\pi^2\varepsilon_1\varepsilon_2}\int x_i x_j \exp\left(-\frac{1}{2}\mathbf{x}^{\mathrm{T}}\boldsymbol{\Xi}\mathbf{x}\right)dx^4. \tag{2.35}$$

To carry out the integration, one can perform a coordinate transform, $\mathbf{y} = \mathbf{V}^{-1}\mathbf{x}$, which reduces matrix $\boldsymbol{\Xi}$ to its diagonal form. Taking into account that

$$\frac{1}{4\pi^2\varepsilon_1\varepsilon_2}\int y_i y_j dy^4 \exp\left(-\frac{1}{2}\mathbf{y}^{\mathrm{T}}\hat{\boldsymbol{\Xi}}\mathbf{y}\right) = [\mathrm{diag}(\varepsilon_1, \varepsilon_1, \varepsilon_2, \varepsilon_2)]_{ij} \equiv \hat{\boldsymbol{\Sigma}}_{ij}, \tag{2.36}$$

one obtains that the matrix of the second-order moments is

$$\boldsymbol{\Sigma} = \mathbf{V}\hat{\boldsymbol{\Sigma}}\mathbf{V}^{\mathrm{T}}. \tag{2.37}$$

Using Eqs. (2.25) and (2.37), one can easily prove that matrix $\boldsymbol{\Sigma}$ is the inverse of matrix $\boldsymbol{\Xi}$. Consequently, a symplectic transform $\mathbf{VU}$ reduces matrix $\boldsymbol{\Sigma}$ to its diagonal form. Applying a similar scheme as above for obtaining emittances and eigenvectors from matrix $\boldsymbol{\Xi}$, one finds that the mode emittances ε_1 and ε_2 can be computed from matrix $\boldsymbol{\Sigma}$ as roots of its characteristic equation,

$$\det(\boldsymbol{\Sigma}\mathbf{U} + i\lambda\mathbf{I}) = 0, \quad \varepsilon_l = \lambda_l, \tag{2.38}$$

while the equations for the eigenvectors are

$$(\mathbf{\Sigma U} + i\varepsilon_l \mathbf{I})\mathbf{v}_l = 0. \tag{2.39}$$

It also follows from Eq. (2.37) that the total beam emittance is equal to

$$\varepsilon_{4D} = \varepsilon_1 \varepsilon_2 = \sqrt{\det(\mathbf{\Sigma})}. \tag{2.40}$$

Taking into account that the beam motion from point s to point s' results in the matrix Ξ transformation so that $\Xi(s') = \mathbf{M}(s, s')^{\mathrm{T}}\Xi(s)\mathbf{M}(s, s')$ and using Eq. (2.29) and the motion symplecticity, one can easily prove that the mode emittances ε_1 and ε_2 are the motion invariants, i.e., there is no configuration of linear electric and magnetic fields which can change them. Consequently, each mode emittance is bound to the corresponding betatron mode. If the beamline is built so that the motion is decoupled at some point, then the mode emittances coincide with conventional horizontal and vertical emittances.

2.1.5 Beta-Functions of Coupled Motion

Employing the previously introduced notation, one can describe a single-particle phase-space trajectory along the beam orbit as

$$\begin{aligned} \mathbf{x}(s) &= \mathbf{M}(0,s)\mathrm{Re}\left(\sqrt{2I_1}\mathbf{v}_1 e^{-i\psi_1} + \sqrt{2I_2}\mathbf{v}_2 e^{-i\psi_2}\right) \\ &= \mathrm{Re}\left(\sqrt{2I_1}\mathbf{v}_1(s) e^{-i(\psi_1+\mu_1(s))} + \sqrt{2I_2}\mathbf{v}_2(s) e^{-i(\psi_2+\mu_2(s))}\right), \end{aligned} \tag{2.41}$$

where the vectors $\mathbf{v}_1(s) \equiv e^{i\mu_1(s)}\mathbf{M}(0,s)\mathbf{v}_1$ and $\mathbf{v}_2(s) \equiv e^{i\mu_2(s)}\mathbf{M}(0,s)\mathbf{v}_2$ are the eigenvectors of the matrix $\mathbf{M}(0,s)\mathbf{M}\mathbf{M}(0,s)^{-1}$, ψ_1 and ψ_2 are the initial phases of betatron motion and I_1 and I_2 are the corresponding actions, and $\mathbf{M} = \mathbf{M}(0, L)$ is the transfer matrix for the entire ring. The terms $e^{-i\mu_1(s)}$ and $e^{-i\mu_2(s)}$ are introduced to bring the eigenvectors to the following form:

$$\mathbf{v}_1(s) = \begin{bmatrix} \sqrt{\beta_{1x}(s)} \\ -\dfrac{iu_1(s) + \alpha_{1x}(s)}{\sqrt{\beta_{1x}(s)}} \\ \sqrt{\beta_{1y}(s)}e^{i\nu_1(s)} \\ -\dfrac{iu_2(s) + \alpha_{1y}(s)}{\sqrt{\beta_{1y}(s)}}e^{i\nu_1(s)} \end{bmatrix}, \quad \mathbf{v}_2(s) = \begin{bmatrix} \sqrt{\beta_{2x}(s)}e^{i\nu_2(s)} \\ -\dfrac{iu_3(s) + \alpha_{2x}(s)}{\sqrt{\beta_{2x}(s)}}e^{i\nu_2(s)} \\ \sqrt{\beta_{2y}(s)} \\ -\dfrac{iu_4(s) + \alpha_{2y}(s)}{\sqrt{\beta_{2y}(s)}} \end{bmatrix}, \tag{2.42}$$

so that $\mu_1(s)$ and $\mu_2(s)$ would be the phase advances of betatron motion. Here $\beta_{1x}(s)$, $\beta_{1y}(s)$, $\beta_{2x}(s)$, and $\beta_{2y}(s)$ are the beta-functions; $\alpha_{1x}(s)$, $\alpha_{1y}(s)$, $\alpha_{2x}(s)$, and $\alpha_{2y}(s)$ are the alpha-functions which, as will be shown in the next section, coincide with the beta-functions' negative half-derivatives at regions with zero longitudinal magnetic field; and six real functions $u_1(s)$, $u_2(s)$, $u_3(s)$, $u_4(s)$, $\nu_1(s)$, and $\nu_2(s)$ are determined by the orthogonality conditions of Eq. (2.14). Below we will be omitting their dependence on s where it does not cause an ambiguity. Two eigenvectors $\mathbf{v}_1$ and $\mathbf{v}_2$ were chosen out of two pairs of complex conjugate eigenvectors by selecting u_1 and u_4 to be positive.

The first orthogonality condition of Eq. (2.14),

$$(\mathbf{v}_1{}^{+}\mathbf{U}\mathbf{v}_1) = -2i(u_1 + u_2) = -2i,$$

yields $u_1 = 1 - u_2$, and similarly for the second eigenvector, $u_4 = 1 - u_3$. The next two equations, $\mathbf{v}_1{}^{\mathrm{T}}\mathbf{U}\mathbf{v}_1 = 0$ and $\mathbf{v}_2{}^{\mathrm{T}}\mathbf{U}\mathbf{v}_2 = 0$, are identities.

Taking into account the above relations for u_1 and u_4, the remaining two nontrivial orthogonality conditions can be written as follows:

$$\begin{aligned}(\mathbf{v}_2{}^{+}\mathbf{U}\mathbf{v}_1) = &-\left(\sqrt{\frac{\beta_{2x}}{\beta_{1x}}}[i(1-u_2)+\alpha_{1x}] + \sqrt{\frac{\beta_{1x}}{\beta_{2x}}}[iu_3 - \alpha_{2x}]\right)\mathrm{e}^{-i\nu_2} \\ &-\left(\sqrt{\frac{\beta_{1y}}{\beta_{2y}}}\left[i(1-u_3)-\alpha_{2y}\right] + \sqrt{\frac{\beta_{2y}}{\beta_{1y}}}\left[iu_2 + \alpha_{1y}\right]\right)\mathrm{e}^{i\nu_1} = 0,\end{aligned} \tag{2.43}$$

$$\begin{aligned}(\mathbf{v}_2{}^{\mathrm{T}}\mathbf{U}\mathbf{v}_1) = &-\left(\sqrt{\frac{\beta_{2x}}{\beta_{1x}}}[i(1-u_2)+\alpha_{1x}] - \sqrt{\frac{\beta_{1x}}{\beta_{2x}}}[iu_3 + \alpha_{2x}]\right)\mathrm{e}^{i\nu_2} \\ &-\left(\sqrt{\frac{\beta_{1y}}{\beta_{2y}}}\left[i(u_3-1)-\alpha_{2y}\right] + \sqrt{\frac{\beta_{2y}}{\beta_{1y}}}\left[iu_2 + \alpha_{1y}\right]\right)\mathrm{e}^{i\nu_1} = 0.\end{aligned} \tag{2.44}$$

Multiplying both terms in Eqs. (2.43) and (2.44) by their complex conjugate values, one obtains

$$\begin{aligned}A_x{}^2 + \left(\kappa_x(1-u_2) + \kappa_x{}^{-1}u_3\right)\right)^2 &= A_y{}^2 + \left(\kappa_y(1-u_3) + \kappa_y{}^{-1}u_2\right)^2, \\ A_x{}^2 + \left(\kappa_x(1-u_2) - \kappa_x{}^{-1}u_3\right)\right)^2 &= A_y{}^2 + \left(\kappa_y(1-u_3) - \kappa_y{}^{-1}u_2\right)^2,\end{aligned} \tag{2.45}$$

where

$$A_x = \kappa_x \alpha_{1x} - \kappa_x^{-1} \alpha_{2x},$$
$$A_y = \kappa_y \alpha_{2y} - \kappa_y^{-1} \alpha_{1y}, \tag{2.46}$$
$$\kappa_x = \sqrt{\frac{\beta_{2x}}{\beta_{1x}}}, \quad \kappa_y = \sqrt{\frac{\beta_{1y}}{\beta_{2y}}}.$$

Subtracting Eq. (2.45) yields $u_2 = u_3$. Substituting $u_2 = u_3 = u$ into the first equation of Eq. (2.45), one obtains the following expression for u:

$$u = \frac{-\kappa_x^2 \kappa_y^2 \pm \sqrt{\kappa_x^2 \kappa_y^2 \left(1 + \frac{A_x^2 - A_y^2}{\kappa_x^2 - \kappa_y^2}\left(1 - \kappa_x^2 \kappa_y^2\right)\right)}}{1 - \kappa_x^2 \kappa_y^2}. \tag{2.47}$$

By definition u_k ($k = 1, \ldots 4$) are real functions[2] and u_1 and u_4 are positive. That sets a constraint for possible values of beta- and alpha-functions,

$$\frac{A_x^2 - A_y^2}{\kappa_x^2 - \kappa_y^2}\left(1 - \kappa_x^2 \kappa_y^2\right) \geq -1, \tag{2.48}$$

and a constraint on a value of u, $u \leq 1$ (see also Sect. 2.1.6).

Knowing u makes it easy to find $\nu_1 + \nu_2$ and $\nu_1 - \nu_2$ from Eqs. (2.43) and (2.44):

$$e^{i\nu_+} \equiv e^{i(\nu_1+\nu_2)} = \frac{A_x + i\left(\kappa_x(1-u) + \kappa_x^{-1} u\right)}{A_y - i\left(\kappa_y(1-u) + \kappa_y^{-1} u\right)},$$
$$e^{i\nu_-} \equiv e^{i(\nu_1-\nu_2)} = \frac{A_x + i\left(\kappa_x(1-u) - \kappa_x^{-1} u\right)}{A_y + i\left(\kappa_y(1-u) - \kappa_y^{-1} u\right)}, \tag{2.49}$$

and, consequently, ν_1 and ν_2:

$$\nu_1 = \frac{1}{2}(\nu_+ + \nu_-) + \pi(n + m),$$
$$\nu_2 = \frac{1}{2}(\nu_+ - \nu_-) + \pi(n - m). \tag{2.50}$$

Here n and m are arbitrary integers. Equation (2.49) results in that ν_- and ν_+ are determined modulo 2π which, consequently, yields that ν_1 and ν_2 are determined modulo π (see Eq. (2.50)) resulting in additional solutions. Actually there are only two independent solutions for ν_1 and ν_2. The first one corresponds to the case when

---

[2] Equation (2.47) also demonstrates that if beta- and alpha-functions are chosen incorrectly, such that the value of the discriminant is negative, u becomes imaginary, thus redetermining the alpha-functions.

both n and m have the same parity, which is equivalent to $m+n=m-n=0$. The second one corresponds to different parity of m and n, which is equivalent to $m+n=m-n=1$. Thus, in a general case, one has four independent solutions for u and ν_1 and ν_2 set by symplecticity conditions: two solutions for u and two solutions for ν_1 and ν_2 for each u.

Finally, we can express the eigenvectors in the following form:

$$\mathbf{v}_1 = \begin{bmatrix} \sqrt{\beta_{1x}} \\ -\dfrac{i(1-u)+\alpha_{1x}}{\sqrt{\beta_{1x}}} \\ \sqrt{\beta_{1y}} e^{i\nu_1} \\ -\dfrac{iu+\alpha_{1y}}{\sqrt{\beta_{1y}}} e^{i\nu_1} \end{bmatrix}, \quad \mathbf{v}_2 = \begin{bmatrix} \sqrt{\beta_{2x}} e^{i\nu_2} \\ -\dfrac{iu+\alpha_{2x}}{\sqrt{\beta_{2x}}} e^{i\nu_2} \\ \sqrt{\beta_{2y}} \\ -\dfrac{i(1-u)+\alpha_{2y}}{\sqrt{\beta_{2y}}} \end{bmatrix}. \tag{2.51}$$

That yields the following expression for matrix $\mathbf{V}$ (see Eq. (2.16)):

$$\mathbf{V} = \begin{bmatrix} \sqrt{\beta_{1x}} & 0 & \sqrt{\beta_{2x}}\cos\nu_2 & -\sqrt{\beta_{2x}}\sin\nu_2 \\ -\dfrac{\alpha_{1x}}{\sqrt{\beta_{1x}}} & \dfrac{1-u}{\sqrt{\beta_{1x}}} & \dfrac{u\sin\nu_2-\alpha_{2x}\cos\nu_2}{\sqrt{\beta_{2x}}} & \dfrac{u\cos\nu_2+\alpha_{2x}\sin\nu_2}{\sqrt{\beta_{2x}}} \\ \sqrt{\beta_{1y}}\cos\nu_1 & -\sqrt{\beta_{1y}}\sin\nu_1 & \sqrt{\beta_{2y}} & 0 \\ \dfrac{u\sin\nu_1-\alpha_{1y}\cos\nu_1}{\sqrt{\beta_{1y}}} & \dfrac{u\cos\nu_1+\alpha_{1y}\sin\nu_1}{\sqrt{\beta_{1y}}} & -\dfrac{\alpha_{2y}}{\sqrt{\beta_{2y}}} & \dfrac{1-u}{\sqrt{\beta_{2y}}} \end{bmatrix}. \tag{2.52}$$

Below we will call eleven functions, $\beta_{1x}(s)$, $\beta_{1y}(s)$, $\beta_{2x}(s)$, $\beta_{2y}(s)$, $\alpha_{1x}(s)$, $\alpha_{1y}(s)$, $\alpha_{2x}(s)$, $\alpha_{2y}(s)$, $u(s)$, $\nu_1(s)$, and $\nu_2(s)$, the generalized Twiss functions. Only eight of them are independent. Other three can be determined from the symplecticity conditions. Although for known eigenvectors the Twiss parameters can be determined uniquely, it is not the case if we know only alpha- and beta-functions. In this case an application of symplecticity conditions leaves four independent solutions for the eigenvectors. Two of them are related to the sign choice for u in Eq. (2.47), and other two (for each choice of u) are related to uncertainty of ν_1 and ν_2 in Eq. (2.50). The latter is related to the fact that the mirror reflection with respect to the x or y axis does not change β's and α's but changes the relative signs for the x and y components of the eigenvectors, with subsequent change of ν_1 and ν_2 by π. It can also be achieved by a change of the coupling sign (simultaneous sign change for gradients of all skew quads and magnetic fields of all solenoids), which does not change the beta-functions but does change the ν-functions by π. To choose a unique solution for the eigenvectors, one needs to know which of the two choices for u and ν_1 (or ν_2) needs to be taken in addition to the alpha- and beta-functions.

In the case of weak coupling, one should normally choose $\mathbf{v}_1$ as the eigenvector, which mainly relates to the horizontal motion, and $\mathbf{v}_2$ to the vertical motion. In the case of strong coupling, the choice is arbitrary. As can be seen from Eq. (2.51), in determining beta- and alpha-functions, swapping two eigenvectors causes the following redefinitions: $\beta_{1x} \leftrightarrow \beta_{2x}$, $\beta_{1y} \leftrightarrow \beta_{2y}$, $\alpha_{1x} \leftrightarrow \alpha_{2x}$, $\alpha_{1y} \leftrightarrow \alpha_{2y}$, $u \rightarrow 1-u$, $\nu_1 \rightarrow -\nu_2$, and $\nu_2 \rightarrow -\nu_1$.

2.1.6 Derivatives of the Tunes and Beta-Functions

Let us consider the relations between the beta- and alpha-functions and the beta-functions and the betatron phase advances. A differential trajectory displacement related to the first eigenvector can be expressed as follows:

$$x(s+ds)=x(s)+x'(s)ds=x(s)+\left(p_x(s)+\frac{R}{2}y\right)ds$$
$$=\sqrt{2I_1}\mathrm{Re}\left(\left(\sqrt{\beta_{1x}(s)}+\left[-\frac{i(1-u(s))+\alpha_{1x}(s)}{\sqrt{\beta_{1x}(s)}}+\frac{R}{2}\sqrt{\beta_{1y}(s)}\mathrm{e}^{i\nu_1(s)}\right]ds\right)\mathrm{e}^{-i(\mu_1(s)+\psi_1)}\right). \tag{2.53}$$

Alternatively, one can express particle position through the beta-functions at the new coordinate $s+ds$:

$$\begin{aligned} x(s+ds) &= \mathrm{Re}\left(\sqrt{2I_1\beta_x(s+ds)}\mathrm{e}^{-i(\mu_1(s+ds)+\psi)}\right) \\ &= \sqrt{2I_1}\mathrm{Re}\left(\left[\sqrt{\beta_{1x}(s)}+\frac{d\beta_{1x}}{2\sqrt{\beta_{1x}(s)}}-i\sqrt{\beta_{1x}(s)}d\mu\right]\mathrm{e}^{-i(\mu_1(s)+\psi)}\right). \end{aligned} \tag{2.54}$$

Comparing both the imaginary and real parts of Eqs. (2.53) and (2.54), one obtains

$$\begin{aligned} \frac{d\beta_{1x}}{ds} &= -2\alpha_{1x}+R\sqrt{\beta_{1x}\beta_{1y}}\cos\nu_1, \\ \frac{d\mu_1}{ds} &= \frac{1-u}{\beta_{1x}}-\frac{R}{2}\sqrt{\frac{\beta_{1y}}{\beta_{1x}}}\sin\nu_1. \end{aligned} \tag{2.55}$$

Similarly, one can write down equivalent expressions for the vertical displacement,

$$y(s+ds) = y(s) + y'(s)ds = y(s) + \left(p_y(s) - \frac{R}{2}x\right)ds$$

$$= \sqrt{2I_1}\mathrm{Re}\left(\left(\sqrt{\beta_{1y}(s)}e^{i\nu_1(s)} - \left[\frac{iu(s)+\alpha_{1y}(s)}{\sqrt{\beta_{1y}(s)}}e^{i\nu_1(s)} + \frac{R}{2}\sqrt{\beta_{1x}(s)}\right]ds\right)e^{-i(\mu_1(s)+\psi_1)}\right),$$

and

$$y(s+ds) = \sqrt{2I_1}\mathrm{Re}\left(\left[\sqrt{\beta_{1y}(s)} + \frac{d\beta_{1y}}{2\sqrt{\beta_{1y}(s)}} + i\sqrt{\beta_{1y}(s)}(d\nu_1 - d\mu_1)\right]e^{-i(\mu_1(s)+\psi-\nu_1(s))}\right),$$

which yields

$$\begin{aligned} \frac{d\beta_{1y}}{ds} &= -2\alpha_{1y} - R\sqrt{\beta_{1x}\beta_{1y}}\cos\nu_1, \\ \frac{d\mu_1}{ds} - \frac{d\nu_1}{ds} &= \frac{u}{\beta_{1y}} + \frac{R}{2}\sqrt{\frac{\beta_{1x}}{\beta_{1y}}}\sin\nu_1. \end{aligned} \tag{2.56}$$

Similar calculations carried out for the second eigenvector yield

$$\begin{aligned} \frac{d\beta_{2y}}{ds} &= -2\alpha_{2y} - R\sqrt{\beta_{2x}\beta_{2y}}\cos\nu_2, \\ \frac{d\mu_2}{ds} &= \frac{1-u}{\beta_{2y}} + \frac{R}{2}\sqrt{\frac{\beta_{2x}}{\beta_{2y}}}\sin\nu_2, \\ \frac{d\beta_{2x}}{ds} &= -2\alpha_{2x} + R\sqrt{\beta_{2x}\beta_{2y}}\cos\nu_2, \\ \frac{d\mu_2}{ds} - \frac{d\nu_2}{ds} &= \frac{u}{\beta_{2x}} - \frac{R}{2}\sqrt{\frac{\beta_{2y}}{\beta_{2x}}}\sin\nu_2. \end{aligned} \tag{2.57}$$

One can see that in the absence of longitudinal magnetic field, the alpha- and beta-functions are related the same way as for the uncoupled case ($\alpha = -(d\beta/ds)/2$) and the derivatives of the phase advances $d\mu_1/ds$ and $d\mu_2/ds$ are proportional to $(1-u)$ and are positive. That explains the selection rule for the eigenvectors formulated in Sect. 4.1.5 which requires u_1 and u_4 being positive ($u_1 = u_4 = 1 - u \geq 0$). Note that there is no formal requirement for $d(\mu_1 + \nu_1)/ds$

and $d(\mu_2+\nu_2)/ds$ being also positive, and therefore u can be negative,[3] while in the majority of practical cases, it belongs to the [0, 1] interval.

2.1.7 Representation of Transfer Matrix and Second-Order Moments in Terms of Generalized Twiss Functions

One can derive a useful representation of the transfer matrix $\mathbf{M}_{1,2} \equiv \mathbf{M}(s_1, s_2)$ between two points of a transfer line in terms of the generalized Twiss functions. Using the definitions of eigenvector and matrix $\mathbf{V}$ (see Eqs. (2.16) and (2.41)), one can derive the following identity:

$$\mathbf{V}_2\mathbf{S} = \mathbf{M}_{12}\mathbf{V}_1. \tag{2.58}$$

Here $\mathbf{V}_1$ and $\mathbf{V}_2$ are the $\mathbf{V}$-matrices given by Eq. (2.52) for the initial and final points. The matrix $\mathbf{S}$ is

$$\mathbf{S} = \begin{bmatrix} \cos\Delta\mu_1 & \sin\Delta\mu_1 & 0 & 0 \\ -\sin\Delta\mu_1 & \cos\Delta\mu_1 & 0 & 0 \\ 0 & 0 & \cos\Delta\mu_2 & \sin\Delta\mu_2 \\ 0 & 0 & -\sin\Delta\mu_2 & \cos\Delta\mu_2 \end{bmatrix}, \tag{2.59}$$

where $\Delta\mu_{1,2}$ are the betatron phase advances between points 1 and 2 for the first and second modes. Multiplying both sides of Eq. (2.59) by the inverse matrix, $\mathbf{V}_1^{-1} = -\mathbf{U}\mathbf{V}_1^{\mathrm{T}}\mathbf{U}$, as given by Eq. (2.9), allows one to express the transfer matrix, $\mathbf{M}_{12}$, in the form

$$\mathbf{M}_{12} = -\mathbf{V}_2\mathbf{S}\mathbf{U}\mathbf{V}_1^{\mathrm{T}}\mathbf{U}. \tag{2.60}$$

In the case of the one-turn transfer matrix, the matrices $\mathbf{V}_1$ and $\mathbf{V}_2$ are equal and Eq. (2.60) simplifies. That results in the following expressions for the matrix elements of diagonal 2×2 sub-matrices:

$$\begin{aligned} M_{11} &= (1-u)\cos\mu_1 + \alpha_{1x}\sin\mu_1 + u\cos\mu_2 + \alpha_{2x}\sin\mu_2, \\ M_{12} &= \beta_{1x}\sin\mu_1 + \beta_{2x}\sin\mu_2, \\ M_{21} &= -\frac{(1-u)^2+\alpha_{1x}^2}{\beta_{1x}}\sin\mu_1 - \frac{u^2+\alpha_{2x}^2}{\beta_{2x}}\sin\mu_2, \\ M_{22} &= (1-u)\cos\mu_1 + u\cos\mu_2 - \alpha_{1x}\sin\mu_1 - \alpha_{2x}\sin\mu_2, \end{aligned}$$

[3] The Tevatron lattice is based on the detailed optics measurement and takes into account large coupling terms coming mainly from the skew-quadrupole components of the superconducting dipoles. If the coupling corrections are adjusted to minimize the tune split, the value of coupling parameter u varies along the lattice in the range of about [−0.002, 0.04].

$$
\begin{aligned}
M_{33} &= u\cos\mu_1 + (1-u)\cos\mu_2 + \alpha_{2y}\sin\mu_2 + \alpha_{1y}\sin\mu_1, \\
M_{34} &= \beta_{1y}\sin\mu_1 + \beta_{2y}\sin\mu_2, \\
M_{43} &= -\frac{u^2+\alpha_{1y}^2}{\beta_{1y}}\sin\mu_1 - \frac{(1-u)^2+\alpha_{2y}^2}{\beta_{2y}}\sin\mu_2, \\
M_{44} &= u\cos\mu_1 + (1-u)\cos\mu_2 - \alpha_{1y}\sin\mu_1 - \alpha_{2y}\sin\mu_2,
\end{aligned}
\tag{2.61}
$$

where μ_1 and μ_2 are the betatron tunes of two betatron modes. The elements for off-diagonal sub-matrices can be found in [3].

We also present here the elements of matrix $\mathbf{\Sigma}$ used in other chapters of this book (see Eq. (2.25)):

$$
\begin{aligned}
&\Sigma_{11} \equiv \langle x^2\rangle = \varepsilon_1\beta_{1x} + \varepsilon_2\beta_{2x}, \quad \Sigma_{33} \equiv \langle y^2\rangle = \varepsilon_1\beta_{1y} + \varepsilon_2\beta_{2y}, \\
&\Sigma_{12} \equiv \langle xp_x\rangle = \Sigma_{21} = -\varepsilon_1\alpha_{1x} - \varepsilon_2\alpha_{2x}, \quad \Sigma_{34} \equiv \langle yp_y\rangle = \Sigma_{43} = -\varepsilon_1\alpha_{1y} - \varepsilon_2\alpha_{2y}, \\
&\Sigma_{13} \equiv \langle xy\rangle = \Sigma_{31} = \varepsilon_1\sqrt{\beta_{1x}\beta_{1y}}\cos\nu_1 + \varepsilon_2\sqrt{\beta_{2x}\beta_{2y}}\cos\nu_2, \\
&\Sigma_{22} \equiv \langle {p_x}^2\rangle = \varepsilon_1\frac{(1-u)^2+{\alpha_{1x}}^2}{\beta_{1x}} + \varepsilon_2\frac{u^2+{\alpha_{2x}}^2}{\beta_{2x}}, \\
&\Sigma_{44} \equiv \langle {p_y}^2\rangle = \varepsilon_1\frac{u^2+{\alpha_{1y}}^2}{\beta_{1y}} + \varepsilon_2\frac{(1-u)^2+{\alpha_{2y}}^2}{\beta_{2y}}, \\
&\Sigma_{24} \equiv \langle p_xp_y\rangle = \Sigma_{42} = \varepsilon_1\frac{\left(\alpha_{1y}(1-u)-\alpha_{1x}u\right)\sin\nu_1 + \left(u(1-u)+\alpha_{1x}\alpha_{1y}\right)\cos\nu_1}{\sqrt{\beta_{1x}\beta_{1y}}} \\
&\qquad + \varepsilon_2\frac{\left(\alpha_{2x}(1-u)-\alpha_{2y}u\right)\sin\nu_2 + \left(u(1-u)+\alpha_{2x}\alpha_{2y}\right)\cos\nu_2}{\sqrt{\beta_{2x}\beta_{2y}}}.
\end{aligned}
\tag{2.62}
$$

For other elements of matrix $\mathbf{\Sigma}$ and the expression of matrix $\mathbf{\Xi}$, we refer reader to [3].

2.1.8 *Edwards–Teng Parameterization*

The material presented in Sects. 2.1.5, 2.1.6, and 2.1.7 is based on the extension of the Mais–Ripken parameterization presented in [3]. However, the consideration of coupled motion would be incomplete without a discussion of the Edwards–Teng parameterization [6], which was proposed earlier and is presently one of the most popular parameterizations for description of coupled optics. It is based on a canonical transform $\mathbf{R}$ which reduces a 4×4 transfer matrix

$$\mathbf{M} = \begin{bmatrix} \mathbf{P} & \mathbf{p} \\ \mathbf{q} & \mathbf{Q} \end{bmatrix} \tag{2.63}$$

to its normal mode form

$$\widetilde{\mathbf{M}} = \mathbf{RMR}^{-1}, \tag{2.64}$$

where

$$\widetilde{\mathbf{M}} = \begin{bmatrix} \mathbf{A} & \mathbf{0} \\ \mathbf{0} & \mathbf{B} \end{bmatrix} \tag{2.65}$$

and **P**, **p**, **Q**, **q**, **A**, and **B** are 2×2 matrices. Edwards and Teng suggested parameterizing a symplectic matrix **R** as follows:

$$\mathbf{R} = \begin{bmatrix} \mathbf{E}\cos\phi & -\mathbf{D}^{-1}\sin\phi \\ \mathbf{D}\sin\phi & \mathbf{E}\cos\phi \end{bmatrix}, \tag{2.66}$$

where **E** is the unit 2×2 matrix and **D** is a 2×2 symplectic matrix,

$$\mathbf{D} = \begin{bmatrix} a & b \\ c & d \end{bmatrix}. \tag{2.67}$$

Thus, matrix **R** is parameterized by four parameters: a, b, c, and ϕ. Matrix $\widetilde{\mathbf{M}}$ describes the particle motion in new coordinates and can be parameterized by six Twiss parameters: $\beta_1, \alpha_1, \mu_1, \beta_2, \alpha_2$, and μ_2 which are called the Twiss parameters of the decoupled motion. Edwards and Teng expressed them through the transfer matrix elements. Here we present their connection to the extended Mais–Ripken parameterization considered above:

$$\begin{aligned} &\sin\phi = \pm\sqrt{u}, \\ &\beta_1 = \frac{\beta_{1x}}{1-u}, \quad \alpha_1 = \frac{\alpha_{1x}}{1-u}, \quad \beta_2 = \frac{\beta_{2y}}{1-u}, \quad \alpha_2 = \frac{\alpha_{2y}}{1-u}, \end{aligned} \tag{2.68}$$

$$\mathbf{R} = \frac{\pm 1}{\sqrt{1-u}} \begin{bmatrix} 1 & 0 & -d_t & b_t \\ 0 & 1 & c_t & -a_t \\ a_t & b_t & 1 & 0 \\ c_t & d_t & 0 & 1 \end{bmatrix}, \quad \begin{aligned} a_t &= \sqrt{\beta_{2y}/\beta_{2x}}\left(\alpha_{2x}\sin\nu_2 + u\cos\nu_2\right), \\ b_t &= \sqrt{\beta_{1x}\beta_{1y}}\sin\nu_1, \\ d_t &= -\sqrt{\beta_{1x}/\beta_{1y}}\left(u\cos\nu_1 + \alpha_{1y}\sin\nu_1\right), \\ c_t &= (a_t d_t + u - 1)/b_t. \end{aligned} \tag{2.69}$$

Details of calculations can be found in [3]. Although, the top Eq. (2.68) yields four different solutions for angle ϕ, there are unique solutions for the beta- and alpha-functions of the decoupled motion and matrix **R**. Note that the choice of sign for matrix **R** in Eq. (2.69) is determined by the requirement that $\sqrt{\beta_{1,2}}$ are positive. However, a problem appears if the value of u is negative somewhere in the lattice. That results in ϕ being purely imaginary. The solution considered in [6] suggests a replacement of sin(ϕ) and cos(ϕ) by sinh(ϕ) and cosh(ϕ) with an appropriate sign changes in the symplectic transform of Eq. (2.66). It formally addresses the issue but still requires a redefinition of Eq. (2.66) symplectic transforms every time u changes its sign.

Edwards and Teng determined the phase advance of the betatron motion using a standard recipe for the uncoupled motion:

$$\widetilde{\mathbf{v}}_i(s)\mathrm{e}^{-i\mu_i(s)} = \widetilde{\mathbf{M}}(0,s)\widetilde{\mathbf{v}}_i(0), \tag{2.70}$$

where $\widetilde{\mathbf{v}}_i(s)$ are the eigenvectors of decoupled motion. It is important to note that the betatron phase advances of both parameterizations are equal; i.e., the betatron phase advance for the Edwards–Teng representation is directly related to particle oscillations in the x or y plane, depending on which plane a particular eigenvector is referenced to.

As will be shown in the next section, the value of u is changing fast if a system approaches the coupling resonance. Consequently, the beta-functions of extended Mais–Ripken parameterization also change fast although the sums $\beta_{1x}+\beta_{2x}$ and $\beta_{1y}+\beta_{2y}$ stay approximately constant. In contrast, the Edwards–Teng beta-functions are insensitive to the coupling resonance, while ϕ and matrix **R** are sensitive.

2.1.9 Betatron Tunes

Substituting the ring transfer matrix expressed in the block form of Eq. (2.63) into the symplecticity conditions of Eqs. (2.8) and (2.10) and performing matrix multiplication, one obtains

$$\begin{aligned}
&\begin{bmatrix} \mathbf{P}^{\mathrm{T}}\mathbf{U}_2\mathbf{P}+\mathbf{q}^{\mathrm{T}}\mathbf{U}_2\mathbf{q} & \mathbf{P}^{\mathrm{T}}\mathbf{U}_2\mathbf{p}+\mathbf{q}^{\mathrm{T}}\mathbf{U}_2\mathbf{Q} \\ \mathbf{p}^{\mathrm{T}}\mathbf{U}_2\mathbf{P}+\mathbf{Q}^{\mathrm{T}}\mathbf{U}_2\mathbf{q} & \mathbf{p}^{\mathrm{T}}\mathbf{U}_2\mathbf{p}+\mathbf{Q}^{\mathrm{T}}\mathbf{U}_2\mathbf{Q} \end{bmatrix} = \begin{bmatrix} \mathbf{U}_2 & 0 \\ 0 & \mathbf{U}_2 \end{bmatrix}, \\
&\begin{bmatrix} \mathbf{P}\mathbf{U}_2\mathbf{P}^{\mathrm{T}}+\mathbf{p}\mathbf{U}_2\mathbf{p}^{\mathrm{T}} & \mathbf{P}\mathbf{U}_2\mathbf{q}^{\mathrm{T}}+\mathbf{p}\mathbf{U}_2\mathbf{Q}^{\mathrm{T}} \\ \mathbf{q}\mathbf{U}_2\mathbf{P}^{\mathrm{T}}+\mathbf{Q}\mathbf{U}_2\mathbf{p}^{\mathrm{T}} & \mathbf{q}\mathbf{U}_2\mathbf{q}^{\mathrm{T}}+\mathbf{Q}\mathbf{U}_2\mathbf{Q}^{\mathrm{T}} \end{bmatrix} = \begin{bmatrix} \mathbf{U}_2 & 0 \\ 0 & \mathbf{U}_2 \end{bmatrix},
\end{aligned} \tag{2.71}$$

where $\mathbf{U}_2$ is the two-dimensional unit symplectic matrix.

Expanding the diagonal sub-matrices, one obtains four scalar equations:

$$\begin{aligned}
&\det(\mathbf{P})+\det(\mathbf{q})=1, \quad \det(\mathbf{p})+\det(\mathbf{Q})=1, \\
&\det(\mathbf{P})+\det(\mathbf{p})=1, \quad \det(\mathbf{q})+\det(\mathbf{Q})=1.
\end{aligned} \tag{2.72}$$

The solution of above equations yields that the determinants for two diagonal and two off-diagonal matrices are equal:

$$\kappa = \det(\mathbf{p}) = \det(\mathbf{q}), \quad 1 - \kappa = \det(\mathbf{P}) = \det(\mathbf{Q}), \tag{2.73}$$

where parameter κ characterizes the coupling strength. The off-diagonal sub-matrices in each matrix equation of Eq. (2.71) are related by matrix transpose, and the matrix equations are linearly dependent leaving only four independent scalar equations. That bounds up matrices $\mathbf{p}$ and $\mathbf{q}$:

$$\mathbf{p} = \mathbf{U}_2\left(\mathbf{P}^{\mathrm{T}}\right)^{-1}\mathbf{q}^{\mathrm{T}}\mathbf{U}_2\mathbf{Q} = \frac{1}{1-\kappa}\mathbf{P}\mathbf{U}_2\mathbf{q}^{\mathrm{T}}\mathbf{U}_2\mathbf{Q}, \tag{2.74}$$

where we took into account that $\mathbf{P}^{-1} = -\mathbf{U}_2\mathbf{P}^{\mathrm{T}}\mathbf{U}_2/(1-\kappa)$.

To separate coupling effects from effects of uncoupled betatron motion, let us consider the betatron motion in the normalized coordinates, so that in the new coordinates the ring transfer matrix would be presented in the following form:

$$\begin{gathered}\mathbf{M}_c = \begin{bmatrix} \mathbf{P}_c & \mathbf{p}_c \\ \mathbf{q}_c & \mathbf{Q}_c \end{bmatrix}, \quad \mathbf{P}_c = \sqrt{1-\kappa}\begin{bmatrix} \cos\mu_x & \sin\mu_x \\ -\sin\mu_x & \cos\mu_x \end{bmatrix}, \\ \mathbf{Q}_c = \sqrt{1-\kappa}\begin{bmatrix} \cos\mu_y & \sin\mu_y \\ -\sin\mu_y & \cos\mu_y \end{bmatrix},\end{gathered} \tag{2.75}$$

where we explicitly took into account that $\det(\mathbf{P}_c) = \det(\mathbf{Q}_c)$. Here and below we denote by index c the vectors and matrices in the normalized coordinate frame. A symplectic matrix $\mathbf{R}_c$ performs transformation from the old to the new coordinates, $\mathbf{x}_c = \mathbf{R}_c\mathbf{x}$, and can be chosen in the following form:

$$\begin{gathered}\mathbf{R}_c = \begin{bmatrix} \mathbf{R}_{cx} & 0 \\ 0 & \mathbf{R}_{cy} \end{bmatrix}, \quad \mathbf{R}_{cx} = \begin{bmatrix} 1/\sqrt{\beta_{xc}} & 0 \\ \alpha_{xc}/\sqrt{\beta_{xc}} & \sqrt{\beta_{xc}} \end{bmatrix}, \\ \mathbf{R}_{cy} = \begin{bmatrix} 1/\sqrt{\beta_{yc}} & 0 \\ \alpha_{yc}/\sqrt{\beta_{yc}} & \sqrt{\beta_{yc}} \end{bmatrix}.\end{gathered} \tag{2.76}$$

Then, the ring transfer matrix in the normalized coordinates is $\mathbf{M}_c = \mathbf{R}_c\mathbf{M}\mathbf{R}_c^{-1}$. Performing matrix multiplication and requiring the resulting matrix to be in the form of Eq. (2.75), one obtains the parameters of matrix $\mathbf{R}_c$:

$$\begin{gathered}\beta_{xc} = \sqrt{\frac{-4M_{12}^2}{4M_{12}M_{21} + (M_{11} - M_{22})^2}}, \quad \alpha_{xc} = \beta_{xc}\frac{M_{11} - M_{22}}{2M_{12}}, \\ \beta_{yc} = \sqrt{\frac{-4M_{34}^2}{4M_{34}M_{43} + (M_{33} - M_{44})^2}}, \quad \alpha_{yc} = \beta_{yc}\frac{M_{33} - M_{44}}{2M_{34}}.\end{gathered} \tag{2.77}$$

The symplectic transform does not change the trace of the diagonal sub-matrices. That yields

$$\cos\mu_x = \frac{\mathrm{tr}(\mathbf{P}_c)}{2(1-\kappa)} = \frac{\mathrm{tr}(\mathbf{P})}{2(1-\kappa)} = \frac{M_{11}+M_{22}}{2(1-\kappa)},$$
$$\cos\mu_y = \frac{\mathrm{tr}(\mathbf{Q}_c)}{2(1-\kappa)} = \frac{\mathrm{tr}(\mathbf{Q})}{2(1-\kappa)} = \frac{M_{33}+M_{44}}{2(1-\kappa)}. \tag{2.78}$$

The off-diagonal sub-matrices of matrix $\mathbf{M}_c$ are

$$\mathbf{q}_c \equiv \begin{bmatrix} a_c & b_c \\ c_c & d_c \end{bmatrix} = \mathbf{R}_{cx}\mathbf{q}\mathbf{R}_{cy}^{-1}, \quad \mathbf{p}_c \equiv \begin{bmatrix} \hat{a}_c & \hat{b}_c \\ \hat{c}_c & \hat{d}_c \end{bmatrix} = \mathbf{R}_{cx}\mathbf{p}\mathbf{R}_{cy}^{-1}. \tag{2.79}$$

Note that Eq. (2.74) uniquely couples a_c, b_c, c_c, and d_c with $\hat{a}_c, \hat{b}_c, \hat{c}_c$, and $\hat{d}_c$. That leaves only four independent parameters for coupling characterization out of eight parameters of off-diagonal sub-matrices. There is another useful property of matrices $\mathbf{P}_c$, $\mathbf{Q}_c$, $\mathbf{p}_c$, and $\mathbf{q}_c$:

$$\mathrm{tr}\left(\mathbf{P}_c{}^{\mathrm{T}}\mathbf{P}_c\right) = \mathrm{tr}\left(\mathbf{Q}_c{}^{\mathrm{T}}\mathbf{Q}_c\right), \quad \mathrm{tr}\left(\mathbf{p}_c{}^{\mathrm{T}}\mathbf{p}_c\right) = \mathrm{tr}\left(\mathbf{q}_c{}^{\mathrm{T}}\mathbf{q}_c\right). \tag{2.80}$$

The first equation follows from the definitions of $\mathbf{P}_c$ and $\mathbf{Q}_c$. The following string of conversions proves the second equation:

$$\begin{aligned}
\mathrm{tr}\left(\mathbf{p}_c{}^{\mathrm{T}}\mathbf{p}_c\right) &= \mathrm{tr}\left(\left(\mathbf{U}_2\left(\mathbf{P}_c{}^{\mathrm{T}}\right)^{-1}\mathbf{q}_c{}^{\mathrm{T}}\mathbf{U}_2\mathbf{Q}_c\right)^{\mathrm{T}}\mathbf{U}_2\left(\mathbf{P}_c{}^{\mathrm{T}}\right)^{-1}\mathbf{q}_c{}^{\mathrm{T}}\mathbf{U}_2\mathbf{Q}_c\right) \\
&= \mathrm{tr}\left(\mathbf{Q}_c{}^{\mathrm{T}}\mathbf{U}_2\mathbf{q}_c\mathbf{P}_c{}^{-1}\mathbf{U}_2\mathbf{U}_2\left(\mathbf{P}_c{}^{\mathrm{T}}\right)^{-1}\mathbf{q}_c{}^{\mathrm{T}}\mathbf{U}_2\mathbf{Q}_c\right) \\
&= -\mathrm{tr}\left(\mathbf{Q}_c{}^{\mathrm{T}}\mathbf{U}_2\mathbf{q}_c\mathbf{P}_c{}^{-1}\left(\mathbf{P}_c{}^{\mathrm{T}}\right)^{-1}\mathbf{q}_c{}^{\mathrm{T}}\mathbf{U}_2\mathbf{Q}_c\right) \\
&= \frac{-1}{1-\kappa}\mathrm{tr}\left(\mathbf{Q}_c{}^{\mathrm{T}}\mathbf{U}_2\mathbf{q}_c\mathbf{q}_c{}^{\mathrm{T}}\mathbf{U}_2\mathbf{Q}_c\right) = -\mathrm{tr}\left(\mathbf{U}_2\mathbf{q}_c\mathbf{q}_c{}^{\mathrm{T}}\mathbf{U}_2\right) = \mathrm{tr}\left(\mathbf{q}_c\mathbf{q}_c{}^{\mathrm{T}}\right).
\end{aligned}$$

Here we used that $\mathbf{P}_c{}^{-1} = \mathbf{P}_c{}^{\mathrm{T}}/(1-\kappa)$, $\mathbf{Q}_c{}^{-1} = \mathbf{Q}_c{}^{\mathrm{T}}/(1-\kappa)$, $\mathbf{U}_2\mathbf{U}_2 = -\mathbf{I}$, and $\mathrm{tr}(\mathbf{AB}) = \mathrm{tr}(\mathbf{BA})$.

To find the betatron tunes, we follow the standard recipe for finding roots of dispersion equation: $\det(\mathbf{M}_c - \lambda\mathbf{I}) = 0$. Computing the determinant and performing further simplifications, we obtain

$$\begin{aligned}
&\lambda^4 - 2\sqrt{1-\kappa}\left(\cos\mu_x + \cos\mu_y\right)\left(\lambda^3 + \lambda\right) \\
&+ \left(2(1-\kappa)\left(1 + \cos\mu_x\cos\mu_y\right) - \mathrm{tr}(\mathbf{p}_c\mathbf{q}_c)\right)\lambda^2 + 1 \\
&= 0.
\end{aligned} \tag{2.81}$$

In the case of stable motion, the solution of Eq. (2.81) consists of two complex conjugated pairs. The corresponding fourth-order polynomial can be presented in the following form:

$$\prod_{i=1}^{4}(\lambda-\lambda_i)=\left(\lambda^2-2\lambda\cos\mu_1+1\right)\left(\lambda^2-2\lambda\cos\mu_2+1\right)$$
$$=\lambda^4-2\left(\lambda^3+\lambda\right)\left(\cos\mu_1+\cos\mu_2\right)+2(1+2\cos\mu_1\cos\mu_2)\lambda^2+1=0. \tag{2.82}$$

Comparing Eqs. (2.81) and (2.82), we obtain the following system of equations:

$$\begin{aligned}\cos\mu_1+\cos\mu_2&=\sqrt{1-\kappa}\left(\cos\mu_x+\cos\mu_y\right),\\ 1+\cos\mu_1\cos\mu_2&=(1-\kappa)\left(1+\cos\mu_x\cos\mu_y\right)-\mathrm{tr}(\mathbf{p}_c\mathbf{q}_c)/2.\end{aligned} \tag{2.83}$$

The solution is

$$\cos\mu_{1,2}=\sqrt{1-\kappa}\frac{\cos\mu_x+\cos\mu_y}{2}$$
$$\pm\sqrt{(1-\kappa)\left(\frac{\cos\mu_x-\cos\mu_y}{2}\right)^2+\frac{1}{2}\left(\kappa+\frac{1}{2}\mathrm{tr}(\mathbf{p}_c\mathbf{q}_c)\right)}. \tag{2.84}$$

The same as sub-matrices **p** and **q**, the sub-matrices $\mathbf{p}_c$ and $\mathbf{q}_c$ are related by Eq. (2.74). That allows one to express $\mathrm{tr}(\mathbf{p}_c\mathbf{q}_c)$ through matrix $\mathbf{q}_c$. Performing matrix multiplication, one obtains

$$\mathrm{tr}(\mathbf{p}_c\mathbf{q}_c)=\frac{1}{1-\kappa}\mathrm{tr}\left(\mathbf{P}_c\mathbf{U}_2\mathbf{q}_c{}^{\mathrm{T}}\mathbf{U}_2\mathbf{Q}_c\mathbf{q}_c\right)=\sin\mu_x\sin\mu_y\mathrm{tr}\left(\mathbf{q}_c{}^{\mathrm{T}}\mathbf{q}_c\right)-2\cos\mu_x\cos\mu_y\det(\mathbf{q}_c)$$
$$=\sin\mu_x\sin\mu_y\left(a_c{}^2+b_c{}^2+c_c{}^2+d_c{}^2\right)-2\cos\mu_x\cos\mu_y\left(a_cd_c-b_cc_c\right). \tag{2.85}$$

To express the tunes directly through the transfer matrix elements, we take into account that the coefficients in Eq. (2.81) do not change when we perform a transform to the normalized coordinates. In particular it means that coefficients in front of λ and λ^3 are equal. It simplifies the calculations. As a result, we obtain the dispersion equation:

$$\lambda^4-\mathrm{tr}(\mathbf{M})\left(\lambda^3+\lambda\right)+\Lambda\lambda^2+1=0, \tag{2.86}$$

where

$$\begin{aligned}\Lambda=&(M_{11}M_{22}-M_{12}M_{21})+(M_{33}M_{44}-M_{34}M_{43})+(M_{11}M_{33}-M_{13}M_{31})\\ &+(M_{11}M_{44}-M_{14}M_{41})+(M_{33}M_{22}-M_{23}M_{32})+(M_{22}M_{44}-M_{24}M_{42}).\end{aligned} \tag{2.87}$$

Then, the solution is [6]

$$\cos\mu_{1,2} = \frac{1}{4}\left(\mathrm{tr}(\mathbf{M}) \pm \sqrt{(\mathrm{tr}(\mathbf{M}))^2 + 8 - 4\Lambda}\right). \tag{2.88}$$

2.1.10 Coupling Strength, Tune Split, and Width of Resonance Stop Band

There is no single parameter to completely characterize coupling. However, the parameter u (see Eq. (2.51)) is one of the most informative. It characterizes the relative contributions of x and y parts to the eigenvector normalization of Eq. (2.14), so that they are proportional to u or $1-u$. In the absence of coupling, the parameter u is equal to 0 (or 1 if x and y vectors are swapped). Note that, in the general case, the equality $u=0$ does not imply an absence of coupling. As one can see from Eqs. (2.45) and (2.49), the condition $u=0$ requires $A_x^{\,2}+\kappa_x^{\,2}=A_y^{\,2}+\kappa_y^{\,2}$ and yields $e^{i\nu_+}=(A_x+i\kappa_x)/(A_y-i\kappa_y)$ and $e^{i\nu_-}=(A_x+i\kappa_x)/(A_y+i\kappa_y)$. These equations do not require auxiliary beta-functions β_{1y} and β_{2x} to be equal to zero, and, consequently, the condition $u=0$ does not automatically mean an absence of coupling. Although strictly speaking u cannot be considered as a unique coupling parameter, it reflects the strength of coupling and is a good value to characterize it in practice. In particular $u=½$ corresponds to 100 % coupling when the motion for both eigenvectors is equally distributed in both planes. It is also useful to note that u does not change in the part of a beamline without coupling terms. Actually, in the absence of coupling, the x and y parts of the eigenvector, $\mathbf{v}_x$ and $\mathbf{v}_y$, are independent and their normalization, $\mathbf{v}_{x,y}{}^{+}\mathbf{U}_2\mathbf{v}_{x,y}=\{u,\ 1-u\}$, does not change because the determinants of the corresponding 2×2 transfer matrices are equal to 1. Here $\mathbf{U}_2$ is the 2D unit symplectic matrix.

To express the value of u through the betatron tunes, we add up the matrix elements M_{11} and M_{22} of Eq. (2.61). That yields $M_{11}+M_{22}=2(1-u)\cos\mu_1+2u\cos\mu_2$. Taking into account that it is also justified for $M_{c_{11}}+M_{c_{22}}$ and by definition $M_{c_{11}}+M_{c_{22}}=2\sqrt{1-\kappa}\cos\mu_x$, we obtain

$$\sqrt{1-\kappa}\cos\mu_x = (1-u)\cos\mu_1 + u\cos\mu_2.$$

That results in

$$u = \frac{\sqrt{1-\kappa}\cos\mu_x - \cos\mu_1}{\cos\mu_2 - \cos\mu_1}. \tag{2.89}$$

Let coupling be small,[4] $\mathrm{tr}(\mathbf{q}_c{}^{\mathrm{T}}\mathbf{q}_c)\ll 1$, and tunes be located in the vicinity of difference coupling resonance. Then, the fractional parts of the tunes, $\hat{\mu}_x, \hat{\mu}_y, \hat{\mu}_1,$

[4] The condition $\mathrm{tr}(\mathbf{q}_c{}^{\mathrm{T}}\mathbf{q}_c)\ll 1$ also results in that $|\kappa|\ll 1$. Actually, expressing both equations through the matrix elements, one obtains $\mathrm{tr}(\mathbf{q}_c{}^{\mathrm{T}}\mathbf{q}_c)=a^2+b^2+c^2+d^2$ and $\kappa\equiv\det(\mathbf{q}_c)=ad-bc$. Obviously, $|ad-bc|<a^2+b^2+c^2+d^2$.

and $\hat{\mu}_2$, are close to each other, and we introduce the following definitions: $\Delta\mu = \hat{\mu}_x - \hat{\mu}_y \ll 1$, $\hat{\mu}_1 = \hat{\mu}_x + \Delta\mu_1$, and $\hat{\mu}_2 = \hat{\mu}_x + \Delta\mu_2$. Consequently, we expand Eq. (2.84) into Taylor series:

$$\cos\mu_x - \Delta\mu_{1,2}\sin\mu_x \approx \left(1 - \frac{\kappa}{2}\right)\frac{2\cos\mu_x - \Delta\mu\sin\mu_x}{2} \pm \sqrt{\left(\frac{\Delta\mu\sin\mu_x}{2}\right)^2 + \frac{1}{2}\left(\kappa + \frac{1}{2}\mathrm{tr}(\mathbf{p}_c\mathbf{q}_c)\right)}. \quad (2.90)$$

Simplifying the above equation with the help of Eq. (2.85), leaving only the leading order terms and returning to the tunes from their differentials, we obtain

$$\hat{\mu}_{1,2} \approx \frac{1}{2}\left((\hat{\mu}_x + \hat{\mu}_y) \pm \sqrt{(\hat{\mu}_x - \hat{\mu}_y)^2 + \delta\mu_d{}^2}\right), \quad |\hat{\mu}_x - \hat{\mu}_y|, \delta\mu_d \ll 1, \quad (2.91)$$

where

$$\delta\mu_d = \sqrt{(a_c + d_c)^2 + (b_c - c_c)^2} = \sqrt{(\hat{a}_c + \hat{d}_c)^2 + (\hat{b}_c - \hat{c}_c)^2} \quad (2.92)$$

represents the minimum tune split. Substitution of Eq. (2.91) to Eq. (2.89) results in the coupling strength dependence on the tunes in the vicinity of difference coupling resonance:

$$u \approx \frac{1}{2}\left(1 - \frac{\hat{\mu}_x - \hat{\mu}_y}{\sqrt{(\hat{\mu}_x - \hat{\mu}_y)^2 + \delta\mu_d{}^2}}\right), \quad |\hat{\mu}_x - \hat{\mu}_y|, \quad \delta\mu_d \ll 1. \quad (2.93)$$

If $\hat{\mu}_x = \hat{\mu}_y$, the coupling parameter is equal to 1/2. That corresponds to 100 % coupled motion. In the case of $|\hat{\mu}_x - \hat{\mu}_y| \gg \delta\mu_d$, the coupling strength, u, is small (or close to one if the order of tunes is reversed.) That corresponds to a weakly coupled motion.

If the tunes are located in the vicinity of coupling sum resonance, then $\hat{\mu}_x + \hat{\mu}_y - 2\pi = \Delta\mu \ll 1$. Introducing the following definitions, $\hat{\mu}_1 = \hat{\mu}_x + \Delta\mu_1$ and $\hat{\mu}_2 = 2\pi - \hat{\mu}_x + \Delta\mu_2$, and expanding Eq. (2.84) into Taylor series, we obtain

$$\cos\mu_x - \Delta\mu_1\sin\mu_x \approx \left(1 - \frac{\kappa}{2}\right)\frac{2\cos\mu_x + \Delta\mu\sin\mu_x}{2} - \sqrt{\left(\frac{\Delta\mu\sin\mu_x}{2}\right)^2 + \frac{1}{2}\left(\kappa + \frac{1}{2}\mathrm{tr}(\mathbf{p}_c\mathbf{q}_c)\right)}. \quad (2.94)$$

Simplifying the above equation with the help of Eq. (2.85), leaving only leading order terms, repeating similar expansion for $\Delta\mu_2$, and returning to the tunes from their differentials, we obtain

$$\hat{\mu}_{1,2} \approx \frac{1}{2}\left(\pm(\hat{\mu}_x - \hat{\mu}_y) + 2\pi + \sqrt{(\hat{\mu}_x + \hat{\mu}_y - 2\pi)^2 - \delta\mu_s{}^2} \right), \quad |2\pi - (\hat{\mu}_x + \hat{\mu}_y)|, \delta\mu_s \ll 1, \tag{2.95}$$

where

$$\delta\mu_s = \sqrt{(a_c - d_c)^2 + (b_c + c_c)^2} = \sqrt{(\hat{a}_c - \hat{d}_c)^2 + (\hat{b}_c + \hat{c}_c)^2} \tag{2.96}$$

represents the total width of the resonance stop band. One can see that tunes μ_1 and μ_2 become imaginary if the tunes μ_x and μ_y are located inside the resonance stop band. Consequently, the particle motion becomes unstable.

2.1.11 Perturbation Theory for Coupled Motion

The symplecticity allows one to build an effective perturbation theory for the case of coupled motion. Let the unperturbed motion eigenvalues and eigenvectors be related by Eq. (2.11). Then, for the perturbed motion one can write

$$(\mathbf{I} + \mathbf{\Delta M})\mathbf{M}\tilde{\mathbf{v}}_j = (\lambda_j + \Delta\lambda_j)\tilde{\mathbf{v}}_j, \tag{2.97}$$

where the new transfer matrix, $(\mathbf{I} + \mathbf{\Delta M})\mathbf{M}$, is not necessarily a symplectic matrix. The eigenvectors of perturbed motion can be presented as a sum of the unperturbed ones,

$$\tilde{\mathbf{v}}_j = \mathbf{v}_j + \sum_{i=1}^{4} \varepsilon_{ij}\mathbf{v}_i, \quad \varepsilon_{ij} << 1, \tag{2.98}$$

and without limitation of generality, one can consider that $\varepsilon_{ii} = 0$ for every i. Substituting Eq. (2.98) into Eq. (2.97), linearizing the resulting equation, and using Eq. (2.11), one obtains

$$\sum_{i=1}^{4} (\lambda_i - \lambda_j)\varepsilon_{ij}\mathbf{v}_i = (\Delta\lambda_j\mathbf{I} - \mathbf{\Delta M}\ \mathbf{M})\mathbf{v}_j. \tag{2.99}$$

In the case of stable unperturbed motion, the eigenvalues and eigenvectors represent two complex conjugate pairs. Taking this into account, $[\mathbf{v}_1\ \ \mathbf{v}_2\ \ \mathbf{v}_3\ \ \mathbf{v}_4] \to [\mathbf{v}_1\ \ \mathbf{v}_1{}^*\ \ \mathbf{v}_2\ \ \mathbf{v}_2{}^*]$, and introducing complex matrix $\mathbf{V}_p = [\mathbf{v}_1\ \ \mathbf{v}_1{}^*\ \ \mathbf{v}_2\ \ \mathbf{v}_2{}^*]$, one can rewrite Eq. (2.99) in the form of two matrix equations:

$$\mathbf{V}_p \begin{bmatrix} 1 & 0 & 0 & 0 \\ 0 & \lambda_1 - \lambda_1{}^* & 0 & 0 \\ 0 & 0 & \lambda_1 - \lambda_2 & 0 \\ 0 & 0 & 0 & \lambda_1 - \lambda_2{}^* \end{bmatrix} \begin{bmatrix} \Delta\lambda_1 \\ \varepsilon_{21} \\ \varepsilon_{31} \\ \varepsilon_{41} \end{bmatrix} = \mathbf{\Delta M M v}_1,$$
$$\mathbf{V}_p \begin{bmatrix} \lambda_2 - \lambda_1 & 0 & 0 & 0 \\ 0 & \lambda_2 - \lambda_1{}^* & 0 & 0 \\ 0 & 0 & 1 & 0 \\ 0 & 0 & 0 & \lambda_2 - \lambda_2{}^* \end{bmatrix} \begin{bmatrix} \varepsilon_{12} \\ \varepsilon_{22} \\ \Delta\lambda_2 \\ \varepsilon_{42} \end{bmatrix} = \mathbf{\Delta M M v}_2. \quad (2.100)$$

Matrix $\mathbf{V}_p$ is built from symplectic vectors and its inverse is equal to

$$\mathbf{V}_p{}^{-1} = -\frac{1}{2i}\mathbf{U}\mathbf{V}_p{}^{\mathrm{T}}\mathbf{U}. \quad (2.101)$$

One can verify it by utilizing the eigenvector normalization of Eq. (2.14). Inversion of Eq. (2.100) with the help of Eq. (2.101) finally results in [7]

$$\begin{bmatrix} \Delta\lambda_1 \\ \varepsilon_{21} \\ \varepsilon_{31} \\ \varepsilon_{41} \end{bmatrix} = -\frac{\lambda_1}{2i} \begin{bmatrix} 1 & 0 & 0 & 0 \\ 0 & \lambda_1 - \lambda_1{}^* & 0 & 0 \\ 0 & 0 & \lambda_1 - \lambda_2 & 0 \\ 0 & 0 & 0 & \lambda_1 - \lambda_2{}^* \end{bmatrix}^{-1} \mathbf{U}\mathbf{V}_c{}^{\mathrm{T}}\mathbf{U}\,\mathbf{\Delta M}\,\mathbf{v}_1,$$
$$\begin{bmatrix} \varepsilon_{12} \\ \varepsilon_{22} \\ \Delta\lambda_2 \\ \varepsilon_{42} \end{bmatrix} = -\frac{\lambda_2}{2i} \begin{bmatrix} \lambda_2 - \lambda_1 & 0 & 0 & 0 \\ 0 & \lambda_2 - \lambda_1{}^* & 0 & 0 \\ 0 & 0 & 1 & 0 \\ 0 & 0 & 0 & \lambda_2 - \lambda_2{}^* \end{bmatrix}^{-1} \mathbf{U}\mathbf{V}_c{}^{\mathrm{T}}\mathbf{U}\,\mathbf{\Delta M}\,\mathbf{v}_2. \quad (2.102)$$

Multiplication of Eqs. (2.102) by $[\,1 \quad 0 \quad 0 \quad 0\,]$ and $[\,0 \quad 0 \quad 1 \quad 0\,]$, correspondingly, results in corrections for the eigenvalues:

$$\Delta\lambda_1 = -\frac{\lambda_1}{2i} \quad \mathbf{v}_1{}^{+}\mathbf{U}\,\mathbf{\Delta M}\,\mathbf{v}_1,$$
$$\Delta\lambda_2 = -\frac{\lambda_2}{2i} \quad \mathbf{v}_2{}^{+}\mathbf{U}\,\mathbf{\Delta M}\,\mathbf{v}_2. \quad (2.103)$$

Taking into account the relationship between the eigenvalue corrections and the tune shifts, $\Delta Q_n = i/(4\pi)\ (\Delta\lambda_n/\lambda_n)$, one obtains [8]

$$\Delta Q_1 = -\frac{1}{4\pi}\mathbf{v}_1{}^{+}\mathbf{U}\,\mathbf{\Delta M}\,\mathbf{v}_1,$$
$$\Delta Q_2 = -\frac{1}{4\pi}\mathbf{v}_2{}^{+}\mathbf{U}\,\mathbf{\Delta M}\,\mathbf{v}_2. \quad (2.104)$$

To demonstrate an application of the above formalism, let us find the tune shifts due to a local focusing perturbation. In the general case the perturbation of the

Hamiltonian is proportional to $\Phi_x x^2 + 2\Phi_s xy + \Phi_y y^2$. That results in the transfer matrix of the perturbation:

$$\mathbf{\Delta M} = \begin{bmatrix} 0 & 0 & 0 & 0 \\ -\Phi_x & 0 & -\Phi_s & 0 \\ 0 & 0 & 0 & 0 \\ -\Phi_s & 0 & -\Phi_y & 0 \end{bmatrix}.$$

Substituting it to Eq. (2.104), one obtains [7]

$$\begin{aligned} \Delta Q_1 &= \frac{1}{4\pi}\left(\Phi_x\beta_{1x} + 2\Phi_s\sqrt{\beta_{1x}\beta_{1y}}\cos\nu_1 + \Phi_y\beta_{1y}\right), \\ \Delta Q_2 &= \frac{1}{4\pi}\left(\Phi_x\beta_{2x} + 2\Phi_s\sqrt{\beta_{2x}\beta_{2y}}\cos\nu_2 + \Phi_y\beta_{2y}\right). \end{aligned} \tag{2.105}$$

One can see that in the case of uncoupled motion, $\beta_{1y} = \beta_{2x} = 0$, the tune shifts coincide with the well-known expression for the tune shift of uncoupled motion. Note that for a quadrupole field $\Phi_x = -\Phi_y$.

2.1.12 Sum and Difference Coupling Resonances

An analysis of the coupled motion using a perturbation theory applied directly to the equations describing initially uncoupled motion is useful in many applications. Let us consider the two uncoupled modes x and y. For each of them the formalism described in Sects. 2.1.1–2.1.7 can be reduced from four dimensional to two dimensional, so that

$$\begin{aligned} \mathbf{x}(s) &\equiv \begin{pmatrix} x \\ x' \end{pmatrix} = \mathrm{Re}\left(A_x\mathbf{v}_1(s)\mathrm{e}^{-i\mu_x(s)}\right) = \tfrac{1}{2}A_x\mathbf{v}_1(s)\mathrm{e}^{-i\mu_x(s)} + \tfrac{1}{2}A_x^*\mathbf{v}_1^*(s)\mathrm{e}^{i\mu_x(s)}, \\ \mathbf{y}(s) &\equiv \begin{pmatrix} y \\ y' \end{pmatrix} = \mathrm{Re}\left(A_y\mathbf{v}_2(s)\mathrm{e}^{-i\mu_y(s)}\right) = \tfrac{1}{2}A_y\mathbf{v}_2(s)\mathrm{e}^{-i\mu_y(s)} + \tfrac{1}{2}A_y^*\mathbf{v}_2^*(s)\mathrm{e}^{i\mu_y(s)}. \end{aligned} \tag{2.106}$$

Here A_x and A_y are the complex amplitudes of horizontal and vertical motion (the same as described by amplitudes A_1, A_2 and phases ψ_1, ψ_2 in Eq. (2.15)) and the eigenvectors $\mathbf{v}_1$, $\mathbf{v}_2$ are two-dimensional vectors:

$$\mathbf{v}_1 = \begin{pmatrix} \sqrt{\beta_x(s)} \\ -\dfrac{i+\alpha_x(s)}{\sqrt{\beta_x(s)}} \end{pmatrix}, \quad \mathbf{v}_2 = \begin{pmatrix} \sqrt{\beta_y(s)} \\ -\dfrac{i+\alpha_y(s)}{\sqrt{\beta_y(s)}} \end{pmatrix}. \tag{2.107}$$

One can express the amplitudes A_x, A_y via $\mathbf{x}(s)$ and $\mathbf{y}(s)$, correspondingly. Indeed, multiplying Eq. (2.106) by $\mathrm{e}^{i\mu_x(s)}\mathbf{v}_1^+U$ or $\mathrm{e}^{i\mu_y(s)}\mathbf{v}_2^+U$ on the left and using the orthogonality conditions in Eq. (2.14), one obtains

$$A_x = \frac{1}{i}e^{i\mu_x(s)}\mathbf{v}_1^+\mathbf{U}_2\mathbf{x}, \quad A_y = \frac{1}{i}e^{i\mu_y(s)}\mathbf{v}_2^+\mathbf{U}_2\mathbf{y}. \tag{2.108}$$

Here $\mathbf{U}_2$ is a 2×2 unit symplectic matrix. Now let us look for the solution of Eq. (2.6) in the form of Eq. (2.106), but with A_x, A_y not being constant. Substituting Eq. (2.106) into Eq. (2.6) and considering N and R as small perturbations, we obtain the equations for A_x, A_y:

$$\begin{aligned} \frac{dA_x}{ds} &= \frac{\mathrm{e}^{i\mu_x(s)}}{2i}\sqrt{\beta_x(s)\beta_y(s)}\left[-\left(N-\frac{R'}{2}\right)\left(A_y\mathrm{e}^{-i\mu_y(s)}+A_y^*\mathrm{e}^{i\mu_y(s)}\right)\right. \\ &\quad \left. -R\left(A_y\frac{\alpha_y(s)+i}{\beta_y(s)}\mathrm{e}^{-i\mu_y(s)}+A_y^*\frac{\alpha_y(s)-i}{\beta_y(s)}\mathrm{e}^{i\mu_y(s)}\right)\right], \\ \frac{dA_y}{ds} &= \frac{\mathrm{e}^{i\mu_y(s)}}{2i}\sqrt{\beta_x(s)\beta_y(s)}\left[-\left(N+\frac{R'}{2}\right)\left(A_x\mathrm{e}^{-i\mu_x(s)}+A_x^*\mathrm{e}^{i\mu_x(s)}\right)\right. \\ &\quad \left. +R\left(A_x\frac{\alpha_x(s)+i}{\beta_x(s)}\mathrm{e}^{-i\mu_x(s)}+A_x^*\frac{\alpha_x(s)-i}{\beta_x(s)}\mathrm{e}^{i\mu_x(s)}\right)\right]. \end{aligned} \tag{2.109}$$

In the vicinity of sum and difference resonances, Eq. (2.109) can be solved by averaging.

Near the sum resonance $\nu_x+\nu_y=k_++\Delta$, the system Eq. (2.109) reduces to

$$\begin{aligned} \frac{dA_x}{ds} &= C_+A_y^*\mathrm{e}^{2i\pi\Delta\cdot s/C}, \\ \frac{dA_y}{ds} &= C_+A_x^*\mathrm{e}^{2i\pi\Delta\cdot s/C}, \end{aligned} \tag{2.110}$$

where the resonance strength C_+ is defined by

$$C_+=\frac{i}{2}\int_0^C\frac{ds}{C}\sqrt{\beta_x\beta_y}\left(2N+R\left[\left(\frac{\alpha_x}{\beta_x}-\frac{\alpha_y}{\beta_y}\right)-i\left(\frac{1}{\beta_x}-\frac{1}{\beta_y}\right)\right]\right)e^{i\left(\mu_x+\mu_y-2\pi\left(\nu_x+\nu_y\right)s/C+2\pi k_+s/C\right)}. \tag{2.111}$$

Here integration is performed over the machine circumference C. For the difference resonance $\nu_x-\nu_y=k_-+\Delta$, the equations are

$$\begin{aligned}\frac{dA_x}{ds}&=C_-A_ye^{i2\pi\Delta\cdot s/C},\\ \frac{dA_y}{ds}&=-C_-^*A_xe^{-i2\pi\Delta\cdot s/C},\end{aligned} \tag{2.112}$$

and the resonance strength is

$$C_-=\frac{i}{2}\int_0^C\frac{ds}{C}\sqrt{\beta_x\beta_y}\left(2N+R\left[\left(\frac{\alpha_x}{\beta_x}-\frac{\alpha_y}{\beta_y}\right)-i\left(\frac{1}{\beta_x}+\frac{1}{\beta_y}\right)\right]\right)e^{i\left(\mu_x-\mu_y-2\pi\left(\nu_x-\nu_y\right)s/C+2\pi k_-s/C\right)}. \tag{2.113}$$

Integration of Eqs. (2.112) over one revolution binds up two complex resonance strengths, C_+ and C_-, to the elements of the off-diagonal sub-matrices $\mathbf{q}_c$ and $\mathbf{p}_c$ of Eq. (2.79).

2.1.13 Emittance Growth at Beam Transfers Due to Optics Mismatch and X–Y Coupling

As an application of the above-developed formalism, we consider here the emittance growth related to an optics mismatch at beam transfer from one ring to another. Let the incoming beam distribution function be Gaussian and be described by bilinear form $\mathbf{\Xi}$ (see Eq. (2.34)). The corresponding eigenvectors and **V**-matrix we denote as $\mathbf{v}_1$, $\mathbf{v}_2$ and $\mathbf{V}$ (see Eqs. (2.16) and (2.32)). The eigenvectors and **V**-matrix of the circulating beam we denote as $\mathbf{v}'_1$, $\mathbf{v}'_2$ and $\mathbf{V}'$. Rewriting Eq. (2.41), we express the coordinates of each particle at the injection point through their new actions (single-particle rms emittances) and new eigenvectors:

$$\mathbf{x}=\frac{1}{2}\left(\sqrt{2I'_1}\left(\mathbf{v}'_1e^{i\psi_1}+\mathbf{v}'_1{*}e^{-i\psi_1}\right)+\sqrt{2I'_2}\left(\mathbf{v}'_2e^{i\psi_2}+\mathbf{v}'_2{*}e^{-i\psi_2}\right)\right). \tag{2.114}$$

Multiplying each side of the above equation by $\mathbf{v}'^{+}_1\mathbf{U}$ or $\mathbf{v}'^{+}_2\mathbf{U}$ and using the orthogonality conditions of Eq. (2.14), we obtain

$$I'_k(\mathbf{x}) = \frac{1}{2}\left|\mathbf{v}_k{}^{+}\mathbf{U}\mathbf{x}\right|^2, \quad k = 1, 2. \tag{2.115}$$

Averaging over all particles yields the new rms emittances:

$$\varepsilon'_k = \int dx^4 I'_i(\mathbf{x}) f(x) = \frac{1}{8\pi^2 \varepsilon_1 \varepsilon_2} \int dx^4 \left|\mathbf{v}'_k{}^{+}\mathbf{U}\mathbf{x}\right|^2 \exp\left(-\frac{1}{2}\mathbf{x}^{\mathrm{T}}\boldsymbol{\Xi}\mathbf{x}\right). \tag{2.116}$$

Similar to Eq. (2.35) a coordinate transform, $\mathbf{y} = \mathbf{V}^{-1}\mathbf{x}$, reduces matrix $\boldsymbol{\Xi}$ to its diagonal form. That results in

$$\varepsilon'_k = \frac{1}{8\pi^2 \varepsilon_1 \varepsilon_2} \int dy^4 \left|\mathbf{v}'_k{}^{+}\mathbf{U}\mathbf{V}\mathbf{y}\right|^2 \exp\left(-\frac{1}{2}\mathbf{y}^{\mathrm{T}}\hat{\boldsymbol{\Xi}}\mathbf{y}\right), \tag{2.117}$$

where matrix $\hat{\boldsymbol{\Xi}}$ is determined by the beam initial emittance in accordance with Eq. (2.28). Taking into account Eqs. (2.36) and (2.37), we finally obtain

$$\varepsilon'_k = \frac{1}{2}\mathbf{v}'_k{}^{+}\mathbf{U}\boldsymbol{\Sigma}\mathbf{U}^{\mathrm{T}}\mathbf{v}'_k, \quad k = 1, 2. \tag{2.118}$$

For initially uncoupled beam characterized by β_x, β_y, α_x, and α_y at the injection point that yields

$$\begin{aligned} \varepsilon_1{}' &= \varepsilon_1 A_{11} + \varepsilon_2 A_{12}, \\ \varepsilon_2{}' &= \varepsilon_1 A_{21} + \varepsilon_2 A_{22}, \end{aligned} \tag{2.119}$$

where

$$\begin{aligned}
A_{11} &= \frac{1}{2}\left(\frac{\beta_x}{\beta_{1x}}\left[(1-u)^2 + \alpha_{1x}{}^2\right] + \frac{\beta_{1x}}{\beta_x}\left[1 + \alpha_x{}^2\right] - 2\alpha_{1x}\alpha_x\right), \\
A_{12} &= \frac{1}{2}\left(\frac{\beta_y}{\beta_{1y}}\left[u^2 + \alpha_{1y}{}^2\right] + \frac{\beta_{1y}}{\beta_y}\left[1 + \alpha_y{}^2\right] - 2\alpha_{1y}\alpha_y\right), \\
A_{22} &= \frac{1}{2}\left(\frac{\beta_y}{\beta_{2y}}\left[(1-u)^2 + \alpha_{2y}{}^2\right] + \frac{\beta_{2y}}{\beta_y}\left[1 + \alpha_y{}^2\right] - 2\alpha_{2y}\alpha_y\right), \\
A_{21} &= \frac{1}{2}\left(\frac{\beta_x}{\beta_{2x}}\left[u^2 + \alpha_{2x}{}^2\right] + \frac{\beta_{2x}}{\beta_x}\left[1 + \alpha_x{}^2\right] - 2\alpha_{2x}\alpha_x\right)
\end{aligned}$$

and β_{1x}, β_{1y}, β_{2x}, β_{2y}, α_{1x}, α_{1y}, α_{2x}, α_{2y}, and u are the generalized Twiss parameters of the ring at the injection point. For uncoupled ring optics, $\beta_{1y} = \beta_{2x} = \alpha_{1y} = \alpha_{2x} = u = 0$, we obtain the well-known expression:

$$\begin{aligned} \varepsilon_x{}' &= \frac{1}{2}\varepsilon_x\left(\frac{\beta_x}{\beta_{1x}}\left[1+\alpha_{1x}{}^2\right]+\frac{\beta_{1x}}{\beta_x}\left[1+\alpha_x{}^2\right]-2\alpha_{1x}\alpha_x\right), \\ \varepsilon_y{}' &= \frac{1}{2}\varepsilon_y\left(\frac{\beta_y}{\beta_{2y}}\left[1+\alpha_{2y}{}^2\right]+\frac{\beta_{2y}}{\beta_y}\left[1+\alpha_y{}^2\right]-2\alpha_{2y}\alpha_y\right). \end{aligned} \tag{2.120}$$

One can see that the emittance growth is absent only if the Twiss parameters of injected and circulating beams are equal.

2.2 Linear Optics Measurements

Linear optics measurements have played an important role for improvement of the Tevatron complex performance. Accurate knowledge of the ring and transfer line optics resulted in the significant reduction of the emittance growth for beam transfers and increased the acceptances of the rings and transfer lines with subsequent reduction of the beam loss at transfers and in the course of other operations. In the case of Tevatron, it resulted in a decrease of the beta-functions at the IPs (with subsequent luminosity increase), helped to maintain equal luminosities at the two experiments, was instrumental in locating magnet misalignments and keeping the machine stable over long periods of time, and provided valuable input for various simulations, including the simulation of beam–beam effects, optimization of helical orbits, and collimation. In the case of Debuncher and Accumulator, an accurate knowledge of magnet focusing allowed us to modify machine optics so as to maximize acceptances of the machines and to improve performance of stochastic cooling systems (see Chap. 7). A number of methods and software tools were developed to streamline the process of data acquisition, processing, and analysis. All methods that were employed for the Tevatron linear optics measurements rely on the beam position information provided by the Tevatron BPM system (see Chap. 9). Later these methods were used to build accurate optics models for Debuncher, Accumulator, and Recycler.

Historically, the first method used for optics measurements in Run II was based on the analysis of orbit response data generated by a small number (usually four) of dipole correctors and an energy change [9]. An automated software program has been used for data acquisition. It makes the corrector current change and records the resulting orbit difference with respect to the nominal orbit (hence, we use the term *differential orbit measurement* to describe the method). The generated orbits were then compared with the model prediction, and the model was corrected to minimize discrepancies between measurements and the model. Later more sophisticated software tools were built for analysis of multiple differential orbits and an automatic correction of optics model for circular machines. Unfortunately this software cannot be used effectively for transfer lines because it requires much more data, and such data acquisition is time prohibitive for the most transfer lines of the Tevatron complex.

The idea of the method is based on an excitation of betatron wave with a single dipole corrector kick. For a transfer line, the wave is propagating downstream of the corrector and corresponding beam displacement is

$$x(s) = \theta\sqrt{\beta(s)\beta(s_0)}\sin(\mu(s) - \mu(s_0)). \tag{2.121}$$

For a ring the closed orbit displacement is

$$x(s) = \frac{\sqrt{\beta(s)\beta(s_0)}}{2\sin(\pi Q)}\theta\cos(|\mu(s) - \mu(s_0)| - \pi Q). \tag{2.122}$$

Here Q is the betatron tune, θ is the corrector kick, β and μ are the betatron function and phase, and index 0 labels the corrector location. A focusing error results in an unaccounted kick with angle

$$\delta\theta = \Delta K l x,$$

where ΔKl is the error in the integrated quadrupole strength and x is the orbit displacement. That affects the phase and amplitude of betatron motion relative to the computer model predictions. As one can see from the above equation, the value of the kick is proportional to the trajectory displacement in the quadrupole, and therefore a single differential orbit has suppressed sensitivity to focusing errors in the vicinity of locations where the differential orbit crosses zero. Therefore, at least two differential orbits (for each plane) are needed to sense all focusing errors. In optimum the betatron phases of these orbits should be shifted by $(n + 1/2)\ \pi$, although deviations from optimum in the range of $\pm\pi/4$ do not introduce large penalty to the measurement accuracy. In the case of energy change, the beam displacement is proportional to the corresponding plane dispersion. This response is "orthogonal" to responses of orbit bumps; it additionally limits possible corrections of quadrupole strength and therefore is extremely helpful in finding an actual machine model. Its usefulness is greatly amplified by limited accuracy of BPM measurements related to the BPM noise and errors of BPMs differential response. Although the noise in principle (but not always in practice) can be reduced by increasing number or duration of measurements, the BPM differential sensitivity is a significant factor affecting the accuracy of the measurements and has to be specially addressed.

The most detailed optics models of rings and transfer lines were built using the OptiM code [10]. The models have included all optics-related elements: magnets, linear and nonlinear correctors, BPMs, profile monitors, aperture limitations, stochastic cooling tanks, etc. All important properties of magnets such as the dependence of strength on the current and nonlinearities derived from magnetic measurements have been also included in the models. Locations of all elements have been cross-checked with machine alignment data. However, comparison of model predictions with measurements showed significant discrepancies pointing out that there are non-negligible errors in the focusing properties of magnets. To address it we added into the model the pseudo-quadrupoles (and if necessary the pseudo-skew-quadrupoles) near each quadrupole which adjustments allowed matching the model to the measurements. Although in the most of cases we do not know origins of focusing errors, such approach allowed us to build credible machine models accurately describing ring or transfer line focusing.

There is significant difference in optics measurements for transfer lines and circular machines. In a ring the beam is permanently present, and therefore an accumulation of multiple BPM measurements can be done fast and accumulation of large datasets is not a problem. In this case multiple correctors are used for optics measurements. That creates a redundancy in the data allowing us to make an automated data analysis resulting in a high-accuracy optics model. Accumulation of differential orbit data for a transfer line happens much slower—once or twice per minute. Usually to achieve a minimally required statistical accuracy, at least four measurements are required for each corrector. Taking into account that at least four correctors and an energy change are required and the reference orbit has to be measured before and after the measurement, one obtains a minimum measurement time of about 15 min. Usually measurements are done for both polarities of differential orbits resulting in 30 min to an hour to acquire good dataset for a transfer line. Therefore, the datasets for transfer lines do not have sufficient redundancy for an automatic data analysis and data are analyzed manually, i.e., a person assigns focusing errors to quads. Normally it takes from a few hours to a day to build a transfer line optics. That is a good compromise between time spent for measurements and the data analysis. Due to lack of data and limited human ability to digest still quite large datasets, the accuracy of obtained optics model is not the same good as for rings. However, we found it adequate to the requirements of Tevatron complex transfers. Together with absolute and differential orbits the software records a beam intensity signals for each BPM. Changes in the beam intensity correlated with beam displacement allow one to identify places of potential beam scraping and, taking into account that the measurements are performed for both polarities of each kick, to formulate suggestions for changes of the beam orbit in the transfer line.

Note also that as part of the effort to improve the status of the accelerator complex, three databases were created: (1) the lattice repository which keeps the optics files, (2) the magnet database which keeps the results of magnetic measurements, and (3) the survey database which keeps results of machine surveys and alignment.

2.2.1 Linear Optics Optimization and Linear Optics Measurements for Transfer Lines

At the Run II beginning improvements of optics for transfer lines was much more important problem than optics improvements for circular machines. There were three main reasons: first, historically much more attention was paid to the optimization of optics for circular machines; second, most transfer lines are between rings belonging to different departments and therefore their ownership was not uniquely determined; third, long transfer lines historically were split into a few segments and optics was designed and supported independently for each segment. The Run II clearly demonstrated that resolving transfer line optics is the same challenging as resolving circular machine optics, and therefore the same attention has to be paid to the transfer line optics design and commissioning. There are ten transfer lines in the Tevatron complex: (1) linac to Booster, (2) Booster to MI, (3) Main Injector to antiproton production target, (4) antiproton production target to Debuncher, (5) Debuncher to Accumulator, (6) Accumulator to Main Injector, (7, 8) two lines (proton and antiproton) from Recycler to Main Injector which are also used for Main Injector-to-Recycler transfers, and (9, 10) two lines (proton and antiproton) from Main Injector to Tevatron. There are also transfer lines for neutrino experiments and experiments with fixed targets which are not discussed here. For almost all lines optics was redesigned to improve transport quality. Optics for all of them was measured and if necessary corrected to meet the design intent.

At the Run II beginning the most outstanding optics problems were related to the Accumulator-to-Main Injector transport of antiprotons at 8 GeV. It is the longest and most complicated transfer line in the Tevatron complex. Therefore, resolving its optics problems is considered here in detail. Optics problems of other transport lines were similar and the same approach and software were used to address them. If not directly mentioned, the discussion in the rest of this section is about the Accumulator-to-Main Injector transport.

The total length of beam transport from Accumulator to Main Injector is more than 900 m. Almost 600 m of this line is also used for the 120 GeV proton beam transport from Main Injector to the antiproton production target. Large difference in the energies results in that the magnetic fields of the low-energy transport are dominated by the residual magnetic field of the magnets. At the Run II beginning the situation was so critical and uncertain that the question of building a new 8 GeV line was seriously discussed. However, the differential orbit measurements proved that the line optics is sufficiently reproducible and the decision was made to redesign and tune optics of the existing line. The main objectives for new optics design were as follows: (1) maximize the line acceptance for existing aperture limitations, (2) match transfer line optics to the optics of both rings including both vertical and horizontal dispersions, and (3) minimize (or better eliminate) any hardware work in the tunnel. The transfer line has 63 quadrupoles connected to 35 quadrupole families. Large number of quadrupole families offers considerable freedom for optics design, but at the same time it greatly complicates finding a good

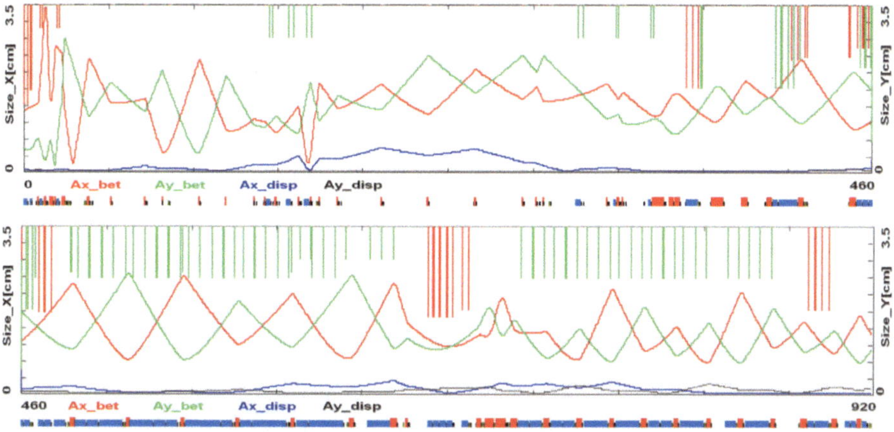

Fig. 2.1 Calculated beam envelopes of the entire Accumulator-to-Main Injector transport for the beam emittances equal to the line normalized acceptances 48 and 42 mm mrad for horizontal (*red*) and vertical (*blue*) planes, correspondingly. *Top* and *bottom* plots present the envelopes for the first and second halves of the beamline. *Blue* and *black* curves present contributions to the beam size coming from the momentum spread of $\Delta p/p = 6 \times 10^{-4}$ corresponding to 2.5σ of typical momentum spread of extracted beam. *Vertical lines* show aperture limitations for horizontal (*red*) and vertical (*green*) planes. *Squares* below the plots present locations of dipoles (*blue*) and quadrupoles (*red*)

solution. Existing aperture limitations were one of the major complications. It forced us to minimize the beta-functions at the aperture limitations and, consequently, yielded an increase of beta-functions in their vicinity and made optics irregular. Although the split of quadrupoles into families was far from optimal, a satisfactory solution was found. Figure 2.1 presents calculated beam envelopes through the entire transport line for the final choice of beamline optics. The horizontal dispersion and both beta-functions were matched to the ring dispersions and beta-functions. The existing quadrupole families did not allow matching the vertical dispersion, but the line optics was designed to minimize the vertical dispersion leakage from the line. Together with a small value of extracted beam momentum spread, it resulted in a negligible contribution of vertical dispersion mismatch to the emittance growth, thus, allowing us to achieve good transfer line performance without reconnecting the quadrupole families.

The optics measurements have been based on the differential orbit measurements. Normally the measurements were performed with reverse protons, where the proton beam is sent from Main Injector to Accumulator. Figure 2.2 presents a typical measurement consisting of five differential orbits representing responses to two horizontal and two vertical correctors and an energy change. To make sure that the BPM response is not affected by beam scraping, the measurements were acquired for positive and negative excitations. It also improves statistical accuracy. Curves present the model predictions after the model was fitted to the data. *X*–*Y* coupling in the line is sufficiently small, and therefore the cross-plane responses for the corrector excited orbits are not presented in the figure. Variations of BPM

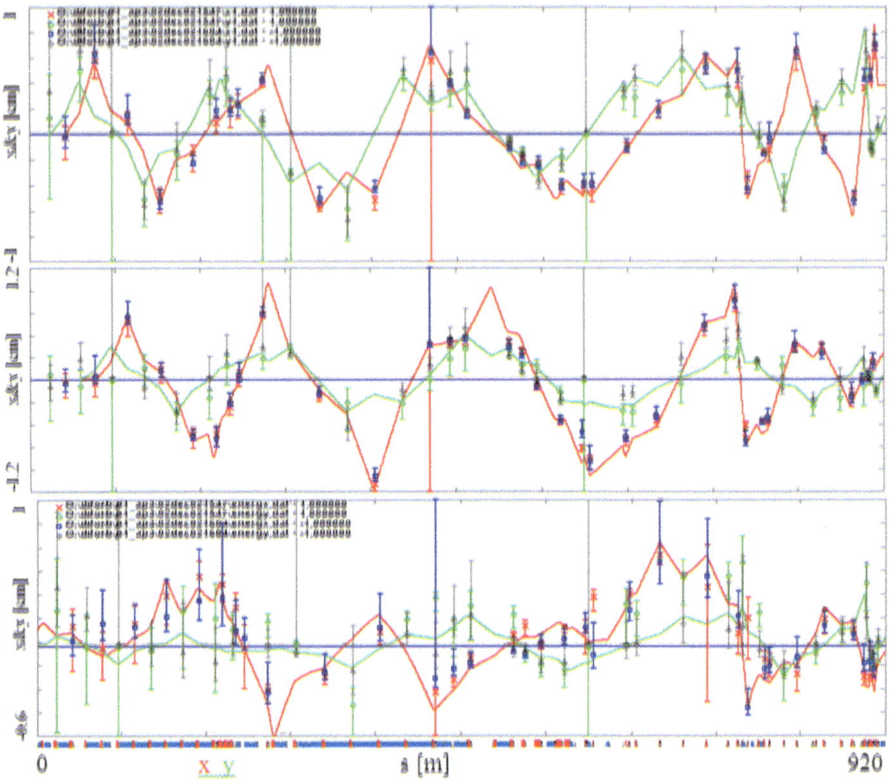

Fig. 2.2 Typical differential obit measurement for the Accumulator-to-Main Injector beam transport and its fitting by refined optics model: *red* and *blue* dots present a horizontal response to a horizontal corrector and *green* and *black* dots a vertical response to a vertical corrector. *Blue* and *black* dots represent inverted values of data acquired with negative corrector excitation. Short error bars present a standard deviation and long error bars present a maximum deviation from the mean value of 3 measurements. Long error bars which cross the entire plot show BPMs with an error status which potentially can have incorrect measurements

differential response significantly complicate finding good solution. In this case the beam displacement reported by a BPM is proportional to the actual displacement but not necessarily with correct coefficient of proportionality.[5] Therefore, the best fit to the data usually yields too large corrections to quadrupole focusing, and,

[5] Experience gained with the upgrade of electronics of Tevatron BPMs carried out in 2004 proved that before the upgrade the major contribution to variations of differential BPM response was related to imperfections of electronics. After the upgrade the spread of variations was reduced from ~10 to ~1 %. Contribution coming from nonlinearity of differential BPM response with coordinate related to the geometry of BPM was much smaller. The imperfection of electronics looks the most probable reason for variations of differential BPM response for the transfer line BPMs.

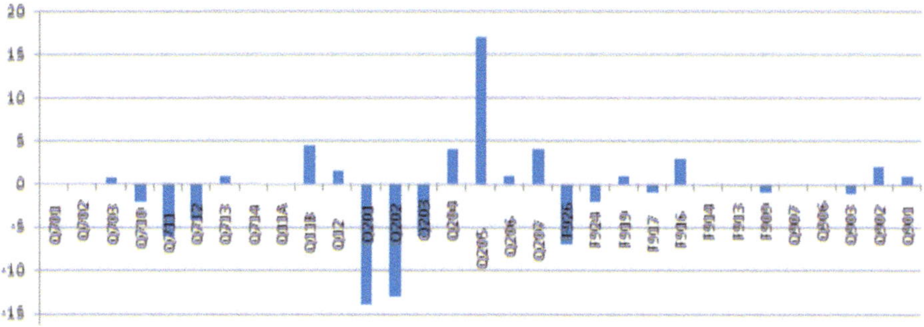

Fig. 2.3 Corrections (in %) to quadrupole power supplies currents for Accumulator-to-Main Injector beam transport. The 120 GeV beamline includes quadrupoles from Q701 to Q207. The rest of the line operates at 8 GeV only

consequently, the model represents poor the actual line focusing. To address this problem we put more trust to the BPMs in which measurements are close to zero, and, consequently, are weakly affected by errors of differential beam response. Normal functioning of these BPMs is verified by measurements with nonzero beam displacements. They come from another corrector of the same plane. Normally optics model update/correction proceeds from the transport line beginning to its end. One corrects quadrupole focusing iterating between different differential orbits. A few rounds of corrections are usually required before a satisfactory match to the measurements is found. Figure 2.3 presents corrections to the power supply currents required to match the model to the differential orbit measurements. One can see significantly larger corrections to the quadrupoles operating both for 8 and 120 GeV transport. Although corrections to quadrupole focusing obtained with this procedure are not unique and do not represent actual errors of the beam transport, they allow one to get an optics model describing the line focusing with satisfactory accuracy. Usually we use the first two correctors of a transfer line, but it does not excite differential orbit in the first quadrupole, and therefore focusing errors of the first quadrupole (closest to Main Injector) are invisible. It can be resolved by an excitation of differential orbit in Main Injector, but it makes both measurements and data analysis more complicated. An upgrade of the transfer line BPM electronics carried out in 2006 made possible accurate position measurements with antiprotons. That allowed us to carry out differential orbit measurements with antiprotons [11]. The measurements were almost not invasive, and because they used antiprotons moving in the opposite to protons' direction, they pointed out a focusing error of the quadrupole closest to the Accumulator (Q901). That resulted in a further improvement of transport quality.

The turn-by-turn measurements of transverse beam sizes of injected antiproton beam offer an independent measurement of the beamline optics. Such measurements became available later in the run when the ion profile monitor (IPM) was commissioned for operations with antiprotons (see Chap. 9). The beam sizes oscillate at the double betatron frequency corresponding to observed frequencies

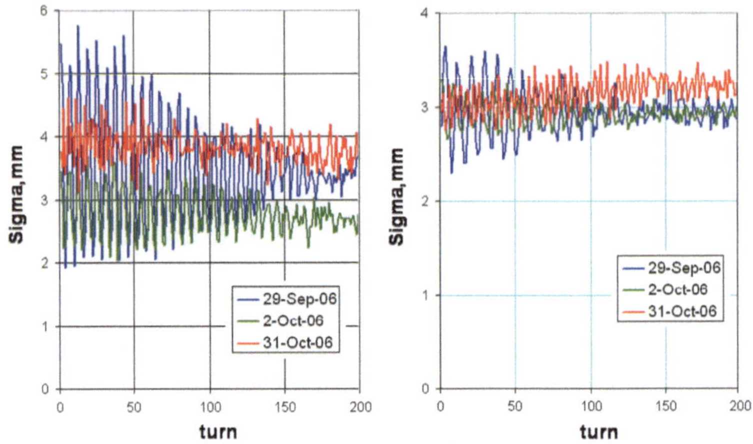

Fig. 2.4 Main Injector IPM measurements of the transverse beam size oscillations at injection. *Left* and *right* plots show horizontal and vertical oscillations, correspondingly, before the optics matching (*blue*), after first correction (*green*), and after second correction (*red*)

$2\nu_x \sim 0.12$ and $2\nu_y \sim 0.16$. Figure 2.4 presents an improvement of IPM signals with improving optics match [11]. As one can see in Fig. 2.4, there was a considerable mismatch between Accumulator and Main Injector before correction. However, the emittance increase was not as bad as the beta-function mismatch because the emittance growth is proportional to $(\Delta\beta/\beta)^2$. The estimate can be obtained from the well-known formula, describing the emittance growth due to optics mismatch:

$$\begin{aligned}\varepsilon' =& \frac{\varepsilon}{2}\left(\frac{\beta_1}{\beta_2}\left[1+\alpha_2{}^2\right]+\frac{\beta_2}{\beta_1}\left[1+\alpha_1{}^2\right]-2\alpha_1\alpha_2\right) \\ &+\frac{\sigma_p{}^2}{2}\left(\beta_2\left(D_0'-D_1'\right)^2+2\alpha_2\left(D_0'-D_1'\right)\left(D_0-D_1\right)+\frac{\left(D_0-D_1\right)^2}{\beta_2}\left(1+\alpha_2{}^2\right)\right),\end{aligned} \tag{2.123}$$

where β_1, α_1, D_1, and D_1' are the beta- and alpha-functions, the dispersion and its derivative for the incoming beam, and β_1, α_1, D_1, and D_1' are the beta- and alpha-functions of circulating beam. Expending this equation for the case of small quadrupole betatron oscillations and taking into account that $\Delta\beta/\beta \approx 2\Delta\sigma/\sigma$, one obtains

$$\frac{\delta\varepsilon}{\varepsilon} \approx \frac{1}{2}\left(\frac{\Delta\beta}{\beta}\bigg|_{\max}\right)^2 \approx 2\left(\frac{\Delta\sigma}{\sigma}\bigg|_{\max}\right)^2. \tag{2.124}$$

For the data presented in Fig. 2.4, it yields $\delta\varepsilon_x/\varepsilon_x \approx 0.5$ (0.03) and $\delta\varepsilon_y/\varepsilon_y \approx 0.07$ (0.015), correspondingly, before and after the final correction.

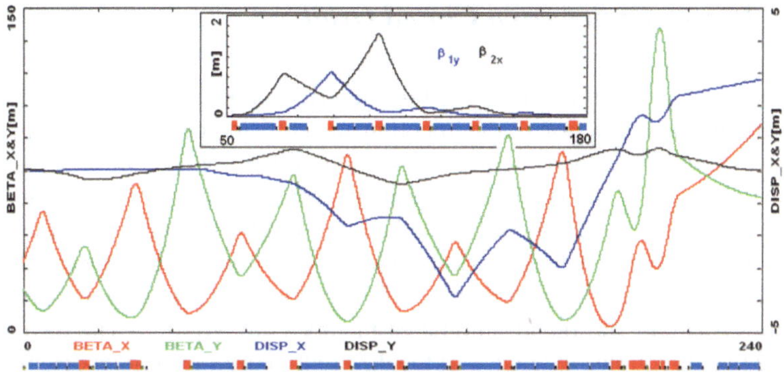

Fig. 2.5 Beta-functions and dispersions for proton transport from Main Injector to Tevatron. *Inset* presents auxiliary beta-functions of coupled motion, β_{1y} and β_{2x}

Standardization of hysteresis protocol and two sets of power supplies (one for 8 and another for 120 GeV) have been required to achieve desired reproducibility of beam transport. However, its seasonal variations still were present in the Accumulator-to-Main Injector line. Other transfer lines operate at the energies which they were designed for. Consequently, they have better reproducibility and have not required additional tuning since the time when their optics was corrected.

As it was already mentioned, Run II inherited a number of problems rooted in Tevatron history. One of them was a vertical dispersion mismatch in the Main Injector-to-Tevatron proton beam transport line. Existing quadrupole powering did not allow matching of vertical dispersion. An independent powering for a string of 7 quadrupoles could address the problem, but it required new power supplies and additional cabling. A simple solution of the problem implied rolling 4 of 7 quadrupoles of the string by small angles. The angles and quadrupole strengths were adjusted to make the uncoupled transport through the string (see inset in Fig. 2.5) and to match beta-functions and dispersions. This line is also used for the antiproton transport described above and the proton beam transport to the antiproton production target. Uncoupled transport through the string simplified their optics correction.

Note that in most cases the emittance increase of about 5 % is acceptable. In this case Eq. (2.124) yields a required transfer line optics match $\Delta\beta/\beta \leq 0.3$. The requirements to optics of circular machines like Tevatron, Debuncher, and Accumulator are much stricter, and therefore a usage of the multi-corrector automated algorithm described in the next section has been absolutely essential to address their optics issues.

2.2.2 Linear Optics from Closed Orbit

The simplified differential orbit method described above was used in 2003 for measurement and correction of the injection and collision lattices for Tevatron. The most spectacular result was achieved for the collision optics—it was discovered that due to the significant (up to 1 %) gradient errors in the final focus quadrupoles, the β^* values were about 30 % larger than the design. Correction of these errors resulted in an immediate increase of the luminosity. However, the data analysis was tedious and the results still were not sufficiently accurate. It was recognized that both more accurate and better-automated methods are required. The development followed two directions: extension of the differential orbit technique discussed here and analysis of the turn-by-turn data discussed in the next section.

Response matrix fitting is a well-known method of calibrating the machine optics. It was first suggested at SLAC [12] and then it was used at NSLS [13] for X-ray ring analysis. Today the method is widely used at many accelerators around the world [14]. At the Tevatron, a modification of the response matrix fitting software developed at Argonne National Laboratory for the Advanced Photon Source (APS) [15] has been used.

The response matrix fitting program SRLOCOFitting [15] written in Tcl/Tk has an extensive graphical user interface, and uses SDDS toolkit [16] for data processing. The code was developed to calibrate the APS model and to provide data for beta-function correction. Coupling correction was not an issue at the APS; therefore, the calculations were limited to the uncoupled case. On the contrary, coupling of horizontal and vertical betatron motion was an important for the Tevatron; therefore, existing analyses had to be expanded to a fully coupled motion. Another important modification was the addition of dispersion to the fit. This allowed us to resolve two issues. First, addition of dispersion adds a constraint on the quadrupole gradients, removing the degeneracy between in-phase quadrupoles. Second, the dispersion can be used to calibrate average gain of BPMs, which otherwise would be a degenerated value. Technically, dispersion is treated as a column of the response matrix. A number of other minor code modifications have been made to ensure that the software could be used with other Fermilab accelerators.

The Tevatron ring has 110 correctors and 118 BPMs in each plane. The response matrix measurement procedure was fully automated and used the following procedure: each steering magnet was excited first with positive current and then with negative current. At each value of current, the orbit was measured 25 times. The total response to the steering magnet excitation was the average positive orbit minus the average negative orbit. The output of the measurement program was an SDDS file containing average orbit responses and their rms deviations in a format readable by SRLOCOFitting. The dispersion measurement was done by scanning the RF frequency, measuring orbit at five points, and fitting a straight line at each BPM. It resulted in an improvement of measurement accuracy. Acquisition of the full response matrix required approximately 2 h of beam time. However, it was

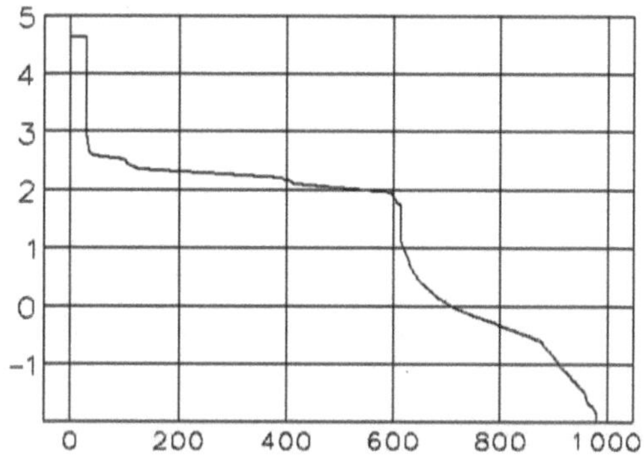

Fig. 2.6 Singular values (logarithmic scale) of the Tevatron response matrix derivative

determined that a good quality fit could be obtained with a smaller dataset, and in normal operations the response matrix was measured using 60 correctors, which took less than 1 h.

The fit produced values of the following variables: quadrupole gradient errors, quadrupole tilts, corrector calibration errors, corrector tilts, BPM gains and BPM tilts, and energy shift due to corrector changes. The total number of unknown variables was about 1,000. Since the LOCO fitting procedure is based on the computation of pseudo-inverse of the response matrix derivative using singular value decomposition, important information is contained in the spectrum of singular values. Figure 2.6 presents a typical plot of the singular values for the Tevatron collision optics. The SVD cutoff was typically chosen at 1, which corresponds to 600–650 singular values.

The main factor limiting accuracy of the LOCO fit is the resolution of the beam position measurement. The BPMs have the resolution of about 10 μm for a single measurement. Besides, the beams oscillate at low (~10 Hz) frequency with the amplitude of about 50 μm. Averaging over 25 measurements has been applied to mitigate the effect of slow oscillations. The overall accuracy of the orbit measurement was then about 15 μm. Figure 2.7 shows the rms difference of the measured orbit and the modeled orbit after the fit for each BPM. In this case, 30 horizontal and 30 vertical orbits were used and the average error was ~14 μm which is close to the orbit measurement accuracy.

The precision of the orbit fitting sets the accuracy of gradient error determination, which in our model is 10^{-3} for the arc quads and 10^{-4} for the final focus quads. The corresponding error in beta-function is about 5 %. In Fig. 2.8 the found quadrupole and skew-quadrupole errors are presented for all locations in the Tevatron.

Two locations with large skew-quadrupole component, D16 and A38, have been identified as tilts of the corresponding quadrupoles. These tilts emerged at the magnet assembly and could not be found by alignment measurements from outside of the magnet.

Table 2.1 summarizes the gradient errors for the final focus quadrupoles. As one can see, the difference from the calibration curve obtained by magnetic measurements can be as high as 1 %.

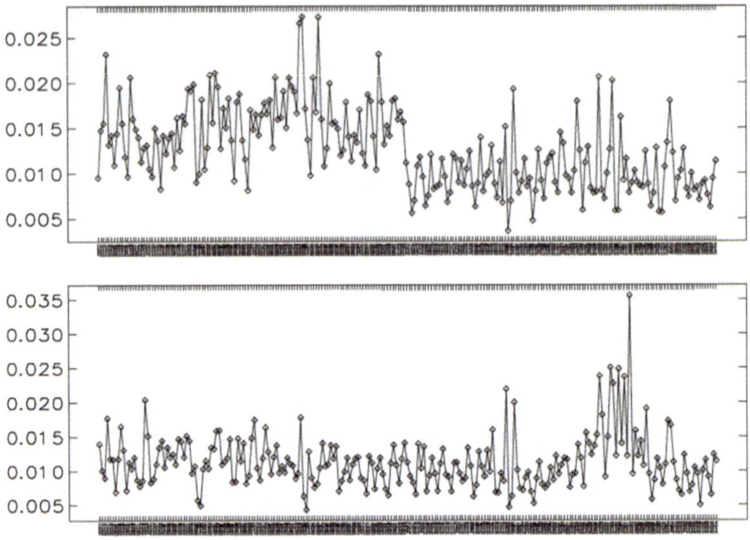

Fig. 2.7 Rms difference (mm) between the measured and modeled orbit vs. BPM name. *Top*, horizontal orbit; *bottom*, vertical orbit

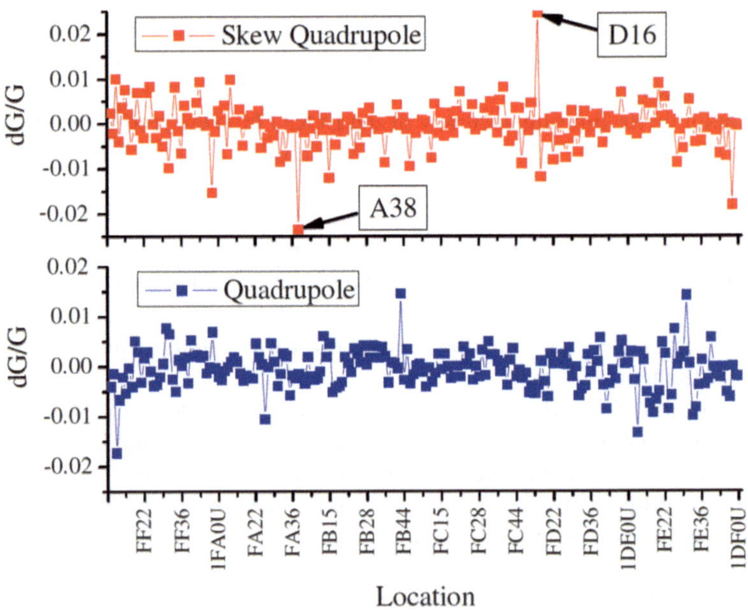

Fig. 2.8 Measured relative quadrupole and skew-quadrupole errors

Table 2.1 Relative quadrupole errors in final focus

Name	Gradient error (10^{-3})
B0Q3	−11.18
B0Q2	−1.87
B0Q3D	−0.09
B0Q3F	−0.47
D0Q3	−9.49
D0Q2	−0.83
D0Q3F	0.24
D0Q3F	−1.84

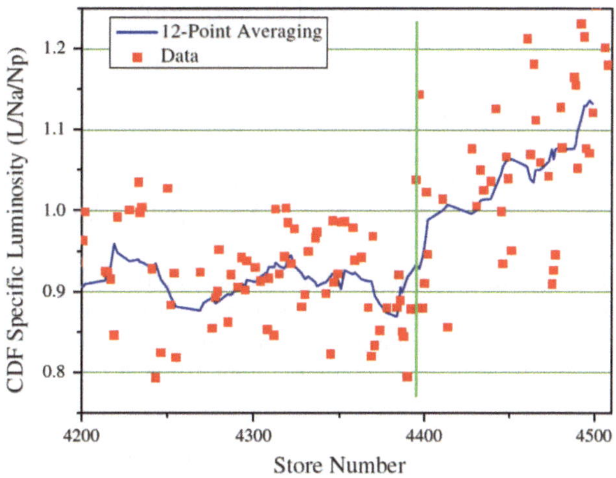

Fig. 2.9 Specific initial luminosity ($L/N_a/N_p$) vs. store number. *Green line* marks the moment when the new optics was put into operation

Based on the knowledge of the lattice details, new collision optics has been implemented in 2005 with the following goals:

1. Eliminate beta-beating in the arcs.
2. Correct the discrepancy in the values of β^* between the two IPs.
3. Decrease the value of β^* from 35 to 28 cm, with an expected gain in luminosity of 11 % (Fig. 2.9). Further decrease of the β^* was not practical because of the growing second-order chromaticity and little gain in luminosity due to the hourglass effect.

Routine optics measurements with LOCO were performed over the entire length of Run II to support collider operations and simulation efforts.

2.2.3 *Turn-by-Turn Measurements*

Orbit response matrix analysis is a powerful tool that supplied precise information about the Tevatron linear lattice imperfections and errors of BPM calibrations.

However, orbit data acquisition is a lengthy process and certainly could not be performed on a daily basis and especially during acceleration. For this regime, the Fourier analysis of turn-by-turn beam position after a single-turn kick proved to be invaluable. This method is fast and offers information about entire machine optics. In particular it allows fast computation of the resonance driving terms for the sum and difference betatron coupling resonances and the location of coupling sources [17].

The distribution of coupling sources around the ring determines the resonance driving terms (see Sect. 2.1.12):

$$w_{\pm}(s) = -\int_0^C ds' \frac{C_{\pm}(s')}{4\sin \pi\nu_{\pm}} e^{-i\nu_{\pm}(2\pi(s-s')/C-\pi\cdot\mathrm{sign}(s-s'))}, \tag{2.125}$$

where $\nu_{\pm} = \nu_x \pm \nu_y$ and

$$C_{\pm}(s) = \frac{\sqrt{\beta_x\beta_y}}{2}\left\{2N + R\left[\left(\frac{\alpha_x}{\beta_x} - \frac{\alpha_y}{\beta_y}\right) - i\left(\frac{1}{\beta_x} \mp \frac{1}{\beta_y}\right)\right]\right\} \times e^{i(\chi_x \pm \chi_y)} \tag{2.126}$$

Here χ_x and χ_y are the periodic phase functions. The functions $w_{\pm}(s)$ are constant in coupling-free regions and experience a discontinuity at the locations of coupling sources. On the coupling resonances $\nu_x \pm \nu_y =$ integer, the functions $w_{\pm}$ are constant. The minimum attainable tune distance is given by

$$\overline{C}_{\pm} = \frac{n_{\pm} - \nu_{\pm}}{\pi} \int_0^C \frac{ds}{C} w_{\pm} e^{in_{\pm}2\pi s/C},$$

with $n_{\pm} = \mathrm{round}(\nu_x \pm \nu_y)$. If the kick occurs in the horizontal plane, the Fourier component $Y_j(\nu_x)$ of $y_j(s)$ is related to the values of $w_{\pm}$ at the j-th BPM via the Twiss functions. When the BPM tilts are negligible or already known (e.g., from the LOCO fit), the number of unknown quantities per BPM is reduced to two and one can retrieve the constant value of $w_{\pm}$ in the region between two BPMs from $Y_j(\nu_x)$ and $Y_{j+1}(\nu_x)$ assuming that there are no strong sources of coupling.

Figure 2.10 presents the vertical injection lattice beta-function reconstructed from turn-by-turn data in comparison with the beta-function obtained by LOCO. One can see that agreement between the two methods is good.

Fig. 2.11 presents the values of real and imaginary parts of $w_{\pm}(s)$ measured at vertical BPMs for horizontal kick and at horizontal BPMs for vertical kick.

An application program for the online turn-by-turn data analysis has been integrated into the Tevatron control system. The program fired the kicker, collected the BPM data, computed Twiss and coupling functions, and, finally, computed and applied the needed corrections to the two main skew-quadrupole circuits SQA0 and SQ. The program was used in routine collider operations during every shot setup and proved invaluable for decoupling on the energy ramp. The time needed to

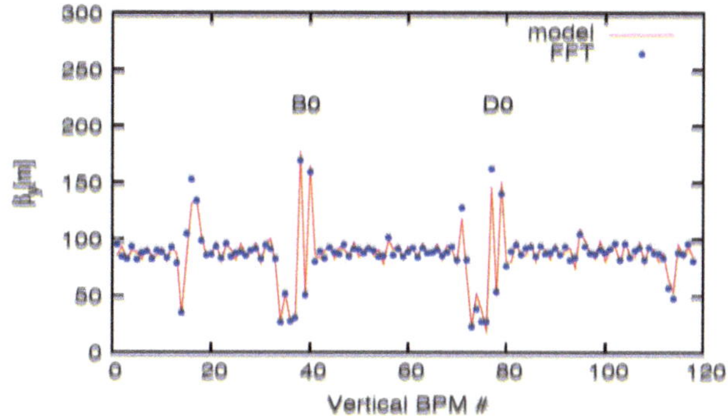

Fig. 2.10 Vertical beta-function at vertical BPMs (injection lattice)

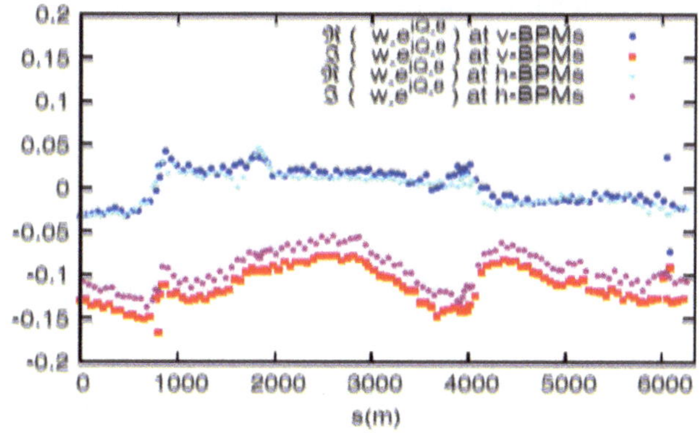

Fig. 2.11 $w_{\pm}(s)$ at vertical BPMs after a horizontal kick (*blue* and *red*), at horizontal BPMs after a vertical kick (*cyan* and *magenta*) as a function of the machine azimuth

retrieve the turn-by-turn data from all Tevatron BPMs was too long for routine use of the method. This limitation could be overcome due to the fact that $w_{\pm}(s) \approx \text{const}$ near the coupling resonances. Since the Tevatron working point at injection ($\nu_x = 20.584$, $\nu_y = 20.574$) is reasonably close to both the difference and the sum resonance, it was possible to use only few BPMs (typically 5 horizontal and 5 vertical) to evaluate the tunes and the functions $w_{\pm}$ at the orthogonal mode BPMs.

2.3 Nonlinear Beam Dynamics

2.3.1 Dedicated Studies of Nonlinear Beam Dynamics in Tevatron

Several important beam studies dedicated to detailed understanding of nonlinear beam dynamics had been carried out at the Tevatron in the late 1980s to early 1990s.

In the E778 beam dynamics experiment, performed in the Fermilab Tevatron, strong nonlinear elements were intentionally added and observations of phase space of nonlinear oscillations were made [18]. For that experiment the Tevatron can be regarded as a linear system on which nonlinearity in the form of 16 sextupole magnets, each of strength S, was intentionally superimposed. The experimental procedure starts with a "needle beam" consisting of some 10^{10} circulating stored protons, to some approximation having essentially the same momentum and to be on the central orbit. Next the same angular deflection D is applied to every particle by a pulsed deflecting magnet. The subsequent beam centroid displacement is sensed for as many as a million turns by beam position monitors (BPM). These measurements are used to generate an experimental Poincaré plot.

To a good approximation, the following equation of motion describes the horizontal particle motion

$$\frac{d^2x}{dt^2} + K(s)x = -\varepsilon(s)x^2. \tag{2.127}$$

Here s is the longitudinal particle coordinate, which advances from 0 to C (the ring circumference). In the course of the experiment the Tevatron tune was about $Q = 19.4$. The anharmonic term in (2.127) is due to sextupole fields of strength $\varepsilon(s)$, proportional to S. That term makes tune of the particle amplitude dependent. Both $K(s)$ and $\varepsilon(s)$ are periodic functions of s with period C. The absence of damping in (1) is valid as the quality factor of these oscillations has a very high value $>10^9$, making this a truly Hamiltonian system. The amplitude (and, thus, the tune) of oscillations can be adjusted by the deflector strength D, so that 20 % of the particles can be trapped in resonance islands. All particles in one of the islands exhibit a tune of exactly 2/5, totally defying decoherence. The resulting BPM signals have been observed to persist for over a minute (approaching a million turns). As an example, Fig. 2.12 shows a "raw" Poincaré plot of transverse beam displacements $x_1(t)$ vs. $x_2(t)$, measured at two positions separated by about one quarter of a betatron wavelength.

The dynamics of a metastable beam of particles "injected" into artificially excited resonance islands in the Tevatron has been further studied in a subsequent experiment [19]. As before, the protons were under the influence of a single dominant nonlinear resonance, caused by the strong excitation of 14 sextupoles in the otherwise nearly linear accelerator. The island location was forced to oscillate at a modulation tune Q_m with an amplitude proportional to the modulation amplitude q:

$$Q(t,a) = Q_{00} + q\sin(2\pi Q_m) + \frac{1}{2}Ua^2. \tag{2.128}$$

Here, the last term represents detuning with amplitude of betatron oscillations a caused by the sextupoles. As in [18], the location and size of the island were adjusted by varying sextupole strengths and the base tune Q_{00} of small-amplitude

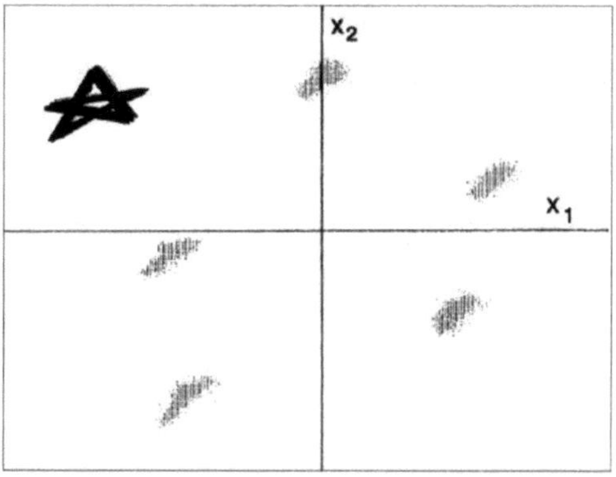

Fig. 2.12 "Raw experimental Poincaré map" exhibiting a metastable state of the accelerator. The logo in the corner of the plot is a demagnified view of the same data with successive points joined by straight lines. The point lands only on every second island, confirming the 2/5 identification [18]

particles. The beam was then allowed to circulate for some 10s to allow transients to decay before data were taken. At 9,000 turns after start of data taking, the tune modulation Eq. (2.128) was turned on by sinusoidal driving two weak quadrupoles. The tune modulation strength and tune, q and Q_m, were linearly ramped for 1 s (about 50,000 turns) and then turned off for the last 0.2 s (some 10,000 turns) of data taking. Figure 2.13 shows the BMP signal during such a chirp. One can clearly see an amplitude modulation of the islands at about 28,000 turns. The persistent signal started dropping dramatically at about 32,000 turns, eventually driving all of the trapped beam out of the resonant island.

The tune modulation trajectory that caused this response is drawn as the dashed line labeled "A" in Fig. 2.14, showing that the signal was lost when the boundary between "amplitude modulation" and "chaos" was crossed. Figure 2.14 also summarizes results from a trajectory labeled "B" that had a very weak constant tune modulation strength $q = 0.000204$, smaller or comparable to realistic operation values. Boundaries found experimentally (circles) and theoretically predicated boundaries between four regions are shown in Fig. 2.14. For the latter ones, the island tune (frequency of small oscillations of the particles trapped in the islands) $Q_I = 0.0063$ was the only free parameter used to adjust the location of these boundaries. A detailed discussion on the theory and explanation of the observations can be found in [18].

The effect of nonlinearity on transverse particle distributions has been studied in yet another beam dynamics experiment [20]. It was concentrated on "stochastic" effects, due to the particle dynamics, that cause "diffusive" evolution of the beam distribution even in the absence of external sources of "noise" or random scattering from residual gas molecules. These effects are studied by adding large nonlinearity to the otherwise comparatively linear machine. At the start of each observation period, a needle-shaped single bunch of some 10^{10} circulating protons was kicked horizontally. This yielded displacement of about 3 mm as observed at a downstream point. The resulting transverse beam profile was repeatedly measured by Flying Wires system every minute or so. During a run of (typical) 30 min duration, each

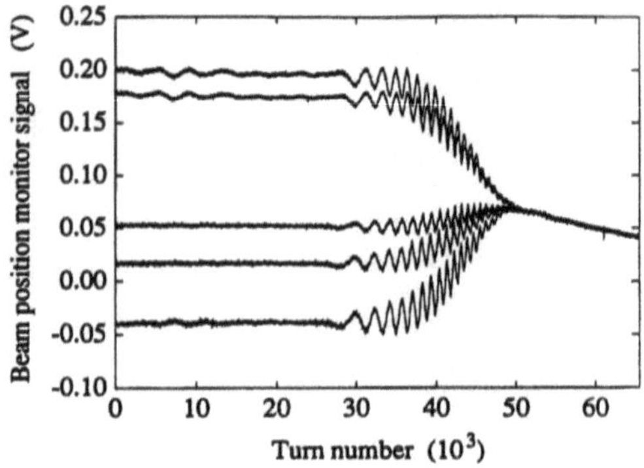

Fig. 2.13 Raw digitized signals of beam position monitor showing a persistent signal and its response to a chirp from $(q, Q_m) = (0, 0)$ to $(q, Q_m) = (0.0102, 0.0031)$. Vertical scale is in volts. Nonzero average initial value is due to a closed orbit offset [19]

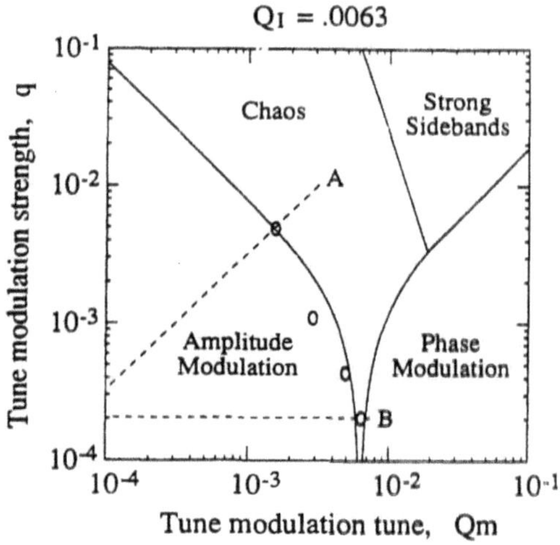

Fig. 2.14 Structure of the tune modulation parameter space (q, Q_m). *A* and *B* correspond to two scans. Four distinct "dynamical phases" are labeled ([19], see in the text)

proton circulates about 10^8 times and executes about 2×10^9 transverse betatron oscillations. The purpose of the kick is to generate a beam in which all the particles are in a region of measurably large diffusion. Individual protons initially oscillate at approximately constant amplitude with damping time equal to many tens of hours, but due to the dynamic diffusion, they start to expand in (initially) void areas of the phase space until they reach a physical aperture (defined by a scraper placed at $x_{\max} = 8$ mm) and get lost. In order to study the influence of resonance, all measurements were performed in the vicinity of the "2/5 resonance" (fractional horizontal tune Q_x close to 0.4).

Contrary to intuition, and unlike multiple scattering, diffusion causes the beam to narrow with time—as shown in Fig. 2.15. That behavior is caused by the sink at the aperture $x_{\max}$ which devours large amplitude particles, reduces the beam intensity, and depletes the tail of the distribution. Time evolution of the beam intensities and full widths, similar to one depicted in Fig. 2.16, was measured at different initial kicks and that allowed to determine the dependence of the diffusion

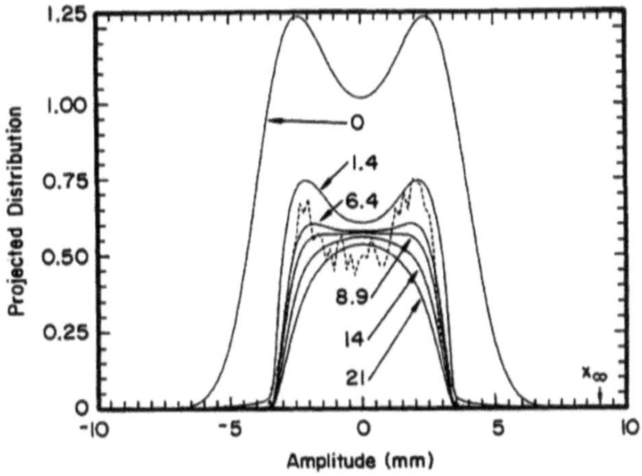

Fig. 2.15 Evolution of the beam profile: the jagged curve is the raw Flying Wire measurement at $t = 6$ min; smooth curves are as predicted by diffusion model. Times (in minutes after hollow beam formation) are indicated [20]

coefficient on the amplitude. Note that similar beam "shaving" phenomena with characteristic intensity decays following $\exp(-t^{1/2})$ law were observed later in the Tevatron Run II when nonlinearity induced by parasitic beam–beam interactions resulted in a significant reduction of a dynamic aperture and beam lifetime degradation—see [21] and discussion in Chap. 8.

2.4 Orbit Motion and Control

Motion of the accelerator components, most notably, quadrupole magnets, results in the beam orbit movements and can lead to a significant deterioration of the collider performance. The mechanism depends on the frequency. At frequencies of betatron sidebands $f_0\,(1-\nu) \approx 19.7$ kHz, fluctuations of the magnetic fields $\delta B(t)$, e.g., due to quadrupole magnet displacements $x(t)$, produce transverse kicks $\delta\theta(t) = \delta B(t)el/Pc = x(t)/F$, where l is the length of the element and F is the focusing length. That leads to the beam emittance growth with the rate of [22]:

$$\frac{d\varepsilon_x}{dt} = \gamma \frac{f_0^2}{F^2} \sum_{k=1}^{N_q} \sum_{n=-\infty}^{\infty} \beta_k S_x(f_0(\nu - n)) \tag{2.129}$$

where f_0 is the revolution frequency, γ is the relativistic factor, ν is the tune, $S_x(f)$ is the power spectral density of the quadrupole motion x, N_q is a total number of quadrupole focusing magnets, and β_k is the beta-function at the k-th quad location. At much lower frequencies, $f \ll f_0$, the kicks lead to a time-dependent displacement of the closed orbit:

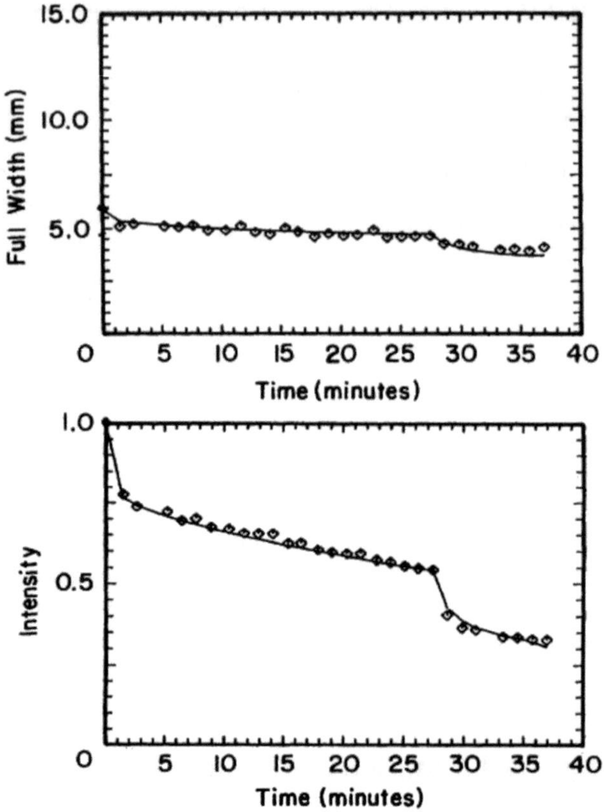

Fig. 2.16 Measured full width and intensity (points) compared to model-derived values (smooth curve). Intensity is normalized to 1 at $t = 0$. The steps at late times are the result of sudden aperture reduction to $x_{max} = 2.7$ mm [20]

$$X_{COD}(s) = \frac{\sqrt{\beta(s)}}{2\sin(\pi\nu)} \sum_{k=1}^{N_q} \sqrt{\beta_k(s)}\theta_k \cos(\varphi(s) - \varphi_k + \pi), \tag{2.130}$$

where s is the location along the ring and $\varphi(s)$, φ_k are betatron phases at the locations of the observation point and at the source of the kth magnet. At very low frequencies, hours to years, the quadrupole magnet displacements are often governed by the "*ATL* law" [23, 24] according to which the mean square of relative displacement dX^2 of the points separated by distance L grows with the time interval between measurements T as

$$< dX^2/dt > = A\ TL \tag{2.131}$$

where A is a site-dependent constant of the order of $10^{-5\pm1}$ μm^2/(s m) and brackets <. . .> indicate averaging over many points of observations distanced by L and over all time intervals equal to T. Such a wandering of the accelerator elements takes place in all directions. Corresponding average closed orbit distortion over the ring with circumference C is equal to [25]

$$< X_{\rm COD}^2(s) > \approx \frac{\beta(s)(\beta_F + \beta_D)}{8F^2 \sin^2(\pi\nu)} ATC \approx \kappa ATC \tag{2.132}$$

where FODO lattice structure is assumed, β_F, β_D are beta-functions at the focusing and defocusing lenses, and numerical coefficient $\kappa \approx 3$ for the Tevatron.

Due to feeddown effects from field non-linearities the Tevatron orbit drifts result in machine optics changes (tunes, coupling, chromaticities). Combined with aperture limitations they lead to increase of beam loss. At the injection energy of 150 GeV when the beams are several mm wide, orbit motion of about a mm leads to losses of the beams at several known places with tight aperture. At the energy of experiment, 980 GeV/beam, beam position in the RF cavities affects stability of high-intensity proton beam, e.g., the power of coherent beam oscillations goes up if the beam is too far off center. Also, oscillations of the RF cavities at synchrotron frequency (85 Hz at 150 GeV and 35 Hz at 980 GeV) are of concern for driving longitudinal emittance growth due to microphonic effects [26]. Large-scale long-term drifts of the orbit can be corrected by dipole correctors, and regular realignment of the magnets—usually during annual shutdown periods—helps to keep the corrector currents under the limit of 50 A.

2.4.1 Measurements of Betatron Oscillations and Orbit Motion

Several instruments were used to detect betatron oscillations in the Tevatron (see Chap. 9). The most challenging were direct measurements of natural betatron oscillations at sub-micron level. Several instruments were built for the detecting such oscillations and measurement of their frequency without additional excitation. Various techniques were employed, including 3D-BBQ (direct diode detector baseband tune) measurement system [27] and the digital tune monitor (DTM) which uses 16 bits 100 MHz ADCs for measuring the tunes on a bunch-by-bunch basis [28]. A very high-precision system employing a fast digital scope (Agilent Acqiris, 10bit, 8GS/s) for measurements of the turn-by-turn vertical centroid positions of individual bunches has been devised and used for digitizing signals from the plates of the VB11 BPM in the large vertical beta-function location that translates into better S/N ratio [29]. The system employs variable attenuators for compensating the beam position offset and phase shifters synchronized within 10 ps to minimize common mode. As a result, subtraction of the two signals by an RF hybrid provides about 44 dB of common mode rejection. Figure 2.17 shows 21,400 turn (0.44 s) record of the vertical beam position at the VB11 location.

The FFT of the data reveal significant excess of the signal at the betatron tune over the noise as shown in Fig. 2.18. The rms amplitude of the betatron oscillations is about 110 nm. Note that the amplitude significantly varies from store to store and often is two to three times smaller. That yields some 10–25 nm range of typical rms

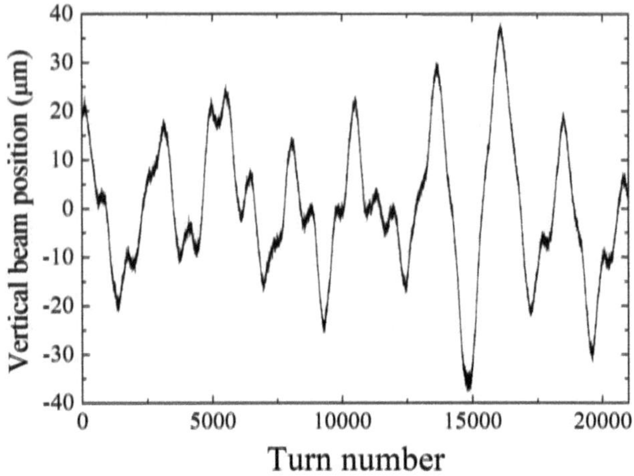

Fig. 2.17 Vertical position of the proton bunch #11 at the beginning of HEP store #6214 (October 2008), measured at the VB11 location with $\beta_y = 900$ m [29]

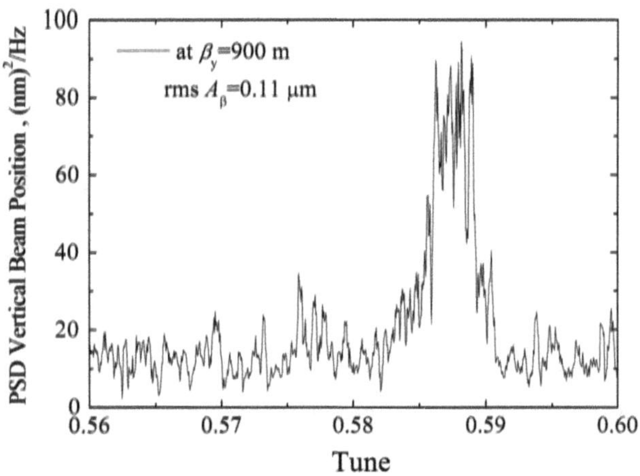

Fig. 2.18 Power spectral density of the vertical betatron oscillations (FFT of the data presented in Fig. 2.17) [29]

betatron motion amplitudes at the average beta-function location with $\beta_y \approx 50$ m. Spectrum of the vertical orbit motion at frequencies 2–1,000 Hz is shown in Fig. 2.19. It scales approximately as $\propto 1/f^3$ and is dominated by the low-frequency beam motion. The strongest lines are the harmonics of 60 Hz main power. The 15 Hz and the 0.45 Hz components can be explained by the effects of the fast cycling Booster synchrotron and the Main Injector on the power distribution systems at FNAL.

At ultralow frequencies, the orbit motion has significant (some 0.1 mm vertical and 0.3 mm horizontal) variation with a period of 12 h, which seems to be associated with Earth tides—see Fig. 2.20 from [30]. The rms of the orbit motion is about 100 μm horizontally and 30 μm vertically. The tide waves are clearly seen in the data from the Hydrostatic Level System (HLS) installed in the MI-8 beamline, the 8 GeV transfer line from the FNAL Booster to the FNAL Main Injector, located within

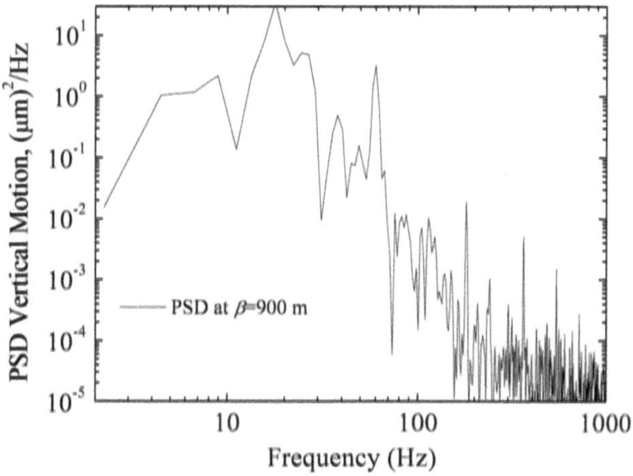

Fig. 2.19 Low-frequency power spectral density of the vertical orbit oscillations [29]

400 m of the Tevatron tunnel. The HLS is described in detail in [24] and consists of 20 submicron resolution water level sensors separated by 15 m from one another. The signal difference for a pair of sensors 135 m apart is plotted at the bottom of Fig. 2.20 and shows some 20 μm peak-to-peak amplitude of the 12 h period component. Slow closed orbit distortions of some 0.5–1 mm rms have been accumulated over 1–2-week intervals and required regular orbit "smoothing," until an automatic orbit stabilization system was introduced in operation in 2005 (see details in Chap. 9) and since then only high-frequency orbit jitter of about 10 μm rms remained.

2.4.2 *Magnet and Ground Motion in the Tevatron Tunnel*

The low-frequency orbit motion has been found correlated with (caused by) the vibrations of the magnets, particularly, strong focusing near interaction regions. Figure 2.21 from [31] demonstrates strong coherence between quadrupole vibrations and the Tevatron orbit motion, especially at certain frequencies. The coherence spectrum $C(f)$ is defined as

$$C_{xy}(f) = \left| \frac{< S_{xy}(f) >}{\sqrt{< S_{xx}(f) >< S_{yy}(f) >}} \right|, \tag{2.133}$$

where $S_{xy}(f)$ is the cross-correlation spectrum of two signals $x(t)$ and $y(t)$.

One can see that the orbit correlates well with the tunnel floor only at low frequencies ~0.1 Hz, while some excessive but small coherence exists at 2–4 Hz. The beam orbit correlates with the quadrupole magnet motion at frequencies of 0.2–2 Hz.

The closed orbit distortions are caused by the displacements of all magnetic elements along the circumference of the Tevatron. The strong coherence between

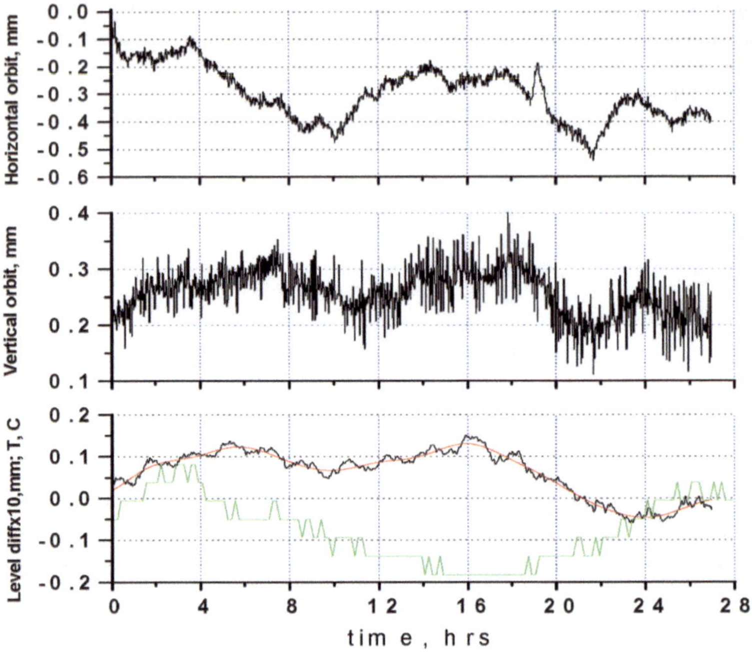

Fig. 2.20 Variation of the Tevatron proton orbits at the F48 location (*top*, horizontal, $\beta_x = 100$ m; *middle*, vertical, $\beta_y = 30$ m); and vertical ground motion as measured in the MI-8 line and temperature (*green* line) measured at the Tevatron sector F48 (both in the bottom plot) during 28 h long collider store #1668 (August 17, 2002)

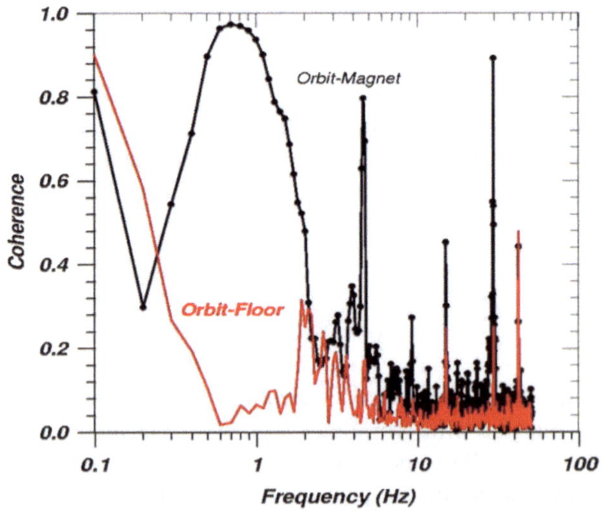

Fig. 2.21 Coherence between signals of the vertical Tevatron beam orbit motion and the F11 magnet vibrations and between the orbit and the tunnel floor (*red line*) [31]

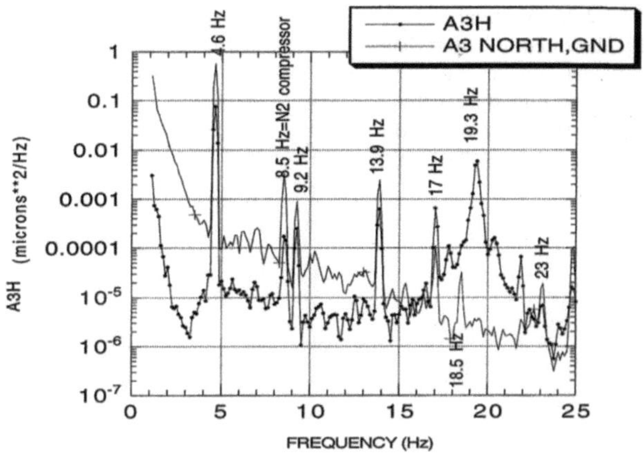

Fig. 2.22 Power spectral density of the A35 quadrupole motion and the ground nearby, from [32]

the magnet and beam vibrations means that there is a common source of vibration along the whole accelerator ring. For example, several remarkable peaks in the orbit–magnet coherence occur at 4.6, 9.2, 13.9 Hz, etc., at the Fermilab site-specific frequencies caused by the Central Helium Liquefier plant operation and well detected everywhere around the ring [32]—see Fig. 2.22.

2.4.3 Slow Diffusion of the Tevatron Tunnel

Analysis of the multiyear Tevatron magnet alignment data shows that in addition to systematic changes due to tides or slow drifts, there is a "random walk" both in time and in space component characterized by the ATL-law of Eq. (4.3) [33].

The alignment system of the Tevatron employs more than 200 geodetic "tie rods" (thick metal rods screwed into the concrete tunnel wall all over the ring and equipped to hold spherical retroreflectors for precise position measurements), each spaced approximately 30 m apart. The positions of the magnets are regularly referenced locally with respect to the "tie rods," while the positions of all the "tie rods" are routinely monitored. The "tie rod" elevation datasets are available for the years of 2001, 2003, 2005, 2006, and 2007. Figure 2.23 shows the change of the elevations $dY(z)$ around the ring accumulated over two intervals—2 years (2003–2005) and 6 years (2001–2007). One can see that longer-term motion has a larger amplitude. The variance $<dY^2(L)> = <(dY(z) - dY(z+L))^2>$ of the elevation difference of the points as a function of the lag (distance between pairs of the measurement points) L has been calculated and averaged over all possible time intervals. That is to say, there are two 1-year intervals (2005–2006, 2006–2007), three 2-year intervals (2001–2003, 2003–2005, 2005–2007), etc., and one for the 6-year interval 2001–2007. The results for the 1-year changes and for the 6-year change are shown in Fig. 2.24. A remarkable difference between the two plots is that 1-year variance scales linearly only up to $L \approx 700$–800 m and does not depend on L beyond that scale, while the 6-year variance grows all the way to distances as

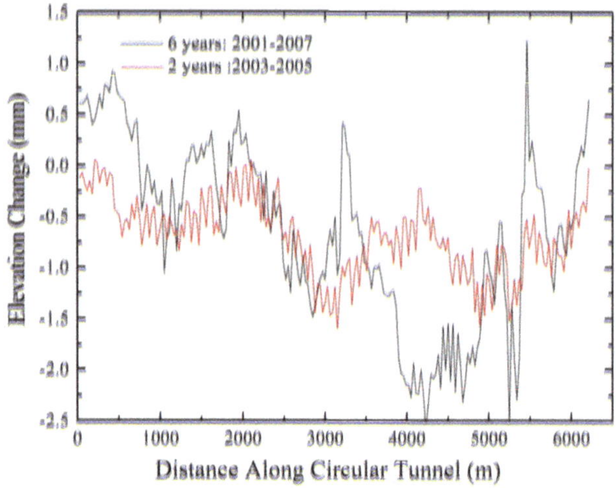

Fig. 2.23 Vertical displacement of more than 200 "tie rods" in the Tevatron tunnel over the period of 2003–2005 and a 6-year period of 2001–2007 [33]

large as 1,800 m. The linear dependence on L is indicative of a significant level of interdependence of the movements of distant points. The calculated variances for all possible time differences can be well approximated by linear fits $<dY^2(L)> = a + bL$ over distances less than 900 m and the slopes (fit parameters b with the error bars) are plotted in Fig. 2.25.

One can see that the variance per unit distance grows with the time interval between the measurements and can be approximated by a linear fit $b(T) = cT$ with $c = 0.153 \pm 0.004$ [mm^2/km/year]. The Tevatron "tie rod" data analysis presented in Figs. 2.24 and 2.25 can be summarized by the ATL law $<dY^2> = ATL$ with coefficient $A_{\text{Tevatron}} = c = (4.9 \pm 0.13) \times 10^{-6}$ μm^2/s/m.

It is to be noted that for small time intervals T the movements of the ground elements are fully uncorrelated if they are separated by a long enough distance $L > L_m$, for example, by more than 800 m for $T = 1$-year intervals as seen in the Tevatron alignment data—see Fig. 2.24—or by more than 120 m for $T = 1$-week intervals as seen in the Tevatron B-sector HLS data [24]. On the basis of these two observations, one can assume that the boundary between totally uncorrelated and the ATL-law regimes scales approximately as $L_m \propto T^{1/2}$.

2.5 Measurements and Corrections of Nonlinear Optics

2.5.1 *First- and Second-Order Chromaticity and Its Correction*

Implementation of electron cooling of antiprotons in the Recycler (see Chap. 7) resulted in a dramatic increase of the antiproton beam brightness. After the 2006 shutdown, the head-on beam–beam tune shift experienced by the proton beam exceeded 0.025 and at times reached 0.03 (see Chap. 8). It was then when beam–

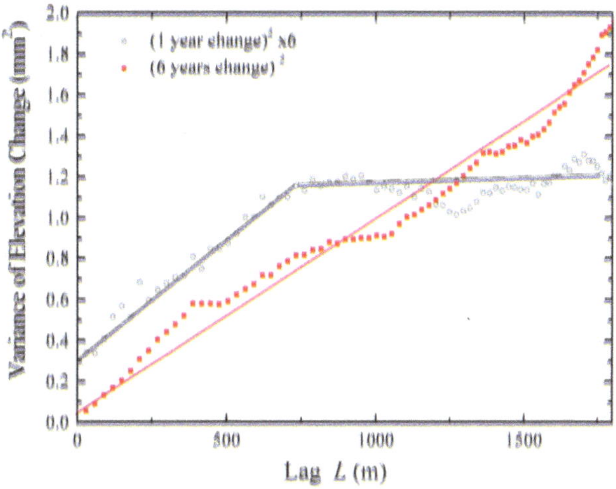

Fig. 2.24 Variances of the Tevatron "tie rod" vertical displacements over time intervals of 1 year (multiplied by 6) and 6 years vs. the distance L [33]

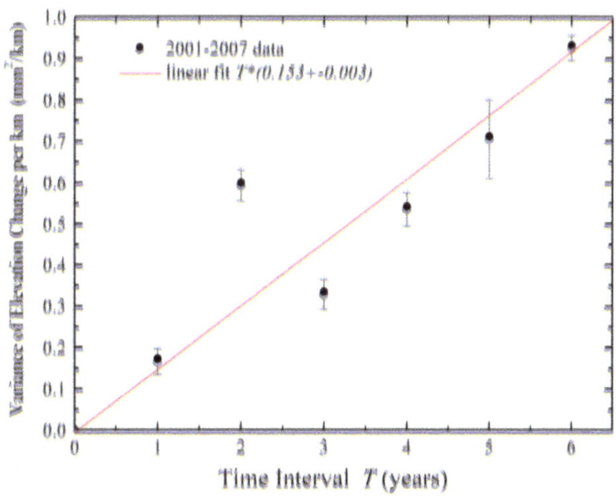

Fig. 2.25 Variances of the Tevatron alignment "tie rods" displacements per unit distance vs. the time interval between the measurements [33]

beam-related losses and emittance blowup started to be observed in protons. It was shown by beam–beam simulations (see Sect. 8.4) that deterioration of the proton lifetime was caused by a decrease of the dynamical aperture for off-momentum particles due to head-on collisions. A contributing factor to this was large chromaticities of the beta-functions at the main IPs.

Initially, a major change of the betatron tune working point was considered as a way to mitigate the beam–beam effect. A possible candidate was the working point near the half-integer resonance, which promised up to 30 % increase of the beam brightness. However, operation near the 1/2 resonance requires careful correction of focusing errors, including the chromatic perturbations. Consequently, it would require machine reproducibility well above achieved and therefore this project was not approved. That motivated the development of a modified chromaticity correction scheme at the Tevatron, aiming at the reduction of the second-order chromaticity by about an order of magnitude.

One can use the well-known perturbation theory approach to describe the distortion of the beta-function caused by chromatic errors (see e.g., [34]). Beta-beating excited by a single quadrupole for an off-momentum particle can be described by the formula:

$$\frac{\Delta\beta}{\beta}(s) = -\frac{\delta}{2\sin(2\pi Q)}\frac{K}{B\rho}\tilde{\beta}\cos\left(2|\tilde{\psi} - \psi(s)| - 2\pi Q\right). \tag{2.134}$$

Here $\delta = \Delta p/p$ is the relative momentum deviation, Q is the betatron tune, K is the quadrupole-integrated strength, $B\rho$ is the magnetic rigidity, ψ is the betatron phase, and ~ denotes values at the location of the quadrupole. In the first-order approximation, the contributions from all quadrupoles are summed to give the total beta-wave. Quadrupoles of the final focus have the largest strength and highest value of beta-function. In the Tevatron, the betatron phase advance between the two interaction regions is close to π; thus, the contributions from the final focus quadrupoles add with the same phase and they dominate the chromatic $\Delta\beta/\beta$.

The contribution to second-order tune chromaticity from a single quadrupole derived from the perturbation theory is given by the following expression:

$$\frac{d^2Q}{d\delta^2} = \left(\frac{1}{4\pi}K\tilde{\beta}\right)\left(\frac{\Delta\beta}{\beta}/\delta\right). \tag{2.135}$$

This effectively means that the second-order chromaticity is proportional to the strength of the final focus quadrupole and to the value of chromatic beta-function at its location. Figure 2.26 shows the comparison of the measured and modeled chromatic beta-function of the Tevatron in the collision mode. The measured parameters were obtained from two orbit response measurements performed at different values of the revolution frequency. The modeled curve was obtained by the perturbation theory. In Fig. 2.27 the same beta-functions are plotted on a zoomed horizontal scale close to the CDF interaction region. One can see that the model gives quite accurate description of the effect.

Sextupole magnets can be used to correct the second-order chromaticity generated by quadrupole focusing errors. Using the same perturbative approach and considering sextupole as a quadrupole at the off-center orbit, we get the expression for the sextupole-driven horizontal chromatic beta-function:

$$\frac{\Delta\beta}{\beta}(s) = \frac{\delta}{2\sin(2\pi Q)}\frac{S\tilde{D}_x}{B\rho}\tilde{\beta}\cos\left(2|\tilde{\psi} - \psi(s)| - 2\pi Q\right) \tag{2.136}$$

where S is the sextupole strength and ~ denotes parameters at the location of the sextupole. Even though the effect of individual sextupoles on the beta-function chromaticity is much less than that of the final focus quadrupoles, one can achieve

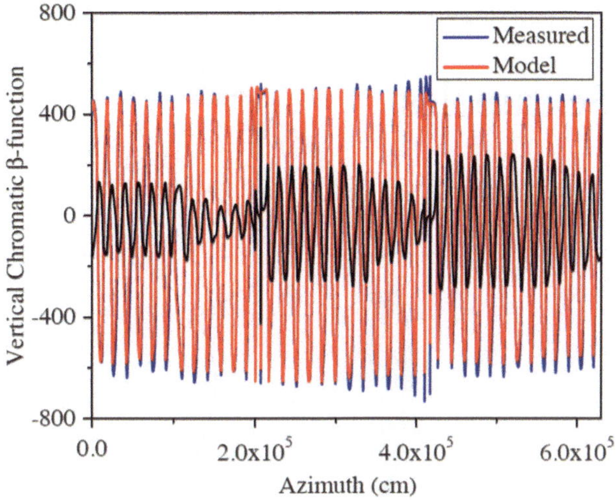

Fig. 2.26 Chromatic beta-function vs. azimuth starting at F0. *Blue line*, measured; *red*, model; *black*, proposed correction

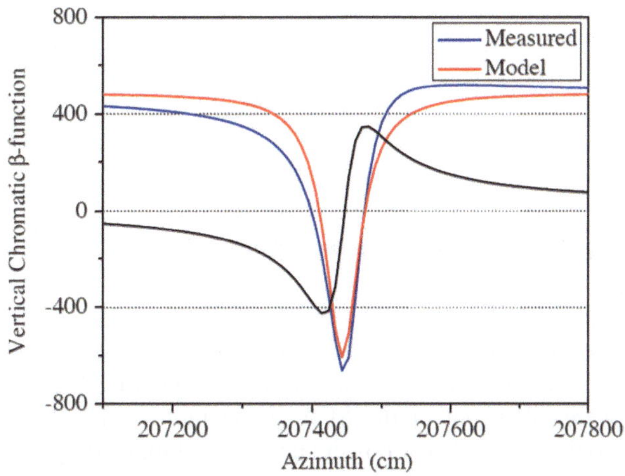

Fig. 2.27 Chromatic beta-function vs. azimuth in the vicinity of CDF IP. *Blue line*, measured; *red*, model; *black*, proposed correction

compensation by selecting many sextupoles with the correct phase advance and combining them into families.

There are 176 chromaticity correction sextupole magnets in the Tevatron. Originally, they were combined into two families SF and SD, each with 88 elements powered in series. Sextupole coils are placed in the so-called spool pieces located next to quadrupoles in the regular FODO lattice of the arcs. The betatron phase advance per FODO cell is close to 60° in both planes. Thus, it was possible to select sextupoles that would have their betatron phase advance with respect to the final focus quads equal to π or $\pi/2$. The total of 46 sextupoles in each family were found to satisfy this condition. However, it was discovered that rewiring them into 4 new circuits would have considerable cost mainly due to the large amount of required new cable. Hence, we had to limit the number of elements in the new circuits and

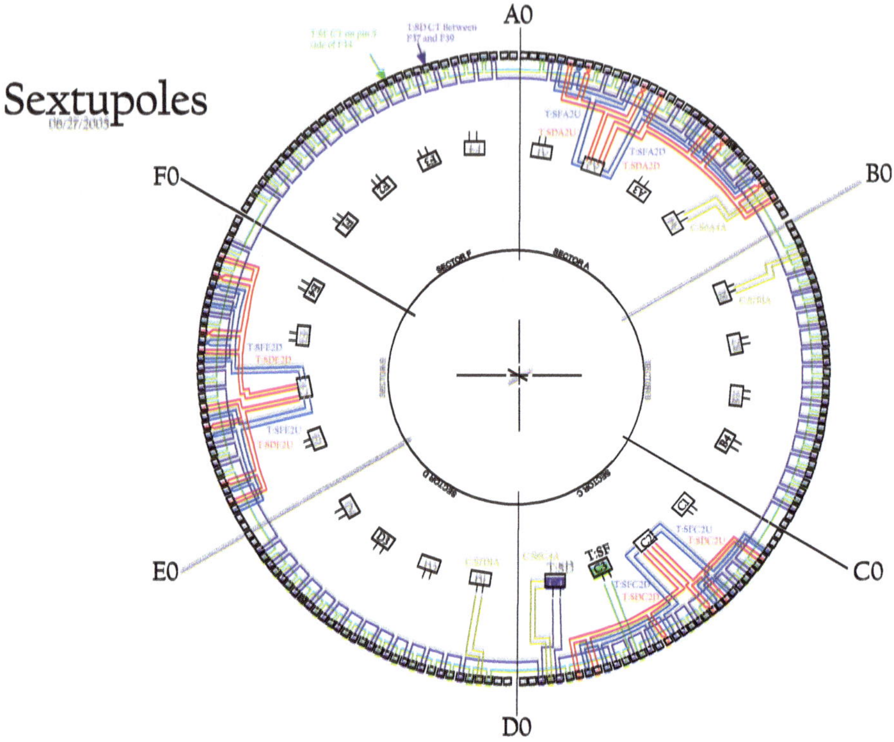

Fig. 2.28 Layout of sextupoles in the Tevatron

group them close to three service buildings which would reduce the length of the new cabling. The final configuration is shown in Fig. 2.28. The total of 44 sextupoles were taken out of the SF and SD families (22 from each). The sextupoles are powered by 12 new power supplies and logically grouped into 4 families. The new circuits are designed to have equal number of elements with positive and negative current. This allows to keep the linear betatron tune chromaticity constant when using the new groups. The disadvantage of this solution is that it breaks the sixfold symmetry of the machine but the expected feed-down effect on the beta-functions is small.

In Figs. 2.26, 2.27, and 2.29, the simulated effect of application of the new families on the chromatic beta-function is plotted for the collision and injection modes. Note that the expected beta-function chromaticity at the CDF IP at collisions is close to zero.

The new sextupole scheme was commissioned in three stages. First, the chosen elements were switched to the new power supplies while being operated at the original current and polarity. This allowed to verify the stability and reliability of electrical components. Second, the new circuits were turned off and the Tevatron was tuned for operating in this new mode. Finally, polarities of some of the

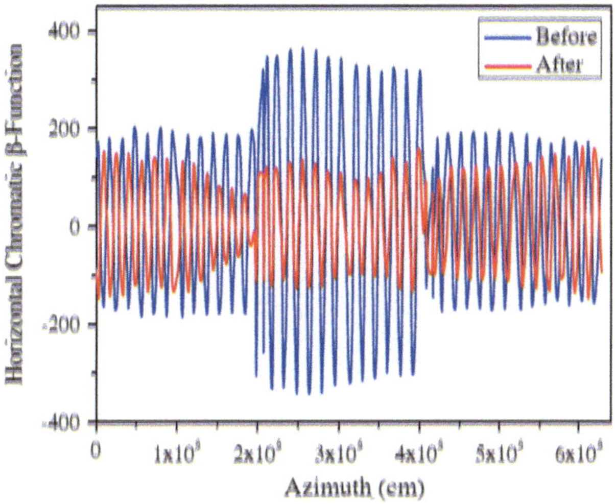

Fig. 2.29 Horizontal chromatic beta-function at the injection energy. *Blue line* is for the original sextupole configuration, *red* for the proposed correction

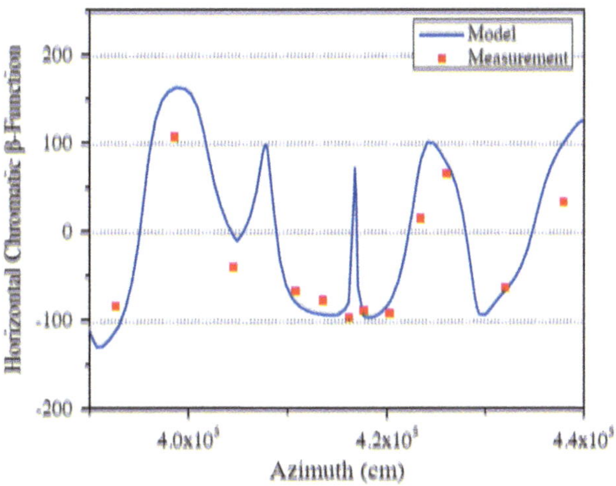

Fig. 2.30 Chromatic beta-function near the D0 IP at the injection energy

sextupoles were flipped and the new families were turned on at the designed strength.

Figures 2.30 and 2.31 show the comparison of expected and measured beta-function chromaticity after implementation of the new circuits in the injection and collision modes, respectively. The corresponding change in the second-order tune chromaticity is presented in Fig. 2.32. Beta-function chromaticities for both the vertical and horizontal planes were corrected.

Measured beta-functions are in good agreement with the calculated values. The second-order tune chromaticity in the collision mode was reduced from −15,000 units to −3,000 units which is close to expected. As a result, we observe a noticeable improvement of the proton beam lifetime at collisions (see Chap. 8).

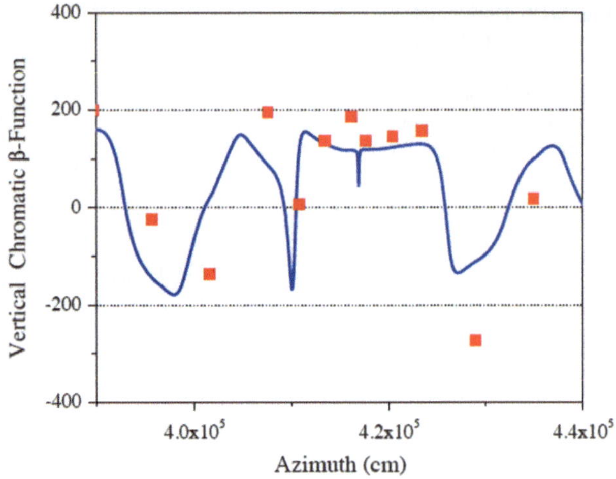

Fig. 2.31 Chromatic beta-function near the D0 IP in the collider mode

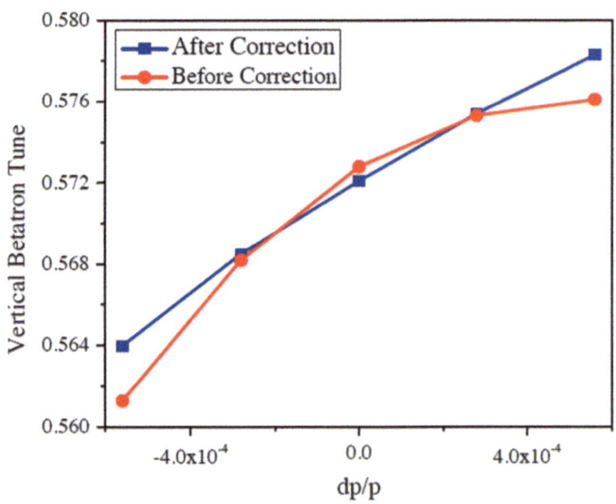

Fig. 2.32 Dependence of the vertical betatron tune on particle momentum in the collider mode

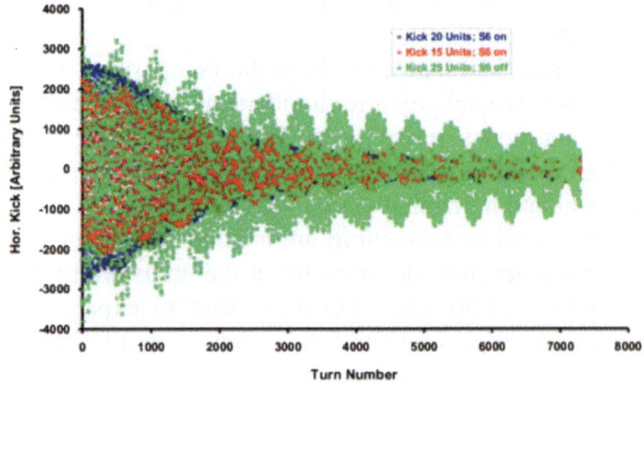

Fig. 2.33 TBT raw data for 7,300 turns after the kick. Three cases are shown: with the sextupoles S6 switched on with $A = 20$ and 15 units and without these special sextupoles for $A = 25$ units

References

1. I. Borchardt, E. Karantzoulis, H. Mais, G. Ripken, Calculation of Beam Envelopes in Storage Rings and Transport Systems in the Presence of Transverse Space Charge Effects and Coupling, DESY 87-161 (DESY, Hamburg, 1987)
2. F. Willeke, G. Ripken, Methods of Beam Optics, in *Proceedings of US Particle Accelerator School (1987 and 1988), AIP Conf. Proc.*, New York, 1989, p. 184
3. V. Lebedev, S. Bogacz, JINST **5**, P10010 (2010)
4. L.D. Landau, E.M. Lifshitz, Mechanics (Course of Theoretical Physics, vol. 1) (1976)
5. A.A. Kolomensky, A.N. Lebedev, *Theory of Circular Accelerators* (Moscow, 1962)
6. D.A. Edwards, L.C. Teng, IEEE Trans. Nucl. Sci. **20**(3), 885–889 (1973); L.C. Teng, Fermilab, FN-229 0100 (1971)
7. A. Burov, V. Lebedev, Coupling and Its Effects on Beam Dynamics, in *Proceedings of Hadron Beam 2008*, Nashville, TN
8. A. Burov, Two-beam instability in electron cooling. Phys. Rev. ST-AB **9**, 120101 (2006)
9. V. Lebedev et al., Nucl. Instrum. Meth. **A558**, 299–302 (2006)
10. V. Lebedev, http://www-bdnew.fnal.gov/pbar/organizationalchart/lebedev/OptiM/optim.htm
11. V. Nagaslaev et al., 8 GeV beam line optics optimization for the rapid antiproton transfers at Fermilab, APAC-07, p. 345 (2007)
12. W.J. Corbett, M. Lee, V. Ziemann, in *Proceedings of PAC'93* (Washington, DC, 1993), p. 108
13. J. Safranek, Nucl. Instrum. Meth. **A388**, 27 (1997)
14. ICFA Beam Dynamics Newsletter No. 44 (2007)
15. V. Sajaev, L. Emery, in *Proceeding of EPAC'02* (Paris, France, 2002), p. 742
16. M. Borland, in *Proceedings of ICAP'98* (Monterey, CA, 1998)
17. Y. Alexahin, E. Gianfelice-Wendt, FERMILAB-PUB-06-093-AD (2006)
18. A. Chao et al., Phys. Rev. Lett. **61**, 2752 (1988)
19. T. Satogata et al., Phys. Rev. Lett. **68**, 1838 (1992)
20. T. Chen et al., Phys. Rev. Lett. **68**, 33 (1992)
21. V. Shiltsev et al., Phys. Rev. ST Accel. Beams **8**, 101001 (2005)
22. V. Lebedev et al., Part. Accel. **44**, 147–164 (1994)
23. B.A. Baklakov et al., Tech. Phys. **38**(10), 894–898 (1993); translated from Sov. Zh. Tech. Fiz., **63**(10), 123–132 (1993)
24. V. Shiltsev, Phys. Rev. ST Accel. Beams **13**, 094801 (2010)
25. V. Parkhomchuk, V. Shiltsev, G. Stupakov, Part. Accel. **46**, 241–258 (1994)
26. J. Steimel et al., in *Proceedings of the 2003 I.E. Particle Accelerator Conference* (Portland, OR, 2003), p. 48
27. C.Y. Tan, in *Proceedings of the 2005 I.E. Particle Accelerator Conference* (Knoxville, TN, 2005), p. 141
28. A. Semenov et al., in *Proceedings of IEEE Particle Accelerator Conference* (Albuquerque, NM, 2007), p. 3877
29. V. Shiltsev, G. Stancari, A. Valishev et al., JINST **6**, P08002 (2011)
30. V. Shiltsev, T. Johnson, X.L. Zhang, Preprint FERMILAB-CONF-02-250 (2002)
31. B. Baklakov et al., Phys. Rev. ST-AB **1**, 031001 (1998)
32. C. Moore, in *Proceedings of the IWAA95, KEK-Proc-95/12* (1996), p. 119
33. V. Shiltsev, Phys. Rev. Lett. **104**, 238501 (2010). doi:10.1103/PhysRevLett.104.238501
34. A.W. Chao, M. Tigner, *Handbook of Accelerator Physics and Engineering* (World Scientific, Singapore, 1999), pp. 263–264

Chapter 3
Magnets and Magnetic Field Effects

J. Annala, D. Harding, V. Shiltsev, M. Syphers, and J. Volk

3.1 Introduction: Magnets of the Tevatron Accelerator Complex

The magnets of the circular accelerators that comprise the Fermilab complex have contributed to, responded to, and solved a string of accelerator physics issues. Most importantly, the magnets are supposed to generate high quality magnetic fields needed for stable long-term dynamics of the particles circulating in the rings. The quality of the transverse magnetic field $\boldsymbol{B}$ is given by the multipole coefficients in the expansion:

$$B_x + i \cdot B_y = B_0 \sum_{n=0} (b_n + ia_n) \left[\frac{x + iy}{R_0} \right]^n, \qquad (3.1)$$

where R_0 is the reference radius (1 in. in the Fermilab accelerators), the pole number is $2(n+1)$ and $b_n(a_n)$ are the normal (skew) multipole coefficients, and b_0 is unity. For example, the multipoles allowed by dipole symmetry, $b_2, b_4, b_6, \ldots$ are designed to be small and would be 0 for a pure $\cos\theta$ coil winding. The precise coil placement, and hence good magnetic field uniformity at the relative level of the multipole coefficients of few 10^{-4}, had the biggest effect on the accelerator performance.

3.1.1 *Booster*

At the heart of the Booster synchrotron are its combined function magnets (dipole, quadrupole, and sextupole), one focusing and one defocusing on each

J. Annala • D. Harding • V. Shiltsev (✉) • M. Syphers • J. Volk
Fermi National Accelerator Laboratory, PO Box 500, Batavia, IL 60510, USA
e-mail: shiltsev@fnal.gov

V. Lebedev and V. Shiltsev (eds.), *Accelerator Physics at the Tevatron Collider*, Particle Acceleration and Detection, DOI 10.1007/978-1-4939-0885-1_3,

girder [1]. Cycling at 15 Hz in a resonant circuit with a bank of capacitors on each girder, the magnets employ an external vacuum skin around both the core and the coils to avoid eddy currents that would be intolerable in a standard beam tube. While the combined function magnets ensure synchronization between the bending and focusing strength through the acceleration cycle, the ability to adjust the tune is limited and variations in the onset of saturation does change the optics during the cycle.

The original air-core correction dipoles and quadrupoles were distributed about the ring. They were weak magnets only designed to run DC and correct at the injection energy. The expectation was that realignment of the gradient magnets would be used to correct for effects at higher energy, and that less steering range would be needed as the accelerated beam shrank. Correction sextupoles were also included, but due to space constraints they were lumped, not distributed, limiting the resonances they could control. Ramped power supplies were eventually added for some correctors, but power supply and heating limitations prevented running them as aggressively as was desired.

In 2009 installation was completed of new set of corrector magnets, power supplies, and controls, with two dipoles, two quadrupoles, and two sextupoles in a single package on every girder [2]. Combined with more powerful beam instrumentation and analysis, the orbit and tunes can now be smoothed through the entire acceleration cycle. The new quadrupoles are strong enough to move the working point in tune space to a much more favorable location near the half integer. Beam losses have been reduced dramatically [3].

Multi-turn injection of 200 MeV protons into the Booster was initially implemented with an electrostatic septum and a pulsed four-bump. Three to four turns filled the aperture. Improvements over the years included the switch to H-injection, with a stripping foil in the middle of the pulsed four-bump [4]; raising of the injection energy to 400 MeV [5]; and finally the change to a three-bump that improves the incoming trajectory and the lattice of the circulating beam, a change made possible by newly designed pulsed injection dipole magnets [6].

Single turn (or partial turn) extraction employs fast kicker magnets that steer the beam into a pulsed septum magnet. Initially the beam was steered around the extraction septum at both extraction points (transfer to the Main Ring and to the beam dump) using just the DC corrector dipoles. These were later supplemented by a stronger, ramped four-bump at each location. In the course of detailed modeling in 2002, it was recognized that the edge focusing of the bumps was having a seriously negative effect on the optics [7]. Over several years the bump strengths were reduced and one extraction point was eventually completely eliminated by relocating the dump. The opened aperture and reduced optical distortion produced significant improvements in the beam transmission efficiency.

3.1.2 Main Ring

The Main Ring, Fermilab's original top accelerator, served as the injector for the Tevatron from the Tevatron's commissioning in 1983 until the commissioning of the Main Injector in 1998. From 1985 onward it also accelerated the protons to produce antiprotons. The Main Ring was an interesting machine in its own right, holding the honor of being the first separated function synchrotron [8]. By separating the bending and the focusing functions, both the bending field and the focusing gradients could be made significantly stronger than in a combined function accelerator, shrinking the circumference of the ring (or increasing the energy) and reducing the beam size. Using H magnets (rather than the C magnets typical of the day) suppressed half the unwanted harmonic components of the field and allowed a smaller, less expensive magnet. Taking the compactness approach to the extreme, there were two dipole geometries, with apertures tailored to the alternating high and low beta functions in the two dimensions.

The Main Ring also pioneered the use of "Lambertson magnets," iron septum magnets for injection and extraction.

The injection energy from the Booster was high enough that the dipole remanent field was not a major problem as it had been in early synchrotrons, but the remanent sextupole field was always an issue.

During Collider Run I, with the detectors in place, the Main Ring was deformed vertically to make a bypass around CDF [9]. With a significant portion of the ring in a different plane from the rest of the ring, some larger gap dipoles were required to bend the beam vertically, as well as some standard dipoles bussed to run at twice the field with the same current.

3.1.3 Main Injector

The Main Injector dipole magnets are a more conservative design than the Main Ring dipoles to maximize reliability [10]. They are robust and energy-efficient. To maximize the useful aperture, the magnets are curved, following the beam sagitta, rather than straight, like the Main Ring dipoles. Injecting at the same energy into a smaller ring (with a higher field) means that the remanent field is less important. More steel in the yoke and a tapered pole reduces the saturation and minimizes the remanent sextupole field. With two lengths of dipole in the lattice and the magnets designed to run into saturation, it was important to ensure that the ends and the body saturated at the same rate. The 3D modeling tools available were insufficient to design the ends, so multiple iterations of prototyping were required to achieve that goal while also eliminating the sextupole component from the end field [11].

The quadrupoles, too, posed matching challenges. The 2.13-m quadrupoles from the Main Ring were to be reused in the Main Injector, along with new quadrupoles of two additional lengths. For the sake of matching the saturation, so that the magnets could run on the same buses, it was agreed that the Main Ring design would be used, just longer. This required ensuring that the effective lengths of the magnets were in the correct ratios, not the iron lengths [12]. Concern about our ability to get this right led to the addition of trim coils to all the new quadrupoles, but they have not proved necessary. Although the spread in magnet strengths was small enough to ignore, having separate focusing and defocusing buses allowed us to narrow the spread of quadrupole strengths even more by putting the stronger ones on one bus and weaker ones on the other bus.

The Main Ring quadrupoles are only twofold symmetric under rotation, so there is an inherent possibility of an octupole component to the field. Indeed, a fabrication error in the original die for stamping laminations led to a significant octupole in some magnets before subsequent cores were machined to compensate and a new die fabricated. This led to the question of how much octupole was appropriate for the Main Injector. Eventually a balance was struck between the needs of a low octupole for injection stability and a high octupole to help with resonant extraction. During installation the old quadrupoles with high octupole were relegated to beam lines.

After several years of operation, it was agreed that larger aperture quadrupoles were needed at the seven extraction and injection points. To track the other magnets, it was agreed to run the magnets on the same buses as the rest of the quadrupoles. The larger aperture for the same gradient raised the pole tip field and drove the magnets slightly farther into saturation than the rest of the magnets. To compensate, independently powered trim coils were included in each magnet. The end field of these magnets was successfully modeled in 3D, a necessity for matching the integrated gradient of the new magnets with the original quadrupoles below saturation [13].

3.1.4 Antiproton Source

The Debuncher and Accumulator rings of the Antiproton Source required very large aperture magnets with storage ring field uniformity [9]. To keep the magnet cost under control, the yoke size was limited and the magnets designed to run with the steel saturated. The poles were tailored to provide a uniform field at this excitation level, with the consequence that the dipoles develop a severe sextupole component below the design excitation. This became an issue when the Accumulator was used to not just store antiprotons but also to decelerate them for use in an internal target experiment. The quadrupole pole shape was critical, as the good field region extended well beyond the pole tip radius on the center plane.

Several different lengths of one dipole magnet style and of both main quadrupole magnet styles were needed, leading to the requirement that the pole ends be contoured to eliminate contributions to the harmonic components of the field integrals. Several prototype iterations were required on the first quadrupole style, but that solution could be scaled up to the larger quadrupoles.

3.1.5 Recycler

The Recycler magnets are treated in detail in Sect. 3.4. We mention here that use of the low-cost permanent magnets, mostly combined function magnets, was only possible because we needed a low energy storage ring and had a large radius tunnel available [14].

3.1.6 Tevatron

Superconducting magnets define the Tevatron, the first synchrotron built with the technology [15]. Subsequent Sects. 3.2 and 3.3 discuss two subtle aspects of the magnet performance, but the most amazing feature is that they worked at all. Issues that had to be addressed included conductor strand and cable fabrication, coil geometry and fabrication, mechanical constraint and support of the coils, cooling and insulation, and protection during quenches. The coil placement, and hence magnetic field uniformity, had the biggest effect on the accelerator performance [16]. Of the allowed (even) harmonics the lowest (6-pole and 10-pole) could be controlled in a straightforward manner, but the higher harmonics could not be controlled in the simple geometry employed (though only the 18-pole had significant magnitude). On the center plane the field integral is nearly constant out to a radius of about 20 mm, as the various multipoles cancel, but moving vertically from the center plane the field uniformity deteriorates.

The simple end geometry of the dipole coils led to a significant sextupole component in the ends. The design sextupole in the body was adjusted to zero the field integral through the whole magnet. There is some evidence that this longitudinal distribution of sextupole field must be accounted for detailed accelerator modeling [17].

Tevatron superconducting magnets, designed in the 1970s, performed beautifully over the years. The Tevatron experience paved the way for other high energy hadron colliders, like HERA (9 m long, 75 mm bore, 5.3 T dipoles made of NbTi conductor, with Al collar and cold iron), RHIC (simple and economical design of 9 m long, 80 mm bore, 3.5 T dipoles), and LHC (15 m long, 56 mm double bore dipoles operating at 2 K super fluid He temperatures with design field of 8.3 T).

3.2 Chromaticity Drifts and Snapback in the Tevatron and Their Compensation

The Tevatron contains 774 superconducting dipole magnets which operate between 0.66 T (corresponding to 150 GeV) at injection and 4.4 T at the peak design field (1 TeV). The Tevatron dipoles are a single aperture cos(θ) (*cosine theta*) design magnet wound from the so-called Rutherford style cable [18]. The magnet is constructed from an upper and lower half, each consisting of a 35 turn inner coil and a 21 turn outer coil. The cable is composed of 23 strands, each of which contains 2,050 separate 9 μm filaments of NbTi embedded in a copper matrix.

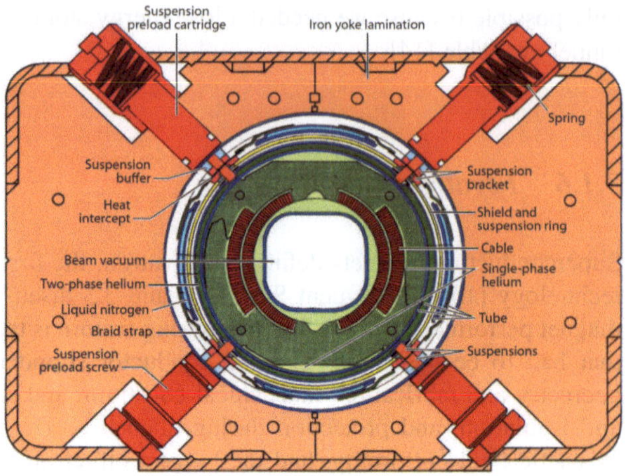

Fig. 3.1 Cross section of the Tevatron superconducting dipole magnet [43]

When the magnets are powered, along with the transport current, there are at least three types of induced currents that can flow in the magnet as well. There are eddy currents between different stands in the cable, and coupling currents between filaments within a strand. Both of these currents decay with time constants of seconds. But there are persistent currents flowing within filaments with extremely long time constants.

The persistent currents exist because of the Meissner effect which causes current to flow on the surface of superconductors in a way that opposes any change in externally applied magnetic field from the interior of the superconductor [19]. In the presence of a changing dipole field, shielding currents form near the surface of the filaments with a magnitude of J_c in a cosine theta pattern that cancels the dipole field in the interior of the filament (Fig. 3.2a). As the external field is increased, the shielding current pattern changes until the filament is penetrated (Fig. 3.2b). This field is known as the penetrating field B_p. The transport current needed to generate the field in real magnets flows at the critical current density in an ellipse interior to the shielding currents (Fig. 3.2c). The transport current is in the same direction as half of the shielding current. Figure 3.1d shows the orientation of four filaments arranged symmetrically around the aperture of a cosine theta dipole magnet, along with their induced dipole moments B_I.

Figure 3.2 shows how persistent shielding currents are set up in a filament to oppose a change in external magnetic field as it (a) starts to rise, (b) fully penetrates the filament, (c) exits with a transport current in the center of the filament and the shielding current induced dipole moments are shown (d) in four filaments symmetrically surrounding a dipole aperture. These induced fields not only have a component opposing the dipole field at the center of the aperture but also have components that lead to a sextupole field, and other allowed (by symmetry) multipoles b_n, $n = 5, 7, \ldots$

The dipole moments induced by the shielding currents can dramatically affect the quality of field in the dipoles. As the transport current reaches its desired current, and then reverses, the shielding currents switch polarity in order to oppose

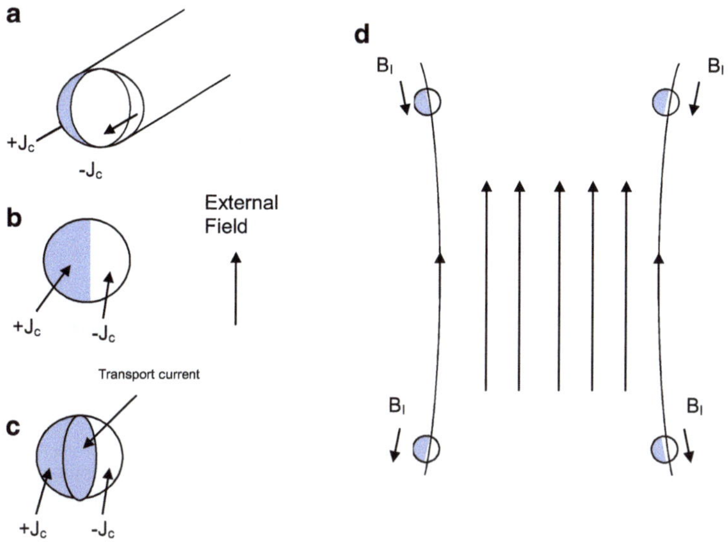

Fig. 3.2 Persistent currents effects in superconducting accelerator magnets: (**a–c**) shielding and transport currents in the filament; (**d**) magnetization currents in four symmetric filaments of a dipole coil

the drop in magnetic field. For this reason, the non-dipole fields (in this case—sextupole) have a large hysteretic component. The persistent currents in the superconductor are most dramatic at low field, with currents well below J_c so the transport current stays near the center of the filament allowing for stronger dipole moments from the shielding pattern. If the magnets are held at a fixed excitation, say, at the injection field, the persistent currents and thus the sextupole fields decay with a logarithmic dependence of time. The source of the decay is the resistive redistribution of Interstrand Coupling Currents (ISCC). These coupling currents flow through a complicated pattern in the copper strands and splices, and as they change, the magnetization of the cable decays.

While the phenomenon of cable magnetization, ISCC, and flux creep was documented long before the Tevatron was built, the implications for the quality of magnetic field was not very well predicted. The hysteretic component of the magnetic multipoles in the dipoles was measured by stopping at a given energy and making careful measurements. It was not known early on that the values of the multipoles would drift when a dwell at a given energy was performed, particularly at lower energies. For this reason, the chromaticity sextupoles had to be adjusted differently, as a function of energy, than predicted by magnetic measurements [20].

The initial operation of the Tevatron was a continuously ramping fixed target machine. Since the injection dwell time was only about 1 s long, the changing sextupole fields at the injection plateau were not an issue. When collider commissioning began, it was noticed that the chromaticity $Q' = dQ/(dP/P)$ on the injection plateau would drift by several tens of units, and the rate of change would decrease

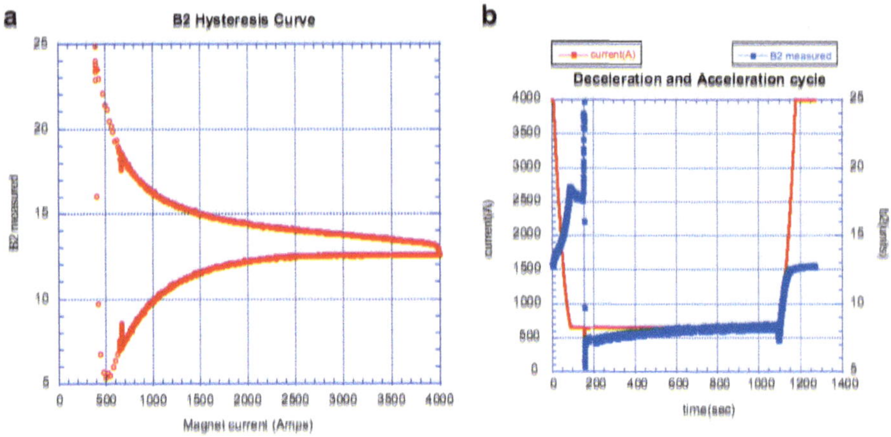

Fig. 3.3 (**a**) Sextupole field coefficient b_2 (multiplied by 10,000) as a function of the current in the Tevatron dipole magnet; (**b**) time dependence of the b_2 and Tevatron bus current for a standard operation cycle

over time. A program was established at the Magnet Test Facility (MTF) to understand the nature of the observed changes in chromaticity. The time resolution of the data acquisition system at MTF was not very good at that point, but it was confirmed that the b_2 component of the dipole magnets would drift away from the hysteresis curve as a function of the log of the time on the injection plateau.

Collider commissioning teams realized that a repeated ramp history would help the chromaticity drift become more repeatable. For these reasons, early collider operation implemented the technique of performing six continuous excitation ramps before stopping at the injection energy. Spending some time at the injection energy then allowed the drift in sextupole fields to be slow enough so that manual chromaticity adjustments could adequately compensate the remaining drift [21].

Figure 3.3a shows the hysteretic b_2 curve along with the drift at the injection level plateaus on both the up ramp and down ramp. The hysteresis curve moves in the counterclockwise direction. This data was taken in 1996 with a data acquisition system that was much better than was available for the original measurements. Figure 3.3b shows the b_2 component of a dipole as a function of time from the end of the deceleration ramp, through the field resetting dip and a 15 min injection porch, followed by an acceleration ramp.

The b_2 drift measurements were made at MTF to verify that operation algorithms were viable. The sextupole field was shown to decay as a function of the log of the time on the injection front porch. There was a faster decay in the first minute, but then both magnet and accelerator measurements displayed the $\ln(t)$ decay. Although the sextupole decay was not explained exactly by theory at the time, the magnet measurement data provided confidence that the operational strategy would be adequate for controlling the chromaticity on the injection porch. The colliding beam sequencer implemented a scheduler function that periodically called

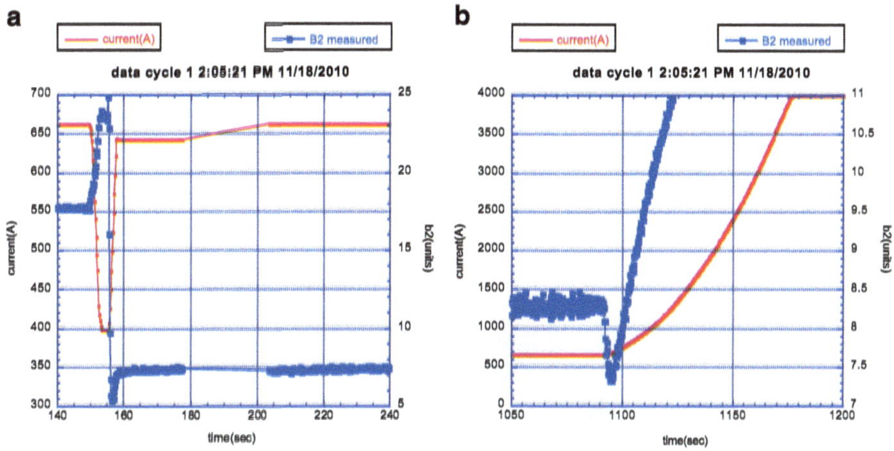

Fig. 3.4 The Tevatron dipole magnet sextupole component b_2 snapback at the end of the back porch (**a**—on the *left*), and at the end of the front porch (**b**—on the *right*)

up the application that loaded the chromaticity tables. Predetermined changes would then be sent to the chromaticity circuits to maintain the desired compensation [20].

The more serious operational issue at that point was the large changes in chromaticity at the start of the energy ramp after a dwell at the injection energy. Measurements showed that the sextupole fields returned to the hysteretic value within the first 30 A of acceleration. A lookup table was developed that calculated an acceleration table based on the length of time spent on the injection porch. This technique allowed acceleration with reduced beam loss and reduced emittance dilution [20].

Figure 3.4a, b shows the b_2 snapback at the end of the back porch, and at the end of the front porch. Again, this data was taken in 1996 with much better data acquisition system than available before the first Collider run.

The Tevatron Collider Run II and the Main Injector era put much more demands on the understanding of dynamic effects in the Tevatron magnets. The plans for maximizing integrated luminosity included the injection of more bunches and the reduction of shot setup times. Procedures involving many pre-cycle ramps and delays on the injection porch were not consistent with efficient luminosity production. Also, the collider plan involved recycling antiprotons, which would require a long dwell time at the injection energy on the down ramp. This extraction dwell would need to vary in time from one store to the next. These considerations necessitated another round of MTF measurements, conducted in the fall of 1996, to understand the effects of different ramp histories. Although the ability to measure and collect harmonic data had improved significantly since the early days of the Tevatron, the power supply regulation at MTF still had limitations and required a modified waveform going into the injection porch to avoid overshoot which would result in reversal of magnetization currents.

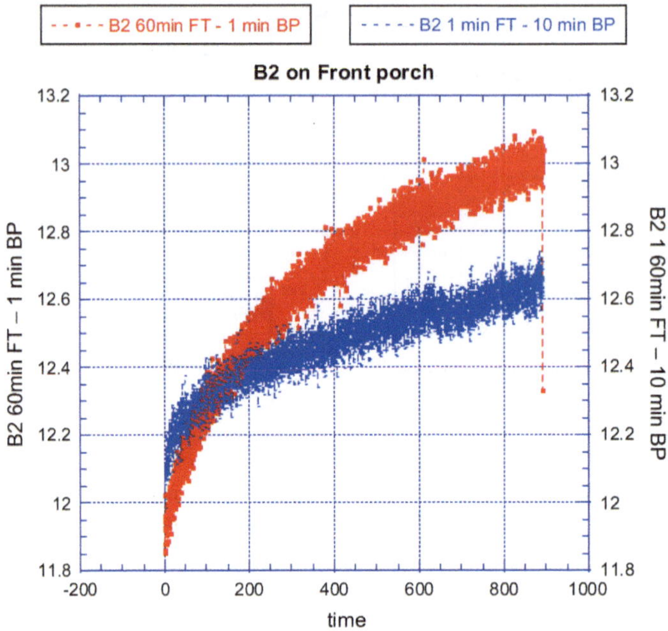

Fig. 3.5 The Tevatron dipole magnet sextupole component b_2 drift at injection after two different current ramps (see details in text)

Figure 3.5 shows the b_2 component of a dipole following two different ramp histories. One of these data sets followed a 60 min flattop with a 1 min back porch. The other data set followed a 1 min flattop with 10 min back porch. These two ramp histories would result in chromaticities that differ by 10 units after the 15 min shown here on the front porch.

Magnet measurements were made to understand how the previous cycles flattop length, as well as the length of the back porch affected the chromaticities that would result on the following front porch. Also, the dependence of ramp energy, ramp rate, and magnet temperature was also investigated. All of these parameters are important parameters in determining the proper compensation for the sextupole drift. Once chosen, the ramp rate, flattop energy, and magnet temperature would not be changing from store to store. It was important to know the functional impact that the flattop length and injection energy dwell times would have on the sextupole fields, as these parameters would necessarily change from store to store. The functional forms of these dependencies were obtained from the 1996 magnet measurements, and then the coefficients were scheduled to be tuned using chromaticity measurements during collider commissioning.

During the very beginning of Collider Run II commissioning, a 30 min pre-cycle and a 90 s back porch was used. Constants for the b_2 drift correction were adjusted to nicely compensate this condition. Since the recycling of antiprotons was not foreseen to take place in the early stages of the run, the need to tune up the drift

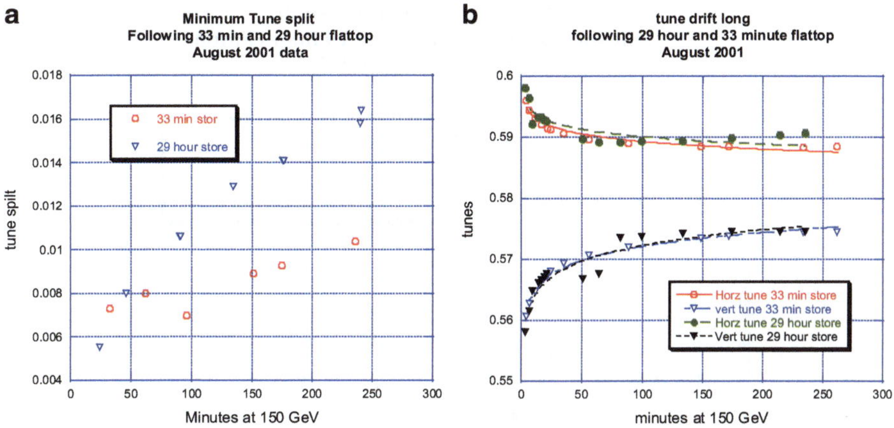

Fig. 3.6 (**a**) The minimum tune split $dQ = |Q_x - Q_y|_{\min}$ as a function of time spent at the injection energy of 150 GeV following two different lengths of flattop—29 h and 33 min at maximum energy of 980 GeV; (**b**) same for the horizontal and vertical tune drifts $Q_x(t)$, $Q_y(t)$

compensation scheme for various lengths of back porch times was not required immediately. As commissioning was rushed it was decided to skip this calibration, and instead stay with a pre-cycle with a 30 min flattop and a 90 s back porch before injecting beam. The run continued until 2004 with this mandatory 30 min pre-cycle before each store.

As the Collider Run II began, new phenomenon of tune and coupling drift were also seen. Prior to Run II, only the chromaticity drifted in a way that needed compensation.

Figure 3.6a, b shows the minimum tune split, and tune drift as a function of time on the front porch following two different lengths of flattop. The source of the tune and coupling drift was not understood, but the time dependence of these fields was similar to the chromaticity drift, so the compensation methods were simply expanded to include quadrupole and skew quadrupole correction circuits as well. It is likely that the source of the tune and coupling drifts is a feed-down effect from orbit offsets in the dipoles and correction sextupoles, but this has not been proven with certainty. All of the measured drifts were compensated fairly well for the first portion of Run II as long as the operation included a 30 min pre-cycle, and a 90 s back porch.

Figure 3.7 shows that the chromaticities are compensated very well for the normal operating scenario of a 30 min pre-cycle. The cycle that follows the long 29 h store shows that the drift compensation was not tuned up for that scenario.

New controls mechanisms were also introduced at the start of the Collider Run II to better automate the compensation of these changing fields. The implementation of the correction scheme was coordinated by a central process known as *TCHROM*. This program ran in the background and kept track of the length of flattop, back porch, and front porch for the previous five cycles. *TCHROM* received knowledge of the state of the Tevatron through a state device that was updated by the Colliding

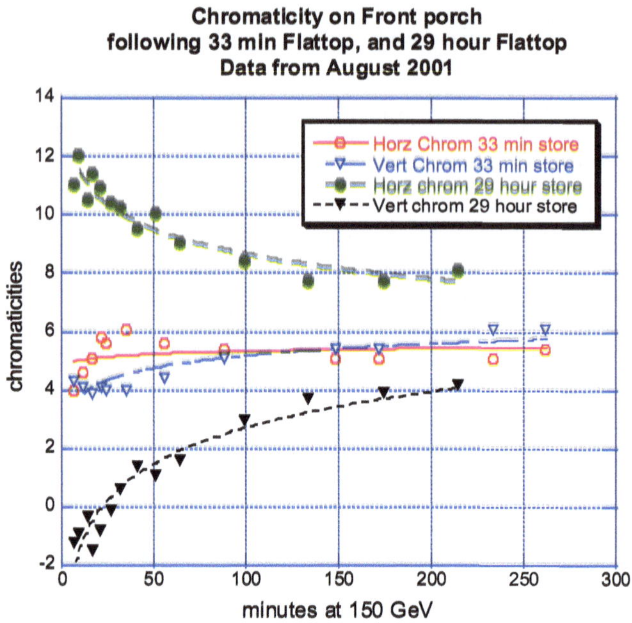

Fig. 3.7 The horizontal and vertical chromaticities $Q_x'(t)$, $Q_y'(t)$ as a function of time spent at the injection energy of 150 GeV following two different lengths of flattop—29 h and 33 min at maximum energy of 980 GeV

Beams Sequencer application program. Before moving to either the front or the back porch, *TCHROM* would calculate compensation curves for the sextupoles, based on the ramp history. The length of time spent on the front or back porch was broadcasted on the machine data link known as *MDAT*. The curves loaded by *TCHROM* played as a function of this time at 150 GeV data frame. At the beginning of the acceleration cycle, the snapback curve is calculated and loaded based on the amount of drift experienced on the front porch. There are a set of 44 constants stored in the database that are used to calculate the proper slope and initial corrections as a function of the previous flattop and back porch length. Since some of the operational flexibility designed into the system was extraneous because of the decision to eliminate recycling of antiprotons, some of the constants are 0, and the equations are more complicated than required for the more simple set of scenarios that are actually used. The controls implementation allows continuous compensation of the field drifts without any interaction from the operators.

There were still occasional issues with compensation of the snapback as a head–tail instability would occasionally arise at the beginning of acceleration causing drastic beam loss. These issues were addressed by running with increased values of chromaticity at the start of the acceleration cycle. This larger than optimal chromaticity did lead to some beam loss, and emittance growth, but these negative effects were fairly minor.

In 2004 there was a very serious effort made within the Fermilab's Technical Division to provide a better model to explain these dynamic effects. Better instrumentation existed by this time so more accurate measurements could be made with much better time resolution. As the behavior of individual magnets was better characterized, a very systematic set of accelerator measurements was planned to implement changes suggested by these measurements. Chromaticity drift measurements were made with various ramp histories. The speed and magnitude of drift were measured as a function of both flattop length and back porch length on the previous cycle. This data along with the magnet measurement data from MTF showing that the dependence on the flattop length saturated at times greater than 1 h lead to great improvements in the compensation equations. The need for a pre-cycle was eliminated for all cases except when the Tevatron ramp was turned off. It was also decided at this time to increase the length of the back porch from 90 s to 5 min as the magnitude and rate of drift on the front porch are less with the longer back porch. The ease of improved compensation was chosen over the decrease in time needed to refill the Collider. The b_2 drift compensation on the front porch always took the form:

$$b_2(t) = b_{20} + m \times \ln[(t + c)/c] \tag{3.2}$$

where both m and c are dependent on the length of the previous flattop up to 1 h.

The 2004 magnet measurements also lead to an improvement in the understanding of the snapback function. Previous to these measurements, the snapback was compensated by a fourth order polynomial function of time from the start of ramp:

$$b_2(t) = b_{2T}\left[1 - \left(t^2/T^2\right)\right]^2 \tag{3.3}$$

where t is time in seconds from the start of ramp, b_{2T} is the total b_2 drift on the front porch, and T is a constant set at 8 s. The function played out completely in 8 s. The improved algorithm was a Gaussian function:

$$b_2(t) = b_{2T} \times \exp\left(-(t/T)^2\right) \tag{3.4}$$

where the length of the snapback compensation T is now a function of the total snapback magnitude b_{2T}:

$$T = [(b_{2T} - 0.061)/0.0682]^{1/2} \tag{3.5}$$

The exact values of the constants in the above equation were obtained by careful beam measurements at the beginning of the ramp. Tune spectra were collected as rapidly as the spectrum analyzer would allow on ramp cycles with different RF frequency offsets. The tune data were collected on three separate ramp cycles with identical ramp histories, but with RF frequency offsets of 0, +40 Hz, and −40 Hz and that was used to calculate the actual chromaticity during the snapback—see Fig. 3.8 from [22].

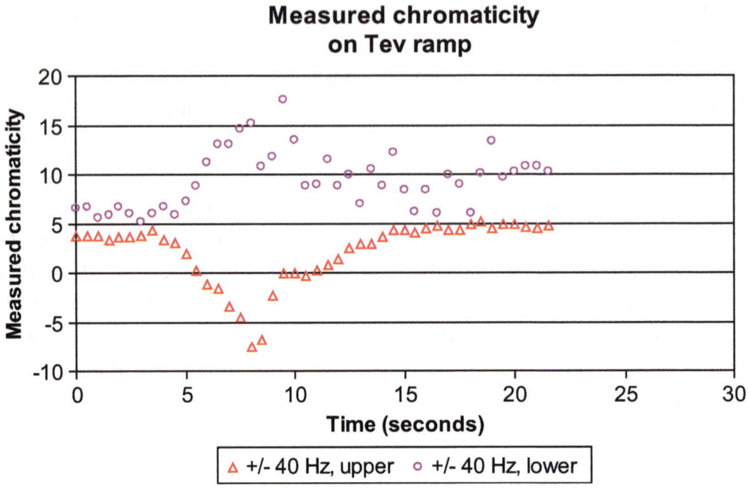

Fig. 3.8 Time evolution of the vertical and horizontal chromaticity values at the beginning of the current ramp. The scatter in data results from difficulty in discerning the correct tune line on the spectrum analyzer when the chromaticity is large

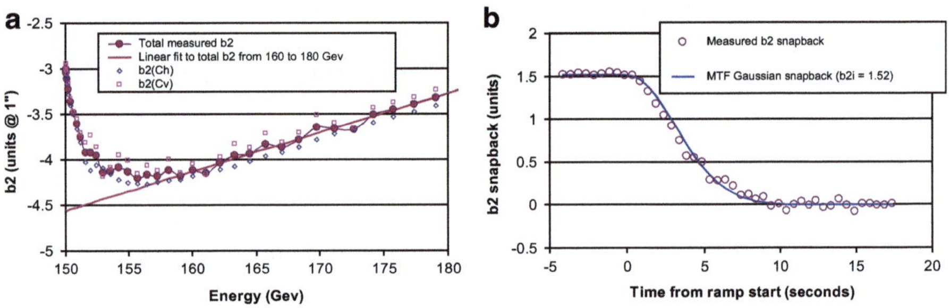

Fig. 3.9 (**a**) The average Tevatron sextupole field component b_2 as a function of energy at the start of the acceleration ramp calculated from the actual chromaticity measurements (see Fig. 3.8); (**b**) Comparison of the b_2 snapback determined from the beam measurements and the snapback calculated on the base of the MTF magnet measurements

Figure 3.9a shows the actual chromaticity measurements converted to units of b_2 as a function of energy at the start of the acceleration ramp. Figure 3.9b shows the agreement between the beam measurements, and the snapback calculated at MTF. One unit of b_2 is equivalent to 26.4 units of horizontal chromaticity and −24.1 units of vertical chromaticity.

The change to a Gaussian compensation curve improved the correction by a couple of units of chromaticity for an 8 s snapback, if the duration of the snapback compensation was unchanged. But the new knowledge that the length of time for the duration of the snapback was a function of the total drift on the injection porch was allowed improving the correction by another 2 units of chromaticity (see Fig. 3.10).

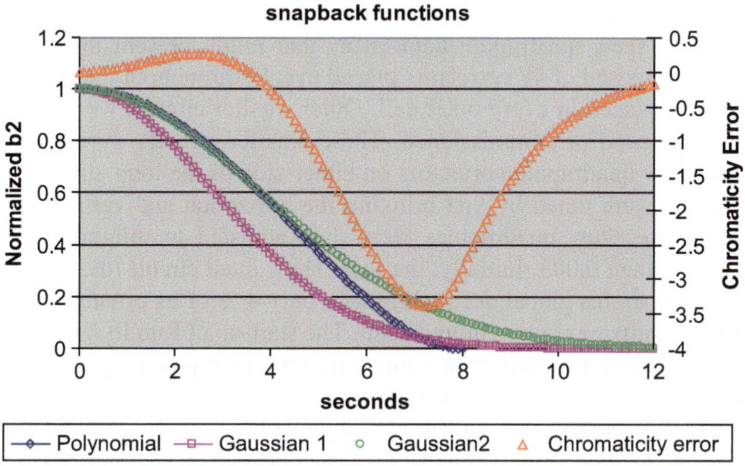

Fig. 3.10 Compensation curves are shown for the early polynomial correction, a Gaussian correction without a modification in the time of the curve, and the final Gaussian correction with a typical time compensation time. Also shown is the chromaticity error resulting from the earlier polynomial correction assuming the *Gaussian2* correction is perfect

Implementing this change allowed the chromaticity tables during acceleration to be changed to values that produced lower chromaticities without the danger of resulting in negative values and a head–tail instability. This resulted in less beam loss and emittance growth during the ramp. Drift compensation tables are calculated for the extraction porch every cycle, but these curves have not been needed since deceleration of antiprotons was eliminated from the operational scenarios.

3.3 Coil Creep and Skew Quadrupole Field Components in the Tevatron

During the start-up of Run II of the Tevatron Collider program, several issues surfaced which were not present, or not seen as detrimental, during Run I. These included the repeated deterioration of the closed orbit requiring orbit smoothing every 2 weeks or so, the inability to correct the closed orbit to desired positions due to various correctors running at maximum limits, regions of systematically strong vertical dipole corrections, and the identification of very strong coupling between the two transverse degrees-of-freedom. It became apparent that many of the problems being experienced operationally were connected to a deterioration of the main dipole magnet alignment, and remedial actions were undertaken [23]. However, the alignment alone was not enough to explain the corrector strengths required to handle transverse coupling.

With one exception, strong coupling had generally not been an issue in the Tevatron during Run I. Based on experience with the Main Ring, the Tevatron was designed with a very strong skew quadrupole circuit to compensate any quadrupole alignment and skew quadrupole field errors that might present themselves. The circuit was composed of 48 correctors placed evenly throughout the arcs, eight per sector, evenly placed in every other cell. Other smaller circuits were installed but not initially needed or commissioned. These smaller circuits were composed of individual skew quadrupole correctors on either side of the long straight sections. These circuits were tuned by first bringing the horizontal and vertical tunes near each other. The skew quadrupoles were then adjusted to minimize tune split, usually to less than 0.003. Initially, the main skew quad circuit (designated T:SQ) could accomplish this global decoupling with only 4 % of its possible current, and the smaller circuits were not required at all. The start-up of Run Ib was complicated by what was later discovered to be a rolled triplet quadrupole magnet in one of the Interaction Regions [24, 25]. This led to a reduction in luminosity of nearly 50 %, as well as operational confusion until it was uncovered.

By the time Collider Run II began, the current needed on the main SQ circuit had increased to 60 % of its maximum value. Some of the smaller circuits were also needed to fully decouple the tunes. With this history, several studies were performed early in Run II to search for strong local coupling sources like the triplet quadrupole, but without success. The strong corrector settings were indicative of a much larger problem than a single-rolled magnet, and the locality of the error was hard to deduce from the setting of a global correction system. Several possible reasons for the increase in coupling were investigated.

3.3.1 Strong Systematic Steering Correction

In late 2002 regions of the Tevatron were found to contain vertical steering magnets whose average strength was required to be nonzero in order to produce a smooth trajectory as seen on the Beam Position Monitors. Compared to the 0.7 μrad average horizontal steering corrector strength, the vertical correctors had a ring-wide average of about 16 μrad, and areas of the Tevatron had strengths of 70–90 μrad averaged over distances of 400 m or so. At 1 TeV, the maximum strength of a corrector is a little more than 100 μrad, so the available correction for general beam steering was limited in these locations. The interpretation of this effect was that these areas contained magnets which were systematically rolled toward the inside of the tunnel. This was verified by magnet roll angle measurements performed in October 2002 and January 2003.

The rolled dipoles and systematic corrections produce a "scalloped" vertical trajectory through the bending regions [26]. Although the distributed beam position monitors read zero displacements, the beam actually underwent ~0.5 mm excursions through these regions, assuming the magnets were rolled about the beam pipe axis. The fact that they are actually rolled about a different axis closer to the floor meant that the beam trajectory was closer to 1 mm or more from the center of the magnet coil.

Since the Tevatron dipoles have a sextupole component, a systematic vertical offset feeds down into a coupling term between the horizontal and vertical motion. The sextupole component is also known to vary as the logarithm of time due to persistent current effects at low magnetic fields; hence the coupling varies with time accordingly. This effect might explain some of the observed tune drift behavior during the Tevatron injection process, but was not nearly enough to explain the large skew quadrupole corrector settings.

In addition to the transverse coupling generated by orbit feed-down through rolled dipole magnets, the coupling due to observed quadrupole magnet rolls was calculated. Taken together, these effects were strong enough to explain a correction of the minimum difference between the two transverse tunes of amount $\Delta Q = |Q_x - Q_y| \approx 0.03$. However, the setting required of the main skew quadrupole correction circuit to decouple the Tevatron was indicative of an order of magnitude larger coupling.

3.3.2 *Injection Experiment*

Beam studies were conducted to see if a local source of the coupling could be pinpointed. All of the skew quadrupoles were set to 0, and then beam position data were collected for the first several turns before dumping the beam. The beam was then mis-steered in the horizontal plane, and orbit data again collected on the first several turns. The orbit differences are shown in Fig. 3.11. The purely horizontal error coupled fully into a vertical error in about 1.5 turns, and then back into the horizontal plane in another half period [27].

The fact that the coupling builds up gradually and not at a few localized sources is indicative of a uniformly distributed source of skew quadrupole fields. The 3-turn period of the coupling is consistent with a tune split of order $\Delta Q \approx 0.3$. It was quickly noted that a systematic skew quadrupole term, $a_1 \equiv (\partial B_x/\partial x)/B_0$, would account for this behavior, and would need to be of order $a_1 = 1.5 \times 10^{-4}$ in.$^{-1}$, or 1.5 "units" [28].

While regions of rolled dipoles existed at the time, the possibility that a systematic roll of the main quadrupoles could give rise to the coupling problem was unlikely, as the focusing and defocusing quadrupoles would have to be rolled in opposite directions. Direct roll measurements of the quadrupoles performed in 2002 also disproved this as the strong coupling source.

3.3.3 *Mechanical Issues*

In January 2003, as plans were being made to correct the rolled Tevatron dipoles discussed above, the question arose of ensuring that the cryostat and coils moved with the warm iron magnet yoke. The hope was that the magnets could be realigned without disturbing the cryogenic and vacuum connections between adjacent magnets.

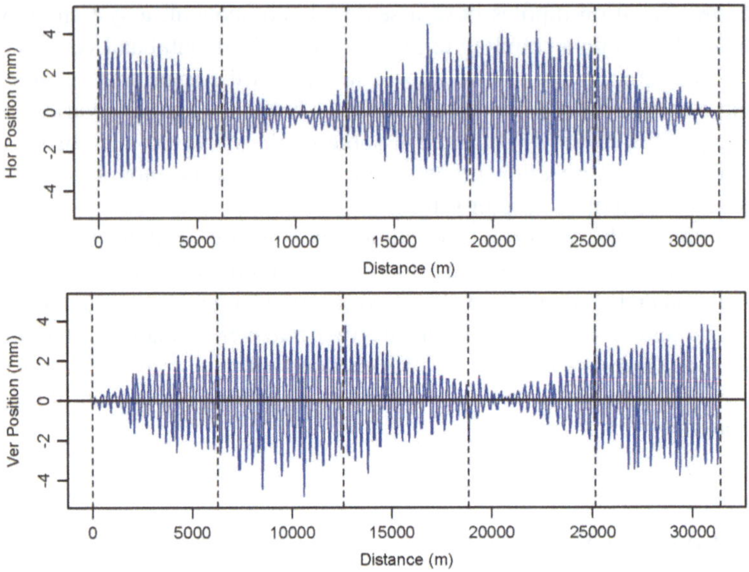

Fig. 3.11 An initial horizontal steering error couples completely to the vertical plane. This data set, taken in February 2003, had all skew quadrupole correctors turned off. *Top*—horizontal beam positions, *bottom*—vertical beam position after traveling distance *s*

The Tevatron dipole cross section is shown in Fig. 3.1. A warm iron yoke surrounds the cryostat tube and the collared coil within it. The collared coil is supported in the cryostat at nine stations down the 6-m length of the magnet. At all but the center station the support is two rectangular blocks of the G11 epoxy-glass plastic, called "suspensions," approximately 8.5 mm × 6.4 mm × 17 mm, aligned with each other, one between the outer cryostat skin and the heat intercept at liquid nitrogen temperature and the other between the heat shield and the phase single helium tube. These suspensions are arranged in grooves that allow the whole cryostat to move longitudinally relative to the yoke as it is cooled. At the middle station more robust G10 tubes, known as "anchors" support the cryostat, anchoring the center against both longitudinal motion and rotation relative to the yoke (a bane of earlier designs of the magnet).

At each station four supports are located at the 45° points. A screw or suspension cartridge through the yoke at each of these points contacts the outside of the cryostat. This allows the cryostat to rest stably in a cradle formed by the lower two screws ("bolts"), while being held down against those supports by two preloaded upper suspensions, spring-loaded cartridges known as "smart bolts." The preloaded springs maintain a force as the collared coil and cryostat shrink during cool down.

As discussed in more detail below, the magnets had been designed to allow adjustment of the coil position while the magnet was cold by adding and removing shims from the cryostat suspension system. During production this feature was used as a standard procedure to compensate for fabrication tolerances and mechanical differences between the top and bottom coils. A depth gauge can be used to measure the distance from the top of the cylinder to the top of the cartridge, a distance that became known as the "lift." For example, as the magnet was cooled, the collared coil shrank, the outer skin of the cryostat deformed slightly, the springs pushed the cylinders inward, and the lift increased. In 2002, lift measurements seemed to have potential for monitoring the movement of the cryostat relative to the yoke during magnet realignment.

To assess the viability of making these measurements in the tunnel and to check the stability of the cryostat over time, technicians measured 18 magnets during a day in February 2003 when the accelerator was down for maintenance. The original (paper) production records of the magnets were retrieved from off-site storage and the lift measurements extracted. By the end of February the data had been analyzed and it was apparent that in those 18 magnets the coil had systematically dropped vertically approximately 0.11 mm relative to the yoke. After a bit of thought it was realized that this was probably due to "creep," slow inelastic deformation under pressure, in the G11/G10 blocks that support the coil within the cryostat. The sizing of the G11 suspensions was a balance between mechanical strength and the heat load a larger conducting cross section would present. The stability of the blocks was the subject of a study at the time [29]. One participant remembers the 1980 guidance to "design the magnets to last for 20 years" [30].

The measured change was enough to produce a skew quadrupole component in the field of about $a_1 = 1.5 \times 10^{-4}$ in.$^{-1}$ or 1.5 "units." If every Tevatron dipole had developed with age a skew quadrupole component, that would explain the observed effects the beam coupling effects.

3.3.4 *Mechanical Issues*

Once the Tevatron dipole magnets were considered to be the main source of the coupling, further verifications were performed using beam measurements as well as magnet measurements. During Run I, upgrades of the low beta insertions required some of the existing spool pieces to be replaced with new devices. As a result, six skew quadrupoles were lost, one upstream and two downstream of each interaction point. Thus, while the distribution of a_1 was essentially uniform around the ring, its correction was no longer uniform.

The evenly spaced coupling errors from the dipole magnets were being compensated with correction circuits that now had gaps at two locations. Since the horizontal and vertical phases advance differently through a half cell, the main skew quad circuit was not able to fully correct the coupling caused by the distributed coupling errors. Powering the small skew quadrupole circuits near the straight sections in conjunction with the main circuit was necessary to fully minimize the global coupling. At injection, the most effective of the smaller circuits was the one surrounding the A0 straight section, and the skew quad at location A49 was also utilized. Since the smaller circuits consisted of only one or two elements, the strength of these correctors needed to be quite high to completely minimize the global coupling. Under these conditions, the nonuniformity of the coupling correction consisted not only of the gaps in the main circuit but also the additional skew quadrupoles around A0 and at A49.

3.3.5 Vertical Dispersion

A horizontal offset in a skew quadrupole results in a vertical kick. A horizontal offset due to momentum in a skew quadrupole will therefore result in a vertical kick, creating vertical dispersion. Given the value of horizontal dispersion at the location of the skew quadrupole elements around A0 and at the gaps in the main skew quad circuit, a significant vertical dispersion resulted. The dispersion at A49 was very small so that a skew quadrupole there did not contribute to the vertical dispersion.

The dispersion can be measured easily by changing the RF frequency and recording the resulting orbit difference. Figure 3.12 shows the dispersion measured in both planes at the injection energy of 150 GeV in early 2002.

The value of the horizontal dispersion function oscillates in the arcs of the Tevatron between about 1 and 5 m, and vanishes at the interaction points, in agreement with the design lattice. By design, the vertical dispersion should be 0, though a small "residual" vertical dispersion is always present due to various error sources, and would be expected to have a random distribution with peaks of approximately 0.2 m in the Tevatron. The vertical dispersion measurements, however, showed a fairly coherent dispersion wave of up to 0.8 m that is largest between D0 and A0, smaller between B0 and D0, and very small between A0 and B0. This pattern is consistent with calculations based on coupling errors due to the strong correction circuits at A0 and A49, and the gaps in the correction circuit near B0 and D0. The vertical dispersion was near maximum value at the injection point at F0 leading to an emittance increase of up to 1π mm mrad or more for injection of coalesced proton bunches [31].

3.3.6 More Magnet Data

The lifts of an additional sample of 66 magnets were measured in the tunnel during a subsequent short maintenance period, confirming the original distribution. Similar changes in the coil position were measured in spare dipoles that had not been in service. The original calculations from the production era were repeated, affirming the linear relationship between displacement and skew quadrupole error term, $a_1 = k{\cdot}y$, where k is 10.6 units/mm. Several spare dipoles were magnetically measured [32], confirming that the long-term change in lift was reflected in the expected change in skew quadrupole. The amount of its life that a magnet had spent cold or

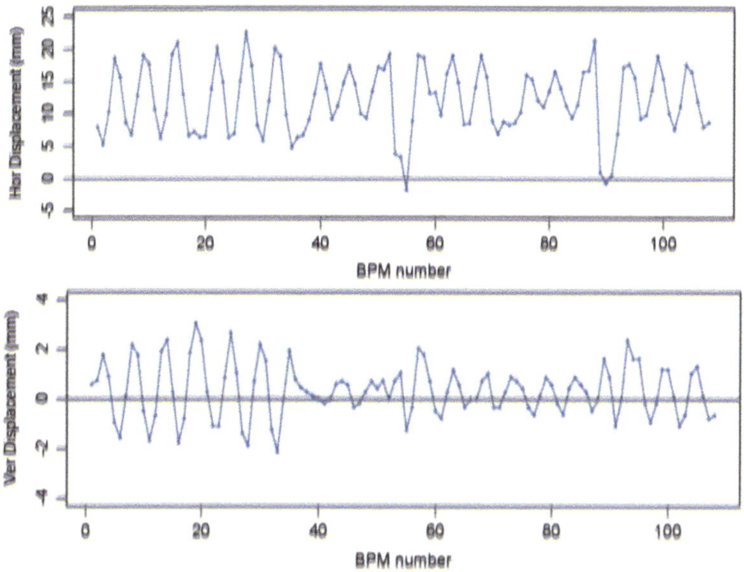

Fig. 3.12 The horizontal (*top*) and vertical (*bottom*) orbit difference given by 500 Hz RF frequency shift, taken in early 2002. The position shift at a BPM is proportional to the dispersion at that location $(x, y) = 3.4$ [mm] $D_{x,y}$ [m]. The vertical displacements clearly show a coherent pattern

warm did not seem to affect its creep. Many more magnets have since been measured during the major shutdowns of 2003 and 2004. The distribution of lift change, averaged over each magnet, is plotted in Fig. 3.13. The average shift of about 0.14 mm (0.0055″) is clearly seen, as is the breadth of the distribution. The difference between the average left and right lift changes is centered on 0, indicating no lateral movement.

The spring force in the smart bolts is sufficiently greater than the weight of the cryostat that we take the compressive force on the upper and lower suspensions to be equal and interpret the change in lift to be creep equally distributed between the upper and lower sets. Note also that an equal creep all around results in a change in the coil position that is the vector sum of the creep in the two lower suspensions.

3.3.7 *Magnet Field Correction During Production*

When there is any asymmetry in the coils of a conductor-dominated dipole magnet, an error field is produced, which we characterize by the coefficients of its harmonic decomposition. For convenience we normalize the error field to the central field, take the reference radius to be 25.4 mm, and quote the coefficients in "units" of parts in 10^4. This means that if we can bring the coefficients into the range of unity, the magnetic field will be acceptable over the useful aperture of the magnet.

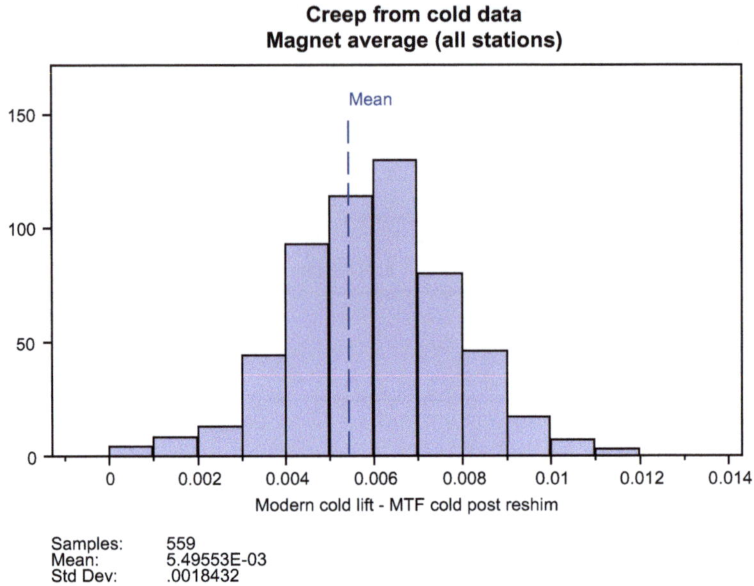

Fig. 3.13 Distribution of the "lift" changes from production to now due to creep (indicative of vertical displacement of the cold mass w.r.t. the magnet iron yoke, in inches). For each magnet the change is averaged over all accessible stations

The asymmetry may be in the geometry of the coils, in their placement in the yoke, or in the superconducting cable. Of particular concern here is the fact that a normal (skew) quadrupole term is produced by the horizontal (vertical) displacement of the coils from the center with a linear dependence. By adjusting the cryostat and collared coil position in the yoke, the quadrupole components of the field could be eliminated or canceled during production.

During assembly brass shims were placed on the end of all four bolts at each station to roughly center the coil in the yoke. Each of the almost 1,000 (including spares) production magnets was measured in detail [33]. After quench testing, the harmonic components of the magnet field integral were measured at one current and the magnet was shimmed according to the guidance provided by a computer calculation. The cryostat support system described above allowed these adjustments while the magnet was cold. While the magnet was still on the test stand, one opposing pair bolts were removed at a time, shims were shifted from one bolt to the other to reposition the coil, and the bolts screwed back in. The magnet was then measured at multiple currents on both the up and down ramps. This data set includes every magnet and provides a measurement of the entire magnet field integral. The specification called for both quadrupole components to be adjusted, if needed, to be within 1 unit of zero. If the first try did not succeed, a new shim adjustment could be calculated and the process repeated. Figure 3.14 shows the final distribution of the skew quadrupole component.

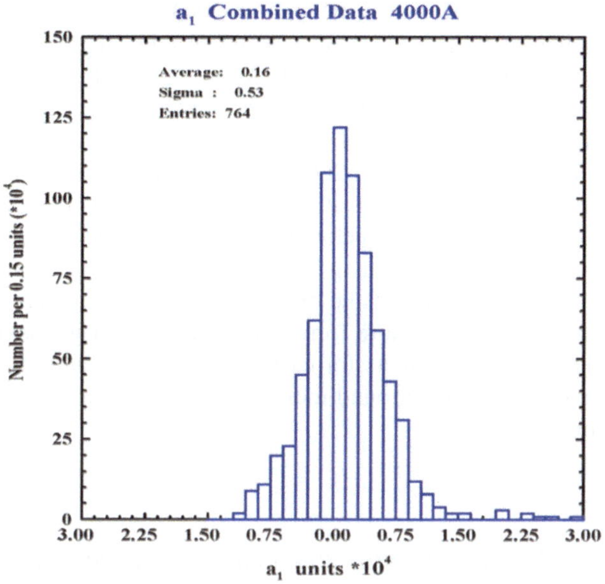

Fig. 3.14 Final distribution of skew quadrupole in Tevatron dipoles after production shimming

3.3.8 The Reshimming Plan

During production the results of shimming could be verified immediately. This would not be possible during reshimming in the tunnel. Examination of the production lift data quickly led us to the conclusion that it was not sufficiently reliable on a magnet-by-magnet basis to warrant the effort of individual treatment. Measuring the lifts of each magnet, comparing them with the production values, and reshimming accordingly would have been fraught with possibilities for error. Instead, it was decided to treat all magnets as though they had changed by the same amount. The addition of a standard shim 3 mil or 0.076 mm (1 mil is one thousandth of an inch, 0.025 mm) was agreed upon, raising the coils by 2.12 mil (0.054 mm).

With 774 dipole magnets in the Tevatron, and with 18 shim locations per magnet, it was clear that to fix the problem completely would take significant tunnel access time. Trial runs above ground had shown that although the job in the tunnel was tedious, requiring the technicians to crawl over the magnets in tight quarters, a crew of three could comfortably reshim two magnets per day. It was not clear, however, how many exceptional cases would arise requiring extra time and how well the crews would hold up with time, even with careful ergonomic planning. So, given some uncertainty in the effectiveness of the reshimming operation, the limited resources, and the uncertainty in the amount of time to reshim the magnets, we decided to start with a modest correction program. We initially focused on the region where the regularity of skew quad correction system had been interrupted, with the idea that the balance of the ring could be reshimmed in future maintenance periods.

For the first run in the fall of 2003 we chose to reshim 106 magnets divided between the two sides of each intersection region, the areas that had their skew correction systems disrupted. After this work, the accelerator performance was significantly improved, so another 12 magnets were reshimmed in March 2004 during a 1-week shutdown. The next major period of repair was the following fall (September–November 2004) when 412 dipoles were reshimmed. This set of dipoles consisted of magnets that were at least four dipole locations from the nearest main skew quadrupole corrector. The performance continued to improve after the second repair, and the remaining 244 dipoles were adjusted in early (March–April) 2006.

The sensitivity of the operation led us to augment our usual quality assurance procedures with more formal analyses [34]. The sheer volume of measurements, 18 lift measurements for each magnet before and after the reshimming, required extensive above ground monitoring in addition to the teams of technicians in the tunnel.

3.3.9 Accelerator Performance Improvements: Reduced Skew Quadrupole Correction, Reduced Vertical Dispersion, Reduced Emittance Growth at Injection

After each round of reshimming the required current in the main skew quadrupole correction circuit decreased, giving clear indication that the reshimming was reducing the skew quadrupole component in the dipoles—see Fig. 3.15 (note that the current is negative and 0 is at the top of the plot). There is only a small change after the first stage, which only addressed the regions where the main skew quadrupoles had been removed. The two stages that addressed the regions covered by the main skew quad circuit were directly effective. The current on the injection "front porch" has decreased by a factor of 6. All other skew quad circuits were reduced to near 0 at the injection energy by the first stage of reshimming, as expected. Stronger coupling corrections are needed at high energy.

The vertical dispersion has been reduced with each phase of the reshimming. A dispersion measurement from 2011, shown in Fig. 3.16, can be compared with the starting dispersion measurement from 2002, shown in Fig. 3.12. The 2011 measurement shows vertical closed obit changes of about 1 mm, oscillating with roughly constant amplitude, where in 2003 the amplitudes varied around the ring, peaking at over three times this amount.

Most importantly, the emittance growth at injection due to the mismatch between the optics of the rings and the injection lines has also decreased dramatically. Figure 3.17 shows the proton injection emittance growth by store number spanning the first two major reshimming periods. Where in early 2003 the rms emittance growth was typically over 1.3π mm mrad, immediately after the first round of reshimming the emittance growth dropped to 0.6π mm mrad. The second major round of reshimming reduced the growth by another 0.4π mm mrad.

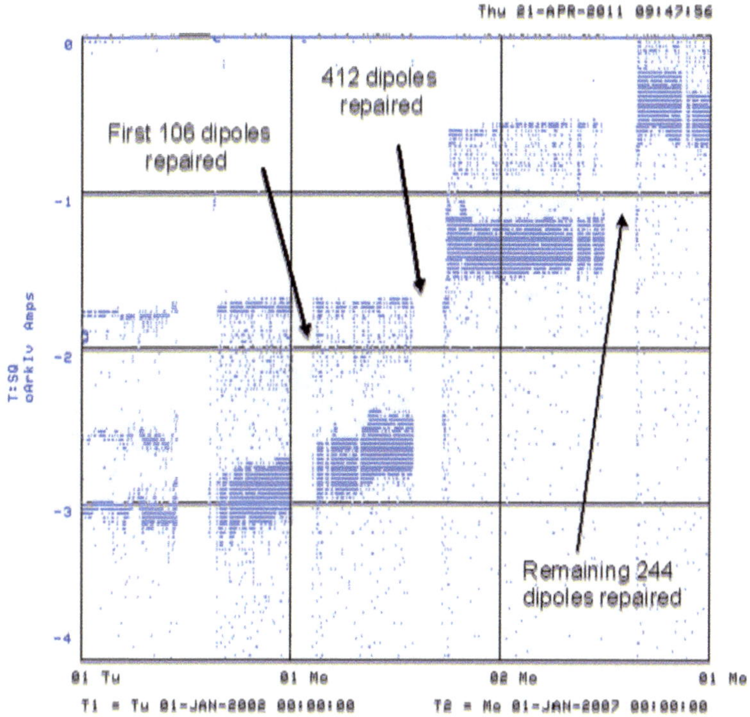

Fig. 3.15 Six-year plot of current in main skew quadrupole circuit T:SQ. The *darkest bands* show the current during the time on the "injection porch"

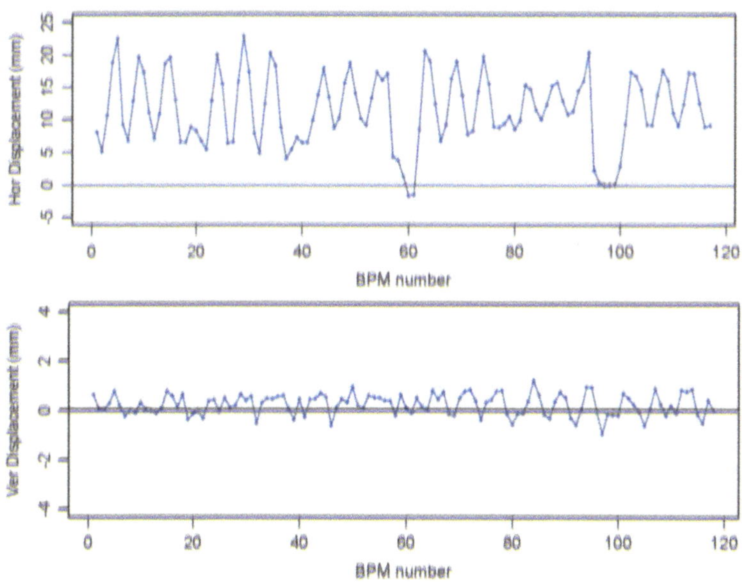

Fig. 3.16 Results of dispersion measurement made in 2011 to be compared with the 2002 measurement shown in Fig. 3.12

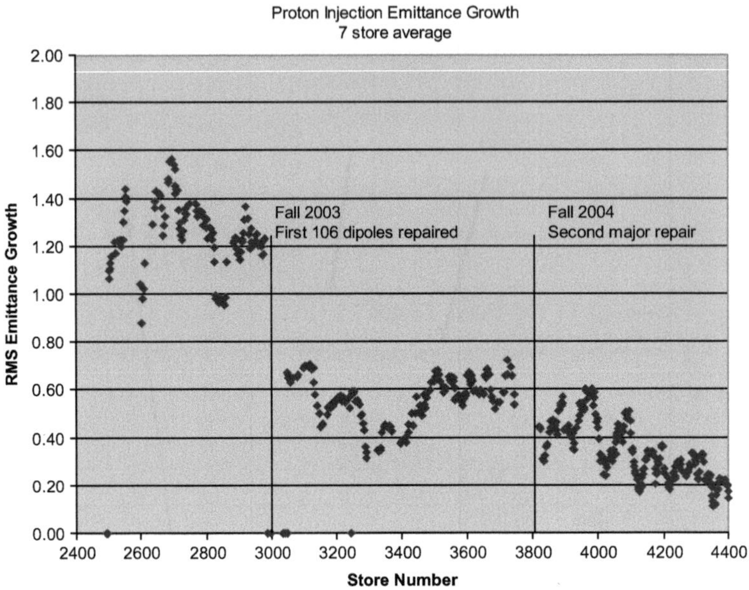

Fig. 3.17 Seven-store average of the proton injection emittance growth spanning the period of the first two major repair periods (April 2003–October 2005)

The reshimming has had an indirect benefit as well. With the major sources of coupling removed, other sources (such as two rolled arc quadrupoles) could be identified and the entire Tevatron lattice restored closer to its design. Correspondingly beam measurements have become much cleaner and easier to interpret, greatly enhancing the operation of the collider.

3.3.10 *Summary*

Many issues plagued the ramp-up of luminosity during Run II of the Tevatron collider program. While adjustments of the global coupling of the accelerator using a systematic skew quadrupole magnet circuit could allow the operators to place the betatron tunes at desirable settings, other confusions ensued. For example, horizontal corrections that seemed to induce vertical oscillations hampered the commissioning of beam damper systems, and unexplained emittance growth—particularly in the vertical degree of freedom—upon injection was difficult to track down. The required settings of the quadrupole corrector circuit indicated that large sources of coupling were present—and beam measurements soon verified that the source was everywhere. Independently but in parallel, the explanation emerged from both the interpretation of beam measurement data and of magnet measurement data taken in the tunnel—the Tevatron dipole magnets were the source and a mitigation scheme was required. The coils of essentially every dipole magnet in the tunnel were shimmed to re-center them in their iron yokes. The process, which took careful planning and special training, was strategically implemented over a period of years during planned shutdowns of the collider program, restoring the Tevatron to its 1980s state of small residual coupling, both globally and locally.

3.4 Recycler Ring Permanent Magnets Design and Stability

The Recycler was designed as a fixed 8 GeV kinetic energy antiproton storage ring [14]. It is the first ever large-scale (3.3 km circumference) accelerator for high energy physics research built with permanent magnets instead of usual electromagnets. It is located in the Main Injector tunnel directly above the Main Injector beamline, near the ceiling, and consists of 362 gradient dipole magnets, and 109 quadrupoles, 8 mirror magnets, and 5 Lambertson magnets. There are four types of gradient magnets: focusing and defocusing, with pole lengths of 3.1 and 4.5 m. Each has different quadrupole and sextupole fields (see Table 3.1). The basic design consists of precision-shaped pole tips and pole spacers with strontium ferrite bricks on the outside of the poles [35, 36]. The entire package is surrounded by 19 mm (0.75 in.) thick flux return. A typical cross section is shown in Fig. 3.18. A standard SrFe brick size of 101.6 mm by 152.4 mm by 25.4 mm (4 in. by 6 in. by 1 in.) was used. The field axis was orientated along the 1 in. thickness. Two bricks were used on each side of the pole pieces to generate the required field.

The variation of the magnetic field of Strontium ferrite with temperature [37] was carefully measured by freezing the magnet to 0 °C then heating the magnet to 40 °C and allowing it to cool. During this process the magnetic field was continually measured with a Morgan (rotating) coil. A least square fit to the data gave the temperature variation B-field coefficient of about −0.2 %/°C. This intrinsic temperature coefficient of the ferrite material is canceled by interspersing a "compensator alloy" between the ferrite bricks above and below the pole tips—see Fig. 3.19. The compensator is an iron-nickel alloy Ni (30 %) Fe (70 %) with a low Curie temperature in the range of 40–45 °C and therefore its permeability depends strongly on temperature. Thin strips of the alloy shunt away flux in a temperature-dependent manner which can be arranged to null out the temperature dependence of the ferrite. The degree of temperature compensation is linearly related to the amount of compensator material in the magnet. Thus the degree of compensation can be "fine tuned" to the required accuracy by adjusting the amount of compensator at the ends of the magnet in a manner similar to the strength trimming with the ferrite. For example, a 20-fold reduction of the temperature coefficient (from 0.2 to 0.01 %/°C) requires that the amount of compensator in the magnet be adjusted correctly to 1 part in 20. This poses no difficulties for

Table 3.1 Recycler gradient magnet parameters

Magnet	Central field (kG)	Bdl (kG/m)	Quadrupole units	Sextupole units
RGF	1.3752	6.182	619.7	8.7
RGD	1.3752	6.183	−598.1	−15.1
SGF	1.330	4.121	1,276.0	0.0
SGD	1.330	4.121	−1,303.0	0.0

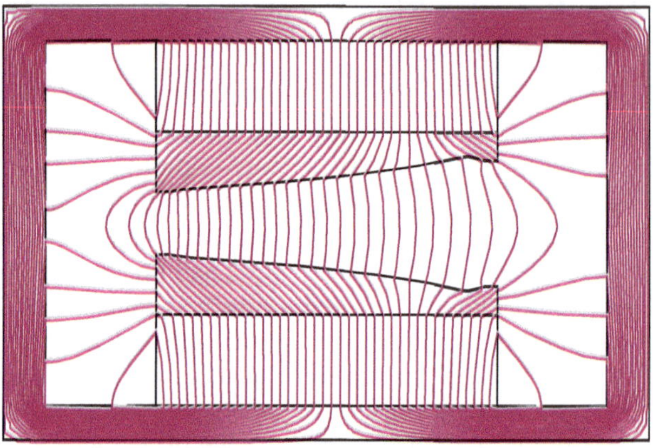

Fig. 3.18 Cross section of the Recycler ring gradient dipole magnet. Each of the magnet types was modeled using PANDRIA code from LANL

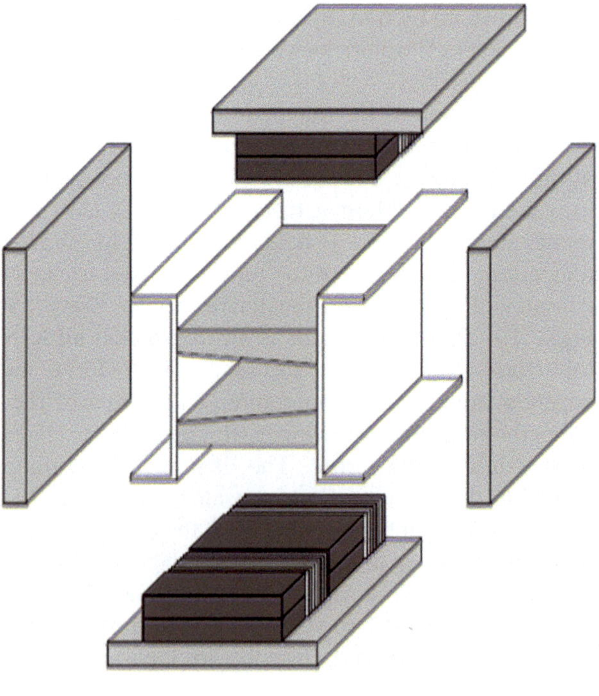

Fig. 3.19 Recycler permanent magnet gradient dipole components shown in an exploded view. For every 4″ wide brick there is an 0.5″ interval of temperature compensator material composed of 10 strips [14]

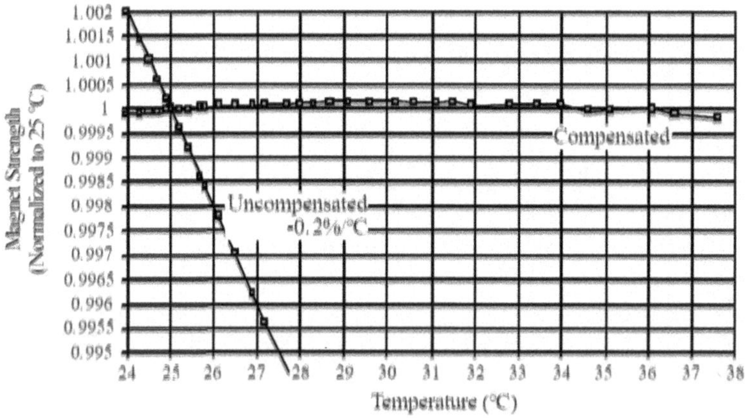

Fig. 3.20 Field strength vs. temperature for a stability test magnet using pre-production samples from the selected vendors for compensator alloy and ferrite

production. Due to the variable nature of the Curie point from heat to heat, the steel strips from different heats were randomly mixed. All Recycler magnets were frozen to 0 °C during the manufacturing process. The field for each magnet was measured at 0 °C and at 22 °C the amount of compensator steel was adjusted to keep the magnetic field in the 0.01 % tolerance. So, as a result the average magnetic field temperature stability of the Recycler magnets was about 0.004 %/°C rms over the operating range of 24–38 °C—see Fig. 3.20.

Long-term field stability was studied using the first production gradient magnet [38]. This magnet was measured with a Morgan (rotating) coil regularly over the first 2 years after its fabrication in November 1997. As seen in Fig. 3.21, the magnet lost a total of 0.1 % of initial strength in the first 2 years due to approximately logarithmic decay and has remained stable. The fact that over more than 6 years of the Recycler operations, its RF frequency of some 52 MHz has changed by less than 200 Hz is further proof that the magnetic field has remained stable within 0.04 %.

The longitudinal field for each gradient magnet was measured during production. It was determined that the field could be made more uniform by varying the number of compensator strips between each set of permanent magnets. More compensator strips were added at the quarter points of the magnet and fewer at the ends and middle. This served to keep the field flatter across the length of the magnet and to maintain the temperature compensation within the specified value of 0.01 %/°C.

To deal with the variation of the azimuthal magnetic field harmonics inevitable due to construction imperfections, each gradient magnet was measured with a Morgan (rotating) coil. Harmonics up to the 12-pole were measured. A FORTRAN program was written to calculate the shape of an end shim for each magnet that would adjust the quadrupole, skew quadrupole, sextupole, and octupole [39]. The program generated machine code for a wire EDM (Electric Discharge Machine) that cuts the exact shape required for each magnet. The shims were ready within

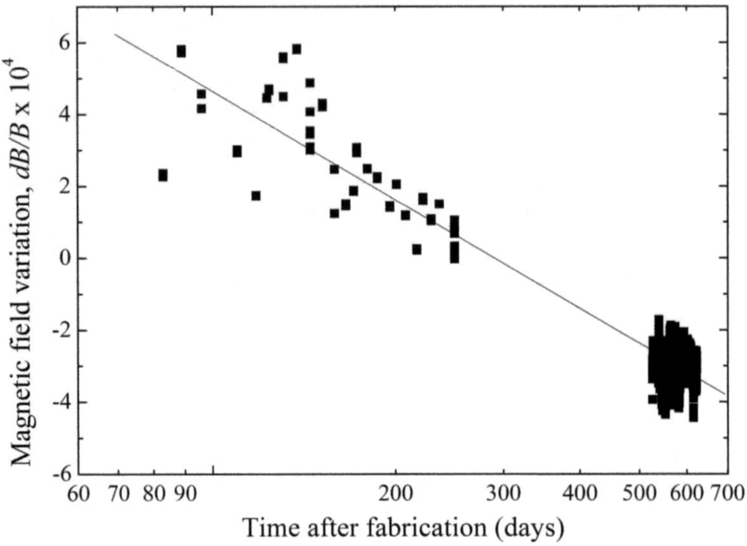

Fig. 3.21 Time evolution of the magnetic field in the Recycler gradient magnet (in the units of 10^{-4} of the main dipole field)

Table 3.2 Quadrupole magnet parameters

Magnet	Gradient (kG/m)	Focal length (1/m)	Gradient × length (kG)
RQMF	26.27	0.045	13.35
RQMD	−25.32	−0.043	−12.86
RQME	−21.97	−0.038	−11.16
RQEB	18.46	0.063	18.76
RQTF	16.68	0.029	8.47
RQTD	−16.87	−0.029	−8.57

24 h after the first measurement. In general, one iteration was required for each magnet. In rare cases two iterations were needed.

The poles of the Recycler quadrupole magnets [40] – see Table 3.2 – were made of type 1008 carbon steel. The hyperbolic shape of the poles was formed by cold extruding, no other machining of the poles was required. Strontium ferrite bricks were placed behind the poles. In the corners, steel washers were added to adjust the gradient, the skew quad, sextupole, and octupole field components. Steel washers were attached to the face of the poles to eliminate the 12-pole component (Table 3.2).

Special magnets needed for the beam injection and extraction—Lambertsons magnets and mirror magnets—were also made with strontium ferrite [41]. The base of the Lambertson magnets was made of solid steel. The hole for the field free region was gun bored then through entire 4 m length. This provided for a field free region of less than 5 G with a field region of 0.16 T. The use of solid base plates was allowed for a common design for all five Lambertson magnets but different configuration of the bending field.

With experience of operation it became clear that small corrector magnets such as skew quadrupoles and sextapoles were required. Several of these magnets were fabricated to solve specific problems [42]. Both strontium ferrite and samarium cobalt permanent magnets were used.

References

1. E.L. Hubbard (ed.), *Booster Synchrotron*, FERMILAB-TM-0405, Batavia, IL, 1973
2. E.J. Prebys et al., New corrector system for the Fermilab booster, in *Proceedings of IEEE PAC07*, Albuquerque, NM, 2007, p. 467
3. W. Pellico, New booster multi-pole correctors commissioning. Presentation to the Fermilab all experimenters meeting, Batavia, IL, 11 October 2010
4. J. Lackey et al., Operation and performance of the new Fermilab booster H-injection system, in *Proceedings of IEEE PAC07*, Albuquerque, NM, 2007, p. 1709
5. C. Hojvat et al., The multiturn charge exchange injection system for the Fermilab booster accelerator. IEEE Trans. Nucl. Sci. **NS-26**(3), 3149 (1979)
6. D. McGinnis, Status of the Fermilab booster after the 400 MeV upgrade, in *Proceedings of EPAC1994*, London, England, 1994, p. 497
7. W. Chou et al., Fermilab booster modeling and space charge study, in *Proceedings of IEEE PAC03*, Portland, OR, 2003, p. 2925
8. F.T. Cole (ed.), *Design Report National Accelerator Laboratory*, Oak Brook, IL, January 1968
9. F.T. Cole (ed.), *Design Report Tevatron 1 Project*, Batavia, IL, 1984
10. P.S. Martin (ed.), *The Fermilab Main Injector Technical Design Report*, Batavia, IL, 1995
11. D.J. Harding et al., Strength and shape of the magnetic field of the Fermilab main injector dipoles, in *Proceedings of IEEE PAC09*, New York, NY, 1999, p. 3318, and references therein
12. D.J. Harding et al., Magnetic field strength and shape measurements of the Fermilab main injector quadrupoles, in *Proceedings of IEEE PAC97*, Vancouver, BC, 1997, p. 3269
13. W. Chou et al., Operational aspects of the main injector large aperture quadrupole (WQB), in *Proceedings of IEEE PAC07*, Albuquerque, NM, 2007, p. 1685
14. G.P. Jackson (ed.), *Fermilab Recycler Ring, Technical Design Report*, FERMILAB-TM-1991, 1996
15. F.T. Cole et al. (eds.), *A Report on the Design of the Fermi National Accelerator Laboratory Superconducting Accelerator*, Batavia, IL, 1979
16. A.V. Tollestrup, *The Amateur Magnet Builder's Handbook*, Fermilab-UPC-086, Batavia, IL, 1979
17. N.M. Gelfand, Tune drifts on the Tevatron front porch, in *Proceedings of PAC07*, Albuquerque, NM, 2007, p. 1691
18. R. Hanft, B. Brown, D. Herrup, M. Lamm, A. McInturff, M. Syphers, Fermilab Preprint TM-1542, 1988
19. K.-H. Mess, P. Scmusser, S. Wolff, *Superconducting Accelerator Magnets* (World Scientific Publishing Co. Pte. Ltd., Singapore, 1996)
20. D. Herrup et al., Phys. Rev. E **49**(6), 5660 (1994)
21. D. Finley, D. Edwards, R. Hanft, R. Johnson, A. McInturff, J. Strait, Fermilab Preprint FN-451, 1987
22. J. Annala et al., in *Proceedings of 2005 IEEE Particle Accelerator Conference*, Knoxville, TN, 2005, p. 54
23. J. Volk et al., Tevatron alignment issues 2003–2004, in *Proceedings of 2005 IEEE Particle Accelerator Conference*, Knoxville, TN, 2005, p. 2146
24. J. Holt et al., Calculating luminosity for a coupled Tevatron lattice, in *Proceedings of 1995 IEEE Particle Accelerator Conference*, Dallas, TX, 1996, p. 456

25. N. Gelfand, Coupling in the Tevatron, in *Proceedings of 1995 IEEE Particle Accelerator Conference*, Dallas, TX, 1996, p. 452
26. M. Syphers, Strong systematic steering correction in regions of the Tevatron, Fermilab Internal Report Beams-doc-491, 2003
27. M. Syphers, D. Harding, Deterioration of the skew quadrupole moment in Tevatron dipoles over time, in *Proceedings of 2005 IEEE Particle Accelerator Conference*, Knoxville, TN, 2005, p. 1967
28. M. Syphers et al., Observations of strong transverse coupling in the Tevatron, in *Proceedings of 2005 IEEE Particle Accelerator Conference*, Knoxville, TN, 2005, p. 2029
29. E. Caulfield, *Investigation of Creep Behavior of the Glass Reinforced Composite Stand-Offs* (Packer Engineering Associates Inc., New York, NY, 1981)
30. G. Kobliska to D. Harding, private communication
31. M. Syphers, Skew Quadrupole tuning and vertical dispersion in the Tevatron, Fermilab Beams-doc-611, 2003
32. G. Ambrosio et al., Results of an investigation of the skew quadrupole issue in Tevatron dipoles, Fermilab Internal Note TD-03-045, 2003
33. R. Hanft et al., Magnetic field properties of Fermilab energy saver dipoles, in *Proceedings of 1983 IEEE Particle Accelerator Conference*, Santa Fe, NM, 1983, p. 3381
34. J. Blowers et al., Reshimming of Tevatron dipoles: a process-quality and lessons-learned perspective, in *Proceedings of 2005 IEEE Particle Accelerator Conference*, Knoxville, TN, 2005, p. 3156
35. G.W. Foster, in *Proceedings of 1995 IEEE Particle Accelerator Conference*, Dallas, TX, 1995, pp. 1298–1300
36. G.W. Foster et al., in *Proceedings of 1998 European Particle Accelerator Conference*, Stockholm, Sweden, 1998, pp. 189–193
37. H.D. Glass et al., in *Proceedings of 1997 IEEE Particle Accelerator Conference*, Vancouver, BC, 1997, p. 3260–3262
38. J. Volk, J. Instrum. **6**, T08003 (2011)
39. G.W. Foster et al., in *Proceedings of 1999 IEEE Particle Accelerator Conference*, New York, NY, 1999, pp. 3324–3326
40. H.D. Glass et al., in *Proceedings of 1997 IEEE Particle Accelerator Conference*, Vancouver, BC, 1997, pp. 3286–3287
41. G.W. Foster et al., in *Proceedings of 1997 IEEE Particle Accelerator Conference*, Vancouver, BC, 1997, pp. 3280–3282
42. J. Volk et al., in *Proceedings of 2003 IEEE Particle Accelerator Conference*, Portland, OR, 2003, pp. 1766–1768
43. S.D. Holmes, V.D. Shiltsev, The legacy of the Tevatron in the area of accelerator science. Annu. Rev. Nucl. Part. Sci. **63**, 435–465 (2013)

Chapter 4
Longitudinal Beam Manipulations

C. Bhat, K. Seiya, and V. Shiltsev

The RF systems installed in the synchrotrons can be used not only for beam acceleration but also to change the longitudinal beam characteristics. Below we present two nontrivial methods of "RF gymnastics" championed at the Fermilab accelerators to satisfy the needs of the Collider Run II—the so-called slip stacking and RF barrier buckets techniques.

4.1 Slip Stacking

Higher proton intensities are required to produce more antiprotons at the target and higher stacking rates in the Antiproton Accumulator. The Run II upgrade had a luminosity goal which required 9.6×10^{12} total antiprotons in the Tevatron. In order to achieve the goal, the Main Injector (MI) was supposed to send 8.0×10^{12} protons in 84 bunches for the antiproton production. Since the maximum beam intensity from the Booster was limited around 5.0×10^{12} particles per pulse (ppp), we planned to increase the intensity in the MI with the scheme called "slip stacking." With the slip stacking, the intensity of the bunch can be doubled by injecting one bunch train at slightly lower energy, another train at slightly higher energy, and bringing them together [1].

The MI accelerates protons to 120 GeV for antiprotons production. In the stacking cycle, 84 bunches were injected from the Booster to the MI, accelerated from 8 to 120 GeV and extracted to production target. Two bunch trains were

C. Bhat • K. Seiya
Accelerator Physics Center, Fermi National Accelerator Laboratory, PO Box 500, Batavia, IL 60510, USA

V. Shiltsev (✉)
Accelerator Physics Center, Fermi National Accelerator Laboratory, MS221, PO Box 500, Batavia, IL 60510, USA
e-mail: shiltsev@fnal.gov

V. Lebedev and V. Shiltsev (eds.), *Accelerator Physics at the Tevatron Collider*, Particle Acceleration and Detection, DOI 10.1007/978-1-4939-0885-1_4,

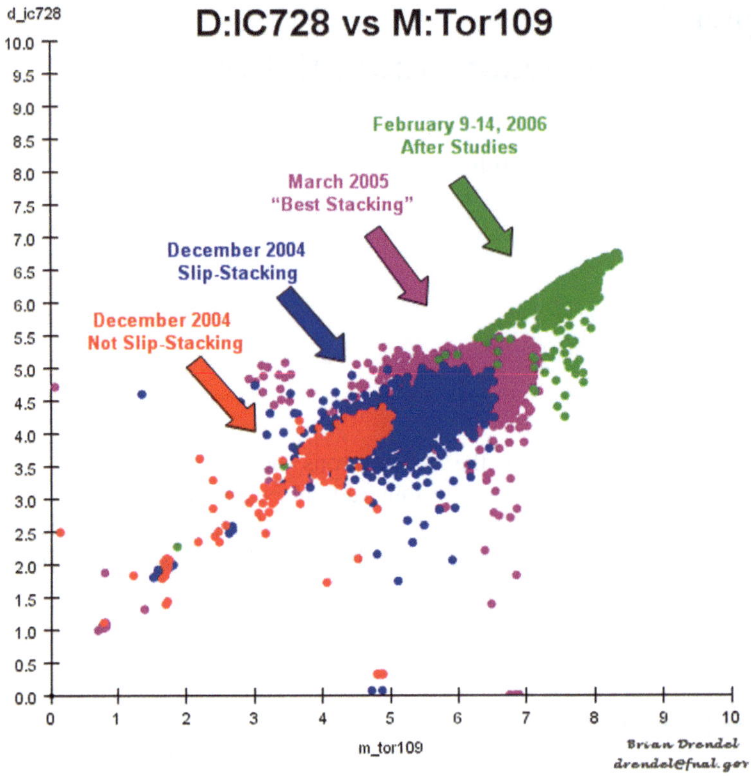

Fig. 4.1 Antiproton production rate from December 2004 to February 2006 as a function of proton intensity on the target (Courtesy B. Drendel, Fermilab)

injected from the Booster and merged to one batch at the injection energy of 8 GeV using slip stacking, and then accelerated to 120 GeV. The intensity was almost doubled from the value before the slip stacking.

The beam studies for slip stacking were started in 2001. Slip stacking became operational in December 2004. This was the first time it had been implemented in any accelerators. Slip stacking increased the beam intensity to the antiproton production target to more than 8.0×10^{12} ppp. The antiproton production rate was improved by 70 % as shown in Fig. 4.1 (the plot shows the increase in the secondary particles flux at the end of the Debuncher injection line D:IC728, in the units of 10^9, vs the proton pulse intensity M:TOR109 on the antiproton production target, in the units of 10^{12}).

The slip stacking scheme was extended to multi-batch stacking for the NuMI neutrino experiment in order to increase the proton intensity at the NuMI target. The MI is sending beam to both antiproton and NuMI targets in one MI cycle. Multi-batch slip stacking has been in operation since January 2008. The number of protons at the NuMI target was increased from 2.25×10^{13} with six Booster pulses to 3.6×10^{13} with nine pulses. The weekly proton intensity on the NuMI target is shown in Fig. 4.2.

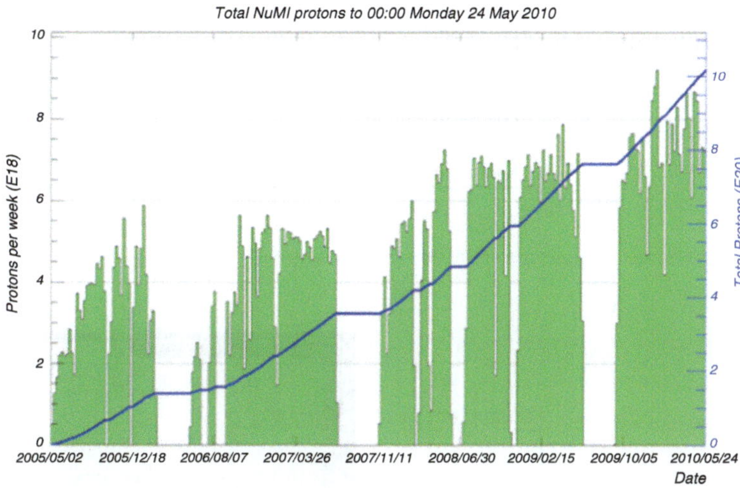

Fig. 4.2 Weekly proton intensity at the NuMI target from May 2005 to May 2010 (courtesy B. Drendel, Fermilab)

4.1.1 RF Parameters and Frequency Curves Optimization at Low Intensity Studies

The scheme of slip stacking uses three different radio frequency (RF) systems and follows four steps. Step 1: The first bunch train is injected from the Booster on the central orbit and captured by the first RF system. To make a room for the second bunch train, the first bunch train is then decelerated until it circulates on the inside of the central orbit. Step 2: The second bunch train is injected on the central orbit and captured by the second RF system. Step 3: As the two bunch trains have slightly different energies, they can move relative to each other. Both bunch trains are accelerated till they are separated by the same amount from the central energy. Step 4: When the two bunch trains coincide at the same longitudinal location, they are captured by the third RF system [2].

Since two bunch trains have different energies, the MI must have an enough momentum aperture to accept both. The momentum aperture of the MI is ± 0.7 % at injection, corresponding to the frequency separation of $\pm 3{,}000$ Hz from the central value of 52.8114 MHz. The main parameters of MI are shown in Table 4.1.

The MI has 18 53 MHz cavities. These are separated into the two groups, each with nine cavities. Low level RF (LLRF) signals go to two groups of cavities individually and produce independent voltages and frequencies. For the slip stacking, we are using three out of nine cavities for each frequency at injection. The RF voltage at injection was adjusted to 62 kV in order to achieve a low momentum spread [3].

The Booster extraction RF voltage was 380 kV and matched to a MI injection voltage of 1 MV for the nonslip stacking normal operation. Since the MI injection

Table 4.1 Main Injector parameters on stacking cycle

RF frequency @ injection	52 811 400 Hz
Proton energy @ injection	8 GeV
Proton energy @ extraction	120 GeV
Harmonic number	588
Transition gamma	21.6
Mean radius	528.3 m

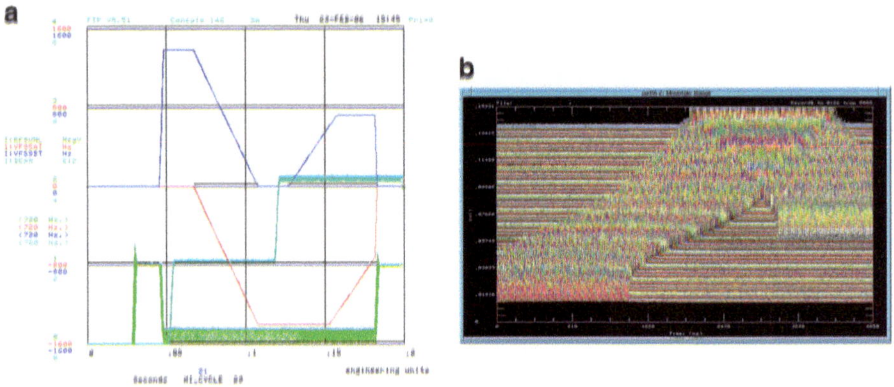

Fig. 4.3 (**a**) Variations of the frequencies of the first RF system (*red*) and the second RF system (*blue*) the total RF voltage (*green*), and the beam intensity (*cyan*) in the MI. (**b**) Mountain range plot of the Main Injector's wall current monitor signals every 1.42 ms for 0.18 s

RF voltage was lowered to 62 kV, a voltage modulation was carried out at the Booster extraction and the bunch shape matched to the MI injection bucket.

Since there were two RF voltages during stacking, they acted on both bunch trains. The bunch shape has been measured to demonstrate that the frequency separation between the first and the second RF systems is adequate. The required separation was estimated from the measurements that had one bunch with the emittance of 0.08 eV s and two RF voltages. The bunch shapes were compared at injection and at 150 ms after injection for the frequency separation from 400 to 1,200 Hz. The frequency separation of 1,200 Hz was enough to keep the bunch shape unchanged.

The frequency as a function of time is shown in Fig. 4.3a for the first and the second RF systems. The first bunch train was injected on the central orbit with nominal frequency at 0.052 s and captured by the first RF system of 62 kV. At this time, the frequency of the second RF system was 1,200 Hz higher than the first RF system. The first bunch was then decelerated to a frequency of 1,200 Hz lower than the original value. After one Booster cycle of 66.7 ms, the second train was injected on the central orbit and captured by the second RF system. After slipping, both bunch trains were captured by all 18 cavities with 1 MV and central frequency. Since we have to inject additional bunch trains from the Booster to the MI, the whole slip stacking process has to be completed in less than two 15 Hz Booster cycles. Total time of the slip stacking process was 133 ms. The slip stacking worked as expected during studies with low intensity and small emittance beams (Fig. 4.3b).

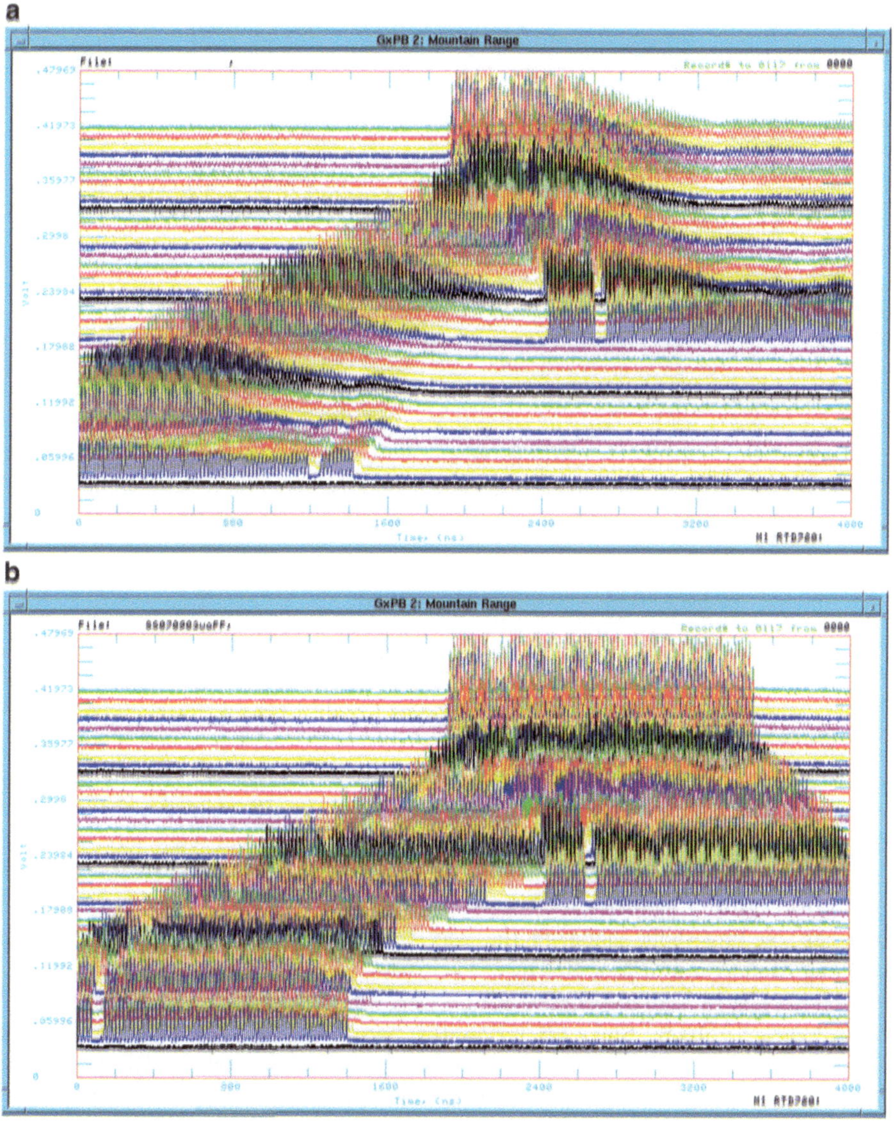

Fig. 4.4 (**a**, **b**) Mountain range plots of the wall current monitor signal. The *upper* and *lower* plots show the signals without and with feedforward beam loading compensation, correspondingly. The signal from the wall current monitor (WCM) reveals the progress of slip stacking from the beginning to the end. The signal was measured every 2.1 ms for 0.24 s [4, 5]

4.1.2 RF Stations Upgrades for the High Intensity Operation

After the beam intensity was increased to 4×10^{12} ppp, the beam loading effects were observed in the bunch signal, as shown in the mountain range plot Fig. 4.4a.

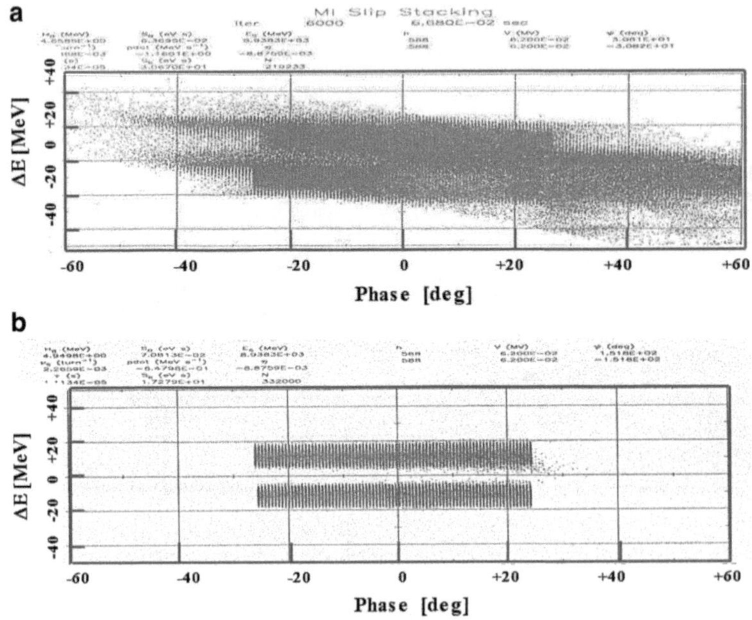

Fig. 4.5 (**a**, **b**) Simulation results in the longitudinal phase space plots. The vertical axis is 20 MeV/div and horizontal is 20 degree/div. The *upper plot* is for the case without the beam loading compensation. The *lower* shows the effect of a 14 dB feedback and 20 dB feedforward beam loading compensation

In presence of a longitudinal feedback system with 14 dB gain and absence of a feedforward beam loading compensation system some particles left their RF buckets and the length of the bunch train exceeded the standard 84 RF buckets. Application of the feedforward beam loading compensation of 14 dB resulted in bunches keeping their shape all the way from injection to the end of the slip stacking as shown in Fig. 4.4b [4].

The MI has 18 high impedance RF cavities which can produce high unexpected beam loading voltage. Since all 18 cavities had to be used for the recapture and acceleration after 130 ms, it was impossible to put shorts on the RF gap in the remaining 12 cavities and reduce the beam loading voltage although just 6 cavities out of 18 were used for the slip stacking.

4.1.3 Simulation with Beam Loading

In order to estimate the required gain for beam loading compensation, computer simulations with beam loading effects were carried out using the ESME code [5]. In the simulation studies, two bunch trains of 84 bunches each with total intensity of 1×10^{13} ppp were set one at the energy slightly above and another slightly below the central energy of the Main Injector. Two RF waves with different frequencies were applied to the trains and evolution of the beam shape after 70 ms is shown in Fig. 4.5a.

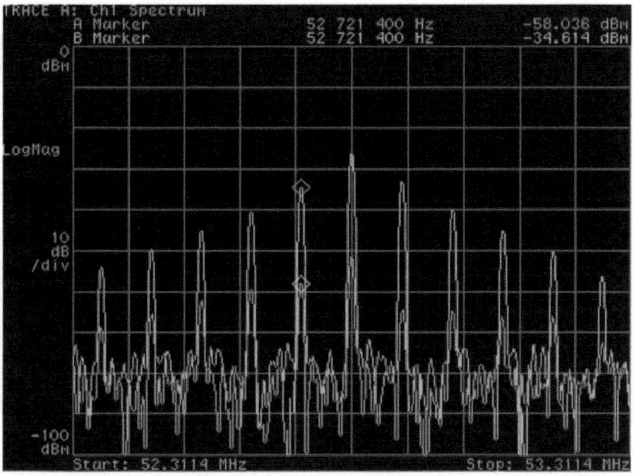

Fig. 4.6 Frequency spectra of the gap voltage monitor on 1 of the 18 cavities. *Blue* and *green traces* show signals with and without feedforward beam loading compensation. The vertical scale is 10 dB/div

As one can see, significant number of particles left their assigned RF buckets due to the beam loading effects and the total length of the bunch train exceeded 84 RF buckets.

Figure 4.5b shows the simulation results with beam loading compensation. Most particles are now captured in the RF buckets and the total train length remains 84 buckets. In these simulations, feedback and feedforward compensation systems applied with gains of 20 and 14 dB, respectively. From these results we estimated that we needed to increase the feedforward compensation gain by 6 dB.

4.1.4 RF Stations Upgrades

In order to increase the gain of the RF feedforward loop, higher RF current was required. The combination of increasing the power and changing the PA operation point from class AB to class A at injection energy was needed to increase the current.

During machine shut down, the solid state power amplifier driver was upgraded from 4 to 8 kW by adding four more modules for each RF station. Figure 4.6 shows a frequency spectra of the gap voltage with and without feedforward compensation on one of the 18 cavities and the reduction was 24 dB. The reduction varied from cavity to cavity the cavities but the average value was about 20 dB.

4.1.5 High Intensity Operation

The MI operated with slip stacking since December 2004. The total beam intensity on target has been increased from 4.5×10^{12} ppp to 8.0×10^{12} ppp as shown in Fig. 4.7a. Figure 4.7b shows mountain range plot with a wall current monitor

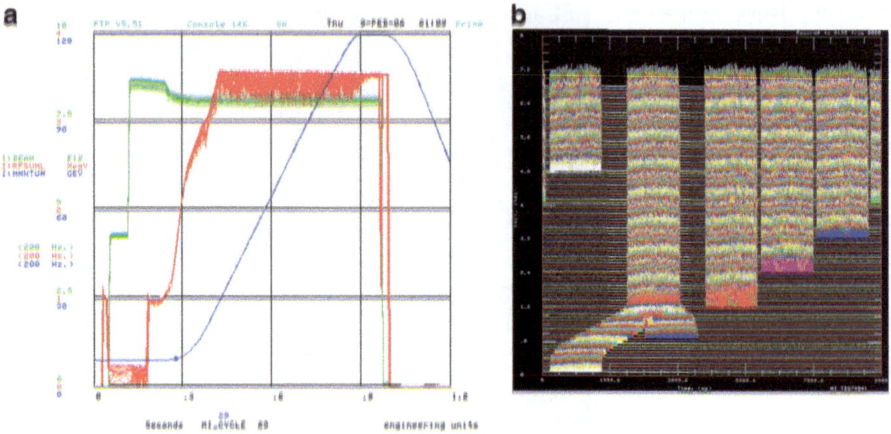

Fig. 4.7 (**a**) The MI cycle from injection to the top energy: beam intensity (*green*), RF voltage (*red*) and particle momentum (*blue*). (**b**) Mountain range plot with wall current monitor signals during slip stacking process. Horizontal scale is 10 μs (one revolution period in the Main Injector)

(WCM) signal for a slip stacking process from injection. The intensity of one Booster pulse was 4.25×10^{12} ppp and the total intensity at injection was 8.5×10^{12} ppp. Two batches were combined to one within two 15 Hz Booster cycle and third batch was injected after that.

There were beam losses of 5×10^{11} ppp (6 %) at the beginning of acceleration, around 9 GeV, which were minimized later by tuning of the frequency separation and RF voltages.

4.1.6 Beam Loss Simulations

The frequency separation and RF voltage for slipping were optimized to minimize beam losses at high intensity and they were 1,400 Hz and 90 kV. Beam studies and simulation studies showed that the beam loss depends on the longitudinal emittance at injection.

In order to estimate longitudinal acceptance at the MI injection with two frequencies used for slip stacking, multi-particle simulation was performed. Figure 4.8a shows the particles in the longitudinal phase space from injection to the point after 12,000 turns which is about as long as the duration of the slip stacking process (181 ms). Figure 4.8b shows an injection distribution of particles which could stay around synchronous phase after 12,000 turns. The simulations indicated that the maximum momentum acceptance is about ± 8 MeV under the MI operation parameters.

The longitudinal beam dynamics at injection energy was experimentally studied using phase space tomography method. The tomography data of four different

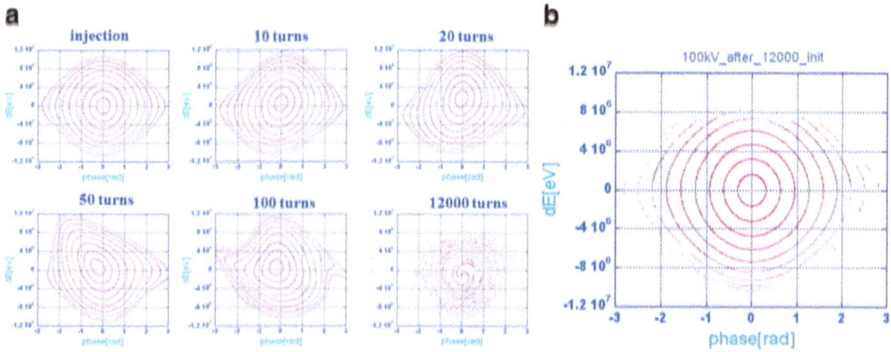

Fig. 4.8 (**a**) Longitudinal phase space evolution over 12,000 turns simulated with two different frequencies. (**b**) Injection distribution of the particles which could stay around synchronous phase after 12,000 turns

Booster bunches are shown in Fig. 4.9a. One can see that the longitudinal phase space distributions vary from bunch to bunch and a 95 % momentum spread was larger than ±15 MeV. Figure 4.9b shows the tomography data at recapture time which shows particles are filling the recapture RF bucket.

The phase space tomography results were used in the simulation as an input distribution. The particles were slipped, recaptured, and accelerated from 8 to 9 GeV. The results of the modeling indicated the particle which were to hit momentum aperture of the MI and then became beam losses at 9 GeV.

4.1.7 Multi-Batch Slip Stacking

The MI used to send one slip stacked batch to antiproton production and five normal batches to the NuMI beam line on the mixed mode operation cycle. The multi-batch slip stacking has been in operation since January 2008. A total of 11 Booster batches were injected in the MI. Two batches were sent to antiproton production and nine to the NuMI beam line. The intensity on the antiproton target stayed at 8–8.5 $\times$ 10^{12} ppp, but the intensity to the NuMI target was increased from 22.5 $\times$ 10^{12} to 35–36 $\times$ 10^{12} ppp. The total intensity at 120 GeV was 45 $\times$ 10^{12} ppp with a cycle efficiency of 95 %.

Figure 4.10a shows a mountain range plot during the 11 batch slip stacking process. The total time for slip stacking, merging 10 batches into 5 batches and injection of the 11th batch is 0.733 s. The beam was accelerated to 120 GeV with a MI cycle time of 2.2 s as shown in Fig. 4.10b. Figure 4.11a, b shows the total beam power in the MI was increased from 250 to 370 kW with 11 batch slip stacking on the mixed mode operation [6–8].

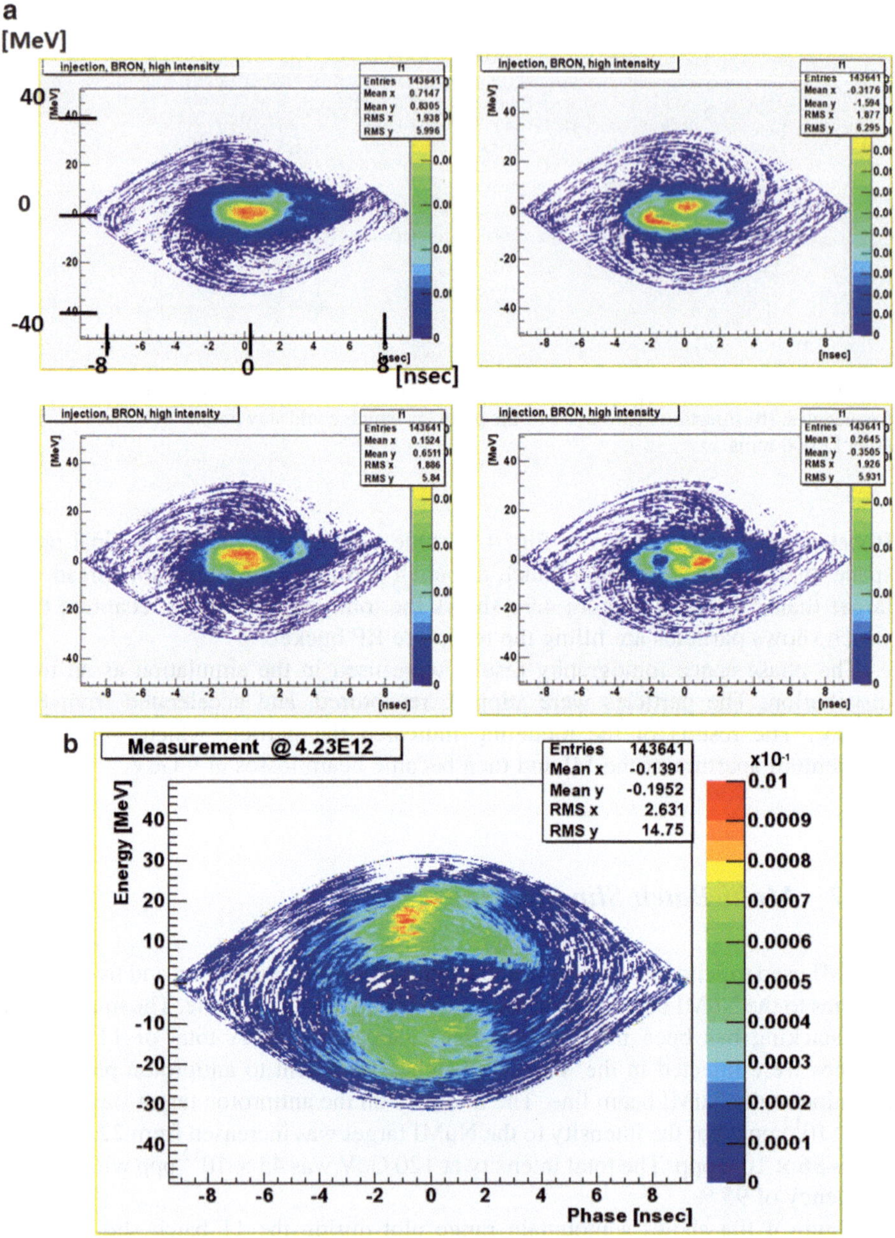

Fig. 4.9 (**a**) Longitudinal phase space tomography at injection for four different bunches. (**b**) Longitudinal phase space tomography at recapture

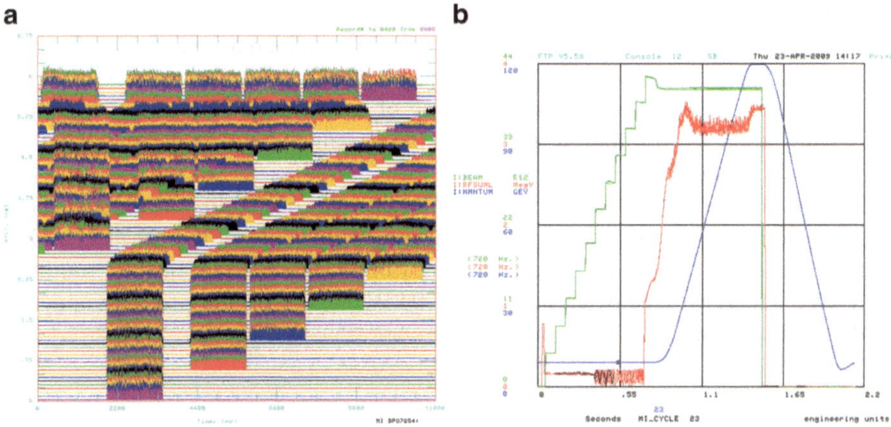

Fig. 4.10 (**a**) Mountain range plot showing the 11 batch slip stacking process in the mixed mode cycle. Horizontal scale is one MI revolution period of ~11 μs. (**b**). The total intensity (*green*), RF voltage (*red*), and momentum in the MI (*blue*) on the mixed mode cycle

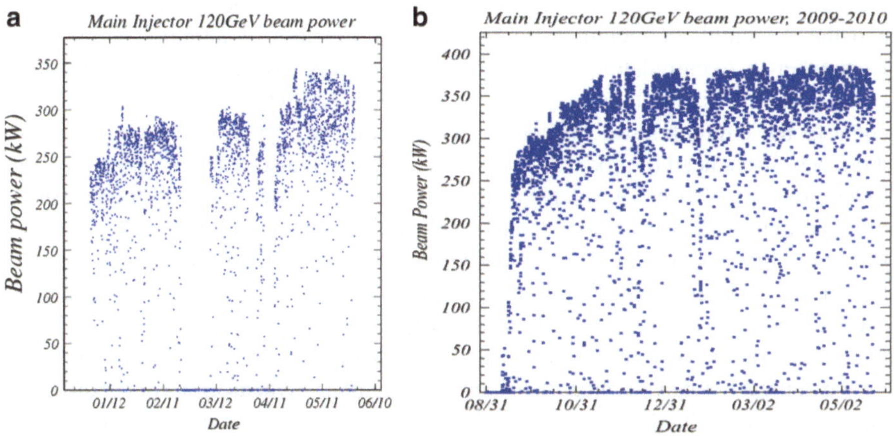

Fig. 4.11 (**a**) The total beam power of mixed mode cycle from January to July 2008. Plot shows before and after the 11 batch slip stacking process became operational. (**b**) The total beam power of mixed mode cycle from August 2009 to May 2010

4.1.8 Summary

Beam studies have been carried out with low intensity proton beams in the MI and they verified that the slip stacking technique works as expected.

We estimated by numerical simulations the required beam loading compensation needed for the high intensity slip stacking operation. The gain of feedforward system was increased by doubling the power of solid state driver and by programming of the PA grid bias. The effects were observed in the beam and gap voltage signals.

Since December 2004, the slip stacking has been operational and intensity provided by MI on the antiproton production target was increased to 8.0×10^{12} ppp. The antiproton production rate was increased by 70 %.

There were beam losses at the beginning of acceleration. The loss mechanism was understood by simulation and beam measurements and it depended on the longitudinal emittance at injection.

The multi-batch (11 Booster batches of 84 bunches each) slip stacking has been in operation since January 2008. The total proton intensity in the MI at 120 GeV reached 4.5×10^{13} ppp with a cycle efficiency of 95 %. The Main Injector routinely delivered a total average beam power of 370 kW on both the antiproton production and NuMI targets.

4.2 Recycler Barrier RF Buckets

Barrier bucket RF systems in synchrotrons were not invented by accident or serendipity but by their sheer necessity. At the very early stages of the Tevatron I Project, it was realized that the circumference difference of about 30 m between the Debuncher and the Accumulator would result in an antiproton loss of about 7 % for every Debuncher-to-Accumulator transfer. Therefore, it was essential to develop a technique to preserve a minimum gap in the antiproton beam in the Debuncher with length equal to the circumference difference between the two synchrotrons. Furthermore, there was also a need to produce an isolated 2.5 MHz sinusoidal wave corresponding to harmonic number $h = 4$ to have a single RF bucket in the ring, with the rest of the buckets suppressed in order to extract a single antiproton bunch from the Accumulator ring for collider operation. These two requirements led to the initial development of barrier RF technology at Fermilab [9]. The concept of a "suppressed bucket" in synchrotrons has been addressed earlier [10]. However, a significant research was undertaken on the barrier RF (also called wide-band RF) at Fermilab. As a result, it has been used in many applications and presently many synchrotrons at Fermilab are equipped with such RF systems [11]. Note that the Recycler uses only a barrier RF system in all its beam manipulations, unlike any other storage ring in the world.

4.2.1 Barrier RF Waveforms

One can imagine a variety of barrier RF waveforms. A schematic view of typical barrier RF voltage excursions for a synchrotron operating below transition energy γ_T (like the Recycler [12]) is shown in Fig. 4.12. The barrier combinations shown in the first two examples are used for confining a parcel of beam in a synchrotron. Figure 4.12d is an example of gap producing barrier combination, also referred to as "anti-buckets." Electronic generation of the barrier waveforms illustrated in Fig. 4.12 is a straightforward LLRF electronics problem. However, it requires

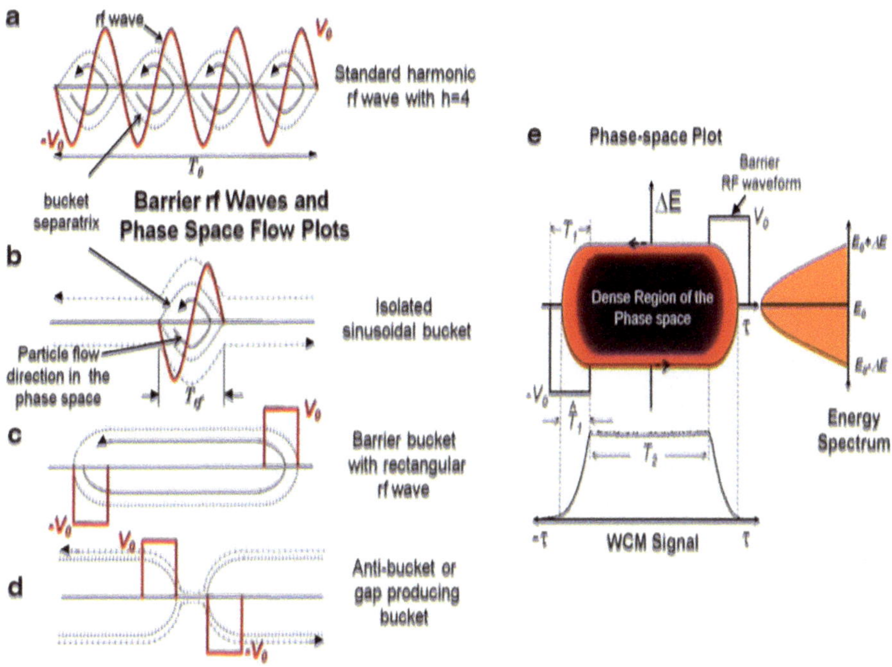

Fig. 4.12 A schematic view of standard and barrier RF wave forms and RF buckets (*left*). The *dashed*, *solid*, and *dashed-dot curves* indicate bucket boundaries, particle flow directions in longitudinal phase space, and flow directions of particles outside the buckets, respectively. A schematic of $(\Delta E, \tau)$-phase space distribution of particles in a rectangular barrier RF bucket with its line charge distribution (wall current monitor-WCM signal) and energy distribution are shown on the *right*

much larger power to feed a cavity than in the case of a resonance RF cavity to produce the same peak voltage. A barrier RF wave of an arbitrary shape in a circular accelerator can be presented as superposition of Fourier harmonics of the revolution frequency. The RR barrier RF system uses >100 Fourier components to generate any waveform which is more than enough for all practical needs. The LLRF signals synthesized by this method are fed to a broadband RF cavity after an amplification using a broadband power amplifier. Below we give a general overview of dynamics of a charged particle in the presence of such a barrier RF wave in a synchrotron.

4.2.2 Longitudinal Beam Dynamics of Charged Particle in the Barrier RF Buckets

The motion of a particle relative to a synchronous particle is governed by [13, 14]

$$\frac{d\tau}{dt} = -\eta \frac{2\pi \Delta E}{T_0 \beta^2 E_0} \quad \text{and} \quad \frac{d(\Delta E)}{dt} = \frac{eV(\tau)}{T_0}, \tag{4.1}$$

where ΔE is an energy deviation from the synchronous energy E_0, e and βc are the particle electric charge and velocity, τ is the time difference between the arrivals of a particle and the synchronous particle located at the center of the RF bucket, η is the ring slip-factor, and T_0 is the revolution period. $V(\tau)$ is the amplitude of the RF voltage waveform. Using the above equations, one can get the half bucket height ΔE_b, given by

$$\Delta E_b = \sqrt{\frac{2\beta^2 E_0}{|\eta|} \frac{\left| \int_{T_2/2}^{T_2/2+T_1} eV(\tau)d\tau \right|}{T_0}}. \tag{4.2}$$

Since the bucket height depends upon $\int eV(\tau)d\tau$ the exact shape of the wave form is not critical. The quantities T_1 and T_2 denote barrier pulse width and gap between RF pulses, respectively, as indicated in Fig. 4.12e.

In the case of Recycler, one mostly deals with rectangular barrier buckets:

$$V(\tau) = \begin{vmatrix} -V_0 \text{ for } -T_1 - T_2/2 \le \tau < -T_2/2, \\ 0 \quad \text{for } -T_2/2 \le \tau < T_2/2, \\ V_0 \text{ for } T_2/2 \le \tau < T_1 + T_2/2. \end{vmatrix} \tag{4.3}$$

The advantage of a rectangular waveform is that for a given maximum RF voltage V_0, the available bucket area will be maximized. A schematic view of the beam phase space distribution in a rectangular barrier RF bucket along with the definition of various parameters relevant to this chapter is shown in Fig. 4.12e.

It has been shown [14–16] that the energy offset $\Delta\hat{E}$ of a particle is related to its depth of penetration $\hat{T}_1$ into a rectangular barrier by

$$\Delta\hat{E} = \sqrt{\frac{2\beta^2 E_0}{|\eta|} \frac{eV_0 \hat{T}_1}{T_0}}, \tag{4.4}$$

and the total longitudinal emittance ε_l of the beam particles with $\Delta E \le \Delta\hat{E}$ is

$$\varepsilon_l = 2T_2\Delta\hat{E} + \frac{8\pi|\eta|}{3\omega_0\beta^2 E_0 eV_0} \Delta\hat{E}^3, \tag{4.5}$$

where $\omega_0 = 2\pi/T_0$. The synchrotron oscillation period of the particle is

$$T_s = \frac{2T_2}{|\eta|} \frac{\beta^2 E_0}{|\Delta\hat{E}|} + \frac{4|\Delta\hat{E}|T_0}{eV_0}. \tag{4.6}$$

The Eq. (4.6) has two components; one for the region with $V(\tau) = 0$ and the second for $V(\tau) \ne 0$. Figure 4.13 displays a comparison between the calculated

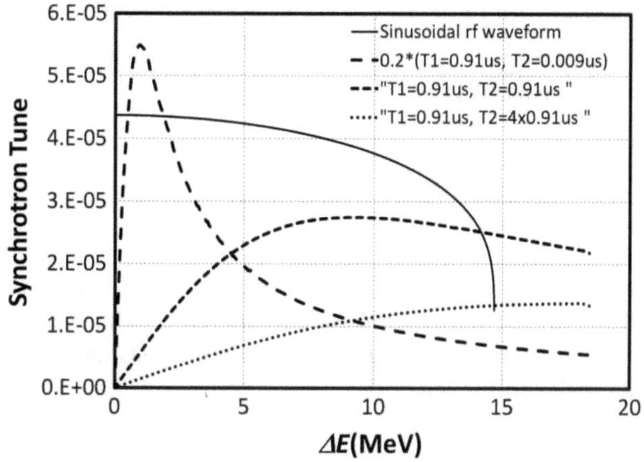

Fig. 4.13 Recycler synchrotron tune versus energy offset of the particle in rectangular barrier buckets for three scenarios of T_2 and their comparison with a sinusoidal RF bucket of length $2T_1$ (*black solid line*)

synchrotron tune ($Q_s = T_0 f_s$) in a sinusoidal RF bucket in Recycler with a period of $2T_1$ and that in a barrier bucket with $T_1 = 0.91$ μs ($\Delta E_b = 18.4$ MeV) and $T_2 \approx 0.0$, 0.91 and 3.64 μs. In all these cases $V_0 = 2$ kV. Note that the dependence of synchrotron frequency on the energy deviation, $\Delta\hat{E}$, in a rectangular barrier bucket is quite different from that for a sinusoidal RF bucket. In the case of the barrier buckets, the peak energy offset at which the synchrotron period reaches minimum is $\Delta\hat{E}\,|_{T_s(\mathrm{Min})} = \sqrt{T_2/4T_1}\Delta E_b$. Consequently, if $T_2 \leq 4T_1$, then the $df_s/d(\Delta E) = 0$ lies well inside the bucket boundary. The particles in the vicinity of $Q_{s\mathrm{Max}}$ lose Landau damping [9] and the beam become susceptible to longitudinal collective instabilities.

All RF manipulations in barrier buckets explained above were modeled using a computer code ESME [17] prior their use in the Recycler.

4.2.3 Recycler RF System

The Recycler RF system is comprised of a LLRF and a high level RF (HLRF) system. The RR LLRF [18] is a very advanced and versatile control system consisting of a Super Harvard Architecture Computer (SHARC) 8-bit digital signal processors (DSPs). The SHARC derives much of its performance like generating data address, hardware loop control, and all multifunction instructions in parallel. The DSP along with its software controls the frequency and phase registers of three direct digital synthesizer (DDS) channels. These DDS modules provide nine RF clock inputs to a new SHARC-based module called the Recycler Bucket Generator (RBG). The RBG contains eight arbitrary waveform generator channels (often called ARB). The DSP controlled ARBs are summed to form the LLRF output that drives Recycler HLRF system. Each arbitrary waveform generator has an independent table, phase, and amplitude, all of which can be changed in real time to perform a specific type of RF manipulation. Any such change may be carried out

Table 4.2 RR LLRF cog rate for the barrier pulses [17]

Cogging param	Medium	Slow	Fast
Maximum rate (Bkt[a]/s)	39.98	4.86	8,400
Maximum derivative (Hz/s)	12.5	1.94	3,024,000
Minimum cogging time (s)	6.9	5.08	0.01389
Maximum Bkts cogged in 1 min	137.996	12.355	~44,000

[a]Bkt = (53 MHz RF bucket length) = 18.935 ns

at a 720 Hz rate. The LLRF commands for barrier RF manipulations are passed from the CPU over a VXI bus to RBG. The RBG then smoothly controls any changes to bucket parameters. Table 4.2 lists available *cogging*[1] rates for a barrier RF pulse in the RR.

In the current RR LLRF architecture ARB5 (out of ARBn, $n = 0, 1, \ldots 7$) is used as RR reference marker for synchronous beam transfers to and from the RR and for cogging the rest of the ARBs. Each of the ARBs are assigned with a predefined RF wave form which can be turned on and off in 128 steps of equal amplitudes at variable rates. The maximum width of a waveform is limited to 256 Bkt (4.84 μs). Sixteen different waveforms similar to those shown in Fig. 4.1 are in use.

The RR LLRF architecture explained above is quite general. Its full potential was realized by developing new sets of commands specific to the RR operation [19]. For example, command like "GrowCoolBucket" is used to expand a "cold bucket" by a certain amount; here a cold bucket is a group of ARB6 and ARB7 which acts like a single entity. Currently, there are about 75 such commands in use, each for a specific purpose. Each RF manipulation scheme, like antiproton beam stacking, is treated as a group of such commands put together in a module or multiple modules.

The RR HLRF [20] consists of four cavities, each driven individually by a broadband solid state amplifier (Amplifier Research Model 3500A100). Each amplifier, operating in push–pull mode, can supply a minimum of 3,500 W output power over the frequency range of 10 kHz to 100 MHz. The amplifiers are connected to the cavities in the tunnel by 20 m of 7/8″ diameter coaxial line.

Figure 4.14 shows an assembly drawing of a RR barrier RF cavity. Each cavity consists of 25 Mn–Zn ferrite (Ceramic Magnetics MN60) rings of dimension 11.5″ OD × 6″ ID × 1″ thick with a spacing of 0.5″ between each ring with three additional rings of NiZn ferrite (CMD10) at its one end. Each ring is supported by Kapton spacer blocks. The whole assembly is enclosed between concentric aluminum cylinders encircling a 4″ diameter stainless steel beam pipe with a 1″ ceramic gap. This gap is electrically connected to the cavity with beryllium–copper finger stock which couples the developed electric field to the beam. To complete the electrical connection between the beam pipe and the cavity outer shell, a thin

[1]This is a quasi-adiabatic process that involves azimuthal displacement of a bunch in a synchrotron. Cogging is carried out by a continuous change in phase and/or frequency of an RF wave relative to a reference marker.

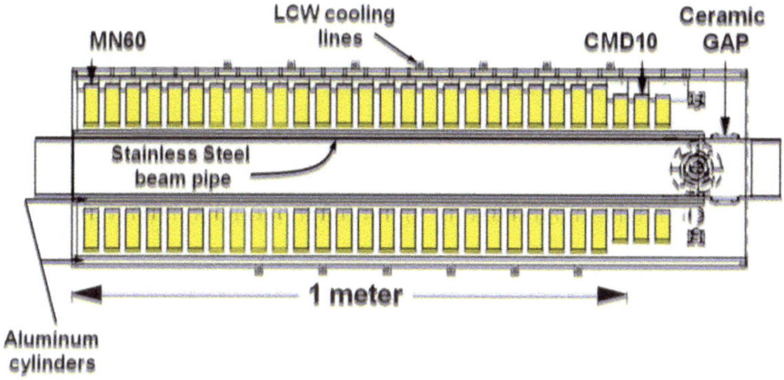

Fig. 4.14 Assembly drawing of RR barrier RF cavity [20]

copper end plate is used. The cavity is cooled using low conductivity water through a coil around the outer shell whereas the ferrite core is air cooled. The impedance of the cavity is approximately 50 Ω over a frequency range of 100 kHz to 20 MHz. The designed peak voltage per cavity is 500 V.

The Recycler RF has the capability to form RF pulses of different wave shapes, lengths and voltages and capture beam in one or more parcels azimuthally distributed around the ring. Also, these parcels can be moved relative to one another and can be independently expanded or compressed.

4.2.4 *RF Linearization*

It is extremely important to have a flat longitudinal line charge distribution in the region with $V(\tau)=0$ for the RF barrier bucket. But the beam profiles in the RR barrier buckets with $T_2 \neq 0$ (in absence of any compensation) showed clear unevenness as shown in Fig. 4.15a even at low intensity (~0.1×10^{12} particles). This lead to unequal intensity and emittance for bunches sent to the Tevatron. The main causes for unevenness in the line charge distribution are (1) nonzero harmonic contents of the revolution frequency between the positive and the negative barrier pulses defining the bucket, (2) polar asymmetry in the fan-back signals of barrier pulses, (3) potential well distortion and beam loading, and (4) RF imperfections. Each of these effects was observed in the RR as beam cooling was improved. Subsequently, correction systems were developed and added to the RR.

A nonzero harmonic component of the revolution frequency was seen for the first time [21, 22] during beam measurements on an isolated rectangular barrier bucket similar to one in Fig. 4.15a. A linearizing circuit [23] was added between the LLRF output and the cavity amplifiers. A block diagram of such a linearizing circuit is shown in Fig. 4.15b. The current system linearizes the frequency in the region from

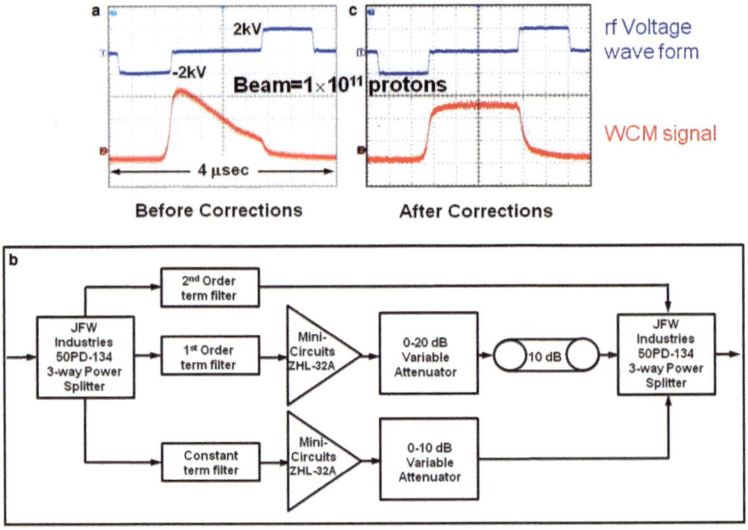

Fig. 4.15 RR barrier bucket (*blue*) and beam profile (*red*) (**a**) before correction (**b**) block diagram of the linearizing circuit (**c**) after the corrections [23]. The horizontal axis is time (400 ns/div)

90 kHz to 1 MHz (up to ten harmonics) with flatness in amplitude better than 0.26 dB and phase 1.8°. Figure 4.15c shows the WCM signal after the corrections.

As the e-cool came online beam got colder and the observed distortion of the beam profile between barriers found to resemble the inverse of the potential well $\int V(\tau)d\tau$ as shown in Fig. 4.16a [24]. A theoretical study of this unevenness arising from potential well distortion was carried out using Haissinski equation [24, 25] which describes the observed beam profile as a function of time according to

$$\rho(\tau) = \rho_0 \exp\left[-\frac{|e|\beta^2 E_0}{|\eta| T_0 \sigma_E^2} \int_0^\tau V_{\text{eff}}(t)dt\right]. \tag{4.7}$$

Here ρ_0, σ_E and $V_{\text{eff}}(t)$ are the ideal profile of the beam, measured root mean square energy spread and fan-back voltage, respectively. The net distortion of the beam current distribution, $(\rho(\tau) - \rho_0)$ can be obtained by expanding Eq. (4.7) as

$$\rho(\tau) - \rho_0 = \rho_0 \frac{|e|\beta^2 E_0}{|\eta| T_0 \sigma_E^2} \int_0^\tau V_{\text{eff}}(t)dt. \tag{4.8}$$

This equation clearly explains one of the important features of experimental observations viz., inverse dependence of distortion on σ_E^2. Figure 4.16b shows a comparison between predicted and measured beam intensity profiles for 0.51×10^{12} antiprotons in a rectangular barrier bucket with $T_2 = 5.8$ µs indicating almost all of the observed unevenness was due to the RF imperfections. The beam loading effect showed up [25] at nearly twice the higher beam intensity under similar conditions.

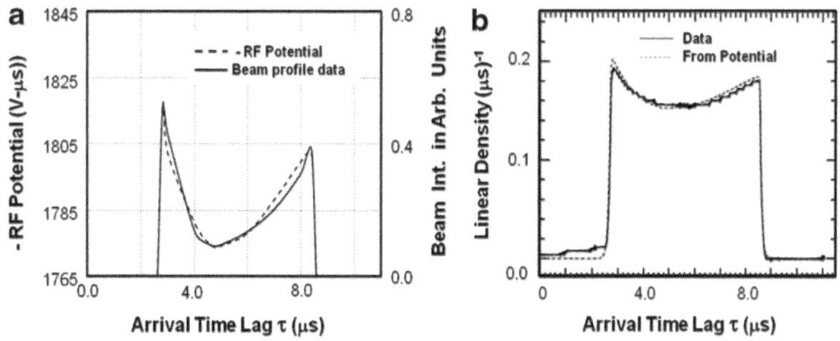

Fig. 4.16 (**a**) Line charge distribution of cold antiproton beam (0.51×10^{12} antiprotons) in a rectangular barrier bucket overlapped on the inverse of the RF potential; (**b**) comparison between computed beam profile using Haissinski equation and measurements [24] for the same beam

To address these problems once and for all, an FPGA-based adaptive correction system was developed [26]. The correction system uses an average error signal obtained by taking the difference between the fan-out and fan-back signals over 128 turns. This signal contains the required information on the observed RF imperfection and beam loading. The final corrections generated using a FPGA firmware is then summed back into the LLRF reference signal by an external summing amplifier board. Figure 4.17a shows block diagram representing the adaptive correction system implemented in the RR. Figure 4.17b shows a typical case of beam profiles before and after the FPGA corrections. This correction system has become an integral part of the RR operation and was found to be very effective for all RF manipulations carried out in the RR.

4.2.5 Measurements of Longitudinal and Transverse Emittance

The Recycler uses Schottky detectors [27] (see Chap. 9) to monitor the emittance in the longitudinal and transverse planes. Originally, this technique was used for a coasting beam. However, it has been shown that for a bunched beam if $T_M f_s << 1$ where T_M is duration of measurement (characteristic time of the band-pass filter), the longitudinal Schottky spectrum represents an instantaneous "snapshot" of the momentum distribution of beam particles [28]. A detailed account of various longitudinal emittance measurement methods available for beam in barrier RF bucket is described in [29]. Currently the longitudinal emittance is obtained using Eq. (4.5); the energy spread of the beam $\Delta\hat{E}$ is measured using the longitudinal Schottky spectrum and T_2 is read from the LLRF fan-out signals. The technique allows emittances measurements for a segmented beam by gating the Schottky signals [30]. The accuracy of the measured longitudinal emittance by this method on a bunch with $T_2 \geq 4T_1$ is $\approx \pm 15$ %. More accurate offline methods based on

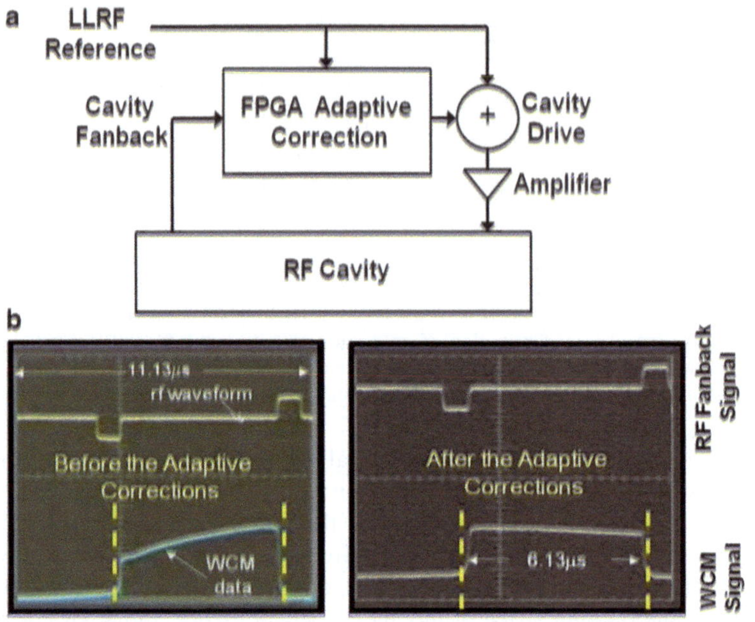

Fig. 4.17 (**a**) Block diagrams [26] for (**a**) the adaptive correction system implemented in the RR LLRF; (**b**) 3.1×10^{12} antiproton beam before and after the FPGA adaptive corrections

beam tomography using Schottky data and/or WCM data [29, 31] have been developed for bunches of any length.

Flying wires (see Chap. 9) and transverse Schottky detectors are used to measure the transverse emittance (ε_T) for the beam in the barrier buckets.

4.2.6 Antiproton Stacking in RR

The antiprotons from the Accumulator Ring are transferred to the RR via the MI in the form of four 2.5 MHz bunches for every transfer. There is a small emittance dilution at the level of a few percent which inevitable during the beam transfers between the Accumulator Ring and the RR because of a frequency mismatch between the Accumulator Ring and the MI [32] and, energy mismatch between the Accumulator Ring and the RR. As a consequence of this, there is a correlation between the overall longitudinal emittance dilution and Accumulator Ring to RR stacking efficiency $E_{AR\rightarrow RR}$, given by

$$E_{AR\rightarrow RR} = \frac{(\text{RR Beam})_{\text{Final}} - (\text{RR Beam})_{\text{Initial}}}{\text{Antiprotons from the Accumulator Ring}}. \qquad (4.9)$$

The correlation arises due to the limited area of the barrier bucket in the RR available for stacking. If the beam particles fall out of the barrier bucket in any stages of the beam stacking, they will become DC and almost 30 % these antiprotons will be lost during each transfer. Therefore, the total Accumulator to RR antiproton transfer efficiency may be looked upon as a product of efficiencies between (1) Accumulator Ring to MI and (2) MI to RR.

Over the years, a number of quasi-adiabatic RF stacking techniques have been developed [33] with an emphasis on maximizing the overall stacking efficiency. The technique currently being used is explained below.

The antiproton stacking in RR is carried out under two different initial conditions; one without any beam or in the presence of partly cooled beam in the RR. In both cases the beam is stacked using rectangular barrier buckets with $T_1 = 0.91$ μs and $\Delta E_b = 18$ MeV. Prior to a new beam transfer an anti-bucket is grown adiabatically to keep the injection region out of any antiprotons with 18 MeV $\leq |\Delta E| \leq$ 34 MeV; the particles with $|\Delta E| <$ 18MeV are confined to the old stack bound by barrier pulses with $V_0 \approx 2$ kV. Nearly a second before the beam transfer to the RR a set of four 2.5 MHz RF buckets of length ≈1.59 μs is superimposed inside the anti-bucket. Following the beam injection, the anti-bucket is replaced by a standard capture barrier bucket and the newly arrived beam is debunched by removing the 2.5 MHz RF buckets. Prior to the beam merging, the bucket height $|\Delta\hat{E}|$ is matched to that of the old stack by using a *morph*-merging technique [33].

Average stacking efficiencies $E_{MI \rightarrow RR}$ and $E_{AR \rightarrow RR}$ for the past three and half years are found to be about 95 and 92 %, respectively. A few percent difference between $E_{MI \rightarrow RR}$ and $E_{AR \rightarrow RR}$ is mainly dominated by the Accumulator Ring to MI beam transfer efficiency as explained earlier.

Antiproton stack size in the RR is built up to a required size over several beam transfers by extracting almost all of the newly accumulated antiprotons from the Accumulator Ring. A set of transfers (consisting of two to three transfers each) will be carried out once for every 30 min and will have an average of about 0.25×10^{12} antiprotons with large emittance—transversely about 6–7 π mm-mr and longitudinally about 25 eVs. Between two sets of transfers, the antiproton beam in the barrier bucket made of old stack and newly arrived antiprotons is cooled using Stochastic cooling systems (rarely with e-cool; e-cool is generally used just before beam transfer to the Tevatron). Simultaneously, new antiprotons are accumulated in the Accumulator Ring for the next set of transfers. Typically, 15–20 sets of transfers are required to reach an optimum stack size[2] of 3.5×10^{12} antiprotons in the RR.

The emittance of the beam in the RR during stacking is shown in Fig. 4.18. The initial emittance of the leftover antiproton beam from the previous Tevatron load was quite large for the example shown here. Each step in the beam intensity represents a new set of transfers. The measured step increment in the beam emittance at the end of each set of transfer is approximately the same as that measured in the MI.

[2] An optimum RR stack size is determined by several factors like the Tevatron luminosity life time, integrated luminosity delivered to collider detectors over a period of a week and also the rate at which the antiprotons are accumulated in the Accumulator Ring (see also discussion in Chap. 1).

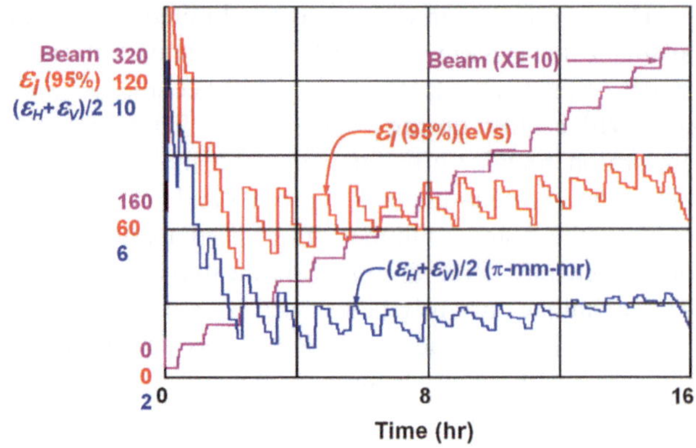

Fig. 4.18 A typical example of RR antiproton stack buildup along with measured longitudinal and transverse emittances. The scales for the ordinate for the three parameters are also shown

With strong electron cooling transverse instability was observed in the antiproton stacks at very high phase space densities. To control the observed transverse instability a damper system is installed with an initial bandwidth of 30 MHz and upgraded to a bandwidth of 70 MHz [34]. The maximum stack size of the antiprotons in RR with the current transverse damper is $\approx 5.4 \times 10^{12}$ with ε_l (95 %) ~ 70 eVs, average ε_T (95 %) ~ 3.4 π-mm-mr.

4.2.7 Antiprotons for the Collider Operation: Longitudinal Momentum Mining

The steps involved in filling up the Tevatron with antiprotons are (1) extract four 2.5 MHz bunches of antiprotons at a time from the RR (or from the Accumulator Ring or a mixture [35]) and send them to the MI and (2) accelerate them from 8 to 150 GeV before transferring to the Tevatron synchronously. These steps are repeated nine times to fill the Tevatron with 36 antiproton bunches. The main requirements for Run II [36] are (1) an efficient and robust way to extract low emittance antiproton bunches from the dense region of the phase space of the cold beam stack and leave behind the particles with very large $\pm \Delta E$ for use after cooling (called "antiproton economy") and (2) all 36 antiproton bunches in the Tevatron should have the same emittance and same number of particles per bunch, i.e., there should be no bunch to bunch variation in the proton–antiproton collider luminosity. To meet these requirements a new mining technique called longitudinal momentum mining (LMM) was developed [37]. One of the limitations to the proton–antiproton luminosity delivered to the high energy physics program prior to the implementation of LMM was significant longitudinal emittance dilution (~300 % from the first to ninth transfer) in the antiproton stack in the Accumulator Ring (e.g., we had this problem throughout the Run I of the collider operation) and in the RR. LLM was crucial to the success of the RR even before the electron cooling was commissioned [35, 37].

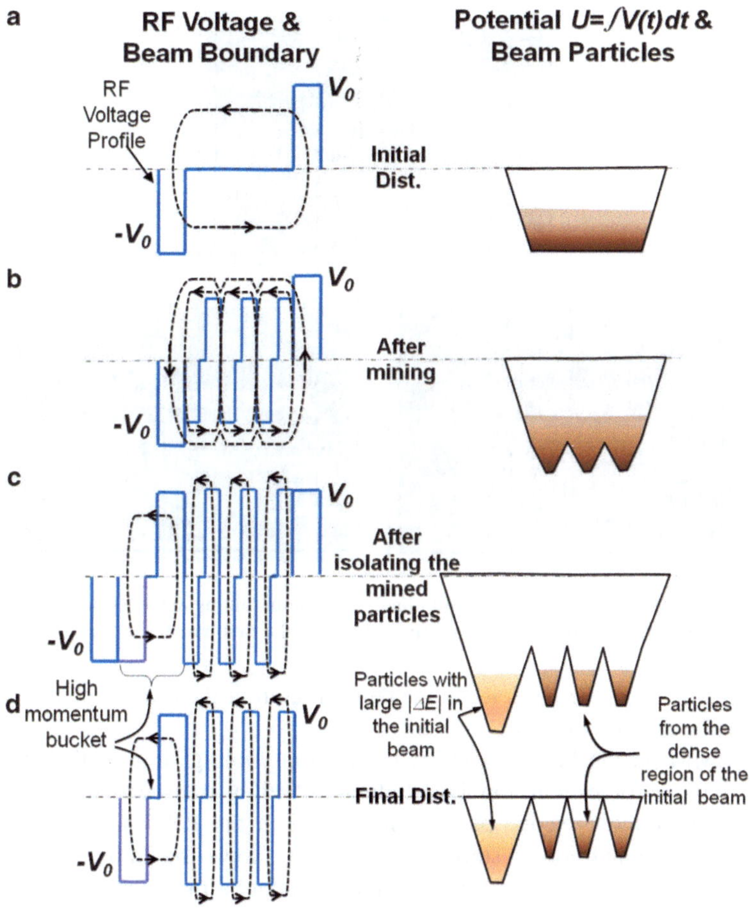

Fig. 4.19 Schematic illustrating the principle of longitudinal momentum mining. Beam particle boundaries in $(\Delta E, \tau)$-phase space (*dashed lines*) are also shown. The potential $\int V(t)dt$ and the beam particles in it in each case are shown in the cartoon on the *right*

The general principle of LMM is illustrated in Fig. 4.19. The RF waveform along with the beam phase space boundary (dashed lines in left figures) and the corresponding potential well containing beam particles are shown for various stages of the mining processes. The objective of LMM is to isolate particles closer to E_0 (i.e., dense region near the bottom of the potential well, Fig. 4.19a) from the rest without any emittance growth. This is accomplished by adiabatically inserting a set of mining buckets (the illustration in figure is for three parcels). The particles which cannot be bound by these "mini" barriers are still bound by the larger barriers shown in Fig. 4.19b and are executing synchrotron oscillations at a relatively higher rate than the ones captured in the mini-buckets. Finally, the un-captured particles are isolated in another RF bucket (high momentum bucket) as shown in

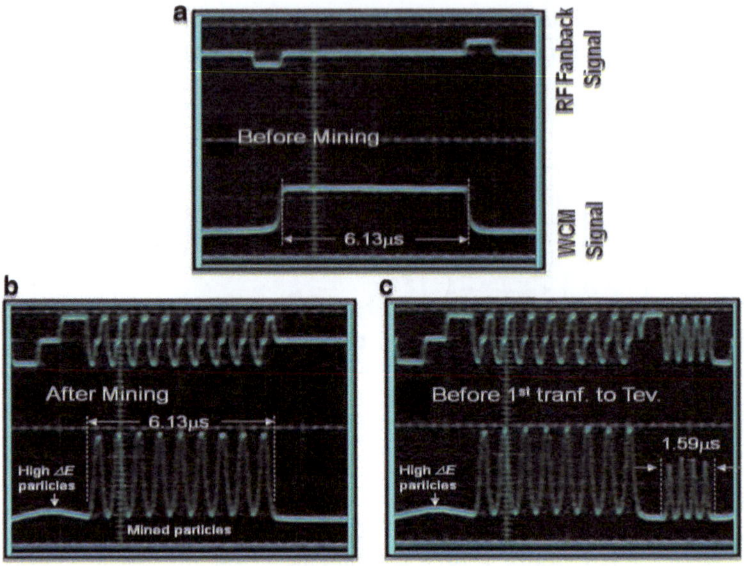

Fig. 4.20 The experimental data on longitudinal momentum mining in the RR with the antiproton beam (**a**) before mining, (**b**) after mining, (**c**) just before the first transfer to the Tevatron

Fig. 4.19c, d. Thus, the particles in the dense region of the $(\Delta E, \tau)$-phase space are mined while leaving the rest.

The LMM on the beam cooled with electron cooling showed that the mined buckets have particles with low longitudinal as well as transverse momenta.[3]

Figure 4.20 shows scope pictures for LMM in the RR just before a typical Tevatron proton–antiproton store. Figure 4.20a shows the initial cold stack of about 3.5×10^{12} antiprotons captured in a rectangular barrier bucket of $T_2 = 6.13$ μs, $T_1 = 0.903$ μs and with RF pulse height of about 1 kV. The measured longitudinal emittance of the stack was about 68 eVs (95 %). The mining is carried out with nine mining buckets with a predetermined area. The size of a mining bucket is decided based on the needs of the Tevatron and MI acceptance. The area of the mini-bucket was chosen to be about 8 eVs. The beam after mining and separating particles with high $|\Delta E|$ is shown in Fig. 4.20b. Subsequently, the first parcel from the right is further divided into four 2.5 MHz bunches after moving it to the extraction region of the RR. Just a few seconds before extraction to the Tevatron, an anti-bucket is grown adiabatically to leave behind any un-captured antiprotons and keep them away from the extraction region as shown in Fig. 4.20c. The anti-bucket is quite important to help sending clean bunches to the Tevatron by eliminating any undesirable DC beam to the MI. During the entire barrier RF manipulations importance is given to preserve the emittance as well as antiproton economy.

[3] In the absence of the electron cooling this is not true (A. Burov, private communications). In any case, the LMM guarantees mining of the particles with low longitudinal momenta.

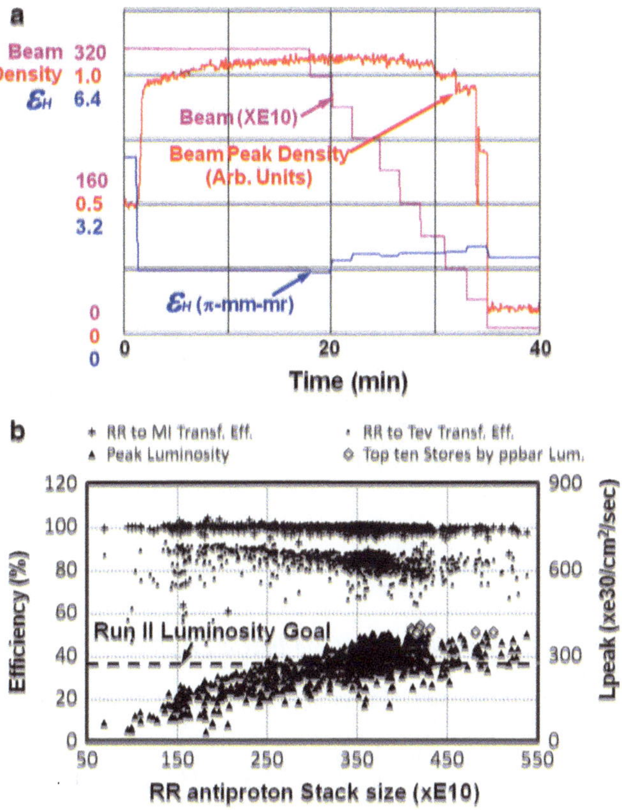

Fig. 4.21 Measured (**a**) antiproton beam intensity, ε_H measured using flying wires and relative peak density during mining and transfers to the Tevatron, (**b**) mid-2007 to present transfer efficiencies and the initial luminosities at the Tevatron. The top ten proton–antiproton stores in the Tevatron by their peak luminosity are also shown (by *diamonds*)

A few improvements have been added to the LMM explained above. For example, we found that it may be more advantageous to eliminate the high momentum bucket and keep a small amount of un-captured antiprotons freely moving in the ring to help transverse stability of the mined beam.

Another improvement, strip mining, is used whenever the Tevatron demands only a part of the antiproton stack. The stack is divided into two parts prior to the mining and the rest is quite similar to the process explained above.

Figure 4.21a shows the measured antiproton intensity, beam peak density, and ε_H with flying wires in the RR for a typical mining case during beam transfer to the Tevatron. The sudden increase in initial peak density represents the transition from the un-mined state to the mined state (see Fig. 4.20a, b). Then, the beam is held in the mined state for a minimum of 2 min with the electron cooling system turned on to cool the beam further. The cooling continues until the last transfer is taken out of the RR. A slow increase in the transverse emittance (see, e.g., Fig. 4.21a) is often observed—as possible cause of this emittance growth is due to a small transverse kick on the beam from the RR extraction kickers. However, the measurements on the extracted beam showed that the ε_T (95 %) ≈1.0 eVs and ε_T (95 %) ≈2.5 π-mm-mr for the 2.5 MHz bunch and the intensity is ~1/36 of the total extracted beam within a fluctuation of about 2.5 %.

Measured RR to MI and RR to Tevatron beam transfer efficiencies as a function of antiproton stack size for all the proton–antiproton stores since August 2007 are shown in Fig. 4.21b. The average transfer efficiency between the RR and MI was close to 100 % while the RR to Tevatron efficiency was about 83 % (with the highest efficiency of about 96 % from an initial stack size around 1.0×10^{12}antiprotons). A major contribution to this antiproton loss in the MI came from the bunch coalescing efficiency at 150 GeV. The initial proton–antiproton luminosities thus obtained in the Tevatron are shown by triangles. Data clearly show that during these years the Tevatron performance exceeded the Run II design luminosity goal of 270×10^{30}/cm^2/s by about 50 %.

References

1. K. Koba, *Status of the Fermilab Main Injector*, in Proceedings of the 2001 IEEE Particle Accelerator Conference, Chicago, 2001, p. 27
2. K. Koba, *Slip Stacking at Low Intensity – Status of the Beam Studies*. Fermilab Internal Note MI-0294 (February 2003)
3. K. Koba, *Slip Stacking Experiments at Fermilab Main Injector*, in Proceedings of the 2003 IEEE Particle Accelerator Conference, Portland, 2003, p. 1736
4. K. Seiya et al., *Status of Slip Stacking at the Fermilab Main Injector*, in Proceedings of the 2005 IEEE Particle Accelerator Conference, Knoxville, 2005, p. 347
5. J. Dey et al., *53 MHz Beam Loading Compensation for Slip Stacking in the Fermilab Main Injector*, in Proceedings of the 2005 IEEE Particle Accelerator Conference, Knoxville, 2005, p. 1958
6. I. Kourbanis, *Present and Future High Energy Accelerator for Neutrino Experiment*, in Proceedings of the 2007 IEEE Particle Accelerator Conference, Albuquerque, 2007, p. 731
7. K. Seiya et al., *Multi-Batch Slip Stacking in the Main Injector at Fermilab*, in Proceedings of the 2007 IEEE Particle Accelerator Conference, Albuquerque, 2007, p. 742
8. K. Seiya, *Slip Stacking*, in CARE-HHH-APD Workshop "BEAM'07", Geneva, 1–5 October 2007
9. J.E. Griffin et al., IEEE Trans. Nucl. Sci., **NS-30**, 3502 (1983)
10. A.G. Ruggiero, Preprint CERN 68–22, Intersecting Storage Rings Division, 4 June, 1968 (and references there in); A.G. Ruggiero, ISR-Th/67-68
11. C.M. Bhat, Preprint FERMILAB-CONF-06-102-AD (2006)
12. G. Jackson (Ed.). *Fermilab Recycler Ring Technical Design Report*. Fermilab-TM-1991 (1996)
13. G. Dome, CERN SPS/84-13(ARF) (1984)
14. S.Y. Lee, *Accelerator Physics (Chap. V)*, 1st edn. (World Scientific, Singapore, 1999), p. 305
15. S.Y. Lee, K.Y. Ng, Phys. Rev. E. **55**, 5992 (1997)
16. T. Sen, C.M. Bhat, J.-F. Ostiguy, Preprint FERMILAB-TM-2431-APC
17. J.A. MacLachlan, in *Proceedings of the International Conference on High Energy Accelerators (HEACC'98)*, XVII, Dubna, 184, 1998; J.A. MacLachlan, Fermilab Report No FN-529 (1989)

18. B. Chase et al., in *Proceedings of the ICALEPS'97* (1997) (and private communications); P. Joireman, B. Chase, K. Meisner (private communications)
19. J. P. Marriner (2001–2002, private communications)
20. J.E. Dey, D.W. Wildman, in *Proceedings of the 1999 IEEE Particle Accelerator Conference*, 1999, p. 869
21. C.M. Bhat, Fermilab Internal Notes MI Note 287, August 2002; Beams-Doc-376-v1 (2002)
22. J.P. Marriner, C.M. Bhat (private communications, 2002)
23. J. Dey et al., in *Proceedings of the 2003 IEEE Particle Accelerator Conference,* 2003, p. 1204
24. J. Crisp et al., in *Proceedings of the HB2006,* 2006, p. 244
25. C.M. Bhat, K.Y. Ng, Preprint FERMILAB-CONF-03-395-T (2003)
26. M. Hu et al., in *Proceedings of the 2007 IEEE Particle Accelerator Conference,* 2007, p. 458; N. Eddy et. al. (private communications, 2007)
27. D. Boussard, Preprint CERN SPS/86-11(ARF) (1986)
28. V. Balbekov, S. Nagaitsev, in *Proceedings of the EPAC04,* 2004, p. 791
29. C.M. Bhat, J. Marriner, in *Proceedings of the 2003 IEEE Particle Accelerator Conference,* 2003, p. 514
30. D.R. Broemmelsiek et al., in *Proceedings of the EPAC2004,* 2004, p. 794
31. A. Burov, Fermilab Internal Notes Beams-doc-2866 (2007); C.M. Bhat, Beams-doc-2884 (2007)
32. B. Chase (private communications, 2002)
33. C.M. Bhat, in *Proceedings of the 2003 IEEE Particle Accelerator Conference,* 2003, p. 2345; C.M. Bhat, *PAC2005*, p 1093; C.M. Bhat, Beams-Doc-2057-V1, December 2005; J. MacLachlan, FERMILAB-Conf-00/117; C.M. Bhat, *EPAC2008*, p. 307; "Stacking of antiprotons by morph-merging," P. Joireman, B. Chase (private communications 2007)
34. N. Eddy, J. Crisp, Preprint FERMILAB-Conf-06-100-AD; N. Eddy et al., AIP Conference Proceedings 868, p. 293 (2006); N. Eddy et al., in *Proceedings of the EPAC2008,* 2008, p. 3254
35. C.M. Bhat et al., in *Proceedings of the 2005 IEEE Particle Accelerator Conference,* 2005, p. 1763
36. P.C. Bhat, W.J. Spalding, in *Proceedings of the 15th Topical Conference on Hadron Collider Physics*, East Lansing, 2004, *AIP Conference Proceedings* 753:30 (2005); Run II Handbook, http://www-bd.fnal.gov/lug/runII_handbook/RunII_index.html
37. C.M. Bhat, Phys. Lett. A **330**, 481–486 (2004); C.M. Bhat, in *Proceedings of the EPAC2004,* 2004, p. 236; C.M. Bhat, Preprint FERMILAB-FN-746 (2004)

Chapter 5
Collective Instabilities in the Tevatron Collider Run II Accelerators

A. Burov, V. Lebedev, L. Prost, A. Shemyakin, V. Shiltsev, J. Steimel, and C.Y. Tan

5.1 Phenomenology of Beam Instabilities in the Tevatron Complex

High luminosity operation of the Tevatron during Collider Run II required high beam intensities all over the accelerator complex, and as a result, five out of six rings (except the Debuncher) had notable problems with beam stability. The instabilities of almost every type were present there: single and multibunch, transverse and longitudinal, due to electromagnetic interaction with vacuum chamber and due to interaction with ions stored in the beam, instabilities happened in both proton and antiproton beams. In many cases, various methods to suppress the instabilities have been implemented, including various damping systems—see Table 5.1. The most severe issues with serious impact on operations were related to transverse head-tail instability in the Tevatron, transverse beam instability in the Booster, instabilities in the Recycler antiproton beams, and longitudinal instabilities in the Tevatron.

5.1.1 Transverse Head-Tail Instability in the Tevatron

Transverse bunch weak head-tail instability was a serious limitation on the maximum bunch current in the Tevatron [1]. It manifested itself as a very fast (50–100 turns) development of vertical or horizontal oscillations and consequent beam loss on the aperture—see Fig. 5.1—accompanied by simultaneous emittance blowup of many bunches in the bunch train. For a long time, the only way to stabilize it was to

A. Burov • V. Lebedev (✉) • L. Prost • A. Shemyakin • V. Shiltsev (✉) • J. Steimel • C.Y. Tan
Fermi National Accelerator Laboratory, PO Box 500, Batavia, IL 60510, USA
e-mail: val@fnal.gov; shiltsev@fnal.gov

V. Lebedev and V. Shiltsev (eds.), *Accelerator Physics at the Tevatron Collider*, Particle Acceleration and Detection, DOI 10.1007/978-1-4939-0885-1_5,

Table 5.1 Instabilities and cures in the Tevatron Run II accelerators

	Instability nature		Dampers	
	Longitudinal	Transverse	Longitudinal	Transverse
Booster	Multibunch, cavities	Laminated wall, space charge	Narrow band	–
Main injector	Multibunch	Multibunch	Bunch by bunch	Bunch by bunch
Accumulator	Stochastic cooling	Due to stores ions	–	Wideband
Recycler	–	Head-tail, resistive wall	–	Wideband
Tevatron	"Bunch dancing"	Head-tail, resistive wall	Bunch by bunch, protons only	Bunch by bunch, protons only

operate Tevatron with high linear chromaticity in both planes $Q'_{x,y} > 10$. High chromaticity values led to short beam lifetime especially in the presence of opposite beam (see Sect. 8.2).

Transverse bunch-by-bunch dampers (vertical and horizontal) were built and commissioned in 2002–2003 [2] and allowed to keep the beam stable at lower chromaticities at the injection energy of 150 GeV (see table below).

Originally	2002	$Q' \sim 10–16$
V/H dampers installed	2003	$Q' \sim 5–8$
Lambertsons liner	2004	$Q' \sim 3–5$
V-damper fights H-damper	Dec 2004	$Q' \sim 8–10$
Octupoles commissioned	Feb 2005	$Q' \sim 0–3$

As seen from the above, other less operationally challenging methods were later employed to secure the beam stability, namely, (a) 0.4 mm thin conductive CuBe liners being installed inside Lambertson magnets that reduced the total Tevatron transverse impedance from $Z_\perp \approx 5–2.4$ MOhm/m to about 1 MOhm/m [1] and (b) commissioning of new circuits of octupoles [3] which generated additional tune spread in the beams and eventually allowed the reduction of chromaticity to a few (0–3) units at 150 GeV and improved beam lifetime to better than 20 h. In December 2004, it was observed that vertical and horizontal dampers "fought" each other—so, one of them had to be turned off (and correspondingly, Q' in that plane had to be increased) in order to let the other one work. A lot of effort was put into investigation of the phenomena—the leading hypothesis was that it is due to local coupling—but there was no satisfactory resolution. So, as soon as the new octupole circuits were operational, the dampers were disabled.

5.1.2 Coherent Synchro-betatron Resonance in the Booster

Booster is a fast cycling proton synchrotron operating at 15 Hz. To exclude the eddy currents excited in the vacuum chamber by fast changing magnetic field,

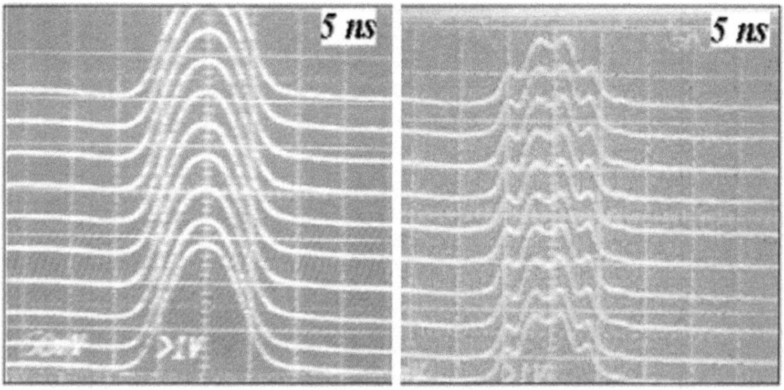

Fig. 5.1 Oscilloscope traces of the longitudinal density profiles of (**a**) the initial ($N_p = 2.6 \times 10^{11}$) and (**b**) remaining ($N_p = 2.6 \times 10^{11}$) 150 GeV proton bunches before and after vertical $l = 2$ weak head-tail instability [1]

its vacuum chamber is formed by poles of laminated combined function dipoles. That addresses the problems related to the eddy currents but greatly contributes to the transverse and longitudinal impedances. In particular, the transverse impedances achieve values of about 100 MΩ/m (see details in Sect. 5.2) adversely affecting beam stability. In operation, the instabilities are suppressed by large chromaticities, $Q'_{x,y} \sim 10$–16. That results in deterioration of the dynamic aperture and the beam lifetime. Our attempts to stabilize the instability with transverse feedback system carried out before and in the course of Tevatron Run II were unsuccessful.

The first detailed studies of beam stability were carried out in 2005 and were mostly devoted to the beam stability at the injection energy [4]. They exhibited that at reduced chromaticity the head-tail motion develops extremely fast with growth time of about 12 turns at nominal Booster intensity and about 14 turns at the half of nominal intensity. For both cases the fractional part of head-tail betatron frequency was close to zero.

Figure 5.2 presents parameters of bunches for the first 150 turns after injection starts at the nominal beam intensity. The process looks as following. First, an injection orbit bump is created just before the injection. Then, the linac beam is injected over several turns (1–11). When the injection is finished the orbit bump is switched off (it takes 10–20 turns) and RF voltage is adiabatically increased causing the beam to be bunched at turn 70 with RF voltage continuing to grow. The first sign of the instability appears at turn 80 causing the beam intensity drop after turn 100. The results of the measurements [4] demonstrated that the instability develops only after bunches are formed, and that its growth rate is weakly dependent on the beam intensity. Later analysis [5] showed that the observed behavior corresponds to the coherent synchro-betatron resonance which develops when the synchro-betatron tune is close to an integer: $Q_{x,y} + 3Q_s \approx 7$. Note the synchrotron tune at injection is quite high $Q_s \approx 0.07$ and the resonance happens even if working tunes are quite

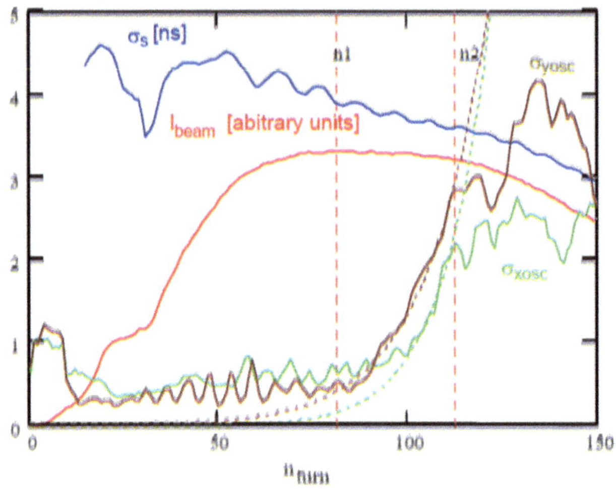

Fig. 5.2 Changes of beam parameters during first 150 turns after injection in the Booster; *red line* AC beam intensity, *blue line* rms bunch length, *vertical* (*brown line*) and *horizontal* (*green line*) rms transverse dipole moments. *Dotted lines* present exponential fits to the rms dipole moments inside a bunch. All signals are averaged over all 84 bunches; total proton intensity is $4.5 \cdot 10^{12}$ [4]

far from the integer resonance. Large value of incoherent betatron tune shift due to space charge tune ($\delta Q_{SC} \approx -0.35$) pushes the bare tunes being above 6.85 making impossible to avoid the resonance in the course of adiabatic bunching. The variation of the growth time from 18 to 12 turns for intensity variation in the range $[1\text{–}5] . 10^{12}$ protons per beam is related to the interference between the synchro-betatron coherent resonance and the head-tail multi-bunch instability related to the large transverse impedance. In the absence of the synchro-betatron resonance, the instability growth time in vicinity of injection energy and small chromaticity is about 100 turns.

5.1.3 Transverse Instability in High-Brightness Antiproton Beam in Recycler

The Recycler ring (RR) is the last (third) ring in a chain of antiproton cooling and stacking stages. Transverse instabilities in RR have been theoretically studied during its design but were deemed a marginal issue for the maximum number of antiprotons that were expected to be stored at any time ($<250 \times 10^{10}$). With strong electron cooling and up to 5×10^{12} stored antiprotons, much brighter beams than initially anticipated are generated. As a result, emittances of the cooled beam are limited by a transverse resistive wall instability. (An ion-capture-driven instability was identified very early in the Recycler operation and was eliminated with clearing

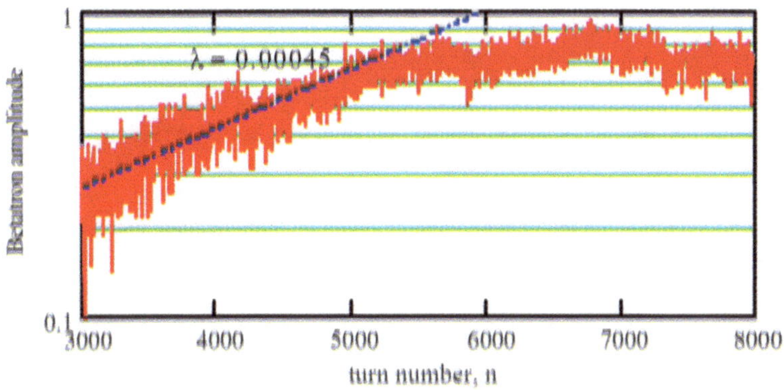

Fig. 5.3 Growth of the betatron oscillations in high-brightness antiproton beam in Recycler without dampers [8]

electrodes and the fact that the stored beam was bunched.) A damper system was installed in 2005 with an initial bandwidth of 30 MHz and eventually upgraded to 70 MHz. Nevertheless, several instabilities were observed during normal operation and prompted studies to better understand their nature and characteristics, as well as to limit their occurrences [6].

A typical instability is characterized by three phenomena: a large and sharp increase of the damper kickers' amplitudes (in particular, the vertical damper kicker), a fast increase of the emittances (mostly vertical) as measured by the Schottky detectors, and a relatively slow beam loss. The instability lasts for 5–15 s, and accordingly, the beam loss is slow. The antiproton emittances measured by the flying wires are almost unaffected by the instability, thus, indicating that this is mostly the tail particles that suffer from the instability and are being lost to the aperture. That observation is also consistent with the general picture of the Landau damping [7]. Without the dampers (or with malfunctioning dampers), most of the beam loss and the emittance blowup happen in <0.1 s (see Fig. 5.3). There is very little motion in the head of a ~6 μs bunch and maximum oscillations occur in the trailing half of the bunch (at some ~2/3 of the bunch length).

To support high luminosity operation of the collider complex, the highest possible effective phase density D_{95} of antiprotons is required. It is defined as

$$D_{95} = \frac{N_{\mathrm{a}}}{4\varepsilon_{\mathrm{L}} 6\varepsilon_{\mathrm{T}}}, \tag{5.1}$$

where N_{a} is the total number of antiprotons in the Recycler (in the units of 10^{10}), ε_{L} is the rms longitudinal emittance (in $\mathrm{eV} \cdot \mathrm{s}$), and ε_{T} is the rms normalized transverse emittance (in μm). The numerical factors are chosen to simplify calculations for widely used "95 % emittance" values for the Gaussian distributions. Without

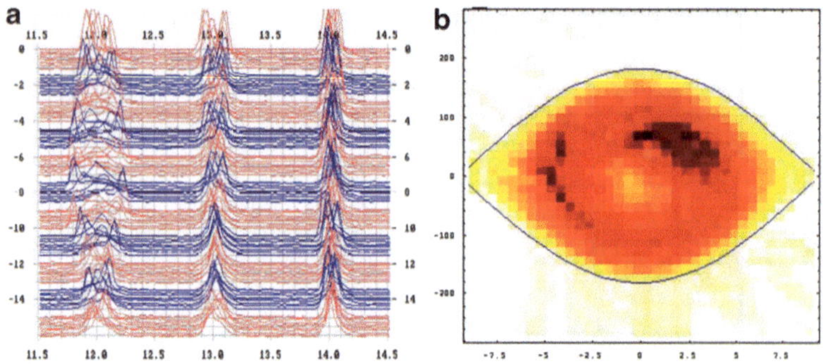

Fig. 5.4 (**a**) "Waterfall" plot of three "dancing" uncoalesced bunches of 150 GeV protons some 19 ns apart; (**b**) tomography of the longitudinal phase space of a high intensity proton bunch in the Tevatron [9]

feedback systems, the operationally achievable factor was limited to $D_{95} < 1.0$, while it did reach as high as 2.6 with 30 MHz bandwidth damper and up to $D_{95} \approx 4.4$ after commissioning of the 70 MHz bandwidth damper system [6].

5.1.4 Longitudinal Instabilities in the Tevatron

The phenomenon of "dancing bunches" in Tevatron refers to notable longitudinal single bunch and coupled bunch instabilities in the proton beam [9, 10]. At the Tevatron injection energy of 150 GeV, large (up to 1 rad) RF beam phase oscillations in high intensity beams can persist for many minutes (see Fig. 5.4, [9]).

The biggest concern for operations was that the "dancing bunches" result in slow bunched beam intensity loss and increase of the "DC beam" intensity (uncaptured particles out of sync with the RF system) which is lost at the start of acceleration. Another manifestation was the regular occurrence of large longitudinal bunch oscillations at 980 GeV energy accompanied by significant longitudinal emittance increase, reduction of luminosity, beam losses, and accumulation of particles in the abort gaps—all of which are very dangerous for operations (see Fig. 5.5). To counteract that, a longitudinal bunch-by-bunch damper was designed, built, installed, and commissioned in the Tevatron in 2002 [11]. Since then, the damper is in operation for every store all the times except the energy ramp. It effectively suppresses both the "dancing bunches" and the single and coupled bunch instabilities. It was found that to be effective, the damper gain should vary slowly during the store in a fashion which tracks proton bunch intensity and bunch length gain ~ N/σ_z. Unfortunately, from time to time, the instability still occurred. The cause of these outbreaks of the instability has not been fully understood. At the later years of the Collider Run II, similar phenomena started to appear in high intensity antiproton beam as well, corresponding damper system has been installed but has not been commissioned due to lack of time.

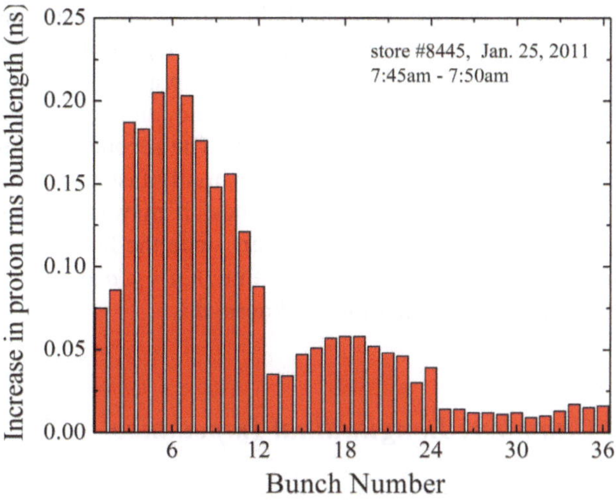

Fig. 5.5 Increase in the rms length of individual proton bunched during the occurrence of longitudinal instability over the course of the HEP store #8445 (from 7:45 am to 7:50 am on January 25, 2011). Before the blowup, the rms bunch length of all the bunches was 2.04 ± 0.03 ns

5.2 Impedances of Laminated Vacuum Chambers

Below, the longitudinal and transverse impedances are derived for round and flat laminated vacuum chambers, and results applied to the Fermilab Booster.

First publications on impedance of laminated vacuum chambers are related to the early 1970s: those are of S.C. Snowdon [12] and of A.G. Ruggiero [13]; 15 years later, a revision paper of R. Gluckstern appeared [14]. All the publications were presented as Fermilab preprints, and that is of no surprise as the Fermilab Booster has its laminated magnets open to the beam. Being in a reasonable mutual agreement, these publications were all devoted to the longitudinal impedance of round vacuum chambers. The transverse impedance and the flat geometry case were addressed in more recent paper of K.Y. Ng [15]. The latest computer calculations of A. Macridin et al. [16] revealed some disagreement with [15] that stimulated further theoretical investigation presented below, which ended up with results in agreement with [16].

Some general conditions are assumed here. First, the frequencies under interest, ω, are supposed to be sufficiently low [17, 18]:

$$\begin{aligned} \omega &<< \gamma\beta c/a \\ \omega &<< 4\pi\sigma/\varepsilon \end{aligned} \tag{5.2}$$

where a is the aperture radius, γ and β are the relativistic factors, c is the speed of light, and σ and ε are the chamber conductivity and dielectric constant. The first condition actually requires the wavelength of the fields to be much longer than the aperture, as they are seen in the beam frame. Note that the specified wavelength parameter $\gamma\beta c/(a\omega)$ is relevant to the wake forces, not to the electric and magnetic fields taken separately. For the separate field components, the relativistic factor does not count; but it does count for the wakes (see, e.g., [19], Eq. (2.41)).

The above condition seems to be satisfied for all practically interesting cases. It allows one to neglect the longitudinal magnetic field, and consequently, the transverse components of the vector potential vanish. The second condition means that the beam electric moments are shielded infinitesimally fast at the chamber surface. While this condition is well satisfied for metals, it may be violated for ferrites [20]. The last case is irrelevant to this chapter, since the laminations are metallic (iron). We also imply that the laminations are thin: $h,d << a$ and that the skin depth, δ, is much smaller than the lamination thickness, d.

5.2.1 Flat Chamber: Longitudinal Impedance

Let the beam current be modulated at a frequency ω:

$$I(\mathbf{r},t) = I_0\delta(\mathbf{r}_\perp)\exp(-i\omega(t - z/v)). \tag{5.3}$$

Due to the horizontal homogeneity, the problem can be solved by the Fourier transform over this coordinate

$$F(x) = \int_{-\infty}^{\infty} F_k \exp(ik_x x)\frac{dk_x}{2\pi}. \tag{5.4}$$

Since only the Fourier components are used below, the subscript k can be safely omitted. For long wavelength, the vector potential reduces to its longitudinal component only. In the free space, it satisfies the transverse Laplace equation and can be presented as

$$A = \frac{I_0 Z_0}{2k_x}\left[\exp(-k_x y) - G\frac{\cosh(k_x y)}{\cosh(k_x a)}\exp(-k_x a)\right]; \quad k_x > 0,\ \ 0 < y < a, \tag{5.5}$$

where $Z_0 = 4\pi/c = 377\ \Omega$, a is the half gap (see Fig. 5.6), $G = G(k_x)$ is the function to be determined from the boundary conditions, and the vector potential is an even function of k_x and y. The first term inside the square brackets describes a direct field of the beam, while the second one is the response due to the induced currents. From here, a ratio of the magnetic fields follows:

$$\left.\frac{H_y}{H_x}\right|_{y=a-0} = i\frac{1 - G}{1 + G\tanh(k_x a)}. \tag{5.6}$$

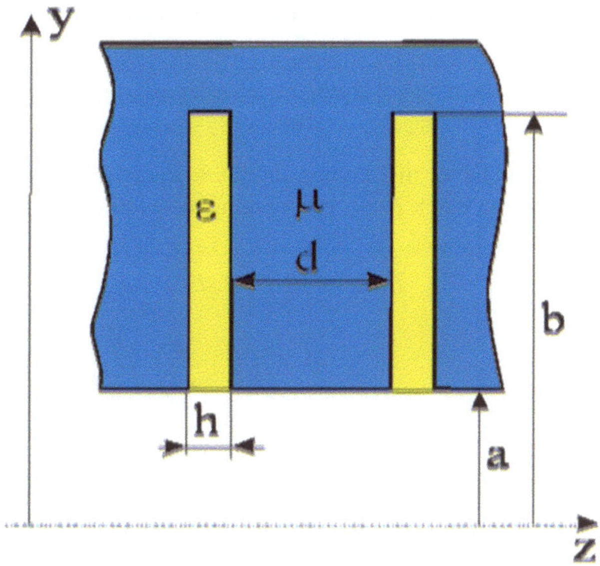

Fig. 5.6 Geometry of the laminated vacuum chamber

Using the boundary conditions at the metal surface, one can easily prove that the vector potential inside a thin crack satisfies the Helmholtz equation:

$$\Delta_{\perp} A^{\text{crack}} = -k^2 A^{\text{crack}}, \tag{5.7}$$

where

$$\begin{aligned} k^2 &\equiv \frac{\omega^2 \varepsilon}{c^2}\left(1+\frac{2\mu}{\kappa h}\right) \equiv \frac{\omega^2 \varepsilon}{c^2} + g^2; \quad \kappa \equiv \frac{1 - i\,\text{sgn}\omega}{\delta} \\ &\equiv (1 - i\,\text{sgn}\omega)\frac{\sqrt{2\pi|\omega|\sigma\mu}}{c}. \end{aligned} \tag{5.8}$$

Note that Eq. (5.7) is only justified if $gh/2 << 1$. In that case the fields inside the crack can be treated as independent from the z-coordinate (coordinate normal to its surface). Otherwise one need to take into account that the fields in the crack are dependent on z as $\cosh(gz)$ or $\sinh(gz)$, resulting in a more complicated form for Eq. (5.7). In most practical cases, the thin crack approximation is valid. Taking into account that the crack is shorted at $y = b$, the fields can be written inside the crack as

$$\begin{aligned} A^{\text{crack}} &= A_0 \sin\left(k_y(b-y)\right); \quad k_y = \sqrt{k^2 - k_x^2}; \\ H_x^{\text{crack}} &= -k_y A_0 \cos\left(k_y(b-y)\right); \\ H_y^{\text{crack}} &= -ik_x A_0 \sin\left(k_y(b-y)\right). \end{aligned} \tag{5.9}$$

A vertical magnetic flux through the metal surface is

$$\int B_y dz\Big|_{\text{metal}} = -2ik_x\mu A_0 \sin\left(k_y(b-y)\right)/\kappa, \tag{5.10}$$

where the factor of 2 comes out due to the two sides of the lamina. Adding the flux through the crack itself, one obtains the average magnetic field, which is

$$\begin{aligned}\overline{B}_y &\equiv \frac{1}{d+h}\left(\int B_y dz\Big|_{\text{crack}} + \int B_y dz\Big|_{\text{metal}}\right) \\ &= -ik_x A_0\left(1+\frac{2\mu}{\kappa h}\right)\frac{h}{d+h}\sin\left(k_y(b-y)\right),\end{aligned} \tag{5.11}$$

yielding

$$\frac{\overline{B}_y}{H_x^{\text{crack}}}\Bigg|_{y=a+0} = \frac{ik_x}{k_y}\frac{h}{d+h}\left(1+\frac{2\mu}{\kappa h}\right)\tan\left(k_y(b-a)\right) \equiv iR_B. \tag{5.12}$$

The condition $y=a+0$ means staying vertically at $y=a+\Delta y$ so that $H,\delta << \Delta y << 1/k$. Similarly, $y=a-0$ means $y=a-\Delta y$. Since both the average magnetic field, Eq. (5.10), and the horizontal field at the crack region are preserved at crossing the magnet border $y=a$, their ratio is preserved as well:

$$\frac{\overline{B}_y}{H_x^{\text{crack}}}\Bigg|_{y=a+0} = \frac{H_y}{H_x}\Bigg|_{y=a-0}. \tag{5.13}$$

Thus, Eqs. (5.11) and (5.5) lead to the induced field amplitude

$$G = \frac{1-R_B}{1+R_B\tanh(k_x a)}. \tag{5.14}$$

At this point, only an average electric field has to be found. To do that, the Maxwell equation

$$-\frac{\partial E_z}{\partial x} + \frac{\partial E_x}{\partial z} = i\frac{\omega}{c}B_y \tag{5.15}$$

can be averaged over a period, yielding

$$\overline{E}_z\Big|_{y=a+0} = -\frac{\omega}{ck_x}\overline{B}_y\Big|_{y=a+0}. \tag{5.16}$$

Average electric fields above and below the boundary (with a thickness of Δy) are related as

$$\overline{E}_z\Big|_{y=a+0} - \overline{E}_z\Big|_{y=a-0} = i\frac{\omega\mu}{\kappa c} H_x\Big|_{y=a-0}. \tag{5.17}$$

Using Eq. (5.14), the horizontal field is found:

$$H_x\Big|_{y=a-0} = \frac{\partial A}{\partial y} = -\frac{Z_0 I_0}{2\cosh(k_x a)} \frac{1}{1 + R_B \tanh(k_x a)}. \tag{5.18}$$

Finally Eqs. (5.18), (5.16), and (5.11) yield the following result:

$$Z_{\|} = -\frac{1}{I_0}\int_0^\infty \overline{E}_z\Big|_{y=0} \frac{dk_x}{\pi} = -i\frac{\omega}{c}\frac{Z_0}{2\pi}\int_0^\infty \frac{dk_x}{k_x} \frac{R_B + \mu k_x/\kappa}{\cosh^2(k_x a)(1 + R_B \tanh(k_x a))}, \tag{5.19}$$

for the longitudinal impedance per unit length. Here we used that

$$\overline{E}_z\Big|_{y=0} = \frac{\overline{E}_z\Big|_{y=a-0}}{\cosh(k_x a)}. \tag{5.20}$$

5.2.2 Flat Chamber: Transverse Impedances

For the horizontal beam oscillations, the vector potential is an even function of the vertical coordinate and odd one of the horizontal; according to [17]:

$$A = -i\frac{D_0 Z_0}{2}\left[\exp(-k_x y) - G\frac{\cosh(k_x y)}{\cosh(k_x a)}\exp(-k_x a)\right];\ k_x > 0,\ 0 < y < a, \tag{5.21}$$

with D_0 as the amplitude of the beam dipole moment oscillations. Note that this field differs from the longitudinal case, Eq. (5.4), only by the amplitude; thus, all the field ratios remain the same. In particular, Eq. (5.14) is valid for this case as well. Using Eqs. (10, 11) of [17], the horizontal impedance follows:

$$\begin{aligned} Z_x &= Z_x^\sigma + Z_x^\infty \\ &= -i\frac{Z_0 \beta}{2\pi}\int_0^\infty \frac{(R_B + \mu k_x/\kappa) k_x dk_x}{\cosh^2(k_x a)(1 + R_B \tanh(k_x a))} - i\frac{Z_0}{2\pi a^2 \beta\gamma^2}\frac{\pi^2}{24}. \end{aligned} \tag{5.22}$$

The vertical impedance can be found from the horizontal by a substitution $\cosh(k_x a) \leftrightarrow \sinh(k_x a)$ in the finite conductance term Z_x^σ and taking twice higher infinite conductivity term [17]

$$
\begin{aligned}
Z_y &= Z_y^{\sigma} + Z_y^{\infty} \\
&= -i\frac{Z_0\beta}{2\pi}\int_0^{\infty}\frac{(R_B + \mu k_x/\kappa)k_x dk_x}{\sinh^2(k_x a)(1 + R_B\coth(k_x a))} - i\frac{Z_0}{2\pi a^2\beta\gamma^2}\frac{\pi^2}{12}.
\end{aligned}
\tag{5.23}
$$

Note that the second terms in the integrand numerator in Eqs. (5.19), (5.22), and (5.23) ($\mu k_x/\kappa$) yield the conventional resistive wall impedances when the crack width approaches zero.

5.2.3 Round Chamber: Longitudinal Impedance

For a round vacuum chamber of radius a and arbitrary walls, the axially symmetric fields in the free space are related so that (Eq. (2.3) in [19])

$$
H_{\varphi} = \frac{2I_0}{rc} - i\frac{\omega r}{2c}E_z,\ r < a. \tag{5.24}
$$

From here, the longitudinal impedance $Z_{\|}$ can be related to the so-called surface impedance R:

$$
Z_{\|} = -\frac{E_z}{I_0} = \frac{Z_0}{2\pi a}\frac{R}{1 - i\omega a R/(2c)};\quad R \equiv -E_z/H_{\varphi}\big|_{r=a-0}. \tag{5.25}
$$

The Maxwell equation $\nabla \times \mathbf{E} = i\omega\mathbf{B}/c$, applied to the azimuthal direction, relates inner and outer average longitudinal electric fields (compare with Eq. (5.17)):

$$
\overline{E}_z\big|_{r=a+0} - \overline{E}_z\big|_{r=a-0} = -i\omega\mu H_{\varphi}/(\kappa c). \tag{5.26}
$$

This can also be written as

$$
R = R_+ - i\frac{\omega\mu}{\kappa c};\quad R_+ \equiv -\frac{\overline{E}_z\big|_{r=a+0}}{H_{\varphi}} \tag{5.27}
$$

Inside the crack, the longitudinal electric field satisfies the Helmholtz equation (compare with Eq. (5.6)):

$$
\begin{aligned}
\Delta_{\perp}E_z^{\text{crack}} &= -k^2 E_z^{\text{crack}}; \\
H_{\varphi}^{\text{crack}} &= i\frac{\omega\varepsilon}{k}\frac{\partial E_z^{\text{crack}}}{\partial(kr)}.
\end{aligned}
\tag{5.28}
$$

From here, the field components are expressed in terms of the Hankel functions:

$$\begin{aligned} E_z^{\text{crack}} &= E_0 \left[H_0^{(1)}(kr) H_0^{(2)}(kb) - H_0^{(2)}(kr) H_0^{(1)}(kb) \right]; \\ H_\varphi^{\text{crack}} &= -i \frac{\omega \varepsilon}{ck} E_0 \left[H_1^{(1)}(kr) H_0^{(2)}(kb) - H_1^{(2)}(kr) H_0^{(1)}(kb) \right]. \end{aligned} \tag{5.29}$$

A factor cosh(gz) is omitted according to the assumption $gh/2 << 1$. Since there is no longitudinal electric field in the metal, only the crack electric field contributes to its average:

$$\overline{E}_z \big|_{r=a+0} = E_z^{\text{crack}} \frac{h}{d+h}. \tag{5.30}$$

Together with Eq. (5.29), this yields

$$\begin{aligned} R_+ &\equiv -\frac{\overline{E}_z}{H_\varphi^{\text{crack}}} \Bigg|_{r=a+0} \\ &= -i \frac{ckh}{\omega \varepsilon (d+h)} \frac{H_0^{(1)}(ka) H_0^{(2)}(kb) - H_0^{(2)}(ka) H_0^{(1)}(kb)}{H_1^{(1)}(ka) H_0^{(2)}(kb) - H_1^{(2)}(ka) H_0^{(1)}(kb)}. \end{aligned} \tag{5.31}$$

With Eq. (5.27), the impedance in Eq. (5.25) follows:

$$Z_{||} = \frac{Z_0}{2\pi a} \frac{R_+ - i\omega\mu/(\kappa c)}{1 - i\frac{\omega a R_+}{2c} - \frac{\omega^2 a \mu}{2\kappa c^2}}, \tag{5.32}$$

where the second term in the numerator is responsible for the conventional resistive wall impedance when the cracks disappear.

5.2.4 *Round Chamber: Transverse Impedances*

For the transverse dipole oscillations, the vector potential in the free space can be written as

$$A = \frac{2D_0}{ca} \left(\frac{a}{r} - G\frac{r}{a} \right) \cos\varphi \equiv A_0 \left(\frac{a}{r} - G\frac{r}{a} \right) \cos\varphi, \tag{5.33}$$

where D_0 is the amplitude of the dipole moment oscillations. In terms of the induced field amplitude G, the transverse impedance is expressed as [18]

$$Z_\perp = Z_\perp^\sigma + Z_\perp^\infty = -i \frac{Z_0 \beta (1-G)}{2\pi a^2} - i \frac{Z_0}{2\pi a^2 \beta \gamma^2}. \tag{5.34}$$

At the inner border, $r = a - 0$, the longitudinal electric and azimuthal magnetic fields follow as

$$\begin{aligned} E_z &= i\omega A/c = i\omega A_0(1-G)\cos\varphi/c; \\ H_\varphi &= -\partial A/\partial r = A_0(1+G)\cos\varphi/a. \end{aligned} \tag{5.35}$$

This relates the surface impedance $R = -E_z/H_\varphi|_{r=a-0}$ and the induced field amplitude G

$$R = -i\frac{\omega a}{c}\frac{1-G}{1+G} \Leftrightarrow 1-G = \frac{2R}{R - i\omega a/c}. \tag{5.36}$$

Note that although the fields E_z, H_φ, etc. and their ratios R, R_+ are denoted by the same symbols for the longitudinal and the transverse cases, they are not the same and should not be confused. Inside the crack, the field components $E_z^{\text{crack}}, H_\varphi^{\text{crack}}$ satisfy Eq. (5.28), leading for the dipole mode to

$$\begin{aligned} E_z^{\text{crack}} &= E_0\left[H_1^{(1)}(kr)H_1^{(2)}(kb) - H_1^{(2)}(kr)H_1^{(1)}(kb)\right]\cos\varphi; \\ H_\varphi^{\text{crack}} &= i\frac{\omega\varepsilon}{ck}E_0\left[H_1^{(1)\prime}(kr)H_1^{(2)}(kb) - H_1^{(2)\prime}(kr)H_1^{(1)}(kb)\right]\cos\varphi. \end{aligned} \tag{5.37}$$

For the calculations, it is useful to remember the derivatives of the Hankel functions are expressed as

$$H_1'(x) = [H_0(x) - H_2(x)]/2. \tag{5.38}$$

Equation (5.37) yields the field ratio

$$\begin{aligned} R_+ &\equiv -\left.\frac{\overline{E}_z}{H_\varphi^{\text{crack}}}\right|_{r=a+0} \\ &= i\frac{ckh}{\omega\varepsilon(d+h)}\frac{H_1^{(1)}(ka)H_1^{(2)}(kb) - H_1^{(2)}(ka)H_1^{(1)}(kb)}{H_1^{(1)\prime}(ka)H_1^{(2)}(kb) - H_1^{(2)\prime}(ka)H_1^{(1)}(kb)}. \end{aligned} \tag{5.39}$$

With Eqs. (5.27) and (5.36), this formula yields the transverse impedance Eq. (5.34)

$$\begin{aligned} Z_\perp &= Z_\perp^\sigma + Z_\perp^\infty = -i\frac{Z_0\beta}{\pi a^2}\frac{R}{R - i\omega a/c} - i\frac{Z_0}{2\pi a^2\beta\gamma^2}; \\ R &= R_+ - i\frac{\omega\mu}{\kappa c} = i\frac{ckh}{\omega\varepsilon(d+h)}\frac{H_1^{(1)}(ka)H_1^{(2)}(kb) - H_1^{(2)}(ka)H_1^{(1)}(kb)}{H_1^{(1)\prime}(ka)H_1^{(2)}(kb) - H_1^{(2)\prime}(ka)H_1^{(1)}(kb)} - i\frac{\omega\mu}{\kappa c}. \end{aligned} \tag{5.40}$$

5.2.5 Impedances of the Booster Laminated Chamber

It would be good to discuss the impedances on a base of real parameters of the Booster magnets. However, some of the important parameters are actually unknown. While the inner and outer aperture a and b as well as the lamina thickness d are perfectly known, we have a poor knowledge of the magnetic permeability μ at the interesting frequency range of hundreds MHz. Moreover, the guiding magnetic field makes that value not just a function of frequency but a tensor function. Another uncertainty relates to the crack width h. Comparison of the average lamina thickness with the entire length of the magnet gives only a magnet-average value for h. There is no reason to assume that these values have a narrow distribution near their average. Ideally, the calculated impedances have to be averaged over this distribution—but it cannot be done even approximately without knowing the rms spread of the crack widths. One more uncertainty relates to thickness of the iron oxide at the lamina surfaces, which may change the crack properties. All these uncertainties can be reduced with a set of dedicated measurements, and some of them are reported in [21]. Longitudinal impedances for the Booster focusing magnet (see its parameters in Table 5.2) calculated using Eq. (5.19), (5.32) are shown in Fig. 5.7 and are in a good agreement with the measurements of [21].

Several features of Fig. 5.7 deserve to be noted:

1. The low limit of the frequency range is determined by the skin depth: at 10 kHz $\delta \approx d/2$.
2. At low frequencies, $f \ll 50$ MHz, a simplistic electrotechnical approximation $Z_{\|} = \frac{\kappa}{\pi d \sigma} \ln(b/a) = Z_{\|}^{\mathrm{conv}} \frac{2a}{d} \ln(b/a) \propto \omega^{1/2}$ for the round geometry coincides with the actual solution. In the case of flat aperture, the low-frequency impedance scaling is different, $Z_{\|} \propto \omega^{3/4}$.
3. Note that impedance of the conventional solid vacuum chamber $Z_{\|}^{\mathrm{conv}} = \frac{\kappa}{2\pi a \sigma}$ exceeds the careless limit $|Z_{\|}/n| \leq Z_0/2$ [19] by a factor of $(\mu\delta/a)\ln(b/a)$. For $\mu >> 1$ this can be a big number. The reason is that the field energy located inside the magnetic chamber grows unlimitedly with the magnetic permeability: $\frac{\mu H^2}{8\pi} 2\pi a \delta \propto \sqrt{\mu}$.
4. A limit for the low-frequency approximation is determined by the field decay along the crack depth, $\mathrm{Im}k \propto \omega^{3/4}$ (see Eq. (5.7)). At sufficiently high frequency, when $\mathrm{Im}\, kb >> 1$, this radial field decay limits the length of the shielding current along the crack surface before it reaches the outer shortcut radius b. At $f > 1$ GHz, $\mathrm{Im}ka \geq 1$, so the path length of the shielding current gets proportional to the field decay length $\mathrm{Im}k$, leading to $Z_{\|} \approx Z_{\|}^{\mathrm{conv}} \frac{2}{d\mathrm{Im}k} \propto \omega^{-1/4}$.
5. For usual, not-laminated vacuum chambers, the longitudinal impedance of the flat chamber is known to be equal to one of the round chamber [22, 23]. In other words, the longitudinal Yokoya factor of the solid flat chamber or the ratio of flat-to-round impedances is 1. As it is seen from Fig. 5.7, the Yokoya factor of the flat laminated chamber is close to 1 at $f \geq 10$ MHz, while at lower frequencies, it may be significantly smaller.

Table 5.2 Accepted parameters of the Booster F-magnet

a	Magnet half gap	2.08 cm
b	Outer shortcut	16.5 cm
d	Lamina thickness	0.064 cm
h	Crack width	0.002 cm
σ	Conductivity	$4.5 \cdot 10^{16}$ 1/s
ε	Dielectric permittivity	4.75
μ	Magnetic permeability	50

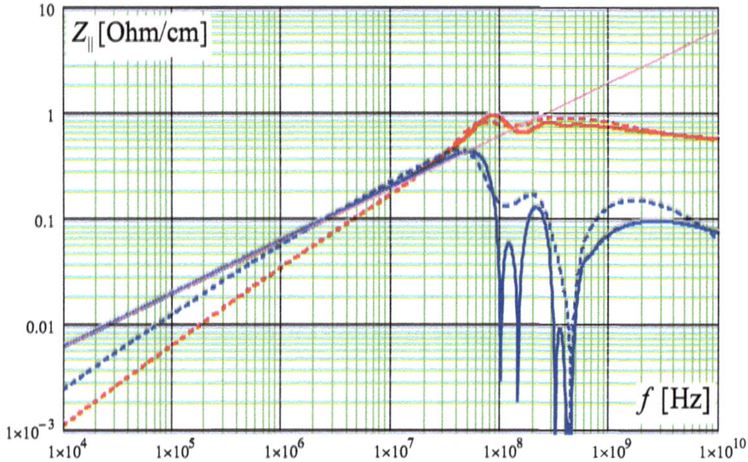

Fig. 5.7 Longitudinal impedances for the round (*solid lines*) and flat (*dash lines*) geometries. *Red lines* are for the real parts, and *blue* for the absolute value of the imaginary parts. The *magenta line* shows the low-frequency approximation $\mathrm{Re}Z_{\|}^{\mathrm{LF}} = \frac{1}{\pi d\sigma\delta}\ln(b/a) = \mathrm{Re}Z_{\|}^{\mathrm{conv}}\frac{2a}{d}\ln(b/a)$ with $Z_{\|}^{\mathrm{conv}} = \frac{\kappa}{2\pi a\sigma}$ as longitudinal impedance of the conventional solid round vacuum chamber of the same metal

The transverse impedances are presented in Fig. 5.8. There are several reasons for the complicated behavior of the transverse impedances. First, the depth of field penetration inside the crack changes at $ka > 1$. Above that frequency (~1 GHz), the shielding current path length is determined by the decay along the crack, while below that it is determined by the aperture a.

The second reason is change of the field structure at $|Rc/(\omega a)| \sim 1$, equivalent to $\mu\delta/d \sim 1$ or $f \sim 10$ MHz. At low frequencies, when $\mu\delta/d > 1$, the fields inside the free space, $r < a$, are of the magnetic type: the magnetic field is almost orthogonal to the magnet surface, $|H_\phi/H_r|_{r=a-0} \sim 1$. In the opposite case, for $\mu\delta/d < 1$, the fields are close to those of the conducting wire: $|H_\phi/H_r|_{r=a-0} >> 1$. Interplay of these and some geometrical factors leads to variety of possibilities for impedance behavior at low frequencies seen in Fig. 5.8. Note, contrary to the longitudinal impedance, the transverse one never exceeds its careless limit $Z_0\beta/(\pi a^2)$. That is why a popular Panofsky-Wenzel estimation of the transverse impedance from the longitudinal is

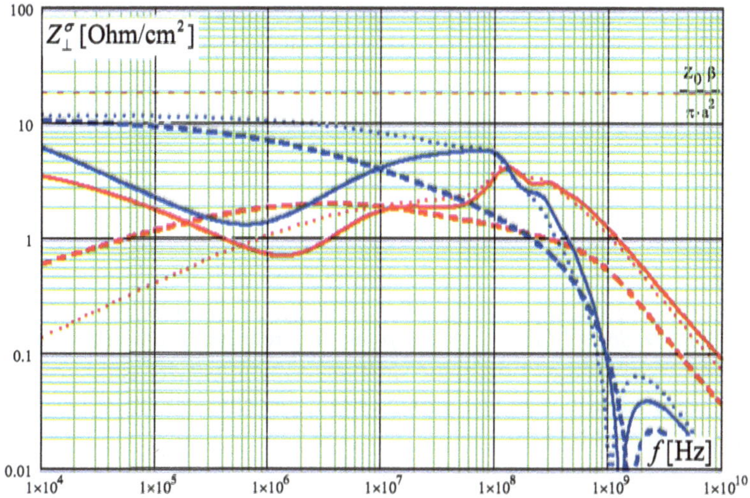

Fig. 5.8 Transverse impedances ($\gamma \to \infty$) for the round (*solid lines*) and flat geometry (*dash lines* for the horizontal and *dot lines* for the vertical). *Red lines* are for the real parts, and *blue* for the absolute value of the imaginary parts

inapplicable here: its use at low frequencies may result in order(s) of magnitude overestimation for the transverse impedance.

5.3 Transverse Instability of Antiprotons in the Recycler

In the course of typical store (of about 16 hours) up to 5×10^{12} antiprotons are accumulated and cooled in Recycler. The reduction of the cooled antiproton beam emittances is limited by a transverse instability [24]. Since the antiprotons are, for the most part, accumulated within long bunches, where synchrotron oscillations are slow enough to be neglected, a coasting beam model [25] appears to be a reasonable first approximation to the stability problem. However, it was realized—both theoretically and experimentally—that the stability thresholds [26] and the spatial behavior of the unstable modes [19, 27] differ from the simplified expectations of the coasting beam model.

5.3.1 Stability of Coasting Beam

A beam stability threshold is determined by an equality of the Landau damping rate Λ and the impedance-driven growth rate: $\Lambda = \Omega_0 \mathrm{Im}\Delta Q_c$. For the coasting beam with space charge-dominated impedance, $|\Delta Q_c| << |\Delta Q_{sc}|$, the growth rate can be easily calculated, since the lattice tune spread can be neglected for that purpose (see, e.g., [19]; Eq. (5.88)). Following [25], the damping rate can be expressed as

$$\Lambda = -\pi\Omega_0\langle\Delta Q_{\rm sc}\rangle\int \Delta Q_{\rm sc} f_x J_x\ \delta(\Delta Q_l + \Delta Q_{\rm sc})d\Gamma\ ,$$
$$\langle\Delta Q_{\rm sc}\rangle \equiv -\left(\int\frac{f_x J_x\ d\Gamma}{\Delta Q_{\rm sc}}\right)^{-1}, \tag{5.41}$$

where $\Delta Q_{\rm sc} = \Delta Q_{\rm sc}(J_x, J_y)$ is the space charge tune shift as a function of the two transverse actions, $f_x = \partial f/\partial J_x$ is a partial derivative of the normalized phase-space density $\int f dJ_x dJ_y dp \equiv \int f d\Gamma = 1$ with p as the relative momentum offset, and $\Delta Q_l = \Delta Q_l(J_x, J_y, p)$ is the lattice tune shift due to nonlinearity and chromaticity ξ. For the Gaussian distribution and a round beam, the chromaticity-related threshold is well approximated by [25]:

$$\frac{|\Delta Q_{\rm sc}(0)|}{\sigma_{\nu p}} = 1.7\ln\left(\frac{|\Delta Q_{\rm sc}(0)|}{{\rm Im}\Delta Q_c}\right), \tag{5.42}$$

where

$$\sigma_{\nu p} = |\xi - n\eta - Q\eta|\sigma_p \equiv \xi_n\sigma_p$$

is the effective chromatic rms tune spread for mode n with effective chromaticity ξ_n, $\Delta Q_{\rm sc}(0)$ is the space charge tune shift at the center of the beam, and σ_p is the rms momentum spread.

Note that the Landau damping rate is determined by the integral over a surface of the resonant particles, whose individual tune shifts $\Delta Q_{\rm sc} + \Delta Q_l$ are equal to the coherent tune shift ${\rm Re}\Delta Q_c$. For the space charge-dominated impedances, $|\Delta Q_c| \ll |\Delta Q_{\rm sc}|$, the coherent tune shift can be neglected in the argument of the delta-function in Eq. (5.41). If the lattice tune spread is determined by the chromaticity only, the resonant surface is represented as

$$\xi_n\frac{\Delta p}{p} + \Delta Q_{\rm sc} = {\rm Re}\Delta Q_c \approx 0.$$

Therefore, the maximum momentum offset of the resonant particles $\Delta p_{\rm res\_max}$ is equal to

$$\Delta p_{\rm res\_max} = \left|\frac{\Delta Q_{\rm sc}(0)}{\xi_n}p\right|. \tag{5.43}$$

For operational purposes, the instability threshold was expressed and measured in terms of the effective phase-space density:

$$D_{95\%} = \frac{N[E10]}{\varepsilon_{(\perp)n,95\%}[\text{mm} \times \text{mrad}] \cdot \varepsilon_{(s)n,95\%}[\text{eV} \times \text{s}]}, \tag{5.44}$$

where N is the number of antiprotons, $\varepsilon_{(\perp)n,95\%}$ is the normalized 95 % emittance, and $\varepsilon_{(s)n,95\%}$ is the longitudinal 95 % emittance for the barrier-bucket RF; the units are shown in the square brackets—see also Eq. (5.1). For the Gaussian distribution, the 95 % emittances are related to the rms emittances as $\varepsilon_{(\perp)n,95\%} = 6\varepsilon_{(\perp)n}$, $\varepsilon_{(s)n,95\%} = 4\varepsilon_{(s)n}$. In terms of the density, Eq. (5.44), the instability threshold, Eq. (5.43), can be expressed as

$$D_{95\%} = 60F\frac{\gamma_0^2\xi_n}{T_0[\text{s}]E_0[\text{eV}]}, \quad F \equiv \ln\left(\frac{|\Delta Q_{\text{sc}}(0)|}{\text{Im}\Delta Q_c}\right); \tag{5.45}$$

units for the revolution time T_0 and the beam energy $E_0 = \gamma_0 m_p c^2$ are shown in the brackets. Since the effective chromaticity and the coherent tune shift depend on the mode frequency, or the harmonic number n, so does the instability threshold. If there are no external feedbacks, the threshold is determined by the mode which gives the lowest density value, Eq. (5.45). For the resistive wall impedance, it is the lowest betatron sideband of the slow waves. When a broadband damper is applied, then for the resistive wall impedance, the beam is most unstable for a wave at the frequency edge of the damper, ~70 MHz for the Recycler. For identical chromaticities and damper bandwidths, the horizontal instability cannot be seen, since the vertical resistive wall impedance is a factor of 2 higher than the horizontal, making the vertical threshold slightly lower due to the logarithmic factor F. However, this slight logarithmic difference can be outweighed by a small difference in the effective chromaticities ξ_{nx} and ξ_{ny} of Eq. (5.42) if the absolute value of the vertical chromaticity sufficiently exceeds that of the horizontal. When the normal chromaticities $\xi_{x,y}$ are small and the effective chromaticities are dominated by the longitudinal factor $n\eta$, polarization of the instability depends on an interplay of these two weak factors and may spontaneously change due to a small uncontrolled variation in the chromaticities.

The threshold expressions in Eqs. (5.42) and (5.45) should be used with some caution. The coherent motion is stabilized by resonant particles, whose individual lattice tune shift compensates their individual space charge tune shift, Eq. (5.41). For the space charge-dominated impedance, these particles are in the far tails—longitudinal and transverse—of the beam distribution. When electron cooling is applied, there is no reason to assume the distribution to be Gaussian, so, strictly speaking, Eqs. (5.42), (5.45) are not applicable. These far tails of the distribution are not measurable, so the general formula in Eq. (5.41) cannot be used either. In this situation, the threshold value of the phase density is found experimentally. It may deviate from the value for a Gaussian distribution by up to a factor of 2 in both directions. Additional reasons for the discrepancy between calculations and measurements are discussed below.

5.3.2 Bunching Effects

Even if the synchrotron tune is much smaller than the coherent tune shift, there are at least three different ways for which the beam bunching may influence the coherent oscillations.

First, for a bunch with a negligible synchrotron tune, the tail-to-head interaction takes place due to a long-range wake field (left from previous revolutions). This leads to a dependence of the coherent tune shift ΔQ_c on the bunching factor $B = T_0/\tau_0 \geq 1$ [27, 28]. The Recycler's wake field is believed to be dominated by the resistive wall contribution; thus, the coherent tune shift slowly grows when the bunch length decreases; for a single bunch in the ring $\Delta Q_c \propto B^{1/3}$ [28], close to a two-particle model where $\Delta Q_c \propto B^{1/4}$ [27]. Note that this leads only to a slow logarithmic growth of the stability threshold in Eq. (5.45), mostly due to $\Delta Q_{\mathrm{sc}} \propto B$.

For a barrier bucket with "infinite walls", the above consideration is the only correction to be applied to the coasting beam model. However, the RF voltage V_{RF} and the barrier width τ_{b} are limited, so a second effect from bunching takes place: particles with sufficient momentum offset $|p| \geq p_{\mathrm{dc}} = \sqrt{2V_{\mathrm{RF}}\tau_{\mathrm{b}}/(|\eta|E_0T_0)}$ are leaving the potential well and spend most of their time outside of the bucket (the so-called DC beam). If the barriers are lower than $\Delta p_{\mathrm{res\_max}}$ in Eq. (5.43), some of the particles responsible for Landau damping do not contribute anymore, and the beam is less stable than it would be with a deeper potential well. Contrary to the first effect of the bunching factor, this one leads to a decrease of the instability threshold the more compressed the bunch is with the same barriers. Indeed, by compressing the bunch, the momentum offset of the AC particles grows, and some resonant particles spill outside the potential well and become DC.

A third factor, which would alter the coasting beam model, is the possibility for the potential well profile to depart from the one resulting from a barrier RF configuration. Before extraction, the beam is kept inside cosine-like potential wells; hence the barrier-bucket theory does not apply. Similar to head-tail modes with strong space charge, where smooth walls of the potential well are better for Landau damping [29], the beam stability threshold for this case can be expected to increase as well.

Finally, it should be mentioned that the presence of multiple bunches around the Recycler also affects the way an instability develops. Indeed, other bunches play the role of "relay stations" for the tail-head signal, thus increasing the coherent growth rate $\Omega_0 \mathrm{Im} \Delta Q_c$ and in turn logarithmically decreasing the instability threshold.

5.3.3 Longitudinal Bunch Tomography

The above arguments indicate that the density threshold in Eq. (5.45) should depend on the shape of RF well. To study this dependence, a longitudinal tomography diagnostic was developed and applied to the Recycler. The idea for that tomography is based on the fact that for a given RF shape, a bunch longitudinal profile provides information about the phase-space density [30]. Thus, measured RF and bunch profiles allow calculating the distribution function of the bunch. This is attained by solving the following set of equations:

$$\begin{aligned}
H(\varepsilon,\tau) &= \frac{\varepsilon^2}{2\mu} + W(\tau);\\
W(\tau) &= -\frac{1}{T_0}\int_0^{\tau} V_{\mathrm{RF}}(t)dt;\\
\lambda(W) &= \int_{-\infty}^{\infty} f(H(\varepsilon,\tau))d\varepsilon = \sqrt{2\mu}\int_W^{H_{\max}} \frac{f(H)}{\sqrt{H-W}}\,dH;\\
I(H) &= \frac{1}{2\pi}\oint \varepsilon(H,\tau)\,d\tau.
\end{aligned} \tag{5.46}$$

Here $H(\varepsilon, \tau)$ is the Hamiltonian as a function of its canonical variables ε (the energy offset) and τ (the timing offset), $V_{\mathrm{RF}}(t)$ is the RF voltage at time t, $W(\tau)$ is the potential energy at position τ, $p_0 = \beta_0 E_0/c$ is the beam momentum, $\mu = p_0 c/|\eta|$ is the effective mass, $\lambda(W)$ is the beam linear density as a function of the potential W (taken from measurements), $f(H)$ is the phase-space density as a function of the Hamiltonian to be found, $I(H)$ is the action variable, $f_I(I) = f(H(I))$, and $H_{\max}$, $I_{\max}$ are the maximal Hamiltonian and action inside the bucket; DC particles are neglected. A solution of Eq. (5.46) can be presented in terms of the integrated distribution or the fraction of particles inside a given action $G(I) = \int_0^I f_I\left(I'\right)dI' / \int_0^{I_{\max}} f_I\left(I'\right)dI'$. Then, its inverse function $2\pi I(G)$ gives the longitudinal emittance, or the phase space, as a function of the percentage of particles contained inside that phase space. Tomography analyses for the Recycler are described in some more details in [31].

5.3.4 *Observations*

Several cases of beam instabilities were observed in the Recycler without external damping. In these cases, the measured instability threshold was in reasonable agreement with Eq. (5.45) (see [32]) and corresponded to $D_{95\%} = 0.5$–0.8. The scatter in the threshold values was due to the limited accuracy of the emittance measurements, the uncertainty of the chromaticity value, and variations in the tail distribution. The other features of the instability were also in line with theoretical predictions. The beam became unstable at the lower betatron sideband primarily in the vertical direction. The coherent oscillations grew for several dozens of turns until a partial beam loss occurred.

To counteract the instability, two transverse dampers (vertical and horizontal) were installed, initially with the bandwidth of 35 MHz [32]. Several studies performed with the antiproton beam in the standard configuration (a "rectangular" barrier bucket with standard barrier height of ~17 MeV/c) clearly showed, as expected, a significant increase of the instability threshold as a result of the dampers installation but again with a sizeable scatter in the threshold $D_{95\%}$. In fact, in one occasion, without turning the dampers off, the instability could not be provoked at all up to $D_{95\%} = 3.1$. Note that turning the dampers off resulted in a fast (<0.1 s) beam loss. However, in two other studies, the beam went unstable at $D_{95\%} = 3.0$ and 2.6.

An important operational limitation was found to be the saturation of the dampers' pickup preamplifiers. It was observed during beam preparation for extraction, when the linear beam density increases by more than a factor of 2. Saturation was effectively turning off the dampers, and the developing instability and accompanying beam loss yielded "clipping" all bunches down to the same peak density.

With advances in the strength of electron cooling and increasing requirements to the beam brightness in the Recycler, the dampers bandwidth became insufficient and started to affect the regular operation of the collider. Therefore, the dampers were upgraded in December 2007. The upgraded version, which has been in use until the end of Run II, had an effective bandwidth of ~70 MHz, and the pre-amplifiers' saturation limit was increased [33]. For this bandwidth ($n \approx 780$) and typical beam parameters ($Q'' = -4$, $\varepsilon_{(a)(\perp)n,95\%} = 2\,\pi$ mm mrad, $B = 0.5$), Eq. (5.45) predicts the threshold phase density of $D_{95\%} = 4.3$. In measurements, the instability threshold was increased to $D_{95\%} = 4.3$–6.9. These numbers show the scatter of several studies carried out with the antiproton beam contained in a rectangular bucket with the standard barrier height (17 MeV/c). In the regular operation, the phase density was kept below 2.7 to guaranty beam stability. However, on a few occasions, the instability still did develop during extraction. The extraction process includes complicated manipulations in the longitudinal phase space, described in detail in Chap. 4. Nevertheless, to illustrate better the extraction process, the three main RF configurations and associated beam longitudinal profiles are shown again in Fig. 5.9. First, the bunch is divided into nine nearly identical pieces by narrow rectangular barriers (called for historical reasons "mined bunches"). Then antiprotons are moved, one mined bunch at a time, into the extraction region. Once there, a mined bunch is adiabatically transformed into four 2.5 MHz smaller bunches, which are then extracted into the matching main injector (MI) RF waveform.

With the dampers in the final configuration (2008–2011), the instability was observed six times in the course of extraction. All of them were similar and had the following main characteristics:

1. The beam loss occurs during the second half of the extraction process.
2. Only one mined bunch at a time goes unstable.
3. Typically, after the first instability, all remaining bunches become unstable as well at later stages. In a couple of exceptions, the very last bunch (#9) remained stable. In those cases, the bunch #9's intensity was ~20 % lower than other bunch intensities because of imperfections of the RF voltage.

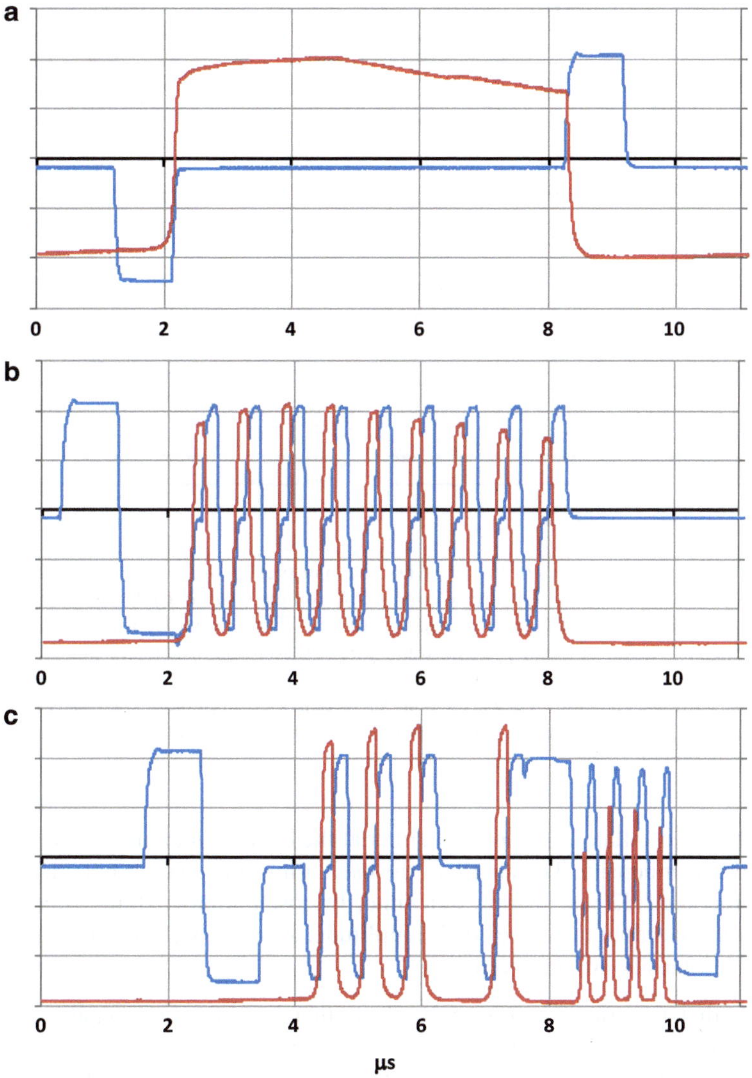

Fig. 5.9 RF voltage (*blue line*) and beam longitudinal profile (measured by the resistive wall monitor, RWM, *red line*) waveforms recorded during an extraction to the Tevatron. (**a**) "Cold" bucket, (**b**) nine mined buckets, and (**c**) mined buckets + 2.5 MHz structure on the bunch which is in the extraction region after having already extracted four "mined" bunches. *Vertical scales* are arbitrary. Note that in case (**a**), the beam longitudinal profile deviates from the rectangular distribution expected for a beam stored within two RF barriers because of the RF imperfection

4. Each beam loss lasts 5–15 s.
5. Traces are recorded with the damper pickups (Fig. 5.10) during 32 ms after detecting the instability. They show that the instability happens at the frequency right outside of the damper bandwidth, ~70 MHz. Only a 100–200 ns portion of

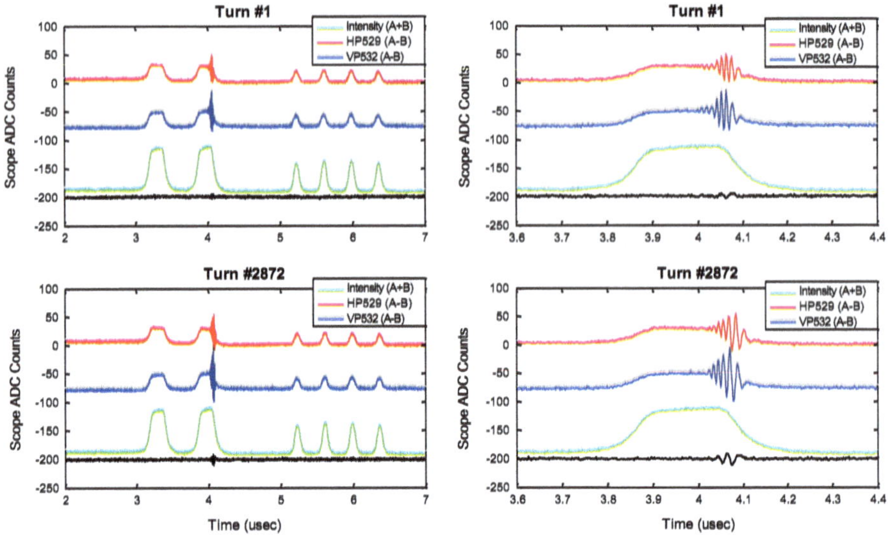

Fig. 5.10 Oscilloscope traces of the transverse damper pickup signals during instability: on the *left* (**a**, **b**), one full revolution for turns #1 and #2872; on the *right* (**c**, **d**), focus on the bunch that went unstable (Bunch #8). The *vertical scale* is arbitrary. The *green trace* is the sum signal and is proportional to the linear density distribution. The *red* and *blue traces* are the difference (not normalized) signals for two damper pickups (*red*: horizontal; *blue* vertical) and reflect the beam transverse position. The *black trace* is the damper vertical kick amplitude. *Top* plots (**a**) and (**c**) are at the early stage of the instability development; *bottom* plots (**b**) and (**d**) are at the end of the recording period of 32 ms

the bunch oscillates. It is located either at the bunch tail or around the maximum of its linear density. The length of the oscillating portion does not change within the recorded 32 ms. During that time, the oscillations exhibit only a modest growth, less than two times their amplitude. Also, no significant changes in the bunch intensity are observed.

Note that the traces in Fig. 5.10 show that both the horizontal and vertical directions went unstable although the largest emittance growth was seen in the vertical direction. This particular case points to the possibility that the instability threshold in the horizontal direction can become very close to (even exceed) the vertical's due to uncontrolled variations of their respective chromaticities. Figure 5.10 also shows almost no response from the dampers kickers because of the frequency response limit discussed above.

The beam phase density was similar at different stages of the extraction process; however, the instabilities occurred only for bunches in one of the RF configurations, the so-called "the mined" bucket. This peculiarity was explained by the combination of the high linear density and low barrier height (8.5 MeV/c vs. standard 17 MeV/c) in this configuration. It leads to an effective exclusion from Landau damping of antiprotons with high longitudinal, low transverse actions. This hypothesis was tested in a dedicated study, where the beam stability was compared in different RF configurations. It was found that lowering the height of the barrier potential by a factor of 2, mimicking what happens for the mined bunches, decreased the threshold phase density $D_{95\%}$ from 6.9 to 4.5.

As in all previous instances, the threshold value of $D_{95\%}$ was calculated using the average 95 %, normalized transverse emittance, $(\varepsilon_{(x)n,95\%} + \varepsilon_{(y)n,95\%})/2$, and the longitudinal emittance. Both measurements are based on signals from the 1.7 GHz Schottky pickups, with the longitudinal emittance calculation done assuming a rectangular bucket and a Gaussian momentum distribution.

The Schottky power is proportional to the square of the rms beam size times the number of particles, i.e., $P_{\text{Schottky}} \propto \sigma^2 \times N$. To calibrate the emittances measured with the Schottky detector, a method that employs beam scrapers with a relatively low intensity beam of antiprotons (~50×10^{10}) in an equilibrium state, i.e., constant distributions, is carried out. First, 5 % of the beam is scraped off in one direction and both the scraper position and the Schottky "emittance" are recorded. Then, the remainder of the beam is scraped away to find the "extinction point" at which the scraper position is also recorded. Assuming that the beta-functions at the scraper and Schottky detectors are known, the scraper travel between the 95 % position and the extinction point gives a measurement of the 95 % emittance at the scraper, which is then computed for the location of the Schottky detector. A calibration factor is thus obtained to convert the beam power of the betatron sidebands measured by the Schottky detector into an emittance number. It should be noted that, strictly speaking, this procedure assumes that the two transverse degrees of freedom are completely uncoupled. An up to ~20 % correction (decrease, i.e., the procedure described *overestimate* the emittance of the beam) needs to be applied when the horizontal and vertical directions are fully coupled. In the Recycler, the coupling is kept to a minimum but cannot be completely eliminated.

In this study, we also used alternate measurements of the transverse and longitudinal emittances. The transverse emittance was measured with the horizontal flying wire, which profile is fitted with a Gaussian function. Note that the flying wire emittance was always lower than the Schottky's by at least a factor of 1.2 (likely a calibration issue), and the ratio was increasing by up to a factor of 2 when the beam was deeply cooled by the electron beam, indicating long non-Gaussian tails.

The longitudinal emittance was calculated with the tomography procedure applied to the longitudinal density profile, acquired with a resistive wall monitor for a fixed measured RF voltage waveform. The tomography approach gave the same qualitative results as the calculation obtained from the Schottky signal for not-too-deeply cooled bunches contained between rectangular barriers. This alternate procedure leads to an even larger stability threshold difference between the two RF configurations described above: for the standard bucket, $D_{95\%} = 10.9$, while $D_{95\%} = 3.8$ for the bucket with barrier height reduced by a factor of 2.

The beam stability was also studied for the beam in the 2.5 MHz RF structure, which mimicked the final stage of the extraction but with significantly stronger electron cooling. Out of four 2.5 MHz bunches, 3 (trailing) bunches went unstable, and their intensity dropped evenly. The oscilloscope traces showed large oscillations of the second bunch, but no oscillations on the others, even at the end of 32 ms recording period. While this structure is clearly far from the coasting beam model considered in Sect. 5.3.1, the threshold density calculated with the flying wire emittance and tomography, $D_{95\%} = 7.2$ (average for the four bunches) was similar to the number found for the long rectangular bunches.

5.3.5 Discussion of the Observations

The observations did not reveal contradictions with the model presented above. In fact, all the quantitative and qualitative features predicted by the model are in agreement with the observations within their accuracy. Without a damper, the instability occurs at the lowest betatron sideband; with the damper, it happens right outside of the damper bandwidth.

In the measurements with the dampers, the value predicted by Eq. (5.45) for the threshold phase density falls into the scatter of experimental observations. Note that in the case of the 70 MHz damper, the logarithm of Eq. (5.45) is large, ~10. The range of the bunch length and transverse emittance variations in the instability studies results in changes of the logarithm by less than 1.5. Correspondingly, the instability threshold predicted by Eq. (5.45) changes also by less than 15 %, i.e., it is almost a constant at a given chromaticity and damper parameters. The large scatter of $D_{95\%}$ observed in the experiments was attributed primarily to the variations of the tail distribution. In addition, the threshold is clearly affected by the finite height of the longitudinal barriers.

Several features such as a slow non-exponential growth of the oscillations and seconds-long times beam losses were originally unexpected. However, a classical exponential growth of an instability describes the behavior of a system sufficiently above the threshold, while in all our experiments, the beam was slowly reaching the threshold density as it was being cooled. Strictly speaking, the instability growth rate at the exact threshold is zero. Then, in this case, it is determined not only by the impedance but also by such factors as beam cooling, synchrotron motion, and all sorts of diffusion for the resonant particles. That is why for that gradual approach of the threshold, the emerging instability can be orders of magnitude slower than the pure impedance-related growth.

5.3.6 Operational Implications

The studies confirmed that the beam configuration most prone to become unstable is the mined bucket. To avoid instabilities caused by overcooling the antiprotons, the electron beam current used in operation was limited to 100 mA, and the offset between the beam centers in the cooling section, which is a manipulation used to reduce the cooling rate, was adjusted in order to stay far from an instability. In addition, we eliminated the step during the preparation of the bunches for extraction that originally separated the portion of antiprotons with the largest momentum offsets (the so-called mining [34]). While no dedicated studies were performed, operationally it allowed applying stronger electron cooling, which led to the highest phase density of extracted beams at the end of Run II.

5.3.7 Note About the Main Injector

The beam extracted from the Recycler is transferred into an identical 2.5 MHz structure in the main injector (MI). Because both machines have similar lattices—hence the beam brightness does not change significantly during the transfer—and transverse dampers were not used at this stage, one may have expected the development of a transverse instability in MI as well. However, no instability-related antiproton beam loss has ever been observed. The reason is that the MI chromaticity was set to a large value, -18, while the measured chromaticity in the Recycler was -2 (horizontal)/-4 (vertical). Operation of MI at a lower chromaticity, -12, resulted in strong horizontal oscillations ([35]) but still not in a beam loss. Note that in the main injector, the beam spends less than 3 s at the injection energy, and the beam lifetime reduction due to high chromaticity is not an issue. Attempts to operate with a similarly large chromaticity in the Recycler were unsuccessful because of the deterioration of the beam lifetime, and in operation, the chromaticity is set to a much smaller value than for MI.

5.3.8 Conclusions

The transverse instability of the antiproton beam in the Recycler was the final limiting factor to the brightness of the extracted beams that could be achieved. Nevertheless, the transverse dampers in conjunction with electron cooling permitted to increase the beam brightness by an order of magnitude.

Qualitative features of the measured instances of the instability fit reasonably well the model developed for a coasting beam. The onset of the instability is determined by the threshold phase density, which value is in agreement with the model within the scatter of experimental data and the precision to which this theoretical threshold can be calculated. The scatter in the data is likely related to variations in the distribution of the tails particles, which determine Landau damping. In particular, lowering the potential depth of the barrier bucket effectively excludes part of the longitudinal tails from damping and may decrease the threshold density by a factor of 2.

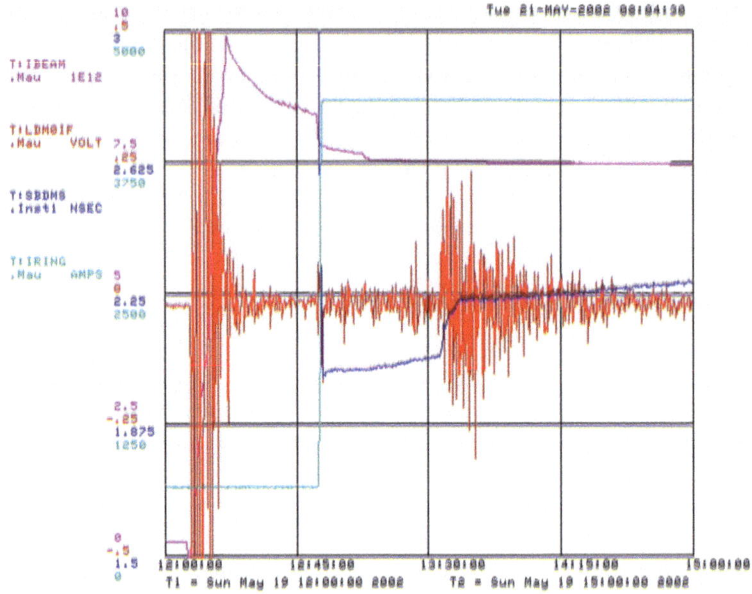

Fig. 5.11 The beam blows up longitudinally (*blue line*, T:SBDMS, the rms bunch length average for all bunches) at about 13:40 h during the store which started at about 13:00 h. We see that when it blows up the phase signal of the bunch oscillates w.r.t. RF (T:LDM0IF). Plotted also are the Tevatron average beam current T:IBEAM and the Tevatron magnet bus current T:IRING [11]

5.4 The Tevatron Wideband Longitudinal Coupled Bunch Mode Dampers

When Run II began in its first year, the high current stored in the Tevatron caused unforeseen problems in the beam dynamics. These needed to be fixed before higher luminosities could be achieved. One of the problems that started to appear at the beginning of 2002 was the rapid blowup of the longitudinal beam size during a store (see Fig. 5.11). Although these blowups did not appear in every store, they seem to be weakly correlated with beam current. There were conjectures that coupled bunch mode instabilities that arose from coupling to the higher-order parasitic modes of the RF cavities were the cause of the instabilities. As the frequency of these higher modes moves as a function of temperature, the coupled bunch modes can be stable or unstable depending on where and how the higher-order parasitic modes line up. Table 5.3 shows 11 stores in the month of May where about 2/3 of the stores were unstable.

The first attempts at controlling this blowup with mode 0 dampers ended in failure. This showed us that the instability may be a longitudinal head-tail or higher-order coupled bunch mode. At the time, we did not have the instrumentation to distinguish between the two types. After much discussion, it was decided that the best course of action was to build a wideband longitudinal damper system which would take care of the coupled bunch mode instabilities. At first glance, the idea of

Table 5.3 Comparison of different stores

Store	Date	Num. protons	Bunch length before blowup (ns)	Bunch length after blowup (ns)	Time before blowup (min)	Bunch lengthening time before blowup (h)	1/e time after blowup (h)	Comments
1302	8 May 2002	1.7e11	2.0	2.3	60	42	67	
1305	9 May 2002	1.67e11	2.0	2.3	6	12	43	
1307	10 May 2002	1.79e11	2.0			53		No blowup
1309	11 May 2002	1.71e11	2.0			42		No blowup
1313	12 May 2002	1.76e11	2.0			40		No blowup
1329	16 May 2002	1.76e11	1.9	2.2	3	No data	77	
1332	17 May 2002	1.78e11	1.9	2.4	6	9	83	
1333	18 May 2002	1.81e11	2.1			50		No blowup
1335	19 May 2002	1.77e11	2.0	2.2	39	40	59	
1337	20 May 2002	1.83e11	2.0	2.2	16	19	56	
1340	21 May 2002	1.94e11	2.0	2.6	2	No data	No data	

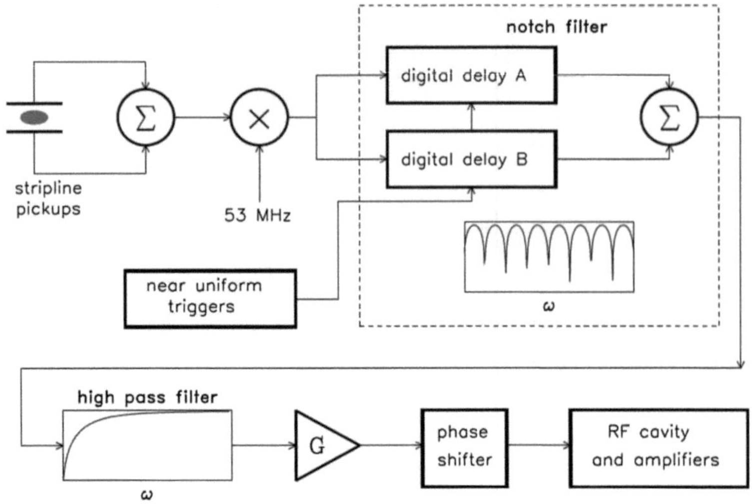

Fig. 5.12 The block diagram of the longitudinal dampers [11]

using the RF cavity themselves as the source of longitudinal kicks on the beam seems to be difficult. This is because each of the four proton RF cavities has a high-quality factor Q ($\approx 10^4$) near its resonance, and thus, its impedance falls off rapidly away from it. Therefore, the amplitude and phase response is not flat at all synchrotron sideband pairs, and the response of the dampers for the mode 1 coupled bunch mode will be an order of magnitude greater than the higher-order coupled bunch modes. It would be impossible to keep the feedback stable for mode 1 and still have useful gain at the higher-order modes. The solution to this problem is to build an equalizer that lifts up the impedance so that it looks constant away from the resonance. Besides the equalizer, the damper also needs a notch filter that suppresses the revolution harmonics (otherwise these harmonics will limit the gain of the loop) and differentiates in time the synchrotron sidebands. Lastly, we also have to time in the system so that the error signal of bunch n is applied exactly one turn later to kick bunch n.

The block diagram of the damper system [11] is shown in Fig. 5.12. The damper system starts at the stripline pickups which sum the beam signals at the two plates to produce a signal that is proportional to the longitudinal position of the beam. This signal is then down converted with the Tevatron RF to produce a phase error (or quadrature) signal w.r.t. it. The error signal is then processed with electronics that perform the following:

(a) Equalize the impedance of the RF cavity
(b) Suppress the revolution harmonics and differentiate the synchrotron sidebands around the revolution lines
(c) One turn delay so that when the dampers pick up the signal of bunch 1, it will kick bunch 1 one turn later

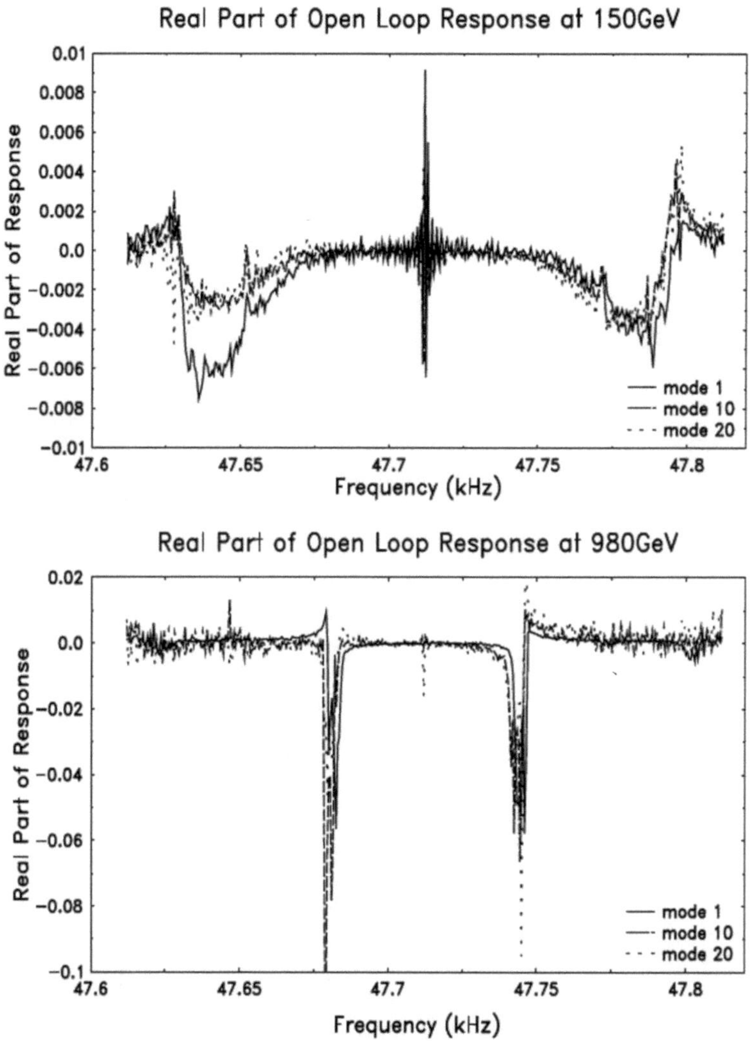

Fig. 5.13 These graphs show the real part of the open-loop response of modes 1, 10, and 20 at 150 and 980 GeV. We have superimposed all the three graphs on top of each other by shifting the frequency of mode 10 by $-10f_0$ and mode 20 by $-20f_0$ [11]

To accomplish (a), we have a high-pass filter (HPF) that equalizes the RF cavity impedance, and for (b) and (c), we have digital notch filters that provide tracking delay and a notch at every revolution harmonic.

The real part of the open-loop transfer measurement of the setup at 150 and 980 GeV is shown in Fig. 5.13. For damping, it is necessary that the real part of the response at $\pm f_s$ be symmetric about the revolution harmonic and negative. A sampling of the open-loop transfer functions of modes 1, 10, and 20 presented in this figure shows that the dampers are phased correctly. Note that the edges of the

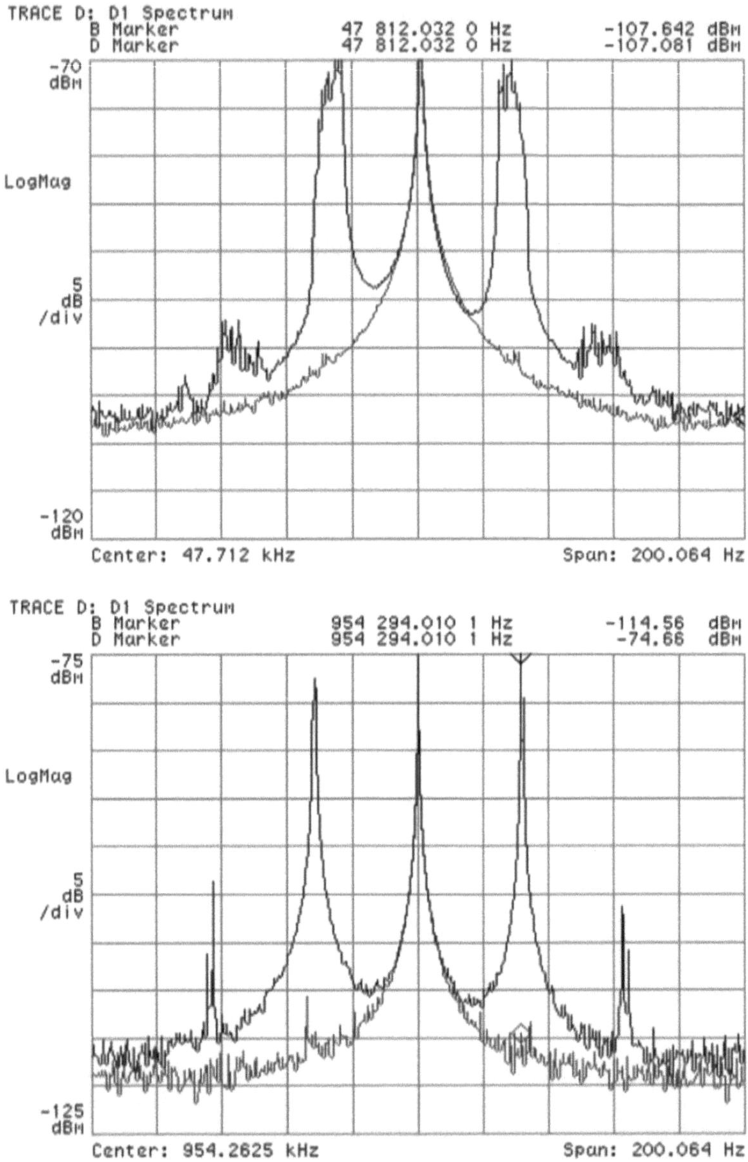

Fig. 5.14 The beam spectra at 980 GeV; top - before the measurement the loop was closed, then the beam was excited by anti-damping; bottom - the damping was turned on resulting in the synchrotron lines of modes 1 and 20 being damped [11]

response are positive rather than negative. This limits the amount of gain that can be set in the damper system.

To test whether the dampers indeed work, we can excite the beam at 980 GeV by switching the sign of the gain. This is a good sign because we can actually excite the beam which means that there is sufficient gain in the loop. When we switch the sign of the gain back to damping, we find that the excitation can be damped. The results of these actions are shown in Fig. 5.14. Although the dampers do perform their job, we find that damping takes 2–3 min in these examples.

After installing the dampers, the problem of sudden beam size growth during a store is rarely observed. To prove to ourselves that the dampers definitely stopped the problem, we deliberately turned the dampers off for one store. In this store the beam blew up longitudinally as before. This conclusively showed us that the longitudinal dampers solved the problem. However, the underlying cause of the blowup is still not understood. There are speculations that higher-order parasitic modes in the RF cavity, phase noise from microphonics, etc. are the source of these blowups. However, operationally, the dampers were a success.

References

1. P.M. Ivanov et al., in *Proceedings of the IEEE PAC'2003*, Portland, OR (2003), p. 3062
2. C.-Y. Tan, J. Steimel, *Fermilab Preprint FERMILAB-TM-2204*, 2003
3. P.M. Ivanov et al., in *Proceedings of IEEE PAC'2005*, Knoxville, TN (2005), p. 2756.
4. V. Lebedev, A. Burov, W. Pellico, X. Yang, *Prepint FERMILAB-CONF-06-205-AD*, 2006
5. A. Burov, V. Lebedev, Phys. Rev. ST Accel. Beams **10**, 054202 (2007)
6. L.R. Prost et al., *Transverse Instabilities in the FERMILAB Recycler*, arXiv:1201.4168 (2012)
7. A. Burov, V. Lebedev, Phys. Rev. ST Accel. Beams **12**, 034201 (2009)
8. A. Burov, V. Lebedev, AIP Conf. Proc. **773**, 350 (2005)
9. R. Moore et al., in *Proceedings of the IEEE PAC'2003*, Portland, OR, p. 1751
10. W. Guo et al., FERMILAB-TM-2203 (2003)
11. C.-Y. Tan, J. Steimel, *FERMILAB-TM-2184*, 2002
12. S.C. Snowdon, *Fermilab TM-277*, 1970
13. A. Ruggiero, Fermilab FN-220 0402 (1971)
14. R. Gluckstern, *Fermilab TM-1374*, 1985
15. K.Y. Ng, *Fermilab FN-0744*, 2004
16. A. Macridin, P. Spentzouris, J. Amundson, L. Spentzouris, D. McCarron, Phys. Rev. ST Accel. Beams **14**, 061003 (2011)
17. A. Burov, V. Lebedev, *FERMILAB-CONF-02-101-T*, 2002
18. A. Burov, V. Lebedev, *FERMILAB-CONF-02-100-T*, 2002
19. A.W. Chao, *Physics of Collective Beam Instabilities in High Energy Accelerators* (Wiley, New York, 1993)
20. A. Burov, *LHC Project Note 353*, 2004
21. J. Crisp, *Fermilab-TM-2145*, 2001
22. K. Yokoya, H. Koiso, Part. Accel. **27**, 181 (1990)
23. R. Gluckstern, J. van Zeijts, B. Zotter, Phys. Rev. E **47**, 656–663 (1993)
24. L.R. Prost et al., in *Status of Antiproton Accumulation and Cooling at Fermilab's Recycler.* Proceedings of the COOL'09, Lanzhou (2009) MOM1MCIO01
25. A. Burov, V. Lebedev, PRSTAB **12**, 034201 (2009)
26. L.R. Prost et al., in *Transverse Instability of the Antiproton Beam in the Recycler Ring.* Proceedings of the PAC'11, New York (2011)
27. A. Burov, V. Lebedev, AIP Conf. Proc. **773**, 350–354 (2005)
28. V. Balbekov, PRSTAB **9**, 064401 (2006)
29. A. Burov, Phys. Rev. ST Accel. Beams **12**, 044202 (2009)

30. L. Michelotti, Phys. Rev. ST Accel. Beams **6**, 024001 (2003)
31. A. Burov, in *Instabilities and Phase Space Tomography in RR*. Fermilab Internal Report beams-doc-3641 (2010)
32. N. Eddy, J.L. Crisp, M. Hu, AIP Conf. Proc. **868**, 293–297 (2006)
33. N. Eddy, J.L. Crisp, M. Hu, in *Measurements and Analysis of Beam Transfer Functions in the Fermilab Recycler Ring Using the Transverse Digital Damper System*, Proceedings of the EPAC'08, Genoa (2008)
34. C.M. Bhat, Phys. Lett. A **330**, 481–486 (2004)
35. M.J. Yang, J. Zagel, *Fermilab Internal Note* beam-doc-3649 (2010), http://beamdocs.fnal.gov/AD-public/DocDB/ShowDocument?docid=3649

Chapter 6
Emittance Growth and Beam Loss

R. Carrigan, V. Lebedev, N. Mokhov, S. Nagaitsev, V. Shiltsev, G. Stancari, D. Still, and A. Valishev

A wide range of diffusion and beam loss mechanisms were studied during the Tevatron Run II commissioning and operations. Many of them were well known [Coulomb scattering, residual gas and intrabeam scattering (IBS)] but required substantial studies and deeper theoretical insights because of unique experimental conditions and beam parameters.

6.1 Single and Multiple Scattering

Conventionally, multiple and single particle scattering in a storage ring are considered to be independent. Such an approach is simple and often yields sufficiently accurate results. Nevertheless, there is a class of problems where such an approach is not adequate and single and multiple scattering need to be considered together. This can be achieved by solving an integro-differential equation for the particle distribution function, which correctly treats particle Coulomb scattering as well as the betatron motion. We start our consideration from the Fokker–Planck equation describing the evolution of particle distribution due to multiple scattering, then we consider multiple IBS, and finally we will move to a simultaneous treatment of multiple and single Coulomb scattering, where we will consider two problems: the beam scattering on the residual gas and an evolution of longitudinal particle distribution in a hadron collider due to single and multiple IBS.

R. Carrigan • V. Lebedev (✉) • N. Mokhov • S. Nagaitsev • V. Shiltsev • G. Stancari • D. Still • A. Valishev
Accelerator Physics Center, Fermi National Accelerator Laboratory, MS341, PO Box 500, Batavia, IL 60510, USA
e-mail: val@fnal.gov; shiltsev@fnal.gov

V. Lebedev and V. Shiltsev (eds.), *Accelerator Physics at the Tevatron Collider*, Particle Acceleration and Detection, DOI 10.1007/978-1-4939-0885-1_6,

6.1.1 Diffusion Equation in the Action-Phase Variables

Let $f(x,\theta,t)$ be a one-dimensional beam transverse phase-space distribution function at time t. In the presence of damping and diffusion the evolution of the function f in a ring can be described by a Fokker–Planck equation:

$$\frac{\partial f}{\partial t}+v_0\theta\frac{\partial f}{\partial x}+\frac{\partial}{\partial\theta}(\kappa(x,t)xf)=\frac{\partial}{\partial\theta}(\lambda(x,\theta,t)\theta f)+\frac{1}{2}\frac{\partial}{\partial\theta}\left(D_\theta(x,\theta,t)\frac{\partial f}{\partial\theta}\right). \quad (6.1)$$

Here the functions $\kappa(x, t)$, $\lambda(x, \theta, t)$ and $D_\theta(x, \theta, t)$ describe the focusing, damping, and diffusion in the ring, and $v_0=\beta c$ is the average beam velocity. Making the transition to the action-phase variables (I, ψ), assuming that the distribution function does not depend on the phase ψ, and performing averaging over the ring circumference one obtains the general form of the Fokker–Planck equation for nonlinear oscillator:

$$\frac{\partial f}{\partial t}-\frac{\partial}{\partial I}(\lambda(I)If)=\frac{\omega(0)}{2}\frac{\partial}{\partial I}\left(\frac{D(I)}{\omega(I)}I\frac{\partial f}{\partial I}\right). \quad (6.2)$$

Here $\lambda(I)$ and $D(I)$ are the damping rate and the diffusion averaged over the betatron phase and the ring circumference, $\omega(I)$ is the betatron frequency, and the action for small amplitude betatron oscillations is introduced as follows:

$$I=\frac{1}{2}\left(\beta_x\theta^2+2\alpha_x x\theta+\frac{x^2}{\beta_x}\left(1+{a_x}^2\right)\right). \quad (6.3)$$

In the case of three-dimensional motion the general form of Eq. (6.2) is:

$$\frac{\partial f}{\partial t}-\sum_{k=1}^{3}\frac{\partial}{\partial I_k}(\lambda_k(\mathbf{I})I_k f)=\frac{1}{2}\sum_{k=1}^{3}\omega_k(0)\frac{\partial}{\partial I_k}\left(\frac{D_k(\mathbf{I})}{\omega_k(\mathbf{I})}I_k\frac{\partial f}{\partial I_k}\right),\quad \mathbf{I}$$
$$=\left(I_x,I_y,,I_z\right),k=x,y,z. \quad (6.4)$$

Note that Eqs. (6.2) and (6.4) conserve the number of particles.

In the case of linear focusing and damping, and amplitude independent diffusion the functions $\kappa(x, t)$, $\lambda(x, \theta, t)$ and $D_\theta(x, \theta, t)$ do not depend on x and θ; and they are directly related to the functions of Eq. (6.1) averaged over circumference (revolution time): $\lambda=\langle\lambda(t)\rangle_t$, $D=\langle\beta_x(t)D_\theta(t)\rangle_t$. That yields the following form of Fokker–Planck equation:

$$\frac{\partial f}{\partial t}-\lambda\frac{\partial}{\partial I}(If)=\frac{D}{2}\frac{\partial}{\partial I}\left(I\frac{\partial f}{\partial I}\right). \quad (6.5)$$

Multiplying Eq. (6.5) by I and integrating we obtain

$$\frac{d}{dt}\bar{I} \equiv \int_0^\infty I\frac{\partial f}{\partial t}dI = \int_0^\infty I\left(\frac{\partial}{\partial I}(\lambda If) + \frac{1}{2}\frac{\partial}{\partial I}\left[DI\frac{\partial f}{\partial I}\right]\right)dI = -\lambda\bar{I} + \frac{D}{2}. \tag{6.6}$$

The replacement of action by the betatron amplitude $a = \sqrt{2I}$ converts the right-hand side of Eq. (6.5) being proportional to the two-dimensional laplacian of f:

$$\frac{D}{2}\frac{d}{dI}\left(I\frac{df}{dI}\right) = \frac{D}{4}\Delta_a f \equiv \frac{D}{4}\frac{1}{a}\frac{d}{da}\left(a\frac{df}{da}\right). \tag{6.7}$$

As an example of Eq. (6.5) application we will find the beam lifetime corresponding to the evolution of particle distribution at the stationary stage when all details of initial distribution are smeared out by diffusion and damping.[1] For unlimited aperture and nonzero damping rate the beam life time is infinite and the equilibrium distribution function is:

$$f(I,t) = Ce^{-I/I_0}, \tag{6.8}$$

where $I_0 = D/2\lambda$ is the average beam action. Further we assume that the aperture limitation is important for one dimension only. Then the boundary condition is: $f_0(I_b) = 0$, where I_b determines the ring acceptance. Looking for a solution in the following form:

$$f(I,t) = C\ e^{-t/\tau}f_0(I), \tag{6.9}$$

and substituting it into Eq. (6.5) one obtains

$$f_0{}'' + \left(\frac{1}{I} + \frac{1}{I_0}\right)f_0{}' + \frac{1}{II_0}\left(1 + \frac{1}{\lambda\tau}\right)f_0 = 0. \tag{6.10}$$

where τ is the beam intensity lifetime and C is a constant. The substitutions $\xi = -I/I_0$ and $a = 1 + 1/(\lambda\tau)$ reduce this equation to the equation of the confluent hypergeometric function ${}_1F_1(a, b, \xi)$ with $b = 1$. Thus the solution (for $\alpha, \lambda > 0$) is:

$$f_0(I) = {}_1F\left(1 + \frac{1}{\lambda\tau}, 1, -\frac{I}{I_0}\right). \tag{6.11}$$

[1] This case describes well the beam lifetime in an electron synchrotron when the horizontal beam emittance is set by equilibrium between synchrotron radiation (SR) damping and diffusion due to SR. The vertical emittance is much smaller and therefore the beam loss due to diffusion in vertical plane is negligible. It makes the problem being single dimensional.

where τ is determined by the requirement that the first root is located at $\xi = -I_b/I_0$, i.e., ${}_1F_1(1+1/(\lambda\tau), 1, -I_b/I_0) = 0$. In the practically interesting case of sufficiently large aperture the numerical solution yields approximate equation for beam lifetime:

$$\tau \approx \frac{0.612}{\lambda}\exp\left(0.812\frac{I_b}{I_0}\right), \tag{6.12}$$

which has an accuracy better than 2 % for $2.2 < \sqrt{I_b/I_0} < 8$. In the absence of damping the solution is:

$$f(I,t) = C\ e^{-t/\tau} J_0\left(\mu_0\sqrt{\frac{I}{I_b}}\right), \quad \tau = \frac{8I_b}{D{\mu_0}^2}, \mu_0 = 2.405\ldots, \tag{6.13}$$

where $J_0(x)$ is the Bessel function of zero order and μ_0 is its first root.

In conclusion we will find the beam lifetime in the absence of cooling for the case of equal aperture limitations and equal diffusions for both the horizontal and vertical planes. We assume a round vacuum chamber, equal beta-functions and sufficiently small momentum spread so that its contribution to the beam sizes could be neglected. Then Eq. (6.4) can be rewritten in the following form:

$$\frac{\partial f}{\partial t} = \frac{D}{2}\left(\frac{\partial}{\partial I_x}\left(I_x\frac{\partial f}{\partial I_x}\right) + \frac{\partial}{\partial I_y}\left(I_y\frac{\partial f}{\partial I_y}\right)\right), \tag{6.14}$$

with the boundary condition $f_0(I_x, I_y)\big|_{I_x+I_y=I_b} = 0$. Making transition to amplitudes, $\left(a_x = \sqrt{2I_x},\quad a_y = \sqrt{2I_y}\right)$ and using Eq. (6.7) we can rewrite Eq. (6.14) as follows:

$$\frac{\partial f}{\partial t} = \frac{D}{4}\Delta_4 f, \tag{6.15}$$

where Δ_4 denotes the laplacian in the four-dimensional phase space. An axial symmetric solution can be presented as a function of single variable—the radius in the four-dimensional space $a = \sqrt{{a_x}^2 + {a_y}^2}$. That yields:

$$\frac{\partial f_0}{\partial t} = \frac{D}{4}\frac{1}{a^3}\frac{d}{da}\left(a^3\frac{df_0}{da}\right). \tag{6.16}$$

Looking for solution in the following form $f(a,t) = C\ e^{-t/\tau} f_0(a)/a$ with boundary condition $f_0\left(\sqrt{2I_b}\right) = 0$ one obtains:

$$f(I,t) = C\frac{e^{-t/\tau}}{\sqrt{I/I_b}} J_1\left(\mu_{10}\sqrt{\frac{I}{I_b}}\right), \quad \tau = \frac{8I_b}{D{\mu_{10}}^2}, \mu_{10} = 3.832\ldots, \tag{6.17}$$

where function $J_1(x)$ is the Bessel function of the first order, and μ_{10} is its first root. Comparing Eqs. (6.13) and (6.17) one can see that the beam lifetime in the case of two-dimensional aperture limitation with round vacuum chamber is ≈2.54 times shorter than in the case of single dimensional aperture limitation.

6.1.2 *Multiple IBS*

IBS of charged particles in beams results in an exchange of energy between various degrees of freedom resulting in an increase of average energy of particles in the beam frame and an increase of the total beam emittance in 6D phase space. The total Coulomb cross section of a two-particle scattering in vacuum diverges; however, it has a finite value for collisions in plasma (or beam) due to field screening by other particles [1] or finite beam dimensions. Usually two scattering regimes are considered: (1) single scattering, when a rare single collision produces a large change of particle momentum (the Touschek effect), and (2) multiple scattering, when multiple frequent collisions cause diffusion. The first phenomenon is usually responsible for the creation of distribution tails and the beam loss in electron machines, while the second one for changes in the distribution core. Such separation is useful in many applications. It usually represents a good approximation for electron synchrotrons where the RF bucket length is much larger than the bunch length. However there are cases when it fails to deliver an accurate result. In particular it cannot be used to compute the beam lifetime in Tevatron where particles fill the entire RF bucket.

The IBS in accelerators is already a rather well-understood subject. The first decisive published work appears to be that of Piwinski [2], followed by Bjorken and Mtingwa [3]. These two earlier works were both carried out from first principles of two-body Coulomb collisions and largely ignored multiple scattering prior works on Coulomb scattering in plasma [4, 5] and astrophysics [6]. In what follows we re-derive the results of [3] using the Landau kinetic equation [4]. The new theoretical results include: (1) closed-form IBS rate expressions using symmetric elliptic integrals, (2) new IBS expressions for beams with coupled betatron motion and the vertical dispersion, and (3) a theoretical approach that combines the small and large angle Coulomb scattering. These new results are then compared with experimental measurements in several Fermilab machines.

Multiple Scattering in Single Component Plasma

Consider spatially homogeneous non-relativistic one-component plasma. The evolution of the velocity distribution function, $f(v_x, v_y, v_z)$, in such plasma is described by the Landau kinetic equation [4]:

$$\frac{df}{dt} = -2\pi n r_0^2 c^4 L_c \frac{\partial}{\partial v_i} \int \left(f \frac{\partial f'}{\partial v'_j} - f' \frac{\partial f}{\partial v_j} \right) \omega_{ij} d^3 v', \tag{6.18}$$

where

$$\omega_{ij} = \frac{\left(\mathbf{v} - \mathbf{v}'\right)^2 \delta_{ij} - \left(v_i - v_i'\right)\left(v_j - v_j'\right)}{\left|\mathbf{v} - \mathbf{v}'\right|^3}, \tag{6.19}$$

$$\int f(\mathbf{v}) d^3 v = 1, \tag{6.20}$$

r_0 is the particle classical radius, n is the plasma density, c is the speed of light, $L_c = \ln(\rho_{max}/\rho_{min})$ is the Coulomb logarithm,

$$\begin{aligned} \rho_{min} &= r_0 c^2 / \overline{v^2}, \\ \rho_{max} &= \sqrt{\overline{v^2} / 4\pi n r_0 c^2}, \end{aligned} \qquad \overline{v^2} = \sigma_{vx}^2 + \sigma_{vy}^2 + \sigma_{vz}^2, \tag{6.21}$$

and $\sigma_{vi} \equiv \sqrt{\overline{v_i^2}}, i = (x, y, z)$ are the rms velocity spreads. Equation (6.18) is obtained in the logarithmic approximation and is justified when $\ln(\rho_{max}/\rho_{min}) \gg 1$, where the plasma perturbation theory can be used. The Landau kinetic equation (6.18) is a nonlinear second-order integro-differential equation on the unknown function f, which is assumed to be nonnegative and integrable together with its moments up to order 2. The general time-dependent solution of this equation is not known. However, one can verify by inspection that the stationary solution of Eq. (6.18) is any Maxwellian velocity distribution function. It can be also easily checked that the following quantities are conserved:

$$\int f\, d^3 v, \int f v d^3 v, \text{ and } \frac{1}{2} \int f |v|^2 d^3 v, \tag{6.22}$$

corresponding respectively to the spatial density, momentum, and kinetic energy of plasma. The symmetric form of Eq. (6.18) can be easily rewritten in a Fokker–Planck form:

$$\frac{df}{dt} = -\frac{\partial}{\partial v_i}(F_i f) + \frac{1}{2}\frac{\partial}{\partial v_i}\left(D_{ij}\frac{\partial f}{\partial v_j}\right), \tag{6.23}$$

where

$$\begin{aligned} F_i(v) &= -4\pi n r_0^2 c^4 L_c \int f\left(v'\right) \frac{u_i}{|\mathbf{u}|^3} d^3 v', \\ D_{ij} &= 4\pi n r_0^2 c^4 L_c \int f\left(v'\right) \frac{u^2 \delta_{ij} - u_i u_j}{|\mathbf{u}|^3} d^3 v', \\ \mathbf{u} &= \mathbf{v} - \mathbf{v}', \end{aligned} \tag{6.24}$$

thus demonstrating that the Landau kinetic equation includes both the dynamic friction and the diffusion.

Let us assume a general Gaussian velocity distribution. By appropriate rotation of coordinate frame it can be reduced to the three-temperature distribution function

$$f = \frac{1}{(2\pi)^{3/2}\sigma_{vx}\sigma_{vy}\sigma_{vz}} \exp\left(-\frac{1}{2}\left(\frac{v_x^2}{\sigma_{vx}^2}+\frac{v_y^2}{\sigma_{vy}^2}+\frac{v_z^2}{\sigma_{vz}^2}\right)\right). \tag{6.25}$$

Then let us calculate the growth rate for the second moments of the distribution function

$$\Sigma_{ij} = \int f v_i v_j d^3v. \tag{6.26}$$

First, it is clear from symmetry of Eq. (6.25) that only the diagonal elements of $\boldsymbol{\Sigma}$ are nonzero, $\boldsymbol{\Sigma} = \mathrm{diag}(\sigma_{vx}^2, \sigma_{vy}^2, \sigma_{vz}^2)$. Second, using Eq. (6.18) we can find the rate of change of these second order moments due to Coulomb scattering in plasma

$$\frac{d}{dt}\Sigma_{ij} = \int \frac{\partial f}{\partial t} v_i v_j d^3v. \tag{6.27}$$

The result describes the rate of energy exchange between degrees of freedom in plasma:

$$\frac{d\boldsymbol{\Sigma}}{dt} = \frac{(2\pi)^{3/2} n r_0{}^2 c^4 L_c}{\sqrt{\sigma_{vx}^2+\sigma_{vy}^2+\sigma_{vz}^2}} \begin{pmatrix} \Psi(\sigma_{vx},\sigma_{vy},\sigma_{vz}) & 0 & 0 \\ 0 & \Psi(\sigma_{vy},\sigma_{vz},\sigma_{vx}) & 0 \\ 0 & 0 & \Psi(\sigma_{vz},\sigma_{vx},\sigma_{vy}) \end{pmatrix}. \tag{6.28}$$

The function $\Psi(x, y, z)$ can be expressed [7] through the symmetric elliptic integral of the second kind, $R_D(x, y, z)$, so that:

$$\Psi(x,y,z) = \frac{\sqrt{2}r}{3\pi}\left(y^2 R_D(z^2,x^2,y^2) + z^2 R_D(x^2,y^2,z^2) - 2x^2 R_D(y^2,z^2,x^2)\right), \tag{6.29}$$

$$R_D(u,v,w) = \frac{3}{2}\int_0^\infty \frac{dt}{\sqrt{(t+u)(t+v)(t+w)^3}} \tag{6.30}$$

where $r = \sqrt{x^2+y^2+z^2}$; $x, y, z \geq 0$. An algorithm for fast numerical evaluation of $R_D(u, v, w)$ is discussed in [7]. The function $\Psi(x, y, z)$ is chosen such that it depends on the ratios of its variables but not on r. It is symmetric with respect to the variables y and z, and is normalized such that $\Psi(0, 1, 1) = 1$. The energy conservation yields: $\Psi(1,0,1) = \Psi(1,1,0) = -\frac{1}{2}$ and $\Psi(x,y,z)+\Psi(y,z,x)+\Psi(z,x,y) = 0$. And in a thermal equilibrium, $\Psi(1, 1, 1) = 0$. In a case of two equal

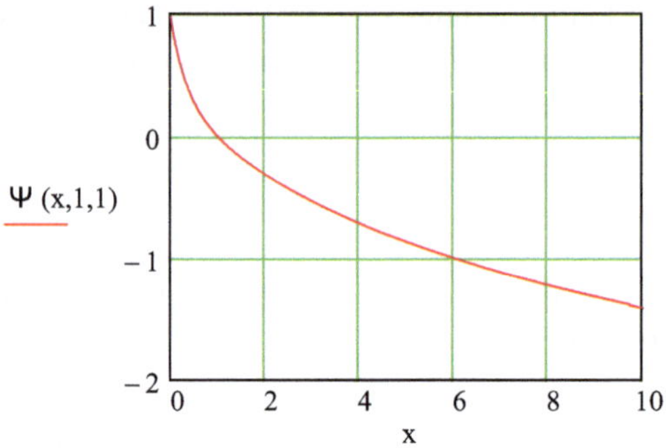

Fig. 6.1 The function $\Psi(x, 1, 1)$ for two equal temperatures

temperatures, the temperature relaxation rate given by (6.28) is identical to that given in [8] and can be expressed in terms of elementary functions:

$$\Psi(x, y, y) = \frac{\sqrt{2}\sqrt{x^2 + 2y^2}}{\pi y^2}\left(F\left(x^2, y^2\right)\left[y^2 + 2x^2\right] - 2x\right), \tag{6.31}$$

where

$$F(u, v) = \begin{cases} \dfrac{1}{\sqrt{(v-u)^3}}\left(v\mathrm{acos}\left(\sqrt{\dfrac{u}{v}}\right) - \sqrt{u}\sqrt{v-u}\right), & u < v, \\ \dfrac{2}{3\sqrt{v}}, & u = v, \\ -\dfrac{1}{\sqrt{(u-v)^3}}\left(v\mathrm{acosh}\left(\sqrt{\dfrac{u}{v}}\right) - \sqrt{u}\sqrt{u-v}\right), & u > v. \end{cases} \tag{6.32}$$

Figure 6.1 shows the function $\Psi(x, 1, 1)$ for x from 0 to 10. If one of the temperatures is zero, a very useful approximation (with a ~0.5 % accuracy) was obtained in [9]:

$$\Psi(0, x, y) \approx 1 + \frac{\sqrt{2}}{\pi}\ln\left(\frac{x^2 + y^2}{2xy}\right) - 0.055\left(\frac{x^2 - y^2}{x^2 + y^2}\right)^2. \tag{6.33}$$

Note that Eq. (6.28) is not self-consistent; i.e., it implies that the distribution function remains Gaussian; however, the diffusion and the friction due to multiple scattering does not allow for the beam distribution to remain Gaussian during the process of temperature exchange until it reaches the thermal equilibrium and the distribution becomes Maxwellian. Additionally, the single large-angle scattering (not considered in this section) creates non-Gaussian tails. However deviations from the Gaussian distribution are comparatively small and Eq. (6.28) represents a good approximation in most practical cases. In particular, it describes well the temperature relaxation in the course of electron beam transport on the case of non-magnetized electron cooling.

Multiple Scattering (IBS) in Accelerators

Although the beam focusing and dispersion effects in accelerators significantly complicate computation of IBS, they do not change the essence of the process. However, the time-dependent nature of transverse focusing results in a continuous growth of the 6-D beam emittance [2, 3]. In difference to plasma where the energy is conserved the binary collisions do not conserve energy in the beam frame resulting in unlimited emittance growth supported by energy transfer from the longitudinal beam motion (with nearly infinite reservoir of energy) to the internal particle motion in the beam frame (BF). The calculations of the IBS rates in accelerators will proceed in the following manner. First, one can verify that for the typical parameters at the Tevatron complex, the beam velocity spread in the BF is far from being relativistic[2] and the particle collision time, ρ_{max}/v, is much smaller than the period of betatron oscillations in the BF. This implies that the results of the previous section can be used in each location and then averaged over the entire length of the accelerator to obtain the overall IBS rates. Second, we will make an assumption that at each location of the accelerator the distribution function, $f(\mathbf{v}, \mathbf{r}, t)$, in the BF, is Gaussian in the 6D phase space. Third, we will calculate the growth rates in the BF using Eq. (6.28). And finally, we will convert these rates into the laboratory frame (LF) emittance growth rates.

Generalizing the Landau kinetic equation [see Eq. (6.23)] for a spatially non-homogeneous distribution function in the BF one obtains:

$$\frac{df}{dt} = -2\pi N r_0^2 c^4 L_c \frac{\partial}{\partial v_i} \int \left(f \frac{\partial f'}{\partial v_j'} - f' \frac{\partial f}{\partial v_j} \right) \omega_{ij} d^3 v' d^3 r' \delta\left(\vec{r} - \vec{r}' \right), \tag{6.34}$$

where N is the number of particles per bunch, the distribution function is normalized to 1

$$\int f(\mathbf{v}, \mathbf{r}, t) d^3 v d^3 r = 1, \tag{6.35}$$

and we assume that in the absence of collisions the beam is in the equilibrium state, i.e., $\partial f / \partial t = 0$. The same as above we choose the initial distribution being Gaussian:

$$f(\mathbf{v}, \mathbf{r}) = \frac{\sqrt{|\Xi|}}{8\pi^3} \exp\left(-\mathbf{X}^T \Xi \mathbf{X} \right), \tag{6.36}$$

where $\mathbf{X}^T = (x, v_x, y, v_y, z, v_z)$, and Ξ is a symmetric positively defined 6×6 matrix determined by ring Twiss parameters and mode emittances of the beam

---

[2] Actually this condition is satisfied for all circular accelerators build to this time with exception of LEP presenting a weakly relativistic case.

(see Chap. 2). By definition its determinant is the squared product of inversed mode emittances, $|\Xi| \equiv \det(|\Xi|) = 1/(\varepsilon_1\varepsilon_2\varepsilon_3)^2$.

Let us define the second moments of the distribution function (6.36) in the BF:

$$\begin{aligned} \Sigma_{ij} &= \int f(\mathbf{v},\mathbf{r}) v_i v_j d^3v d^3r, \\ K_{ij} &= \int f(\mathbf{v},\mathbf{r}) r_i v_j d^3v d^3r, \\ \Lambda_{ij} &= \int f(\mathbf{v},\mathbf{r}) r_i r_j d^3v d^3r. \end{aligned} \tag{6.37}$$

Similar to Eq. (6.27) let us find the rate of change of these moments by using the Landau kinetic equation (6.34) for the distribution function (6.36). Simple integration immediately yields[3]

$$\frac{d\mathbf{K}}{dt} = \frac{d\mathbf{\Lambda}}{dt} = 0. \tag{6.38}$$

To calculate the rate for elements of matrix $\mathbf{\Sigma}$ we will introduce the rotation matrix $\mathbf{T}$, reducing $\mathbf{\Sigma}$ to its diagonal form

$$\mathbf{T}^{\mathrm{T}}\mathbf{\Sigma}\mathbf{T} = \begin{pmatrix} \sigma_1^2 & 0 & 0 \\ 0 & \sigma_2^2 & 0 \\ 0 & 0 & \sigma_3^2 \end{pmatrix}, \tag{6.39}$$

and the matrix function $\mathbf{\Psi}_{\mathrm{IBS}}(\mathbf{T}^{\mathrm{T}}\mathbf{\Sigma}\mathbf{T})$ such that

$$\mathbf{\Psi}_{\mathrm{IBS}}\left(\mathbf{T}^{\mathrm{T}}\mathbf{\Sigma}\mathbf{T}\right) = \begin{pmatrix} \Psi(\sigma_1,\sigma_2,\sigma_3) & 0 & 0 \\ 0 & \Psi(\sigma_2,\sigma_3,\sigma_1) & 0 \\ 0 & 0 & \Psi(\sigma_3,\sigma_1,\sigma_2) \end{pmatrix}, \tag{6.40}$$

with $\Psi(x,y,z)$ given by Eq. (6.29).

Similar to Eq. (6.28) one can now calculate the rate for matrix $\mathbf{\Sigma}$ due to the IBS in the beam frame in the absence of betatron and synchrotron oscillations:

$$\frac{d\mathbf{\Sigma}}{dt} = \frac{N r_0{}^2 c^4 L_c}{4\sqrt{2} a_1 a_2 a_3 \sqrt{\mathrm{Tr}(\mathbf{\Sigma})}} \mathbf{T}\mathbf{\Psi}_{\mathrm{IBS}}\left(\mathbf{T}^{\mathrm{T}}\mathbf{\Sigma}\mathbf{T}\right)\mathbf{T}^{\mathrm{T}}, \tag{6.41}$$

where $a_1 a_2$ and a_3 are the rms sizes of 3D bunch ellipsoid (principle rms beam dimensions). We will now convert this rate to the laboratory frame and calculate the

[3] Note that at this point we consider only variations of f related to the scattering. Effects of betatron motion will be accounted at the next stage of calculations.

emittance growth rates. For accelerators with linear optics, the bunch distribution (6.36) is defined in the LF as:

$$F(\mathbf{r},\boldsymbol{\theta}) = \frac{1}{(2\pi)^3 \varepsilon_1 \varepsilon_2 \varepsilon_3} \exp\left(-\frac{I_1}{\varepsilon_1} - \frac{I_2}{\varepsilon_2} - \frac{I_3}{\varepsilon_3}\right), \tag{6.42}$$

where $\mathbf{r}$ and $\boldsymbol{\theta}$ are the LF canonical coordinates and momenta, I_k are the particle's action variables, and ε_k are the mode emittances of the beam. The mode emittances are defined such that:

$$\varepsilon_k = \int I_k F(\mathbf{r},\boldsymbol{\theta}) d^3\theta d^3 r. \tag{6.43}$$

For a given particle the action variables are bi-linear forms of coordinates and momenta:

$$I_k = B_{ij}^k \theta_i \theta_j + C_{ij}^k \theta_i r_j + Q_{ij}^k r_i r_j, \tag{6.44}$$

where $\mathbf{B}$, $\mathbf{C}$, and $\mathbf{Q}$ are real 3×3 matrices, defined by the accelerator lattice at each location, and summation on the repeated subscripts is implied. It is obvious that (6.36) implies,

$$-\ln(F) = \sum_k \frac{I_k}{\varepsilon_k} = \frac{1}{2}\Theta_{ij}\theta_i\theta_j + \frac{1}{2}H_{ij}\theta_i r_j + \frac{1}{2}M_{ij} r_i r_j, \tag{6.45}$$

where $\boldsymbol{\Theta}$, H, and $\mathbf{M}$ are real 3×3 matrices, uniquely defined by the lattice at each accelerator location and related to BF matrices (6.37). We can now calculate the emittance growth rates as

$$\frac{d\varepsilon_k}{dt} = \left\langle \int I_k \frac{dF}{dt} d^3\theta d^3 r \right\rangle_s, \tag{6.46}$$

where $\langle\rangle_s$ implies averaging over the accelerator circumference. Using (6.40) and (6.41) we obtain

$$\frac{d\varepsilon_k}{dt} = \frac{N r_0^2 c^2}{4\sqrt{2}\beta^2\gamma^4} \left\langle \frac{L_c \sum_{i,j=1}^{3} B_{ij}^k R_{ij}}{a_1 a_2 a_3 \sqrt{\mathrm{Tr}(\boldsymbol{\Sigma})}} \right\rangle_s, \tag{6.47}$$

where β and γ are the usual relativistic Lorentz factors of the bunch in the lab frame, a_i are the principle rms bunch sizes in the laboratory frame, and the 3×3 matrix $\mathbf{R}$ is given by the following expression:

$$\mathbf{R} = \left(\mathbf{G}^{-1}\right)^{\mathrm{T}} \mathbf{T} \boldsymbol{\Psi}_{\mathrm{IBS}} \left(\mathbf{T}^{\mathrm{T}} \boldsymbol{\Sigma} \mathbf{T}\right) \mathbf{T}^{\mathrm{T}} \mathbf{G}^{-1}. \tag{6.48}$$

The BF matrix $\boldsymbol{\Sigma}$ is related to the LF matrix $\boldsymbol{\Theta}$ as follows:

$$\boldsymbol{\Sigma} = (\beta\gamma c)^2 \mathbf{G}^{\mathrm{T}} \boldsymbol{\Theta}^{-1} \mathbf{G}, \tag{6.49}$$

where $\mathbf{G} = \mathrm{diag}(1, 1, 1/\gamma)$. Equation (6.47) is the most general IBS emittance growth rate for an arbitrary Gaussian 6D distribution function in a linear accelerator lattice.

Let us now discuss several specific cases. As a first example consider an accelerator lattice with uncoupled $x - y$ betatron motion. However we will account that both dispersions can be nonzero. Then,

$$\boldsymbol{\Theta} = \begin{pmatrix} \beta_x/\varepsilon_x & 0 & -\beta_x\Phi_x/\varepsilon_x \\ 0 & \beta_y/\varepsilon_y & -\beta_y\Phi_y/\varepsilon_y \\ -\beta_x\Phi_x/\varepsilon_x & -\beta_y\Phi_y/\varepsilon_y & \Theta_{33} \end{pmatrix}, \tag{6.50}$$

where $\Theta_{33} = {\sigma_p}^{-2} + \frac{A_x}{\varepsilon_x} + \frac{A_y}{\varepsilon_y}$, $\Phi_x = D_x' + \frac{\alpha_x D_x}{\beta_x}$, $\Phi_y = D_y' + \frac{\alpha_y D_y}{\beta_y}$, $A_x = \frac{1}{\beta_x}\left({D_x}^2 + (\beta_x\Phi_x)^2\right)$, $A_y = \frac{1}{\beta_y}\left({D_y}^2 + (\beta_y\Phi_y)^2\right)$; $F_D = 1 + \frac{{D_x}^2{\sigma_p}^2}{\varepsilon_x\beta_x} + \frac{{D_y}^2{\sigma_p}^2}{\varepsilon_y\beta_y}$;

β_x, β_y, α_x, and α_y, are the beta-functions and their negative half derivatives; D_x, D_y, D_x', and D_y' are the dispersions and their derivatives; ε_x, ε_y and $\varepsilon_z = \sigma_z\sigma_p$ are the non-normalized transverse and longitudinal rms beam emittances; σ_z is the rms bunch length and σ_p is the relative rms momentum spread. In this case the Eq. (6.47) becomes:

$$\frac{d\varepsilon_k}{dt} = \frac{N{r_0}^2c^2}{4\sqrt{2}\sigma_z\beta^2\gamma^4\sqrt{\varepsilon_x\varepsilon_y}} \left\langle \frac{L_c \sum_{i,j=1}^{3} \mathbf{B}_{ij}^k \mathbf{R}_{ij}}{\sqrt{\beta_x\beta_y F_D \mathrm{Tr}(\boldsymbol{\Sigma})}} \right\rangle_s, \tag{6.51}$$

with

$$\mathbf{B}^x = \begin{pmatrix} \beta_x & 0 & -\Phi_x\beta_x \\ 0 & 0 & 0 \\ -\Phi_x\beta_x & 0 & A_x \end{pmatrix}, \mathbf{B}^y = \begin{pmatrix} 0 & 0 & 0 \\ 0 & \beta_y & -\Phi_y\beta_y \\ 0 & -\Phi_y\beta_y & A_y \end{pmatrix},$$
$$\mathbf{B}^z = \begin{pmatrix} 0 & 0 & 0 \\ 0 & 0 & 0 \\ 0 & 0 & \beta_z \end{pmatrix}, \tag{6.52}$$

where $\beta_z = \sigma_z/\sigma_p$. The Coulomb logarithm is computed similar to the plasma case with the following correction affecting the value of maximum impact parameter in Eq. (6.21):

$$\rho_{\max} = \min\left(\sigma_{\min}, \gamma\sigma_z, \sqrt{\frac{\mathrm{Tr}(\boldsymbol{\Sigma})}{4\pi n r_0 c^2}}\right), \tag{6.53}$$

where $2\sigma_{\min}^2 = \sigma_x^2 + \sigma_y^2 - \sqrt{(\sigma_x^2 - \sigma_y^2)^2 + 4D_x^2 D_y^2 \sigma_p^4}$, $\sigma_x^2 = \varepsilon_x\beta_x + D_x^2\sigma_p^2$, and $\sigma_y^2 = \varepsilon_y\beta_y + D_y^2\sigma_p^2$. If one of the dispersions is equal to zero, then $\sigma_{\min} = \min(\sigma_x, \sigma_y)$.

The above equations can be used for a coasting beam with following substitutions: $\sigma_z \to L/(2\sqrt{\pi})$, $\mathbf{B}_z = \mathrm{diag}(0,0,2)$ implying that $d\varepsilon_3/dt \to d\overline{\sigma_p^2}/dt$. Note that the factor of 2 in matrix $\mathbf{B}_z$ reflects the absence of the synchrotron motion, taken into account in Eq. (6.51).

In many practical applications the longitudinal temperature in the BF is much smaller than the transverse one ($\theta_{||}/\gamma \ll \theta_\perp$) and the vertical dispersion can be neglected. The Eq. (6.51) (bunched beam) then can be reduced to the following:

$$\frac{d}{dt}\begin{pmatrix} \varepsilon_x \\ \varepsilon_y \\ \sigma_p^2 \end{pmatrix} = \frac{N r_0^2 c}{4\sqrt{2}\beta^3\gamma^3\sigma_z}\left\langle \frac{L_c}{\sigma_x\sigma_y\theta_\perp}\begin{pmatrix} \Psi_x(\theta_x,\theta_y) \\ \Psi_y(\theta_x,\theta_y) \\ \Psi(0,\theta_x,\theta_y) \end{pmatrix}\right\rangle_s, \tag{6.54}$$

where $\Psi_y(\theta_y,\theta_x) = \Psi(\theta_y,\theta_x,0)\beta_y/\gamma^2$, $\Psi_x(\theta_x,\theta_y) = \Psi(0,\theta_x,\theta_y)A_x + \Psi(\theta_x,\theta_y,0)\beta_x/\gamma^2$, $\theta_\perp = \sqrt{\theta_x^2 + \theta_y^2}$, $\theta_x^2 = \varepsilon_x/\beta_x(1 + \sigma_p^2(\Phi_x\beta_x)^2/\sigma_x^2)$, $\theta_y^2 = \varepsilon_y/\beta_y$. For an ultra-relativistic machine, $\gamma \gg Q_x$, the terms with $\beta_{x,y}/\gamma^2$ are small and can be neglected.

Similar to the suppression of IBS in an electron beam by magnetic field the multiple IBS is suppressed if σ_p becomes so small that the collision time becomes comparable to the betatron frequency [10]. Such conditions can be achieved in the case of deep beam cooling.

6.1.3 Single and Multiple Scattering at the Residual Gas

It was already mentioned in the previous section that the Coulomb scattering creates non-Gaussian tails. In this section we will consider how Coulomb scattering can be treated so that both single and multiple scattering are correctly accounted. The diffusion coefficient for particle scattering in the medium is well known. For scattering in a ring it can be expressed in the following form:

$$D = 4\pi c\beta\left(\frac{r_0}{\gamma\beta^2}\right)^2 \sum_i Z_i(Z_i+1)L_c^i \oint n_i(s)\beta_x(s)\frac{ds}{L}, \tag{6.55}$$

where the summing is performed over partial densities of residual gas, Z_i is their charges, and the integration over ring circumference averages the gas density, $n_i(s)$, weighted by the horizontal beta-function, $\beta_x(s)$. For simplicity of equation we are considering an evolution of particle distribution in the horizontal plane. The same expressions are applicable to the vertical plane. For the high energy scattering ($\beta > \alpha_{FS}Z_i$, $\alpha_{FS} \approx 1/137$) the Coulomb logarithm is:

$$L_c^i = \ln\left(\frac{\theta_{\max}^i}{\theta_{\min}^i}\right), \theta_{\min}^i \approx \alpha_{FS} Z_i^{1/3} \frac{m_e c}{p}, \theta_{\max}^i \approx \min\left(\frac{274}{A_i^{1/3}} \frac{m_e c}{p}, \theta_b\right), \tag{6.56}$$

where p is the particles momentum and the minimum and maximum angles are determined by the field screening due to atomic electrons and by the diffraction on nuclei.

The solutions of Eq. (6.5) with diffusion of Eq. (6.55) are commonly used to describe the emittance growth in particle accelerators due to various random diffusion processes, including multiple elastic Coulomb scattering. This equation describes the core of the beam distribution well, but completely fails to describe its tails in the case of Coulomb scattering. Far-away tails can be sufficiently well estimated using a single scattering approximation, but in many applications a prediction of tails behavior in the vicinity of the core is required. It is possible to computer-model the distribution function by Monte Carlo methods. However, we found it beneficial to advance the analytical treatment of the Coulomb scattering process to a point, where, for a given residual gas pressure, the distribution function can be obtained by solving an integro-differential equation, proposed below. In what follows we consider Coulomb scattering on the residual gas, but the theory can be easily adapted to other Coulomb scattering phenomena.

To simplify formulas we omit the summation over different gas species below. Neglecting the nuclear form-factor, one can write the differential elastic small angle cross section in the following form [11]:

$$\frac{d\sigma}{d\Omega} \approx 4Z(Z+1)\left(\frac{r_0}{\gamma\beta^2}\right)^2 \frac{1}{\left(\theta_x^2 + \theta_y^2 + \theta_{\min}^2\right)^2}, \tag{6.57}$$

After integrating this over θ_y one obtains the one-dimensional cross section

$$\frac{d\sigma}{d\theta_x} \approx 2\pi Z(Z+1)\left(\frac{r_0}{\gamma\beta^2}\right)^2 \frac{1}{\left(\theta_x^2 + \theta_{\min}^2\right)^{3/2}} \tag{6.58}$$

and the total cross section

$$\sigma_{\text{tot}} \approx \frac{4\pi Z^2}{\theta_{\min}^2}\left(\frac{r_0}{\gamma\beta^2}\right)^2. \tag{6.59}$$

For a combined treatment of both the small- and large-angle elastic scattering let us write the right-hand side of Eq. (6.5) in a general form of the collision integral [3]:

$$\frac{\partial f}{\partial t} - \lambda \frac{\partial (If)}{\partial I} = \left\langle \int_{-\infty}^{\infty} \left(\frac{d\sigma}{d\theta_x}\bigg|_{(\theta-\theta')} - \sigma_{tot}\delta\left(\theta - \theta'\right) \right) n v_o f \delta\left(x - x'\right) d\theta' dx' \right\rangle_{\psi,s}$$
$$= \left\langle \int_{-\infty}^{\infty} \left(2\pi Z(Z+1) \left(\frac{r_0}{\gamma\beta^2}\right)^2 \frac{1}{\left(\left(\theta_x - \theta_x'\right) + \theta_{min}^2\right)^{3/2}} - \sigma_{tot},\delta\left(\theta - \theta'\right) \right) \right.$$
$$\left. n v_o f \delta\left(x - x'\right) d\theta' dx' \right\rangle_{\psi,s}, \qquad (6.60)$$

where $\delta(\ldots)$ is the Dirac delta-function, and averaging is performed over the betatron phase, ψ, and machine circumference. Expressing particle angles and coordinates through the action-phase variables, (I, ψ), one obtains

$$\frac{\partial f}{\partial t} - \lambda \frac{\partial}{\partial I}(If) = \int_0^{I_b} W\left(I, I'\right) f\left(I', t\right) dI', \quad I < I_b, \qquad (6.61)$$

where I_b is the ring acceptance, and we introduced a scattering kernel, $W(I, I')$,

$$W\left(I, I'\right) = n v_o \left\langle \int_{-\infty}^{\infty} \left(\frac{d\sigma}{d\theta_x}\bigg|_{(\theta-\theta')} - \sigma_{tot}\delta\left(\theta - \theta'\right) \right) \delta\left(x - x'\right) d\varphi' \frac{d\varphi}{2\pi} \right\rangle_s . \qquad (6.62)$$

Neglecting θ_{min} in the cross section (6.58), denoting

$$B = 4\pi v_0 \left(\frac{r_p}{\gamma_0 \beta_0^2}\right)^2 Z(Z+1) \oint n(s)\beta(s)\frac{ds}{L}, \qquad (6.63)$$

and temporally omitting the term with $\sigma_{tot}\delta(\theta - \theta')$ one obtains

$$W\left(I, I'\right) = \frac{B}{16\pi} \int_0^{2\pi} d\varphi \int_0^{2\pi} d\varphi' \frac{\delta\left(\sqrt{I}\cos\varphi - \sqrt{I'}\cos\varphi'\right)}{\left|\sqrt{I}\sin\varphi - \sqrt{I'}\sin\varphi'\right|^3}. \qquad (6.64)$$

A lengthy integration yields[4]

---

[4] Method of the integration can be found below in the computation of similar integral for intra-beam scattering [see details further down Eq. (6.80)]. Note also that the Kernel (6.66) can be used without δ-function [like in Eq. (6.65)] in alternative form of integro-differential equation: $\partial f/\partial t - \lambda \partial (If)/\partial I = \int_0^{I_b} W\left(I, I'\right)\left(f\left(I', t\right) - f(I, t)\right) dI'$.

$$W\left(I,I'\right)=\frac{B}{4}\frac{I+I'}{\left|I-I'\right|^3}\ . \tag{6.65}$$

Neglecting $\theta_{\min}$ in our transition from Eq. (6.60) to Eq. (6.64) causes divergence of the integral (6.65) at $I \approx I'$, when the scattering angle is small. Instead of performing the exact integration using Eq. (6.58) (which is already an approximation), one can eliminate this divergence by modifying the kernel of Eq. (6.65) similar to the method used to limit the divergence in Eq. (6.57). Combining Eqs. (6.60) and (6.65) one obtains the kernel:

$$W\left(I,I'\right)=\frac{B}{4}\left(\frac{I+I'+I_{\min}/2}{\left(\left(I-I'\right)^2+\left(I+I'\right)I_{\min}+I_{\min}^2/4\right)^{3/2}}-\frac{2}{I_{\min}}\delta\left(I-I'\right)\right), \tag{6.66}$$

where to find the coefficient in front of δ-function we used the law of particle conservation which requires that

$$\int_0^\infty W\left(I,I'\right)dI'=0. \tag{6.67}$$

The divergence in Eq. (6.65) was eliminated artificially by adding terms containing $I_{\min}$ in a manner similar to $\theta_{\min}$ in Eq. (6.57). Thus, although the new kernel (6.66) is not exact, it, however, has the correct asymptotic. It is symmetric with respect to I and I' as it can be seen from its definition (6.64). At small scattering angles it has the accuracy similar to Eq. (6.5) but it correctly accounts for single and multiple scatterings. Note that the form of the kernel, $W(I,I')$, assumes that the range of beam particle angles is smaller than the maximum scattering angle $\theta_{\max}$, which is well justified in most practical cases. Otherwise $\theta_{\max}$ has to be explicitly taken into account in Eq. (6.57).

The accelerator aperture is always finite. Therefore, the upper limit in the integral of Eq. (6.65) should be replaced with the boundary action value, I_b. This also yields the boundary condition for the distribution function, $f(I_b, t) = 0$.

It is now quite trivial to obtain a Fokker–Planck equation from Eq. (6.61) by expanding the function f in series at $I'=I$, $f\left(I',t\right)\approx f(I,t)+f'(I,t)\left(I'-I\right)+\frac{1}{2}f''(I,t)\left(I'-I\right)^2$, and integrating to $I_b=\overline{\beta}\theta_{\max}^2$, where $\overline{\beta}$ is the average ring beta-function. The integration yields:

$$\frac{\partial f}{\partial t}-\lambda\frac{\partial}{\partial I}(If)\approx\frac{B}{4}\ln\left(\frac{I_{\max}}{I_{\min}}\right)\frac{\partial}{\partial I}\left[I\frac{\partial f}{\partial I}\right]. \tag{6.68}$$

Recalling that $\ln(I_{\max}/I_{\min})=2L_c$ we arrive to Eq. (6.5) with diffusion of Eq. (6.55).

Normally the difference between the maximum and minimum impact parameters is many orders of magnitude. It makes it impossible to solve Eq. (6.61) directly. However considered above diffusion model allows one to create an effective numerical algorithm which solves for the distribution function evolution under single and multiple scattering. For a numerical solution we split the total range of the action variable, $[0, I_b]$, into N equal size cells, $\Delta I = I_b/N$. Then, Eq. (6.61) can be rewritten as

$$\delta f_n = \frac{\delta t}{\Delta I}\left(\lambda \frac{f_{n+1}I_{n+1} - f_{n-1}I_{n-1}}{2} + \sum_{m=0}^{N-1} \widetilde{W}(n,m) f_m\right), \quad I_n = n\Delta I. \tag{6.69}$$

Taking into account that the cell size is much larger than the minimum action, $I_{\min}$, we can write the probability of a particle exchange for two distant cells

$$\widetilde{W}(n,m) = \frac{B}{4}\frac{n+m}{|n-m|^3}, \quad n \neq m, \ \ m \pm 1. \tag{6.70}$$

To find the probability of the particle exchange for nearby cells we use diffusion equation. It yields:

$$\widetilde{W}(n, n \pm 1) = \frac{BL_c}{2}\left(n \pm \frac{1}{2}\right). \tag{6.71}$$

The probability $\widetilde{W}(n,n)$ is determined by the particle conservation so that

$$\sum_{m=0}^{\infty} \widetilde{W}(n,m) = 0 \ . \tag{6.72}$$

The index m in this sum is running to infinity. It takes into account that a particle can be scattered outside of the accelerator aperture. Consequently, the particle number is not conserved in a finite aperture of a ring.

6.1.4 *Single and Multiple Scattering for Longitudinal Degree of Freedom in Hadron Colliders*

The single IBS scattering becomes important when there is a large difference between rms velocities of different degrees of freedom in the BF. In this case a single scattering can result in a momentum transfer significantly exceeding the rms value of the coldest degree of freedom thus creating non-Gaussian tails and particle loss. In ultra-relativistic colliders and storage rings the longitudinal momentum spread in the beam frame, $\Delta p'_{||} = \Delta p_{||}/\gamma$, is much smaller than the transverse one.

In this case a single scattering with large momentum transfer can result in that both particles scatter out of longitudinal machine acceptance. For the case when the longitudinal temperature is much smaller than the transverse one it is called the Touschek effect. It was first investigated in electron storage rings for a flat beam and non-relativistic energies in the BF [12, 13]. The radiation damping in electron synchrotrons typically results in the rms momentum spread being much smaller than the RF bucket height. That allows one to consider single and multiple scattering separately resulting in a simple treatment of the single IBS developed in [13]. However this approximation is usually not valid in hadron colliders where initially the beam is usually well inside of the RF bucket and the beam loss is dominated by single scattering; but shortly later, the multiple scattering results in the beam distribution reaching the RF bucket boundary leading to domination of multiple scattering in the particle loss.

As it was already pointed out the considered above model of IBS assumes that the beam stays Gaussian in the course of its distribution evolution and the focusing is linear. These conditions are satisfied sufficiently well for transverse degrees of freedom. However such approximation is quite coarse for the longitudinal degree of freedom if the bunch length is comparable to the RF bucket length. In further consideration we assume that the longitudinal momentum spread in the BF is much smaller than the transverse one ($\sigma_p/\gamma \ll \theta_\perp$); that the transverse distributions are Gaussian and do not depend on the longitudinal action; and we also take into account that the longitudinal motion is nonlinear and is described by dimensionless Hamiltonian:

$$H = \frac{\hat{p}^2}{2} + U(\phi) \xrightarrow[]{\substack{\text{Single}\\ \text{harmonic RF}}} \frac{\hat{p}^2}{2} + 2\Omega_s^2\left(\sin\frac{\phi}{2}\right)^2, \tag{6.73}$$

where Ω_s is the small amplitude synchrotron frequency.

In the case when multiple scattering is only accounted the evolution of longitudinal distribution and the particle loss from the RF bucket can be described by Eq. (6.2) with $\lambda = 0$:

$$\frac{\partial f}{\partial t} = \frac{1}{2}\frac{\partial}{\partial I}\left(\frac{D_{||}(I)}{\omega(I)} I \frac{\partial f}{\partial I}\right). \tag{6.74}$$

Here in comparison to Eq. (6.2) we redefined diffusion coefficient so that $D_{||}(\omega) = \omega(0)D(\omega)$. It simplifies formulas and looks more natural for description of diffusion in the longitudinal motion. To find $D_{||}(\omega)$ we will follow the following procedure. As one can see from Eq. (6.73) a momentum change $\delta\hat{p}$ results in an energy change: $\delta H = \hat{p}\,\delta\hat{p}$. Taking into account that $dH = \omega(I)\,dI$ we obtain:

$$\frac{d}{dt}\overline{\delta I^2} = \left\langle \frac{\hat{p}^2}{\omega(I)^2}\frac{d}{dt}\overline{\delta\hat{p}^2} \right\rangle_\psi, \tag{6.75}$$

where $\langle \ldots \rangle_\psi$ denotes averaging over synchrotron phase. On the other hand, multiplying Eq. (6.74) by $(I' - I)^2$ and integrating it with $f(I) = \delta(I' - I)$ one obtains

$$\frac{d}{dt}\overline{\delta I^2} = \frac{1}{2}\int \left(I' - I\right)^2 \frac{\partial}{\partial I'}\left(\frac{D_{||}(I')}{\omega(I')} I' \frac{\partial f}{\partial I'}\right) dI' = \\ \int \left(\frac{D_{||}(I')}{\omega(I')} I' + \left(I' - I\right)\frac{d}{dI'}\left(\frac{D_{||}(I')}{\omega(I')} I'\right)\right) f(I') dI' \xrightarrow{f(I')=\delta(I'-I)} \frac{D_{||}(I)}{\omega(I)} I. \tag{6.76}$$

Comparing Eqs. (6.75) and (6.76) one obtains

$$D_{||}(I) = \left\langle \frac{\hat{p}^2}{I\omega(I)} \frac{d}{dt}\overline{\delta\hat{p}^2} \right\rangle_\psi. \tag{6.77}$$

Using cross section of Eq. (6.57) and performing averaging one finally obtains[5]:

$$D_{||}(I) = \frac{\widetilde{A}L_c}{2\pi I\omega(I)} \oint \hat{p}^2 n(\phi) d\psi, \tag{6.78}$$

where we took into account that the local diffusion is proportional to the beam linear density $n(\phi) = \int f(I(\phi, \hat{p}))d\hat{p}$ normalized so that at the process beginning $\int_{-\pi}^{\pi} n(\phi)d\phi = 1$, the choice of parameter

$$\widetilde{A} = 4\pi^3\sqrt{2\pi}\frac{N r_0{}^2 f_{RF}{}^3 \eta^2}{\beta^4\gamma^3}\left\langle \frac{\Psi(0, \theta_x, \theta_y)}{\sigma_x\sigma_y\sqrt{\theta_x{}^2 + \theta_y{}^2}} \right\rangle_s, \tag{6.79}$$

will be clarified later in this section, f_{RF} is the RF frequency, η is the ring slip factor, and N is the initial particle number in the bunch. Substituting diffusion (6.78) into Eq. (6.6) (for $\lambda = 0$) and performing integration with linear RF and Gaussian distribution yield the momentum growth rate of Eq. (6.54).

---

[5] For rectangular distribution with total bunch length ϕ_{tot} ($n(\phi) = 1/\phi_{tot}$ within bunch) and linear RF one obtains $\oint \hat{p}^2 n(\phi) d\psi = 2\pi\omega I/\phi_{tot}$, and, consequently, $D_{||}(I) = \widetilde{A}L_c/\phi_{tot}$.

To take into account the single scattering one needs to replace the Fokker–Planck equation (6.74) by integro-differential equation[6] [14]:

$$\frac{\partial f(I,t)}{\partial t} = \int_0^{a_b} W\left(I,I'\right)\left(f\left(I',t\right) - f(I,t)\right)dI', \tag{6.80}$$

where a_b is the longitudinal acceptance determined by a finite value of the maximum momentum deviation or by RF bucket size. Further we will assume that the upper limit of the integral is equal to infinity but the boundary condition determines that $f(a) = 0$ for $a > a_b$. To proceed further we need to take into account that even strong single scattering which instantly moves a particle to the longitudinal bucket boundary implies a small angle scattering (in the BF) which differential cross section is proportional to $1/q^3$, where $\hat{q} = \hat{p} - \hat{p}'$ is the longitudinal momentum transfer. Integrating over longitudinal distribution and averaging over betatron motion one obtains:

$$\frac{\partial f(I,t)}{\partial t} = \frac{\widetilde{A}}{8\pi}\int_0^{\infty} n(\phi)\frac{f\left(I',t\right) - f(I,t)}{|\hat{p} - \hat{p}'|^3}\delta\left(\phi - \phi'\right)d\psi d\psi' dI'. \tag{6.81}$$

In difference to Eq. (6.60) the local diffusion (scattering) is position dependent, i.e., is proportional to the local density, $n(\phi)$. For now we also neglect the divergence in the integral at $\hat{p} = \hat{p}'$. This deficiency will be addressed later in a manner similar to the one used in the derivation of integro-differential equation (6.61) describing scattering on the residual gas. The parameter $\widetilde{A}$ used in Eq. (6.81) should be determined by averaging over transverse distributions and the ring circumference. As will be shown below that it is determined by Eq. (6.79). Comparing Eqs. (6.80) and (6.81) one can write for the kernel in the integral:

$$\begin{aligned} W\left(I,I'\right) &= \frac{\widetilde{A}}{8\pi}\int_{-\pi}^{\pi}\int_{-\pi}^{\pi} n(\phi)\frac{\delta(\phi - \phi')}{|\hat{p} - \hat{p}'|^3}d\psi d\psi' = \frac{\widetilde{A}\omega\omega'}{8\pi}\oint\oint n(\phi)\frac{\delta(\phi - \phi')}{|\hat{p} - \hat{p}'|^3}\frac{d\phi d\phi'}{\hat{p}\hat{p}'} \\ &= \widetilde{A}\frac{\omega\omega'}{4\pi}\int_{\max\left(a(I),a\left(I'\right)\right)}^{\min\left(b(I),b\left(I'\right)\right)} n(\phi)\left(\frac{1}{|\hat{p} - \hat{p}'|^3} + \frac{1}{|\hat{p} - \hat{p}'|^3}\right)\frac{d\phi}{\hat{p}\hat{p}'}. \end{aligned} \tag{6.82}$$

where $b(I)$ and $a(I)$ determine the range of motion ($\phi \in [a(I), b(I)]$), and we used that $d\psi = \omega d\phi/\hat{p}$. This equality can be obtained by making a ratio of two

[6] Here we choose alternative form of Eq. (6.61).

straightforward equations: $d\psi = \omega\ dt$ and $d\phi = \hat{p}\,dt$. Substituting $\hat{p} = \sqrt{2H(I) - U(\phi)}$ and performing integration one obtains for $I' \geq I$:

$$W\left(I,I'\right) = \frac{\widetilde{A}\omega\omega'}{8\pi\left(H\left(I'\right) - H(I)\right)^3}\left[\left(H\left(I'\right) - H(I)\right)\int_{a(I)}^{b(I)} n(\phi)\frac{d\phi}{\hat{p}} + 2\int_{a(I)}^{b(I)} n(\phi)\hat{p}\,d\phi\right],$$
$$I' \geq I. \tag{6.83}$$

Taking into account that the kernel is a symmetric function, $W(I, I') = W(I', I)$, one obtains the kernel for $I \geq I'$. To remove divergence at $I = I'$ we follow the recipe of Eq. (6.66). That yields:

$$W\left(I,I'\right) = \widetilde{A}\omega\omega' \frac{\left(H\left(I'\right) - H(I)\right)\int_{a(I)}^{b(I)} n(\phi)\frac{d\phi}{\hat{p}} + 2\int_{a(I)}^{b(I)} n(\phi)\hat{p}\,d\phi}{8\pi\left(\left(H\left(I'\right) - H(I)\right)^2 + \left(H\left(I'\right) - H(I)\right)\Delta E_{\min} + \Delta E_{\min}{}^2/4\right)^{3/2}},$$
$$I' \geq I. \tag{6.84}$$

To verify that the parameters $\widetilde{A}$ used in Eqs. (6.78) and (6.80) are equal we obtain the diffusion equation (6.74) from integro-differential (6.80). Expending distribution function into Tailor series one obtains:

$$\frac{\partial f(I,t)}{\partial t} \approx \int_{I_{\min}}^{I_{\max}} \left(W\left(I,I+I'\right)\left(f'(I)I' + f''(I)\frac{I'^2}{2}\right)\right.$$
$$\left. + W\left(I-I',I\right)\left(-f'(I)\Delta I + f''(I)\frac{I'^2}{2}\right)\right)dI', \tag{6.85}$$

where the term with $f(I)$ is equal to zero and was omitted due to particle conservation [see Eq. (6.67)]. Changing the integration variable to $H = I\omega$, introducing functions

$$B(H) = \frac{1}{2}\int_{a(H)}^{b(H)} n(\phi)\frac{d\phi}{\hat{p}} = \frac{1}{4\omega}\oint n(\phi)d\psi \text{ and } C(H) = \frac{1}{2}\int_{a(H)}^{b(H)} n(\phi)\hat{p}\,d\phi$$
$$= \frac{1}{4\omega}\oint n(\phi)\hat{p}^2 d\psi,$$

and expending the kernel to Tailor series we obtain:

$$
\begin{aligned}
\frac{\partial f}{\partial t} &\approx \frac{\widetilde{A}\omega}{4\pi}\int_{H_{\min}}^{H_{\max}}\left((hB(H)+C(H))\left(f'_H(H)h+f''_H(H)\frac{h^2}{2}\right)\right.\\
&\quad\left.+\left(h\left(B(H)-B'_H(H)h\right)+C(H)-C'_H h\right)\left(-f'_H(H)h+f''_H(H)\frac{h^2}{2}\right)\right)\frac{dh}{h^3}\\
&\approx\frac{\widetilde{A}\omega}{2\pi}\int_{H_{\min}}^{H_{\max}}\left(f'_H(H)C'_H+f''_H(H)C(H)\right)\frac{dh}{h}=\frac{\widetilde{A}\omega}{2\pi}\left(f'_H(H)C'_H+f''_H(H)C(H)\right)\\
&\quad\ln\left(\frac{H_{\max}}{H_{\min}}\right),
\end{aligned}
\tag{6.86}
$$

where $h = H(I') - H(I)$ and in the final two equalities we left only logarithmic terms. After simple transformations taking into account that $\ln(H_{\max}/H_{\min}) = 2L_c$ we obtain Eq. (6.74) with diffusion of Eq. (6.78).

In conclusion we note that to find an evolution of bunch population and rms emittances, Eq. (6.80) with kernel (6.84) has to be solved together with the equations describing the evolution of transverse emittances. For numerical solution one can adopt an algorithm considered above in Sect. 6.1.3.

6.2 Diffusion due to RF Noise

Another important mechanism instigating longitudinal emittance growth is related to noise in the RF system of a storage ring. In the absence of longitudinal cooling (which is normally the case for hadron colliders) this effect has to be accounted in computations of particle distribution evolution and all means have to be applied to minimize it to an acceptable level. In this section we generalize the theory which was initially developed in [15] and which describes an effect of RF system noise on the evolution of longitudinal particle distribution.

6.2.1 Noise Induced Diffusion for Nonlinear Oscillator

A particle motion in the field of nonlinear oscillator perturbed by noise can be described by a Hamiltonian $H(\phi, \hat{p}, t)$ consisting of the main unperturbed part and a perturbation driven by a small random value $\alpha(t)$:

$$H(\phi,\hat{p},t) = H_0(\phi,\hat{p}) + \alpha(t)\hat{V}(\phi,\hat{p}). \tag{6.87}$$

Let us apply a canonical transformation to the action and phase variables I, ψ of the unperturbed Hamiltonian: $H_0(\phi,\hat{p}) = H_0(I)$, leading to the oscillation frequency $\omega(I) = dH_0/dI$.

Perturbation changes the action I so that:

$$\frac{dI}{dt} = -\alpha(t)\frac{\partial\hat{V}}{\partial\psi} = -\frac{\alpha(t)}{\omega}\frac{d\hat{V}}{dt} \equiv -\frac{\alpha(t)}{\omega}W(\phi,\hat{p}), \tag{6.88}$$

where $W = -d\hat{V}/dt$ is a power function of the perturbation. The noise $\alpha(t)$ is conventionally described by its correlator $K(\tau)$:

$$\langle\alpha(t)\rangle = 0, \quad \langle\alpha(t)\alpha(t-\tau)\rangle = K(\tau). \tag{6.89}$$

Let Δt be a time interval, large compared with the noise correlation time, but still so small, that action changes only a little for that time.[7] Thus, at lowest order over the small parameter α, the action changes according to

$$\begin{aligned} \langle\Delta I^2\rangle &= \frac{1}{\omega^2}\left\langle\int_t^{t+\Delta t}\int_t^{t+\Delta t}\alpha(t_1)W(t_1)\alpha(t_2)W(t_2)dt_1dt_2\right\rangle \\ &= \frac{\Delta t}{\omega^2}\int_{-\infty}^{\infty}K(\tau)c(\tau)d\tau, \end{aligned} \tag{6.90}$$

where $c(\tau) \equiv \langle W(t)W(t-\tau)\rangle_t$ and $\langle\rangle_t$ denotes time averaging over a period of the nonlinear oscillations $T = 2\pi/\omega$.

Ensemble of these oscillators can be described by a distribution function $f(I, t)$. Its evolution is induced by the noise and satisfies the Fokker–Planck equation (FPE) of (6.74). Taking initial distribution being localized at small interval of actions near I, or $f(I') = \delta(I' - I)$, a change of action is determined by Eq. (6.76). Comparing Eqs. (6.90) and (6.76), one obtains the diffusion coefficient:

$$D_{||}(I) = \frac{1}{I\omega}\int_{-\infty}^{\infty}K(\tau)c(\tau)d\tau. \tag{6.91}$$

To express the diffusion through the noise spectral density we expend the periodic function $c(t)$ into the Fourier series:

[7] It implies that the emittance growth rate is sufficiently small which is true in hadron colliders and storage rings.

$$
\begin{aligned}
c(\tau) &= \sum_{n=-\infty}^{\infty} c_n \exp(in\omega\tau), \\
c_n &= \frac{1}{T}\oint \exp(-in\omega\tau)c(\tau)d\tau = \frac{1}{T^2}\oint\oint \frac{dV}{dt'}\frac{dV}{dt''}\exp\left(-in\omega\left(t' - t''\right)\right)dt'dt'' \\
&= |W_n|^2 = n^2\omega^2|V_n|^2,
\end{aligned} \tag{6.92}
$$

where the Fourier amplitudes are given by

$$
V_n = \frac{1}{T}\oint \hat{V}(\phi(t), \hat{p}(t))\exp(-in\omega t)dt. \tag{6.93}
$$

Usually, the unperturbed Hamiltonian is presented as a sum of kinetic and potential energies: $H_0(\phi, \hat{p}) = \hat{p}^2/2 + U(\phi)$. In this case, it can be more convenient to perform a calculation of the Fourier images (6.93) transforming them to integrals over the coordinate:

$$
W_n = \frac{1}{T}\oint W \exp(-in\omega t)dt = \frac{1}{T}\oint W \exp(-in\omega t(\phi))\frac{d\phi}{\hat{p}(\phi)}, \tag{6.94}
$$

where $t(\phi)$ is a function inverse to $\phi(t)$. Note that the functions $p(\phi)$ and $t(\phi)$ are not single-valued; they require separate analytic equations for every trajectory interval between the turning points in the unperturbed potential $U(\phi)$. In a simplest case when $U(\phi)$ is an even function with a single local minimum at $\phi = 0$ there are only two turning points, $\phi = \pm\alpha(I)$, and

$$
t(\phi) = \begin{cases} t_0(\phi) \equiv \int_0^x d\phi' / \left|\hat{p}\left(\phi'\right)\right|, & \text{for the half cycle with } \hat{p} > 0, \\ T_0/2 + t_0(-\phi), & \textit{otherwise}. \end{cases} \tag{6.95}
$$

These expressions can be further simplified when the perturbing potential $\hat{V}(t) = \hat{V}(\phi(t), \hat{p}(t))$ has a definite parity. If it is odd $\left(\hat{V}(t + T/2) = -\hat{V}(t)\right)$, all even harmonics of Eq. (94) are zeroed, and the result can be presented as:

$$
|W_n| = \begin{cases} \frac{4}{T}\left|\int_0^a \frac{d\phi}{\hat{p}} W \sin\left(n\omega t_0(\phi)\right)\right|, & \text{for odd } n \\ 0, & \text{otherwise} \end{cases}, \quad \text{for odd } \hat{V}, W. \tag{6.96}
$$

In opposite case

$$
|W_n| = \begin{cases} \frac{4}{T}\left|\int_0^a \frac{d\phi}{\hat{p}} W \cos\left(n\omega t_0(\phi)\right)\right|, & \text{for even } n \neq 0 \\ 0, & \text{otherwise} \end{cases}, \quad \text{for even } \hat{V}, W. \tag{6.97}
$$

The diffusion coefficient of Eq. (6.91) can be expressed as a Fourier series of the noise power at harmonics $n\omega$:

$$D(I) = \frac{4\pi}{I\omega} \sum_{n=1}^{\infty} |W_n|^2 P(n\omega), \tag{6.98}$$

where

$$P(\omega) = \frac{1}{2\pi} \int_{-\infty}^{\infty} K(\tau) \mathrm{e}^{-i\omega\tau} d\tau \tag{6.99}$$

is the noise spectral power at n-th harmonic of oscillations, so that $K(\tau) = \int_{-\infty}^{\infty} P(\omega) \mathrm{e}^{i\omega\tau} d\omega$.

The diffusion is driven by the noise power at resonance harmonics. If the noise is white, $P(n\omega) = P$ and $K(\tau) = 2\pi P\delta(\tau)$, the diffusion comes out as

$$D(I) = \frac{2\pi P \langle W^2 \rangle}{I\omega} = \frac{P}{I} \oint W^2 dt. \tag{6.100}$$

The above equations solve a general problem of nonlinear oscillator affected by noise. The distribution function of such oscillators is described by the FPE (6.74), and the diffusion coefficient is calculated by one or another method described above.

6.2.2 *Linear Oscillator*

A simplest application of the suggested approach is a linear oscillator driven by noise. Then $H_0 = \hat{p}^2/2 + U(\phi)$,

$$U(\phi) = {\Omega_s}^2 \phi^2/2, \quad \phi = \sqrt{2I/\Omega_s} \sin\psi, \hat{p} = \sqrt{2I\Omega_s} \cos\psi. \tag{6.101}$$

In the case of phase noise the perturbation is described by $\hat{V}(\phi) = {\Omega_s}^2 \phi$ and $\alpha(t) \equiv \tilde{\phi}(t)$, where $\tilde{\phi}(t)$ is the function describing RF phase fluctuations with corresponding spectral density of $P_\phi(\omega)$. That yields a constant diffusion determined by the noise spectral density at the synchrotron frequency:

$$D = 2\pi {\Omega_s}^4 P_\phi(\Omega_s), \quad \frac{d}{dt}\overline{\phi^2} = \pi {\Omega_s}^2 P_\phi(\Omega_s), \quad \frac{d}{dt}\bar{I} = \pi {\Omega_s}^3 P_\phi(\Omega_s). \tag{6.102}$$

In the case of amplitude noise the perturbation is described by $\hat{V}(\phi) = {\Omega_s}^2 \phi^2/2$ and $\alpha(t) = \tilde{V}(t)/V_0$, where $\alpha(t)$ is the relative RF voltage fluctuations with corresponding spectral density of $P_u(\omega)$. That yields a constant diffusion linearly growing with action determined by the noise spectral density at the double synchrotron frequency:

$$D = \pi \Omega_s^3 I P_u(2\Omega_s), \quad \frac{d}{dt}\overline{\phi^2} = \pi \Omega_s^2 \overline{\phi^2} P_u(2\Omega_s),$$
$$\frac{d}{dt}\bar{I} = \pi \Omega_s^2 I P_u(2\Omega_s). \tag{6.103}$$

6.2.3 Phase and Amplitude Noise in Storage Rings with Single Harmonic RF

Now let us apply a general theory developed above to an analysis of the bunch widening due to noise in a single harmonic RF system. Then the unperturbed longitudinal motion is described by the Hamiltonian:

$$H(\phi, p) = \frac{\hat{p}^2}{2} + \Omega_s^2(1 - \cos\phi), \tag{6.104}$$

where $\Omega_s = \omega_0\sqrt{eV_0h\eta/(2\pi mc^2\gamma\beta^2)}$ is the synchrotron frequency, η is the ring slip-factor, ω_0 is the angular frequency of particle revolution, and h is the harmonic number. The canonical momentum $\hat{p}$ is associated with the energy offset as $\hat{p} = \eta h\omega_0\,\delta p/p$. A noise in the amplitude $\widetilde{V}(t)$ and phase $\widetilde{\phi}(t)$ of RF voltage is accounted by substitution,

$$V_0 \to V_0 + \widetilde{V}(t), \quad \phi \to \phi + \widetilde{\phi}(t),$$

leads to the Hamiltonian similar to (6.87):

$$\begin{aligned} H(\phi, \hat{p}) &= \frac{\hat{p}^2}{2} + \Omega_s^2(1 - \cos(\phi)) - \Omega_s^2\widetilde{u}(t)\cos\phi + \Omega_s^2\widetilde{\phi}(t)\sin\phi \\ &= H_0 + \widetilde{u}(t)V_u + \widetilde{\phi}(t)V_\phi. \end{aligned} \tag{6.105}$$

where $\widetilde{u}(t) = \widetilde{V}(t)/V_0$, and we assume that $\gamma > \gamma_t$.

The solution for unperturbed motion is well known. The turning point phase, the action and frequency are:

$$\begin{aligned} \phi_{\max} &= 2a\sin\left(\sqrt{H/(2\Omega_s)}\right), \\ I &= \frac{1}{2\pi}\oint \hat{p}\,dx = \frac{8\Omega_s}{\pi}(\mathrm{E}(\kappa) - (1-\kappa^2)K(\kappa)) \le \frac{8\Omega_s}{\pi}, \\ \omega &= 2\pi\left(\oint \frac{dx}{\hat{p}}\right)^{-1} = \frac{\pi\Omega_s}{2K(k)} \le \Omega_s, \end{aligned} \tag{6.106}$$

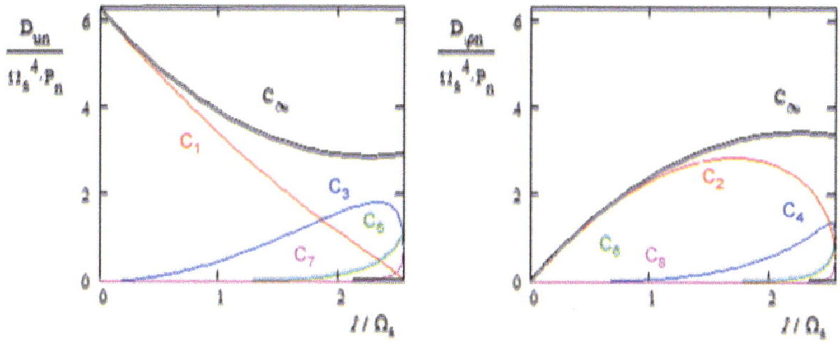

Fig. 6.2 Dimensionless diffusions for the phase (*left*) and amplitude (*right*) RF noise

where $\kappa = \sin(\phi_{\max}/2) = \sqrt{H/(2\Omega_s)}$, and a substitution

$$\sin(\phi/2) = \sin(\phi_{\max}/2)\sin\psi \equiv \kappa\sin\psi \tag{6.107}$$

was used to reduce the integrals to the elliptic functions: $\mathrm{E}(k) = \int_0^{\pi/2}\sqrt{1-k^2\sin^2(\phi/2)}d\phi$ and $K(k) = \int_0^{\pi/2} d\phi/\sqrt{1-k^2\sin^2(\phi/2)}$.

To find the diffusion for the white noise we use (6.100), where we take into account that the correlation functions for phase and amplitude noises are $K_\phi(\tau) = 2\pi P_\phi\delta(\tau)$ and $K_u(\tau) = 2\pi P_u\delta(\tau)$, and P_ϕ and P_u are the spectral powers of the phase and amplitude noise $\tilde{\phi}(t)$ and $\tilde{u}(t)$. Simple calculations yield both diffusion coefficients:

$$D_\phi = \frac{8P_\phi\Omega_s^{\,5}}{I}\int_0^{\phi_{\max}}\cos^2\phi\sqrt{\kappa^2-\sin^2(\phi/2)}d\phi = \frac{8P_\phi\Omega_s^{\,5}}{I}\left(2E(\kappa)-2\left(1-\kappa^2\right)K(\kappa)-\mathrm{R}(\kappa)\right),$$
$$D_u = \frac{8P_u\Omega_s^{\,5}}{I}\int_0^{\phi_{\max}}\sin^2\phi\sqrt{\kappa^2-\sin^2(\phi/2)}d\phi = \frac{8P_u\Omega_s^{\,5}}{I}\mathrm{R}(\kappa), \tag{6.108}$$

where

$$\mathrm{R}(\kappa) = \frac{8}{15}\left[2\left(1-\kappa^2+\kappa^4\right)E(\kappa) - \left(2-3\kappa^2+\kappa^4\right)K(\kappa)\right]. \tag{6.109}$$

The curves marked as C_∞ in Fig. 6.2 present dependencies of these diffusions on the action in a dimensionless form. For small amplitude oscillations these functions coincide with the results for linear approximation of Eqs. (6.102) and (6.103).

For colored noise using Fourier amplitudes of Eq. (6.93) one obtains:

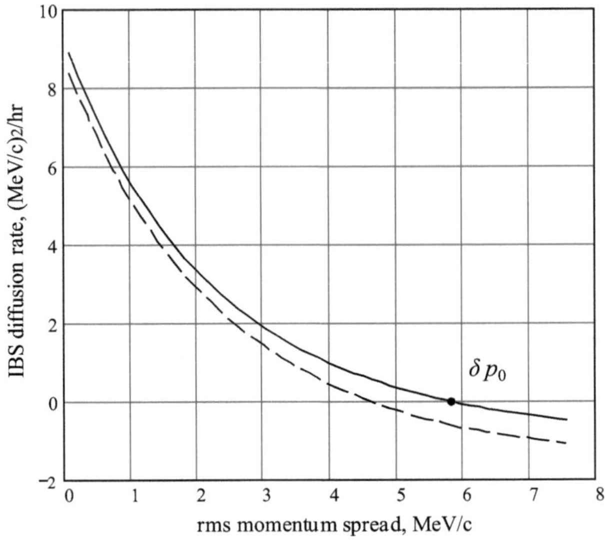

Fig. 6.3 The calculated longitudinal IBS diffusion rate as a function of the rms beam momentum spread in the Recycler for a coasting beam of 100 × 10[12] antiprotons with a constant 5 μm emittance (norm. 95 %). The *solid line* is modeled with realistic beta-functions; the *dashed line* is the round-beam, zero-dispersion, smooth approximation

$$D_\phi = \sum_{n=1}^{\infty} P_\phi(n\omega) C_{\phi n}(I), \quad C_{\phi n}(I) = \frac{\omega \Omega_s{}^4}{\pi I} \left| \oint d\phi \cos\phi \exp(in\omega t(\phi)) \right|^2,$$
$$D_u = \sum_{n=1}^{\infty} P_u(n\omega) C_{un}(I), \quad C_{un}(I) = \frac{\omega \Omega_s{}^4}{\pi I} \left| \oint \frac{d\phi}{2\pi} \sin\phi \exp(in\omega t(\phi)) \right|^2. \tag{6.110}$$

Due to symmetry function $C_{\phi n}(I) = 0$ for even n, and $C_{un}(I) = 0$ for odd. The first few nonzero functions are presented in Fig. 6.3. In the case of white noise the spectral density is the same at all harmonics. Consequently, it can be moved out of sums in Eq. (6.110). Then the coefficients

$$C_{\phi\infty} = \sum_{n=1}^{\infty} C_{\phi n}, \quad C_{u\infty} = \sum_{n=1}^{\infty} C_{2n}, \tag{6.111}$$

describe the diffusion due to white noise so that $D_\phi = PC_{\phi\infty}$ and $D_u = PC_{u\infty}$. These equations are identical to Eq. (6.108).

The integrals in Eq. (6.110) can be calculated using properties of the Jacobian elliptic functions appearing after the substitution (6.107) and leading to:

$$\theta = am(\vartheta, \kappa), \quad \sin(\phi/2) = \kappa sn(\vartheta, \kappa), \quad p = 2\Omega_s cn(\vartheta, \kappa) \text{ with}$$
$$\vartheta = 2K(\kappa)\psi/\pi. \tag{6.112}$$

From here (see [16], p. 911, [17], pp. 292, 295),

$$V_{\phi,n} = {\Omega_s}^2 \oint \frac{d\psi}{2\pi} \sin\phi \exp(in\psi) = -\omega \oint \frac{d\psi}{2\pi} \phi \exp(in\psi) \frac{dp}{d\psi}$$

$$= \begin{cases} \dfrac{2in\omega^2}{\cosh(n\nu)}, & \text{for odd } n \\ 0, & \text{otherwise } n \end{cases} \quad V_{u,n} = {\Omega_s}^2 \oint \frac{d\psi}{2\pi} \cos\phi \exp(in\psi)$$

$$= -2{\omega_s}^2 \kappa^2 \oint \frac{d\psi}{2\pi} sn^2(\vartheta,\kappa) \exp(in\psi) = \begin{cases} \dfrac{2in\omega^2}{\sin h(n\nu)}, & \text{for even } n \\ 0, & \text{otherwise } n \end{cases} \tag{6.113}$$

where $\nu = \pi K'(\kappa)/(2K(\kappa))$.

In conclusion we need to note that in the case when the bandwidth of RF noise is larger than the revolution frequency (like for the LHC RF system) the summing in Eq. (6.110) should also include the revolution frequency harmonics: $\sum_{n=1}^{\infty} C_n P(n\omega_s) \to \sum_{n,m=1}^{\infty} C_n P(n\omega_s + m\omega_0)$.

6.3 Experimental Studies of Single and Multiple Scattering

6.3.1 IBS in the Fermilab Recycler Ring

The Fermilab Recycler is a 3.3-km 8.9-GeV/c fixed momentum storage ring located in the Fermilab Main Injector tunnel. Table 6.1 presents relevant Recycler parameters. On the Run II, the Recycler played a key role as the repository of large stacks of antiprotons (6×10^{12}) with the appropriate phase space characteristics to be used in collider stores. Small-angle multiple IBS is the dominant heating mechanism, which determines the equilibrium emittance achievable in the Recycler.

The longitudinal IBS heating has been of concern because of a small longitudinal emittance required for Tevatron collisions. Figure 6.3 shows the calculated longitudinal IBS diffusion rate. The IBS theory, described above, has been used to make these calculations. Also, the measured Recycler lattice functions have been used. One can see in Fig. 6.3 (solid line) that the longitudinal heating vanishes for a certain rms momentum spread, δp_o, and becomes cooling above this momentum spread. For comparison, also shown in Fig. 6.3 is the heating (cooling) rate calculated in a smooth, round-beam, zero-dispersion approximation using the value of the average β-function, β_{ave} from Table 6.1. One can exploit this feature of the longitudinal diffusion rate to minimize the longitudinal emittance growth by compressing the bunch length, and thus increasing the momentum spread, until the diffusion rate vanishes. In the Recycler ring this is accomplished by employing a barrier-bucket RF system. For a typical 95 % transverse emittance of 5–7 mm mrad the point of vanishing longitudinal heating cannot be reached without losing the "tail" particles at the momentum aperture. A practical solution has been to maintain the momentum spread as high as possible (around 4 MeV/c), while preserving the

Table 6.1 Recycler ring parameters

Parameter	Value	Units
Average β-function, β_{ave}	30	m
Max. dispersion	2	m
Transition, γ_t	20.7	
Typ. transverse beam emittances (n, 95 %), ε_n	3–7	mm mrad
Number of antiprotons	≤ 6	10^{12}

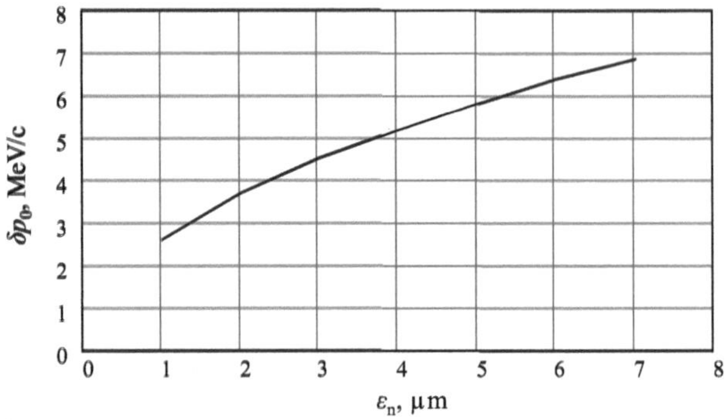

Fig. 6.4 The calculated value of the rms momentum spread in MeV/c, for which the longitudinal IBS heating vanishes, as a function of beam emittance (norm., 95 %)

beam lifetime. Figure 6.4 shows the value of the rms momentum spread, δp_o, for which the diffusion rate vanishes, as a function of the transverse beam emittance.

Below, we describe an experiment in which we created a beam distribution with the momentum spread less than δp_o for a given emittance and observed longitudinal heating, and an experiment in which we created a beam distribution with the momentum spread greater than δp_o for a given emittance and observed longitudinal cooling.

To verify the Recycler IBS model we first created a beam distribution with a large transverse emittance and a small rms momentum spread. According to the IBS model this would correspond to longitudinal heating and small transverse cooling. The vacuum-related transverse heating rate was measured in a separate experiment by recording emittances of a coasting beam with very low current (1×10^{10}) to avoid the IBS contribution. This vacuum-related growth rate was measured to be 0.60 mm mrad/h. Figure 6.5 shows the transverse emittance evolution for a costing beam of 100×10^{10} protons.

One can observe from Fig. 6.5 that the emittance growth rate is in fact smaller than that for a zero beam current. We attribute the difference to IBS-related transverse cooling, which, according to our model, is −0.15 mm mrad/h.

The longitudinal degree of freedom allows us to observe the effect of IBS directly because it is the dominant effect. All other heating mechanisms are negligible for high beam currents. Figure 6.6 shows the evolution of the rms momentum spread for the same beam as in Fig. 6.5. The IBS model uses the transverse emittance growth rate as an input parameter.

The second experiment was to demonstrate the longitudinal IBS cooling. The measurements were conducted with a bunched antiproton beam of 25×10^{10}, which was initially cooled transversely to a very small emittance (<2 mm mrad).

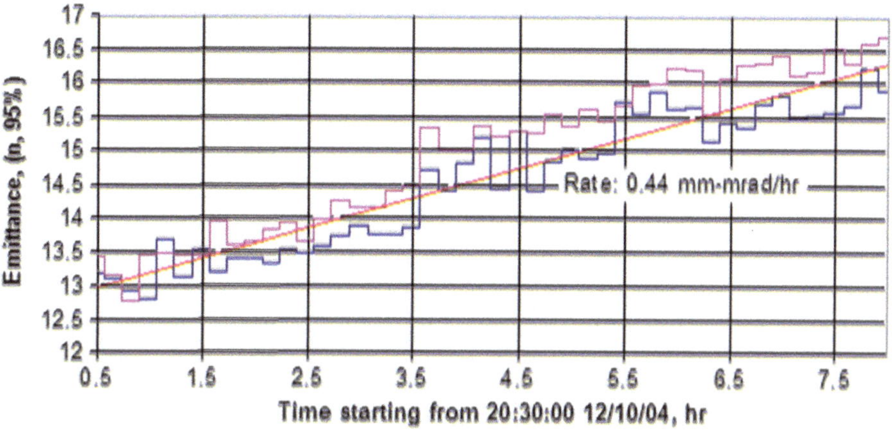

Fig. 6.5 The transverse emittance (horizontal and vertical) evolution for a coasting beam of 100×10^{10} protons in the Recycler. The linear fit yields a growth rate of 0.44 μm/h

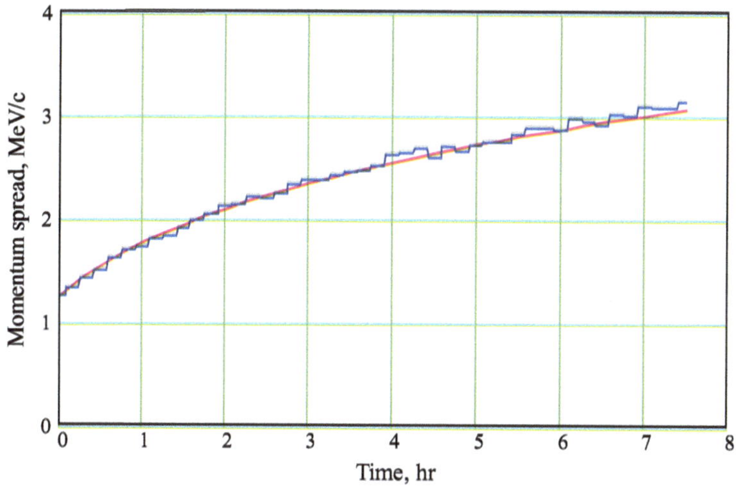

Fig. 6.6 A comparison of the measured rms momentum spread (MeV/c) and the IBS model as a function of time for a coasting beam of 100×10^{10} protons

The rms momentum spread was increased to over 4.5 MeV/c by compressing the beam longitudinally. Figures 6.7 and 6.8 show the measured transverse emittance and longitudinal rms momentum spread evolutions. Also shown is the IBS model. The only adjustable parameter in this model is the vacuum-related emittance growth rate. The best fit corresponds to this rate being 0.55 mm mrad/h—consistent with our previous measurements.

In conclusion, we would like to note that the developed IBS model for the Recycler has been verified with beam measurements made during storage and

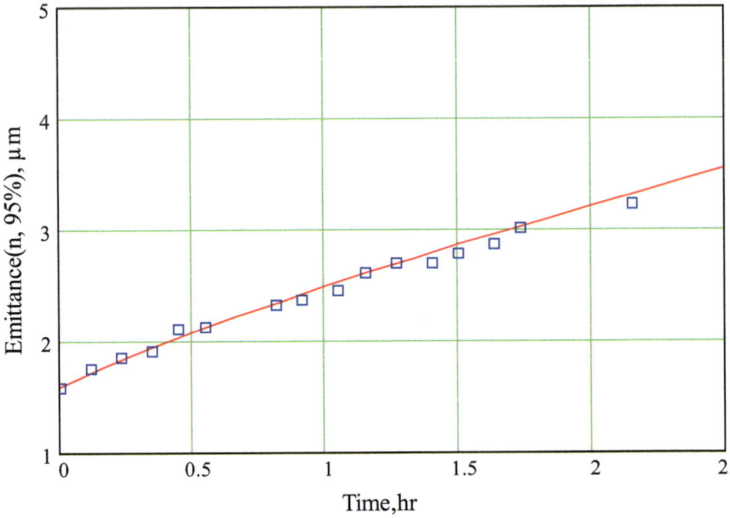

Fig. 6.7 The measured transverse emittance evolution for a bunched antiproton beam (25×10^{10}). Also shown is the IBS model (*solid line*)

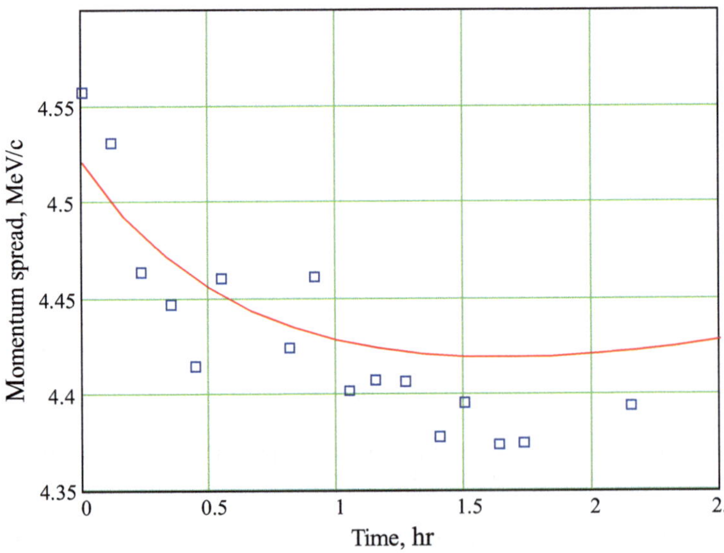

Fig. 6.8 The longitudinal rms momentum spread (MeV/c) as a function of time (hours). The IBS model (*solid line*) has only one adjustable parameter—the vacuum-related transverse emittance growth rate

cooling of the antiprotons before extraction to the Tevatron. We separately verified longitudinal IBS heating and cooling effects by adjusting the beam parameters to demonstrate these effects. Furthermore, the good agreement observed between model and measurement increased our confidence in the beam diagnostics.

6.3.2 Single and Multiple Gas Scattering in Tevatron

The luminosity evolution model described below required separation of different mechanisms affecting the particle loss and emittance growth. In particular we needed to separate the contributions of the gas scattering and the emittance growth driven by external noises, e.g., magnetic field fluctuations, vibrations, etc. Both mechanisms create transverse diffusion but in difference to the magnetic noise the gas scattering creates non-Gaussian tails. This difference was utilized to separate these contributions in the below described experiment [18] carried out in 2002.

First, using the beam scraping it was verified that the particle distribution of a 150 GeV proton beam, injected into Tevatron, is very close to a Gaussian one. Second, a new beam was injected. It was unbunched to exclude the beam heating by the RF noise and the IBS. The beam intensity was sufficiently small to make sure that the coherent effects did not affect the beam dynamics. Third, the beam was scraped horizontally and vertically. The scraping time of a few minutes is much shorter than a characteristic time of the beam evolution. The fraction of particles removed by the vertical scraping (~25 %) allowed us to predict the initial vertical particle distribution with sufficient accuracy. Fourth, the beam scrapers were removed and the beam was left untouched for 1 h. Then, we moved the vertical scraper in, while recording the beam intensity as a function of the scraper position. The vertical scraper was chosen so that the beam momentum spread would not affect the measurements.

The results of the measurements and the comparison with numeric simulations are shown in Fig. 6.9. Taking into account that only one fitting parameter, the unknown average Tevatron beampipe vacuum pressure, is used there is a good agreement between the theory and the measurements. Note that although the Coulomb logarithm is not a well-determined value and depends on Z its uncertainty does not exceed 10–20 %. The experimentally determined value, L_c, coincides with the theoretical one within 5 % for $Z=7$. The measured 0.8 mm mrad/h emittance growth rate (rms normalized) corresponds to an average Tevatron vacuum of 4×10^{-9} Torr (room-temperature N_2 equivalent). Note that high accuracy of the beam current measurements allowed us to measure tiny tails of the distribution function, which could not be seen by regular beam profile monitors. If the large angle scattering is switched off in the simulations, so that particle scattering is described by diffusion only, there is large difference between calculations and measurements as presented in Fig. 6.10.

A good agreement between the observed and the predicted distribution function tails yields an important practical conclusion that the emittance growth at the injection energy is largely due to gas scattering. However the accuracy of the experiments does not exclude an emittance growth rate excited by noise in magnets at the level of about 20 %, i.e., about 0.16 mm mrad/h. Vacuum improvements carried out in 2003–2006 further reduced the emittance growth rate and the beam loss due to scattering on the residual gas by a factor of 5–10. Indirect measurements point out that the emittance growth rate does not exceed 0.1 mm mrad/h at injection.

The normalized emittance growth due to gas scattering is inversely proportional to the beam energy and is about 6.5 times smaller at the collision energy. We have

Fig. 6.9 Dependence of the beam current on the vertical scraper position for the beam core (*top*) and beam tails (*bottom*); *solid line*—measurements, *dashed line*—computer simulations for $L_c = 8.6$, *dotted line*—the dependence which would be measured with the initial distribution; x_0 marks the final scraper position at the initial scraping

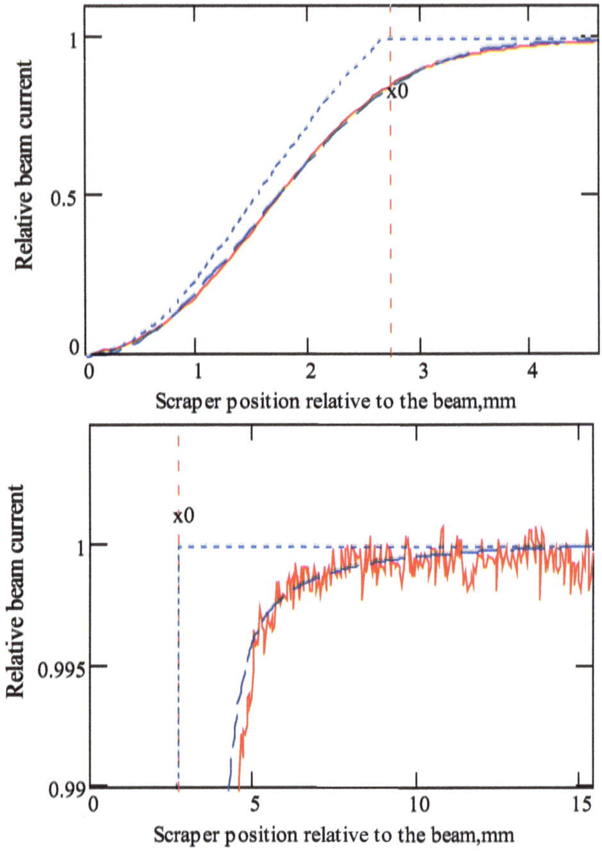

limited knowledge about the magnetic field fluctuations in the Tevatron magnets but have reasons to believe that their contribution to the normalized emittance growth rate at the top energy is bigger than the one due to the vacuum (see next section).

6.3.3 Emittance Growth Mechanisms in the Tevatron Beams

The luminosity in the head-on collisions can be obtained from a well-known formula

$$L = \gamma f_B \frac{N_a N_p}{4\pi\beta^* \varepsilon} H(\sigma_s/\beta^*), \tag{6.114}$$

where ε is the average rms normalized emittances of two round beams $(\varepsilon_a + \varepsilon_p)/2$, $H(x)$ is the "hourglass factor" which depends on the ratio of the RMS bunch length σ_s and beta-function at IPs β^*, γ is the relativistic factor, and f_B is the frequency of

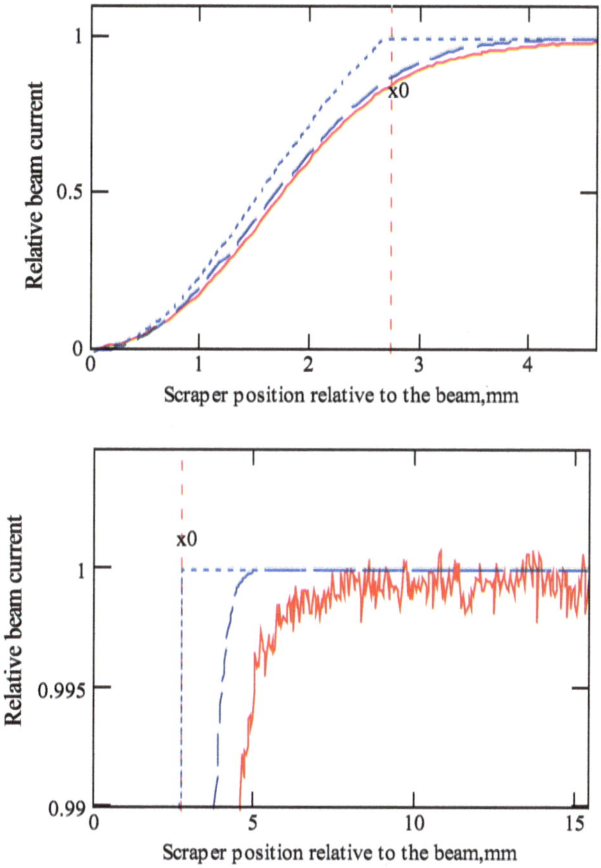

Fig. 6.10 The same as Fig. 6.9, but the large angle scattering is switched off in the model

bunch collisions. Evolution of the Tevatron luminosity over the course of colliding store can be well approximated by a simple empirical fit [19]:

$$L(t) = \frac{L_0}{1 + t/\tau_L}, \tag{6.115}$$

with only two parameters are used: the initial luminosity L_0 and the initial luminosity lifetime τ_L. The luminosity integral $I = \int L dt$ is the most critical parameter for collider experiments. It depends on the product of peak luminosity and the luminosity lifetime, e.g., for a single store with initial luminosity L_0 and duration T, the integral is $I \approx L_0 \tau_L \ln(1 + T/\tau_L)$. From Eq. (6.114), one obtains:

$$\tau_L^{-1} = \frac{dL(t)}{L(t)dt} = \left|\tau_\varepsilon^{-1}\right| + \tau_{N_a}^{-1} + \tau_{N_p}^{-1} + \tau_H^{-1}. \tag{6.116}$$

The luminosity lifetime has four major contributions coming from the growth rates of beam emittances, the beam intensity decay rates, and the hourglass factor

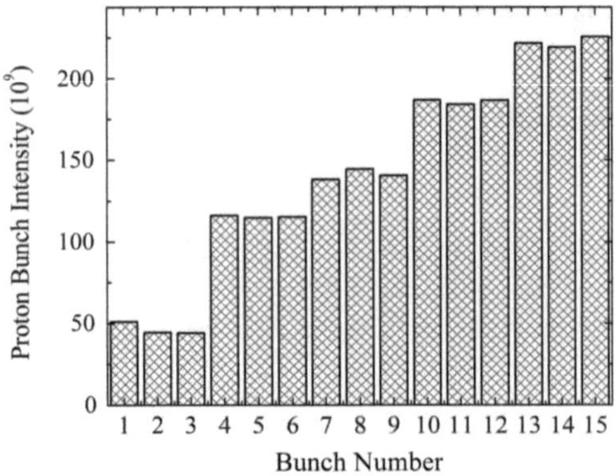

Fig. 6.11 Proton bunch populations at the beginning of the experimental studies store [20]

decay rate. For the end of the Collider Run II operation with range of initial luminosities between 3.0×10^{32} cm^{-2} s^{-1} to 4.0×10^{32}, the largest contribution to the luminosity lifetime of about $\tau_L = 5.2$–5.7 h came from the beam emittance growth with a typical time of $\tau_\varepsilon \sim 10$–13 h. The growth was dominated by IBS in the proton and antiprotons bunches, with small contributions coming from the beam gas scattering and external noises. The beam–beam effects can lead to fast emittance blowups and significant losses (see discussion in Chap. 8) but those were routinely corrected or compensated. The hourglass factor decays with $\tau_H \sim 70$–100 h due to longitudinal IBS heating and a smaller contribution coming from the RF system noises.

Several beam experiments have been conducted to separate contributions of different phenomena to the emittance growth. In one of them [20], 15 proton bunches with various intensities were accelerated to 980 GeV, positioned to the proton helix with the low-beta optics (see Fig. 6.11). The bunches had very different emittances ε varied from 2.3 to 3.6 π mm mrad and rms bunch lengths σ_z in the range from 1.71 to 2.10 ns. The bunches were left in the machine for 3.1 h and their emittances, bunch lengths, and intensities were regularly measured by the Flying Wires system and the SBD system, correspondingly (see details on the Tevatron beam instrumentation in Chap. 9).

The transverse velocity spread is much larger than the longitudinal one for the Tevatron bunches. Then, neglecting the dispersion contribution to the beam size and further simplifying Eq. (6.54) one obtains estimates for the growth rates:

$$\frac{d\varepsilon_T}{dt} \approx \frac{\gamma^{3/4} N_p C_T}{{\varepsilon_T}^{1.5} \varepsilon_L^{0.5}}, \quad \frac{d\sigma_z^2}{dt} \approx \frac{N_p C_L}{\gamma^{1/4} {\varepsilon_T}^{1.5} \varepsilon_L^{0.5}}, \tag{6.117}$$

where C_L and C_T are the constants determined by machine parameters, and both vertical and horizontal emittances are assumed equal, $\varepsilon_x \approx \varepsilon_y \approx \varepsilon_T$. Equation (6.117) suggests that the IBS-driven emittance growth has to be proportional to the factor $F_{IBS} = N_p/({\varepsilon_T}^{1.5}\sigma_s)$. Note that in the described below analysis we used the vertical emittance only because its systematic error and uncertainty for the Tevatron Flying Wires system are much smaller than for the horizontal emittance.

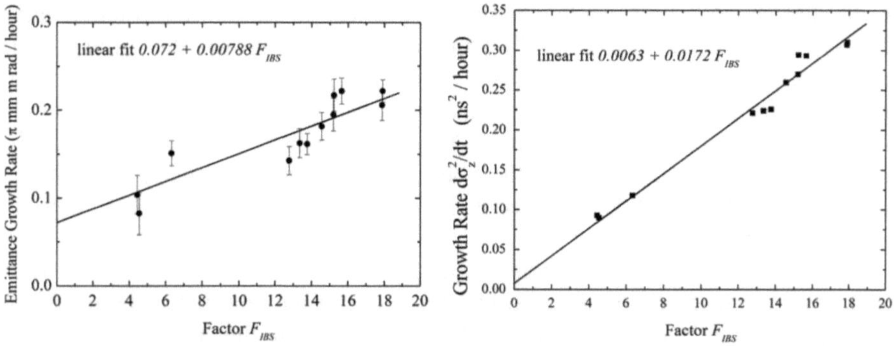

Fig. 6.12 Vertical emittance growth rates (rms, norm.) of proton bunches vs the IBS factor F_{IBS} (*left*); the rms bunch length growth rates vs the IBS factor F_{IBS} (*right*) [20]

Such choice is justified by the presence of strong $x-y$ coupling which is additionally amplified by operation close to the coupling resonance. That results in an equalization of emittances so that they are quite close and are proportional to each other. The experience accumulated in the course of Run II also verifies such choice. Since the bunch emittances are evolving in time, then F_{IBS} is a function of time, too. The observed growth rates for the emittance and the rms bunch length are plotted versus F_{IBS} for all the bunches are presented in Fig. 6.12.

The observed growth rates can be approximated by linear fits:

$$\begin{aligned} \frac{d\varepsilon_V}{dt}[\pi\,\mathrm{mm\,mrad/h}] &= (0.072 \pm 0.02) + (0.0079 \pm 0.0015)\cdot F_{IBS}, \\ \frac{d\sigma_s^2}{dt}\left[\mathrm{ns^2/h}\right] &= (0.0063 \pm 0.0193) + (0.0173 \pm 0.0013)\cdot F_{IBS}. \end{aligned} \tag{6.118}$$

The intercept in the transverse emittance growth of 0.072 π mm mrad/h yields an upper estimate on the growth rate due to intensity independent effects such as scattering on the residual gas and diffusion driven by external noises. The emittance growth due to the gas scattering is equal to

$$\frac{d\varepsilon_{x,y}}{dt} = \frac{2\pi c {r_p}^2}{\gamma\beta^2}\left(\sum_i n_i Z_i (Z_i+1) L_{C_i}\right)\overline{\beta_{x,y}}, \tag{6.119}$$

where $\overline{\beta_{x,y}} = \int \beta_{x,y} ds/C \approx 70$ m. It yields that the observed zero-intensity emittance growth rate corresponds to an equivalent average room-temperature N_2 ($Z_i = 7$) vacuum pressure of $(2.4 \pm 0.7) \times 10^{-9}$ Torr, which, consequently, yields the beam intensity lifetime due to nuclear beam-gas interaction of about 140 ± 40 h. It contradicts to the measured beam lifetime of about $\tau_H \sim 800 \pm 500$ h (confirmed in many other similar measurements), i.e., less than 20 % of the observed low intensity beam emittance growth rate is due to the gas scattering.

Consequently, the corresponding equivalent average pressure of room-temperature N_2 is less than $P \approx 0.5 \times 10^{-9}$ Torr; and that most of the growth rate ($\approx$0.06 π mm mrad/h) is related to other mechanism. The most probable reason is the emittance growth related to the transverse beam shaking at the lowest betatron sideband exited by noise in magnets. Note that the large angle electromagnetic scattering on the residual gas and the Touschek scattering give rise to the lifetimes of about 30,000 and 7,000 h, respectively, and, thus, produce negligible contributions to beam loss.

Indirect confirmation of the noise effect comes from the observed ambient beam betatron motion amplitude of some 110 nm in one of the high-beta (900 m) BPMs during collisions (corresponding to $A_\beta \approx$ 10-25 nm at the arcs, see Chap. 2, Sect. 2.4). That exceeds the Schottky noise by more than an order of magnitude. If that motion is due to the external noise, then corresponding emittance growth rate can be estimated as:

$$\frac{d\varepsilon_T}{dt} \approx \gamma f_0 \frac{A_\beta^2}{2\bar{\beta} N_{dec}}, \tag{6.120}$$

where N_{dec} the decoherence time expressed in the number of turns. That yields the growth rate of about 0.01–0.05 π mm mrad/h in high luminosity stores. Here we assumed that N_{dec} is mostly determined by nonlinearity of the beam–beam forces and is estimated to be about 50–100 turns for protons with beam-beam parameter of $\xi \sim 0.02$.

The intercept in the bunch length growth is proportional to lengthening due to RF noise. Assuming that the phase noise dominates one obtains a spectral density of phase noise at the synchrotron frequency[8] $5 \cdot 10^{-11}$ rad^2/Hz. It coincides with the direct phase noise measurements within measurement accuracy.

6.4 Tevatron Run II Beam Halo Collimation System and Collimation Studies

Even in good operational conditions, a finite fraction of the beam will leave the stable central area of accelerator because of beam–gas interactions, IBS, proton–antiproton interactions in the IPs, RF noise, ground motion, and resonances excited by the accelerator elements imperfection. These particles form a beam halo. As a result of halo interactions with limiting apertures, hadronic and electromagnetic showers are induced in accelerator and detector components causing numerous deleterious effects ranging from minor to severe. The most critical for colliders are beam losses in superconducting magnets and accelerator related backgrounds in the

[8] Note that by definition the spectral density used in measurements is 4π larger than the spectral density used in Sect. 6.2.3.

collider detectors. Only with a very efficient beam collimation system can one reduce uncontrolled beam losses in the machine to an allowable level. Beam collimation is mandatory at any superconducting hadron collider to protect components against excessive irradiation, minimize backgrounds in the experiments, maintain operational reliability over machine life, and reduce the impact of radiation on the environment [21].

During the Collider Run I (1994–1996) the Tevatron halo removal system experienced limitations that prompted a design of a new system for the Collider Run II. The new design specified that the entire halo removal process needed to be more efficient and putting the collimators in after the beam acceleration has to be much faster with the target time of about 5 min. This implied that the halo removal process would have to be based on a two-stage collimation and setting the collimation up have to be automated. A new collimation system [22] has been designed for the Tevatron Run II to localize most of the losses in the straight sections D17, D49, EØ, F17, F48, F49, and AØ. It incorporated four *primary collimators* (targets) and eight newly built 1.5-m long *secondary collimators*. New motion control hardware capable of fast processing of data from beam loss beam intensity monitors was installed. That allowed us to build a computer controlled feedback allowing us to achieve a 2 in. collimator displacement (full travel) within 15 s. A central control software system was also developed to coordinate the global sequence of motion for all 12 collimators while incorporating the halo removal system into the Tevatron Collider sequencer software. At the design stage, a multi-turn particle tracking and beam halo interactions with the collimators were performed with the STRUCT code [23]. Using the STRUCT calculated beam loss distributions and the MARS code [24] we carried out the Monte Carlo hadronic and electromagnetic shower simulation, and the secondary particle transport in the accelerator and detector components, including shielding calculations with real materials and magnetic fields. The Collider Run II halo removal system was installed, commissioned, and became operational since June 2001.

The system was upgraded several times following operational needs. For example, in 2002, the Tevatron Electron Lenses (TEL) have been set up to remove undesirable uncaptured particles from the abort beam gaps and, thus, reduce the risk of damage of high-energy physics particle detectors CDF and D0 during beam aborts. In 2003, following several instances of unsynchronized abort kicker pre-fires in the Tevatron, an additional *tertiary* collimator was installed at the A48 location to protect the CDF detector components [25]. In 2005 the beam scraping procedures were optimized for faster operation and highly efficient repeated scraping (double scraping) of the beams. Also, in 2010 a collimation during the portion of the low-beta squeeze was added in order to reduce losses at CDF and D0 that were causing frequent quenches. Quenching the low-beta quadrupoles during the squeeze became more of a problem once the antiproton intensity and beam brightness became larger. Sensitive steps in the low-beta squeeze, where the beam separation between the proton and antiproton is small, create losses at large beta locations, mainly the cryogenic low-beta quadrupoles. A single collimator at E0 was placed at 5σ to create a limiting aperture moving the loss point of sensitive steps away from CDF and D0 IPs to a region that has robust quench limits.

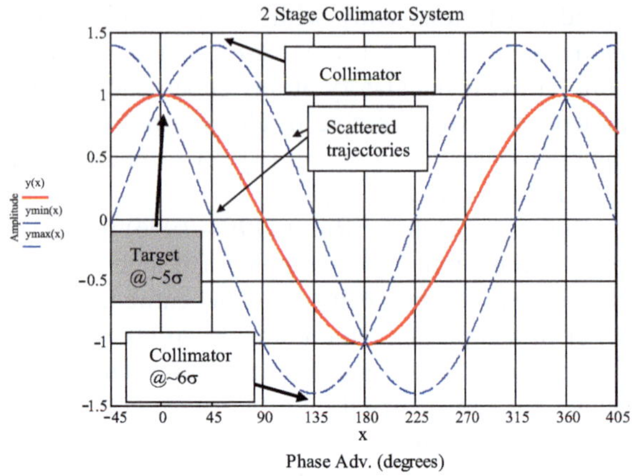

Fig. 6.13 Placement of the target and secondary collimators to produce a two-stage collimator system [27]

This additional collimator has worked well, allowing an increase in number of antiprotons while basically excluding quenches in the low-beta squeeze. Novel ideas to improve the beam collimation efficiency—namely, a bent crystal collimation and hollow electron beam collimation— have been extensively and successfully studied at the end of Run II.

6.4.1 Collimation System Design

The principles of a two-stage collimation system are described in [21]. The system consists of horizontal and vertical primary collimators and a set of secondary collimators placed at an optimal phase advance, to intercept most of the particles out-scattered from the primary collimators during the first turn after beam halo interaction with primary collimators. An impact parameter of multi-GeV and TeV protons on the primary collimators is ~1 μm [26]. The design studies [22] show that in the Tevatron, a 5-mm thick tungsten primary collimator positioned at 5σ (of rms beam size) from the beam axis in both vertical and horizontal planes would function optimally, reducing the beam loss rates as much as a factor of 4–10 compared to the system without such a scatterer. Secondary collimators located at the appropriate phase advances are a 1.5-m long L-shaped steel jaws positioned at 6σ from the beam axis in the horizontal and vertical planes. They are aligned parallel to the envelope of the circulating beam. Figure 6.13 schematically depicts placement of the collimators in such a system.

In the course of Run II the halo removal system consist of 12 collimators. Four of them are primary collimators and eight are secondary collimators. The collimators are arranged in four sets: two proton and two antiproton sets. They are installed around the Tevatron ring as shown in Fig. 6.14. Placement of collimators in the Tevatron is limited to a few locations since there is limited warm space and the proton and antiproton beams are on helical orbits.

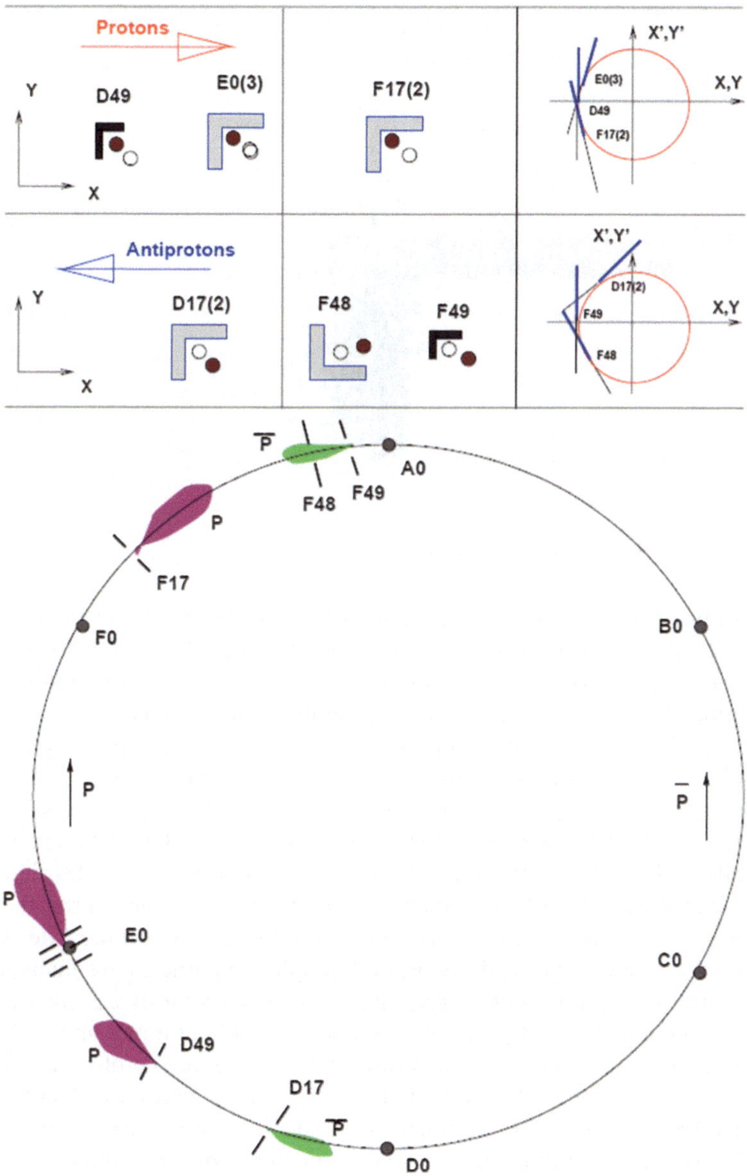

Fig. 6.14 Tevatron Collider Run II Halo Removal Collimator Layout. CDF and D0 detectors are located at B0 and D0, respectively [27]

A proton primary collimator is placed at the beginning of the D17 straight section outward and up of the closed orbit (Fig. 6.14). It intercepts the large amplitude protons and a positive off-momentum beam. Protons scattered from

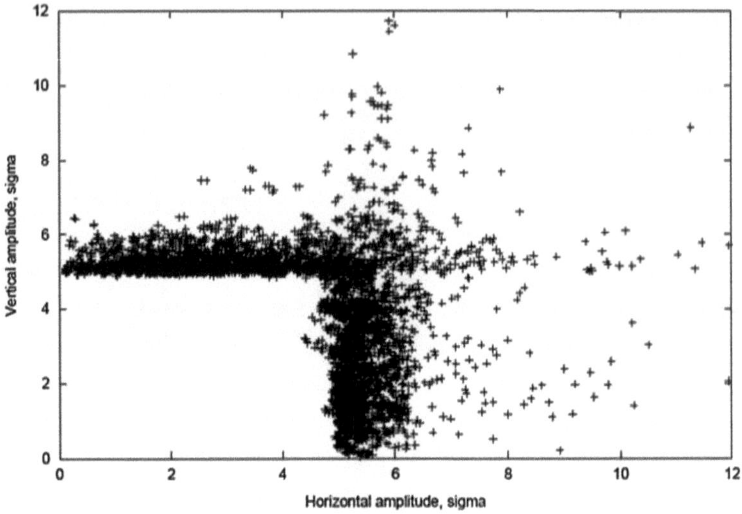

Fig. 6.15 Proton beam halo on a secondary collimator [27]

this collimator are presented by a vertical line in the transverse phase diagram (Fig. 6.14). Protons with a positive angle are intercepted by a D17(3) secondary collimator at the end of the D17 straight section. An A0 secondary proton collimator positioned outward and up of the circulating beam is intended to intercept the negative angle protons emitted from the primary collimator. A primary collimator D49 and secondary collimators E0(1) and F17(2) are used to deal with the protons with negative momentum deviations. Antiproton beam cleaning consists of primary collimators F49, F17(3) and secondary collimators D17(2), F48, F17(1) and E0(2).

Detailed STRUCT/MARS simulations assumed that halo particles first hit the primary collimator with a 1 to 3 μm impact parameter. On the next turns, the impact parameter—as a result of scattering—increases to about 0.3 mm. After the first interaction with a primary collimator, high amplitude particles are intercepted by the secondary collimators, but a large number of particles survive. Some fraction of the halo is not intercepted by a primary/secondary collimator pair and will interact with a primary collimator on the next turns. On average, halo protons interact with the primary collimator 2.2 times. Particles with the amplitudes less than 6σ are not intercepted by the secondary collimators and do survive for several tens of turns until they increase amplitude in the next interactions with the primary collimator. The tail of halo is extended above 6σ (Fig. 6.15). Large-amplitude particles, which escape from the cleaning system at the first turn, are able to circulate in the machine, before being captured by the collimators on the later turns. This defines the machine geometric aperture.

The calculations [22] have shown and later measurements confirmed that the inefficiency of the Tevatron collimation system defined as a leakage of halo protons from its components is ~10^{-3}. At the same time, the most critical function of the system in Run II has been identified as reduction of background rates in the collider experiments. Beam loss in the B0 and D0 depends strongly on the secondary collimator offset with respect to the primary collimators. It has been shown in [28] that a part of the accelerator-related backgrounds in the D0 and CDF detectors is originated from the beam halo loss in the inner triplet region. The studies [29] have revealed that it is mainly due to beam-gas elastic scattering in the regions between the nearest to the IP secondary collimator and the corresponding inner triplet. This process will obviously increase the background rates. In addition to the optically small aperture at β_{max} location, the aperture restrictions in this area are the D0 forward detector's Roman pots placed at 8σ and the B0 Roman pots placed at 10σ at the entrance and exit of the beam separators. Note that the Roman pot systems had been removed in the middle of the Collider Run II. Thus, for the collider detectors, the above-defined inefficiency is not the whole story. The more appropriate definition of collimation inefficiency would be a ratio of backgrounds in the detectors with collimation to that without collimation. For the Tevatron Run II it is calculated as 6.7×10^{-3}, or a factor of 150 reduction of losses. The corresponding measurements are described below.

6.4.2 Collimation System Operation

The collimator hardware consists of a Motorola VME 162 processor and Advanced Controls System Corp. Step/Pac stepping motor drivers that interface to the VME processor [30]. LVDT's (linear voltage differential transformers) are used to read collimator positions. Figure 6.16 is the block diagram for the hardware controls for a single collimator.

The Collider II collimator halo removal system was designed with the capability of incorporating feedback into the motion of a collimator. The system uses two sources for feedback. The first source is feedback from a local beam loss monitor. Four standard Tevatron beam loss monitors and amplifiers are interfaced to the VME processor to provide loss monitor feedback. Two of these loss monitors are used to detect losses in the proton direction and two in the antiproton direction. Two loss monitors for each particle type are used to provide redundant loss monitor signals in case of failure during collimator movement. The second source of feedback comes from a beam intensity signal. A Fast Bunch Integrator system (see Chap. 9) is used to provide beam intensity signals for both proton and antiproton beams at a 360 Hz update rate. Feedback is accomplished by encoding proton and antiproton intensity signals onto the global machine data link (MDAT). The MDAT signal is decoded by each of the VME processors at a 720 Hz rate.

Processing the feedback internal to the VME is accomplished by sampling the loss monitor and/or beam intensity signal periodically while the collimator is moving. The smallest step the collimator can make is 25 μm. This minimum step takes 20 ms to complete. A wait step occurs after the move step to provide more flexibility to timing movements. During this step, loss monitor signals and/or beam intensity signals are sampled every 4 ms and are compared to a loss limit value or

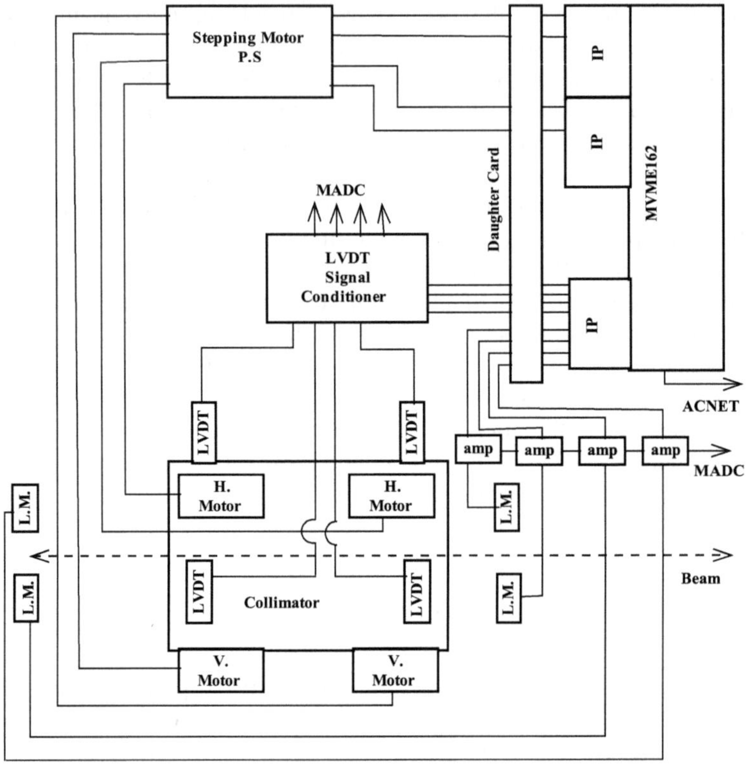

Fig. 6.16 Block diagram of the hardware controls for a single collimator [27]

beam intensity percentage allowed to be removed to decide if the collimator is to be halted for the next step. Each collimator VME frontend has 17 parameters that the user can change to specify details about feedback processing.

The halo removal system also utilizes software that allows global coordination of all 12 local collimator VME front ends. This global coordination software is called an open access client (OAC). An OAC is a central process that runs on a VAX (LINUX after control system change) and has controls hooks into the main Tevatron sequencer software [30]. The OAC employs a finite state machine that is configurable by the user to preprogram one or many collimators to complete a task on a transition of a state. For example, on the state "Goto injection positions" all collimator front ends are preprogrammed with local parameters that define their out of beam positions. The OAC owns a configurable matrix of states of the collimators and the user specifies which collimators are to move when the state is transitioned. Once the state is transitioned, all collimators will be moved back to their injection positions. There were 11 defined collimator states with names like: "Go to injection positions", "Begin halo removal scraping", and "Retract proton collimators". There is one special collimator state which is "Global Collimator Abort". A transition to this state stops all 12 collimators immediately. Figure 6.16 is a block diagram of the OAC.

The halo removal process is conducted in the Tevatron at the flattop energy of 980 GeV after the proton and antiproton beams have been brought into collisions. This process is initiated by the Tevatron sequencer software. There are four sub-sequence operations that are necessary in order to complete the halo removal. (1) *Move Collimators to Initial Positions*—this sub-sequence moves all the collimators at 1.25 mm/s speed into the beam to the "half way" point to the beam. The motivation of this sub-sequence is to speed up the process. (2) *Intermediate Halo Removal*—here each set (proton and antiproton) of collimators and targets are moved together under beam loss monitor feedback until a small loss is detected and all collimators are stopped. This sub-sequence is also preformed in order to reduce the total amount of time the halo removal process takes. (3) *Perform Halo Removal*—each secondary collimator and target is moved serially into the beam. Secondary collimators are moved under loss monitor feedback with a step size of 0.025 mm until they reach the edge of the beam to shadow the losses by the primary collimator. After all secondary collimators are placed next to the beam, each target is moved under loss monitor and beam intensity feedback until 0.4 % of each beam (proton and antiproton) is removed. (4) *Retract Collimators for Store*—after targets and secondary collimators have reached their final assignment, they are retracted approximately 1 mm. This is the position they remain at for the duration of the store. This roughly leaves the targets and secondary collimators at the 5 and 6σ points as specified by the system design. The halo removal system is a necessary and integral part of Tevatron Collider operations. The halo removal system is completely automated and benefits operations with ease of use. The entire process takes as fast as 7 min. Figure 6.17 presents loss rates during the process of beam collimation early in store #8709 (May 2011).

A zero time in Fig. 6.17 corresponds to the moment when two 980-GeV beams are brought to collisions. Over the next six and a half minutes collimators sequentially approach the beams and scrape them. Horizontal position of one of the collimators (D49H) is shown in the bottom plot. One can see that the collimator moved very close to the beam twice—that is intentionally done to repeat the scraping procedure and guarantee lower loss rates afterward. The proton beam intensity on the upper plot shows a number of small drops due to the scraping. The bottom plot shows proton halo loss rate as measured by CDF detector and the antiproton loss rate measured by D0 detector. After the scraping is over—at about 7 min—the CDF detector luminosity monitor starts operation and reports maximum luminosity of about 430×10^{30} cm^{-2} s^{-1}. Note that compared to the first moments after the collisions are initiated, during the luminosity operation the proton halo loss rate drops by factor of 100 from 2–3 MHz to 20–30 kHz, while antiproton rate is down by a factor of 4–5 (from some 8–10 to 2 kHz).

The merit of halo removal efficiency is to simply record the proton and antiproton halo losses at CDF and D0 IPs before halo removal, and, then, to divide it by the same losses recorded at the completion of halo removal. Table 6.2 presents statistics of the reduction in the losses averaged over 100 stores in January–May 2011.

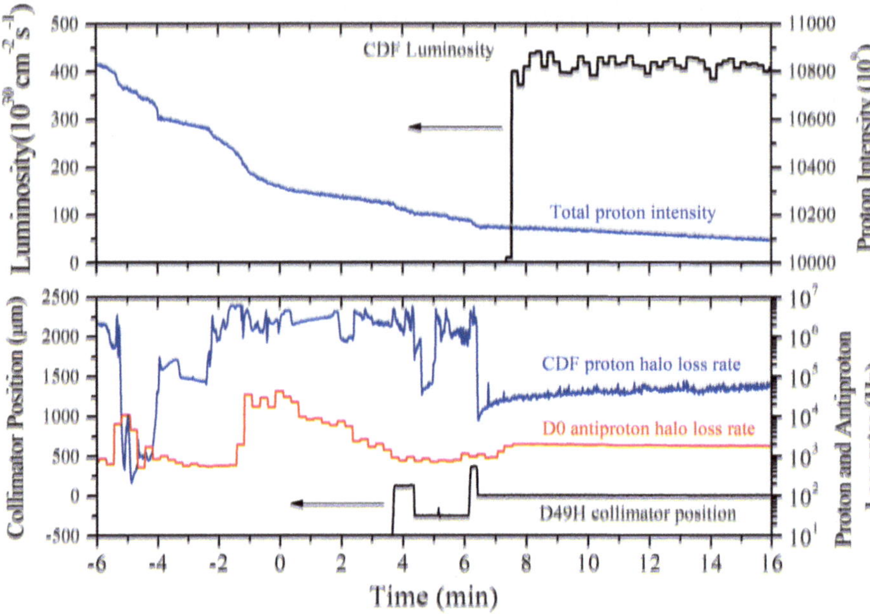

Fig. 6.17 Collimation process early in store #8709 (May 2011). *Upper plot* shows proton beam intensity and CDF luminosity. *Bottom plot* shows proton halo loss rate as measured by CDF detector, D0 antiproton loss rate and horizontal position of one of the collimators (D49H). Zero time corresponds to the moment when two beams are brought to collisions [27]

Table 6.2 Merit of halo removal efficiency (2010–2011)

Halo loss counter at CDF or D0 IP	Factor of reduction of halo losses after halo removal
CDF proton halo loss	112
CDF antiproton halo loss	80
D0 proton halo loss	13
D0 antiproton halo loss	19

One can see that the biggest reduction is seen in the CDF proton rate (over a factor of 100). The reduction in the D0 proton halo loss is relatively small, that can be attributed to the fact that for the proton direction the CDF Interaction Point acts as an additional collimator and, thus, reduces the proton halo losses at D0.

6.4.3 *Abort Gap Beam Removal*

Particles not captured by the Tevatron RF system, and, therefore, not synchronized with it, pose a threat since they can quench the superconducting magnets during acceleration or at beam abort [31]. The mechanisms of such uncaptured beam generation are somewhat different at the injection (150 GeV) and top (980 GeV) energies. Coalescing in the Main Injector typically leaves a few percent of the 150 GeV beam particles outside RF buckets. These particles are transferred together with the main bunches to the Tevatron. In addition, the single IBS (the Touschek effect), diffusion due to multiple IBS, and phase and amplitude noise of the RF voltage, drive particles out of the RF buckets. This is exacerbated by the fact that after coalescing and injection, 95 % of the particles cover almost the entire RF bucket area. The uncaptured beam is lost at the very beginning of the Tevatron energy ramp. These particles are out-of-sync with the Tevatron RF accelerating system, so they do not gain energy and quickly (<1 s) spiral radially into the closest horizontal aperture. If the number of particles in the uncaptured beam is too large, the corresponding energy deposition results in a quench (loss of superconductivity) of the superconducting (SC) magnets and, consequently, terminates the high-energy physics store. At the injection energy, an instant loss of uncaptured beam equal to 3–7 % of the total intensity can lead to a quench depending on the spatial distribution of the losses around the machine circumference.

At the top energy, uncaptured beam generation is mostly due to the IBS and RF noise while infrequent occurrences of the longitudinal instabilities or trips of the RF power amplifiers can contribute large spills of particles to the uncaptured beam. Uncaptured beam particles are outside of the RF buckets, and therefore move longitudinally relative to the main bunches. Contrary to the situation at the injection energy of 150 GeV, when synchrotron radiation (SR) losses are practically negligible, 980 GeV protons and antiprotons lose about 9 eV/turn due to SR. For uncaptured beam particles, this energy loss is not being replenished by the RF system, so they slowly spiral radially inward and die on the collimators, which determine the tightest aperture in the Tevatron during collisions. The typical time for an uncaptured particle to reach the collimator is about 20 min.

The presence of the uncaptured beam is very dangerous not only for the collider elements but also for the high-energy physics particle detectors CDF and D0 as the abort gap particles generate unwanted background and can be kicked onto the detectors' components by the beam abort kickers. A number of ideas have been proposed for elimination of the uncaptured beam in the Tevatron. The TEL has been found to be the most effective. The advantages of the TELs are twofold: (1) an electron beam is in close proximity to the proton or antiproton orbits and generates a quite strong transverse kick; (2) the TELs possess short rise and fall times (~100 ns), so they can be easily adjusted to operate in a variety of different pulsing schemes. Another uncaptured beam removal method tested during machine studies was a transverse strip line kicker operating with a narrow noise bandwidth. The kicker signal was timed into the abort gap to diffuse uncaptured beam particles transversely. With the noise power limited by a 300 W amplifier, that method was found significantly less effective than using the TELs.

The TEL #1 and #2 were installed in the Tevatron in 2001 and 2006, respectively, for compensation of beam–beam effects (see Chap. 8, Sect. 8.3). In early 2002, it was found that TEL-1 can effectively remove uncaptured protons if timed into the abort gap and operated in a resonant excitation regime [31]. TEL-2 is also

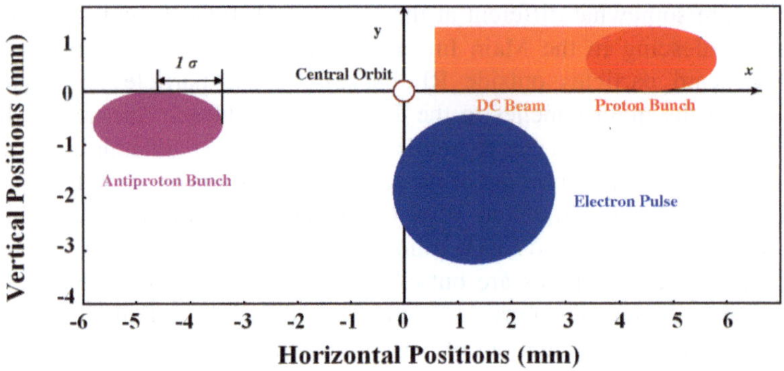

Fig. 6.18 The relative positions of the proton, antiproton, and electron beam during uncaptured beam removal [31]

able to function as an abort gap cleaner. For that, the electron beam pulse is synchronized to the abort gap and positioned near the proton beam orbit. Electric and magnetic forces due to the electron space charge produce a radial kick on high-energy protons depending on the separation *d*:

$$\Delta\theta = \mp\frac{1 \pm \beta_e}{\beta_e} \cdot \frac{2J_e L_e r_p}{e \cdot c \cdot \gamma_p} \cdot \begin{cases} \dfrac{d}{a}, & d < a, \\ \dfrac{a}{d}, & d > a, \end{cases} \tag{6.121}$$

where the sign reflects repulsion for antiprotons and attraction for protons, $\beta_e = v_e/c$ is the electron beam velocity, J_e and L_e are the electron beam current and the interaction length, a is the electron beam radius, r_p is the classical proton radius, and $\gamma_p = 1{,}044$ is the relativistic Lorentz factor for 980 GeV (anti)protons. The factor $1 \pm \beta_e$ reflects the fact that the contribution of the magnetic force is β_e times the electric force contribution and depends on the relative direction of the electron and proton velocities.

For 5 kV electrons with typical peak current of about 0.6 A and 5 mm away from the protons, the estimated kick is about 0.07 μrad. When the pulsing frequency of the TEL is near the proton beam resonant frequency, this beam–beam kick resonantly excites the betatron oscillations of the beam particles.

In the uncaptured beam removal operation, the TEL electron beam is placed 2–3 mm away from the proton beam orbit horizontally and about 1 mm down vertically as depicted in Fig. 6.18. For normal Tevatron operation, the fractional part of the tunes are $Q_x = 0.583$ and $Q_y = 0.579$ for horizontal and vertical planes respectively. These tunes are placed between the strong resonances at $4/7 \approx 0.5714$ and $3/5 = 0.6$. When an uncaptured particle loses energy due to synchrotron radiation, its horizontal orbit is changed proportionally to the lattice dispersion,

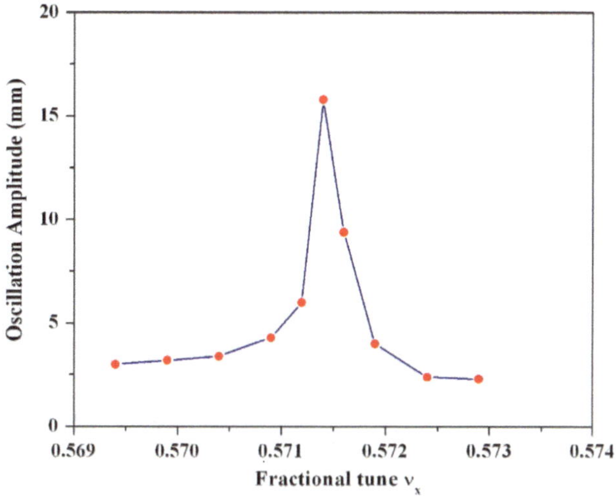

Fig. 6.19 Betatron oscillation amplitude of the particles driven by the TEL in vicinity of the $Q = 4/7$th resonance line (simulations) [31]

$x = D_x(\delta P/P)$, and its betatron tunes are changed due to the lattice chromaticity $Q'_{x,y} = dQ_{x,y}/(dP/P)$ so that

$$Q_{x,y} = Q^0_{x,y} + Q'_{x,y}\left(\frac{dp}{p_0}\right) + \Delta Q_{x,y}(x^2), \tag{6.122}$$

where the third term accounts for slight tune changes due to nonlinear magnetic fields. Typical operational chromaticities of the Tevatron at 980 GeV are set to $Q'_{x,y} \approx +(6\text{–}10)$, so the tune decreases with the energy loss. As the tune, driven by the TEL, approaches one of the resonant lines, the amplitude of the particle betatron oscillations grows, eventually exceeding a few millimeters until the particle is intercepted by the collimators. Figure 6.19 presents one set of the simulation results of the particle amplitude driven by the TEL in the vicinity of the 4/7th resonance. The maximum amplitude is determined by the nonlinearity of the force due to the electron beam and the nonlinearity of the machine. Note that without the TEL, a particle would still be intercepted by a horizontal collimator after its orbit moved about 3 mm inward due to SR. The TEL simply drives it more quickly, preventing the accumulation of uncaptured beam.

The electron beam pulsing scheme is demonstrated in Fig. 6.20, where the green oscilloscope trace is the signal from the TEL Beam Position Monitor (BPM) pickup electrode and the blue trace is the total electron current. In the BPM signal, one can see three negative pulses representing the electron beam pulses in the 3 abort gaps whereas the 36 positive pulses are the proton bunch signals with the small negative adjacent antiproton bunch signals. The intensity of the antiproton bunches was ten times less than that of the proton bunches at the end of that particular store, so they appear only as very small spikes near the large proton bunches. During a typical HEP store, the train of three electron pulses is generated every seventh turn for the purpose of excitation of the 4/7 resonance for the most effective removal of the

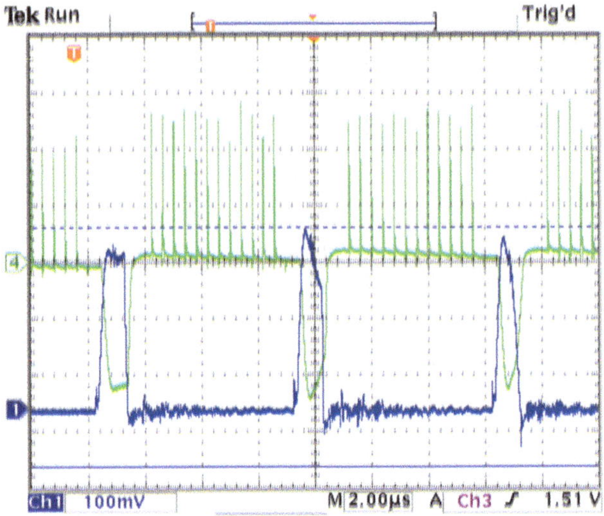

Fig. 6.20 Scope traces of the electron beam pulses (*blue*) and the TEL BPM signal showing electron, proton, and antiproton bunches (*green*). One division of the horizontal axis is 2 μs. About one Tevatron revolution period is shown

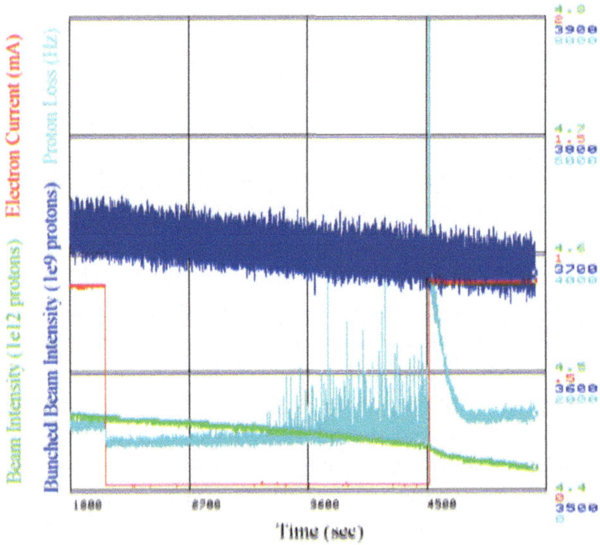

Fig. 6.21 Uncaptured beam accumulation and removal by TEL: the electron current was turned off and turned back on 40 min later again

uncaptured proton beam particles. The electron pulse width is about 1 μs and the peak amplitude is about 250 mA.

The uncaptured beam removal process was demonstrated in an experiment in which the TEL was turned off for about 40 min and then turned on again as shown in Fig. 6.21. The blue trace is the total bunched proton beam intensity measured by the Fast Bunch Integrator (see Chap. 9); the red trace is the average electron current measured at the TEL electron collector; the green trace is the total number of particles in the Tevatron as measured by DCCT; and the cyan trace is the abort gap proton beam loss rate measured by the CDF detector counters. After the TEL was

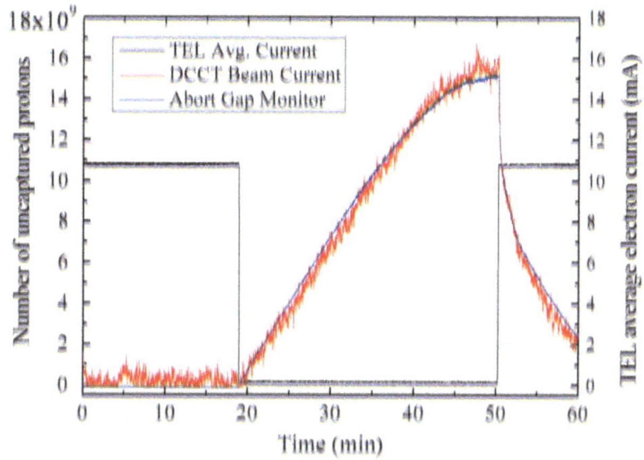

Fig. 6.22 Uncaptured beam accumulation and removal by the TEL. The *black line* represents the average electron current of the TEL; the *red line* is the uncaptured beam estimated from the DCCT measurement; the *blue line* is uncaptured beam in the abort gap measured by the AGM [29]

turned off, the abort gap loss rate was reduced by about 20 % but then started to grow. After about 20 min, the first spikes in the losses started to appear and grow higher. Notably, the bunched beam intensity loss rate (blue line) did not change, so the rate of particles escaping from the RF buckets was about constant. As soon as the TEL was turned on, a large increase in the abort gap losses and reduction of the total uncaptured beam intensity could be seen in Fig. 6.21. About 15×10^9 particles of the uncaptured beam in the abort gaps were removed by the TEL in about $\tau_{\mathrm{TEL}} = 3$ min and the abort gap loss rate went back to a smooth equilibrium baseline.

The calibration of the abort gap monitor (AGM—see details in Chap. 9) has been performed using the TEL as presented in Fig. 6.22. The TEL was turned off during a store (average electron current is shown in black) at about $t = 20$ min. Accumulation of the uncaptured beam started immediately and can be measured as an excess of the total uncaptured beam current with respect to its usual decay. The blue line in Fig. 6.22 shows the excess measured by the Tevatron DCCT, $\delta N_{\mathrm{DCCT}}(t) = N_{\mathrm{TEL\ on}}(t) - N_{\mathrm{decay\ fit\ TEL\ off}}(t)$. The abort gap uncaptured beam intensity measured by the AGM (red line) and the DCCT excess grow for about 30 min before reaching saturation at intensity of about 16×10^9 protons. Then the TEL was turned on resulting in the quick removal of the accumulated uncaptured beam from the abort gaps. This method of calibration of the AGM with respect to DCCT interferes with the collider operation resulting in higher losses (see Fig. 6.21 above and discussion), so this operation is performed only when required. The AGM is used for the routine monitoring of the uncaptured beam. The typical rms error of the uncaptured beam intensity measurement is about 0.01×10^9 protons for the AGM, and some 0.3×10^9 protons for the DCCT.

The amount of the uncaptured beam is determined by the rate of its generation and the removal time τ:

$$N_{\mathrm{DC}} = \left(\frac{dN_{\mathrm{bunched}}}{dt}\right) \times \tau. \tag{6.123}$$

The characteristic time needed for a 980 GeV particle to lose enough energy due to SR is about $\tau_{SR} = 20$ min, so the TEL reduces the uncaptured beam population by about one order of magnitude.

At injection energy, the synchrotron radiation of protons is negligible, so the TEL is the only means to control uncaptured beam. As noted above, one of the TELs is used routinely in the Tevatron operation for the purpose of uncaptured beam removal at 150 and 980 GeV. In 2007, the typical antiproton intensity increased to about a third of the proton intensity, and therefore the antiproton uncaptured beam accumulation started to pose an operational threat. An antiproton AGM, similar to the proton one, has been built and installed. By proper placement of the TEL electron beam between the proton beam and the antiproton beam (illustrated in Fig. 6.18), we are able to remove effectively both uncaptured protons and uncaptured antiprotons. In addition, we have explored the effectiveness of the uncaptured beam removal at several resonant excitation frequencies. For that, we have pulsed the TEL every 2nd, 3rd, 4th, 5th, 6th, and 7th turn. Reduction of the uncaptured beam intensity was observed at all of them, though usually the most effective was the every 7th turn pulsing when the Tevatron betatron tunes were close (slightly above) to $Q_{x,y} = 4/7 = 0.571$ or every 6th turn pulsing when tunes were closer to $Q_{x,y} = 7/12 = 0.583$.

6.4.4 Beam Halo Collimation by Bent Crystals

Since the original suggestion of bent crystal channeling [32] there has been interest in exploiting the technique for accelerator extraction [33] and, later, collimation [34]. Luminosity-driven channeling extraction was observed for the first time in a 900 GeV study at the Fermilab Tevatron during Collider Run I [35]. The experiment, Fermilab E853, demonstrated that useful TeV-level beams can be extracted from a superconducting accelerator during high luminosity collider operations without unduly affecting the background at the collider detectors. Multipass extraction was found to increase the efficiency of the process significantly. The beam extraction efficiency was about 25 %. Studies of time-dependent effects found that the turn-to-turn structure was governed mainly by accelerator beam dynamics. Based on the results of the E853 experiment, it was concluded that it is feasible to construct a parasitic 5–10 MHz proton beam from the Tevatron collider [36].

An efficient beam collimation system is mandatory for any collider or high-power accelerator. The Tevatron Run II approach has been to use a two-stage collimation system in which a primary collimator is employed to increase the betatron oscillation amplitudes of the halo particles, thereby increasing their impact parameters on secondary collimators (see preceding section). A bent crystal can coherently direct channeled halo particles deeper into a nearby secondary absorber. This approach has the potential of reducing beam losses in critical locations and, consequently, to reduce the radiation load to the downstream superconducting magnets.

There are several processes which can take place during the passage of protons through the crystals: (a) amorphous scattering of the primary beam; (b) channeling; (c) dechanneling due to scattering in the bulk of the crystal; (d) "volume reflection" off the bent planes; and (e) "volume capture" of initially unchanneled particles into the channeling regime after scattering inside the crystal. The particle can be captured in the channeling regime, oscillating between two neighboring planes if it enters within crystal's angular acceptance of:

$$\theta < \theta_c = \sqrt{\frac{2U_0}{pc}}, \tag{6.124}$$

where p is the particle momentum and U_0 is the crystal's planar potential well depth. The critical angle θ_c is about 7 μrad for 980 GeV/c protons in the (110) silicon crystal orientation. When the crystal is bent, particles still can be channeled (and thus deflected) if the bend radius R is greater than a critical value $R_c = pv/eE_m$, where E_m is the maximum strength of the electric field in the channel, about 6 GV/cm for the (110) silicon crystal orientation. That yields $R_c \approx 1.6$ m for 980 GeV/c protons. Bending of the crystal decreases the critical channeling angle, the capture probability of particles into the channeling regime and the dechanneling length [33]. If the particle momentum is not within the critical angle but has a tangency point with the bent planes within the crystal volume, almost all particles are deflected to the opposite direction with respect to the crystal bending. The effect is called the volume reflection (VR) [33] and it has very wide angular acceptance equal to the crystal bend angle (of the order of hundreds of microradians compared to several microradians of the channeling acceptance). The drawback of the volume reflection regime is that the deflection angle is small, approximately $(1.5–2) \times \theta_c$. However, this can be overcome by using a sequence of several precisely aligned bent crystals, so that the total deflection angle is proportionally larger.

In the Tevatron beam crystal collimation experiment T980 [37–39] both single crystals (for vertical and horizontal deflection) and multi-strip crystal assemblies (for vertical multiple VR) have been used. Collimation of circulating beams is very different from bent crystal experiments with extracted beams [40] because of smaller initial "impact parameters" and the possibility of interplay of different effects. In an accelerator such as the Tevatron several phenomena determine the impact parameter (the depth of the particle penetration at the first interaction with the crystal). These include four diffusion and two orbit processes. The first two diffusion processes are scattering on the residual gas and noise in magnetic field. They produce the amplitude growth of about 4 nm/√turn. The RF noise results in the diffusion rate of ~12 nm/√turn (hor.) and ~1 nm/√turn (vert.). The beam diffusion due to beam–beam or other nonlinear effects can produce up to ~10–40 nm/√turn; and, important only for the uncaptured DC beam particles located in the abort gaps, the diffusion due to the excitation by the TEL results in ~7 μm/turn. For interaction with amorphous targets, the diffusion rates are ~200 μm/√turn for a 5 mm length of amorphous silicon, and about ~1,200 μm/√turn for a 5 mm tungsten primary target. The two orbit processes

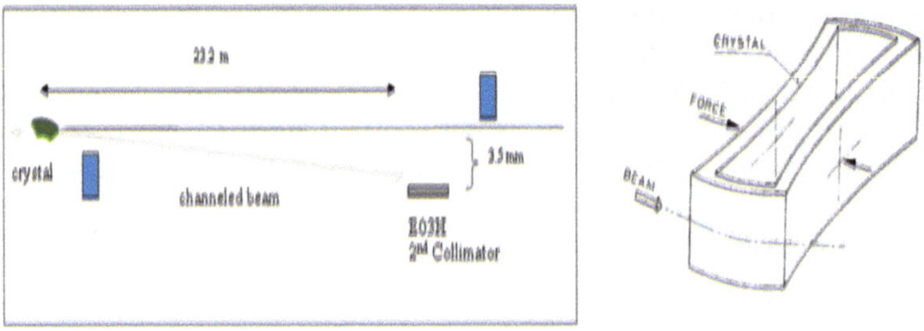

Fig. 6.23 (**a**) (*Left*) General layout of the T980 experiment at E0, the straight section used for the crystal collimation test; (**b**) (*right*) "O"-shape crystal prepared by Petersburg Nuclear Phusics Institute and used at RHIC and Tevatron. The length along the beam is 5 mm

are: transverse orbit oscillations with amplitude of ~20 μm and frequencies of ~15 Hz (i.e., with period equal to some 3,000 revolutions in the Tevatron—see Chap. 2, Sect. 2.4) and synchrotron motion of particles near the boundary of the RF bucket with amplitudes of ~1 mm (hor) and ~70 μm (vert) at ~35 Hz (1,300 turns). The orbit motion due to synchrotron oscillations occurs at nonzero dispersion locations, e.g., at the E03 location of the T980 experiment crystals where the dispersion is about $D_x = 2$ m horizontally and 18 cm vertically. The resulting impact parameters are estimated to be of the order of 0.2–1 μm for transverse halo particles and ~10–30 μm for the particles in the abort gaps which have leaked out of the RF buckets. All that makes the properties of the surface of the crystal (rather than the bulk of the crystal) pivotal for collimation (contrary to the extracted beam studies)—e.g., it's roughness or the miscut angle [40]. These processes are very hard to predict and simulate; hence, the T-980 experimental studies are of great importance for understanding the beam physics aspects and determining feasibility of the method for high energy colliders like the LHC and a Muon Collider.

Figure 6.23a shows a schematic of the T-980 experimental layout. During normal Tevatron operations, a 5-mm tungsten target scatters the proton beam halo into a 1.5-m long stainless steel secondary collimator E03, 50 m downstream of the target. For the bent crystal experiments, a goniometer containing single or multi-strip bent crystals is installed 23.7 m upstream of the E03 collimator. Scintillation counter telescopes detect secondary particles from protons interacting with the target and E03 collimator. An ionization chamber (beam loss monitor LE033) also detects secondary particles scattered from E03. A PIN diode telescope detects the secondaries scattered from the bent crystal. Under the above configuration, channeled beam is signaled by a reduction of the rate in the PIN telescope (channel LE033C) with attendant increases in the rates of the LE033 and E1 counters.

A modified BNL goniometer assembly [41] and an O-shaped 5-mm silicon crystal with a bending angle of 0.44 mrad were originally installed in the Tevatron downstream of the horizontal primary collimator in the fall of 2004. The crystal was set at $5.5\sigma_b \approx 2.5$ mm from the beam center and aligned in the halo by varying the crystal angle in steps of several μrad. The interaction probability in the 5 mm long crystal was monitored by the PIN diode and plotted as a function of the crystal angle as shown in

Fig. 6.24 Crystal angle (see text)

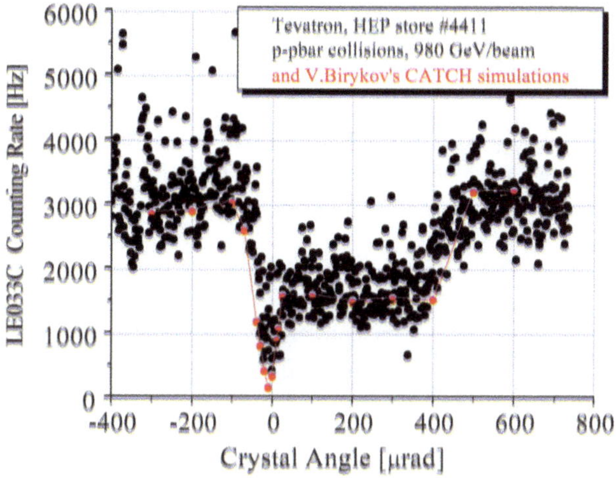

Fig. 6.24. A channeling dip is present at zero angle to the crystal's plane with a width of 22 ± 4 μrad (rms). The width of the channeling dip is a convolution of the beam divergence, the channeling critical angle, multipass channeling effects, and possible crystal distortions. It is difficult to do a deconvolution of the crystal angular scan to get the critical angle. However, the distribution is consistent with the beam divergence and 7 μrad channeling critical angle at 980 GeV. At the bottom of the dip the LE033C signal is 22 % of the signal at a random angular setting. This depth is a measure of the channeling efficiency and gives a channeling efficiency of $\eta_c = 78 \pm 12$ % including the effects of multiple passes. A shoulder extends 460 ± 20 μrad to the right of the channeling dip. This shoulder width is close to the expected magnitude of the crystal bend. The shoulder is a coherent crystal effect acting over the whole arc of the crystal bend due to volume reflection. Like channeling, the volume reflection diminishes nuclear interactions and thereby decreases the LE033 rate. The whole-arc efficiency, η_r, was 52 ± 12 %. The larger red dots and associated curve show the results of Biryukov's CATCH simulation [42] for the conditions in the Tevatron. Note that there are no free parameters in this simulation except average counting rate. Most impressive effect of using the bent crystal at the channeling angle instead of a tungsten primary collimator is the reduction of the CDF beam losses at the opposite side of the ring by a factor of two, in a good agreement with predictions [37].

In 2009 the 0.44-mrad bend O-shaped crystal in the horizontal goniometer was replaced with the new 0.36-mrad O-shaped one with negative 0.12-mrad miscut angle built by IHEP (Protvino), and a new vertical "push–pull"-type goniometer was installed 4-m upstream, housing two crystals—the multiple (eight) strip crystal from IHEP and the old 0.44 mrad O-shaped crystal, so that there were crystals for collimation in both vertical and horizontal planes. Since then crystal collimation has been routinely employed during many collider stores. Additional beam instrumentation was added. Fast automatic insertion of the crystals has been implemented. A vertical multi-strip crystal system has been successfully tested and both multiple-VR beam at the E03 collimator and channeled beam at the F17

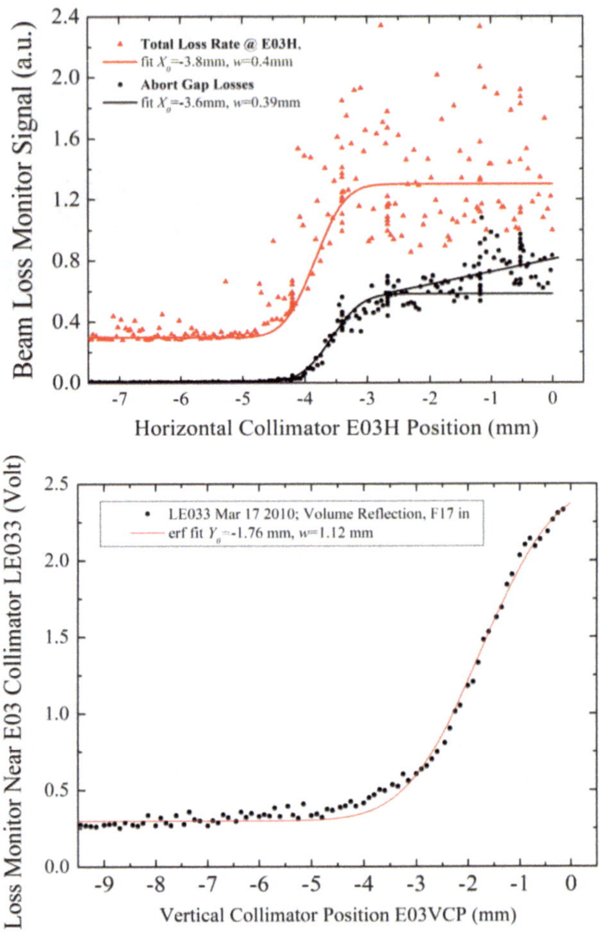

Fig. 6.25 Collimator scan with crystal set at: (**a**) (*top*) the channeling angle; (**b**) (*bottom*) Collimator scan with 8-strip vertical crystal set at the VR angle. *Solid lines* are for "erf" fits of the data [27]

collimator some 1 km downstream of the E03 have been observed. A reduction of ring wide losses was reproducibly obtained along with local loss effects on the collimator due to crystal channeling and VR. The first ever study of two plane crystal collimation was also conducted.

To measure the deflection of the channeled (or VR) particles once the crystal angle is set to the channeling (or VR) peak the position of an appropriate collimator can be slowly scanned, starting from a completely retracted position and moving toward the beam edge. An example of such a scan is shown in Fig. 6.25a for horizontally deflected channeled protons at the E03H collimator. The curves show the total measured loss rate (red dots) as well as the counting rate synchronized to the abort gaps only (black dots). There are three distinct regions: (a) a region of negligible losses, where the collimator does not intercept any beam; (b) a steep increase in the losses, where the collimator intercepts the channeled beam; (c) a region where the losses increase slowly: the collimator is additionally intercepting

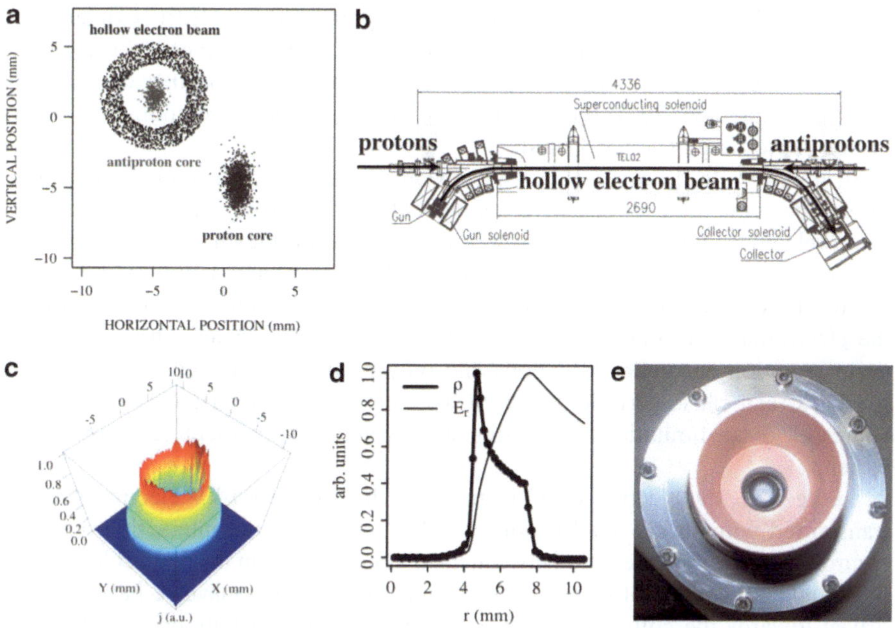

Fig. 6.26 Hollow electron beam collimator: (**a**) transverse beam layout; (**b**) top view of the beams in the Tevatron; (**c**) measured current profile; (**d**) measured charge density ρ and calculated radial electric field E_r; (**e**) photograph of the electron gun [27]

de-channeled and amorphous scattered particles. Both abort gap and total loss signals show a small deflection angle of (3.6–3.8)mm/24 m = 150–160 μrad instead of the expected 360 μrad. Such a difference can either be attributed to the effect of the "miscut angle" [38] or be due to non-ideal crystal surface that becomes important at the very small impact parameters. The angular spread in the channeled beam is about of 0.4 mm/24 m = 17 μrad rms that is larger than the channeling acceptance of $2 \times \theta_c = 13.4$ μrad. A similar scan of the VR beam made with the E03 vertical collimator presented in Fig. 6.25b shows the beam at 1.76 mm/28 m = 63 μrad, i.e., approximately where it is supposed to be, and about 40 μrad rms wide [39].

6.4.5 Hollow Electron Beam Collimator

The hollow electron beam collimator (HEBC) is a novel concept of controlled halo removal for intense high-energy hadron beams in storage rings and colliders [43, 44]. It is based on the interaction of the circulating beam with a 5-keV, magnetically confined, pulsed hollow electron beam in a 2-m-long section of the ring. The electrons enclose the circulating beam, kicking halo particles transversely and leaving the beam core unperturbed (Fig. 6.26a, b). By acting as a tunable diffusion enhancer but not as a hard aperture limitation, the HEBC extends conventional collimation systems beyond the intensity limits imposed by tolerable losses. The concept was tested experimentally at the Tevatron between October 2010 and September 2011. It represents a promising option for scraping high-power beams in the Large Hadron Collider.

In high-power hadron machines, conventional two-stage collimation systems offer robust shielding of sensitive components and are very efficient in reducing beam-related backgrounds at the experiments. However, they have limitations. The minimum distance between the collimator and the beam axis is limited by instantaneous loss rates (especially as jaws are moved inward), radiation damage, and by the electromagnetic impedance of the device. Moreover, beam jitter, caused by ground motion and other vibrations and only partially mitigated by active orbit feedback, can cause periodic bursts of losses at aperture restrictions. The HEBC addresses these limitations, emerging as a viable complement to conventional systems.

In the TEL, the electron beam is generated by a pulsed 5-kV electron gun and transported with strong axial magnetic fields. Its size in the interaction region is controlled by varying the ratio between the magnetic fields in the main solenoid and in the gun solenoid. Halo particles experience nonlinear transverse kicks and are driven towards the collimators. If the hollow current distribution is axially symmetric, there are no electric or magnetic fields inside and the beam core is unperturbed. A magnetically confined electron beam is stiff, and experiments with electron lenses showed that it can be placed very close to and even overlap with the circulating beam. Another advantage is that, contrary to conventional systems, no nuclear breakup is generated in the case of ion collimation. In a setup similar to that of the TEL, with a peak current of 1 A, an overlap length of 2 m, and a hole radius of 3 mm, the corresponding radial kick is 0.3 μrad for 980-GeV antiprotons. The intensity of the transverse kicks is small and tunable: the device acts more like a soft scraper or a diffusion enhancer, rather than a hard aperture limitation. Because the kicks are not random in space or time, resonant excitation is possible if faster removal is desired.

Analytical expressions for the current distribution were used to estimate the effectiveness of the HEBC on a proton beam. They were included in tracking codes such as STRUCT [23], Lifetrac [45], and SixTrack [46] to follow core and halo particles as they propagate in the machine lattice. These codes are complementary in their treatment of apertures, field nonlinearities, and beam–beam interactions. Preliminary simulations suggested that effects would be observable and that measurements are compatible with normal collider operations [47, 48].

A 15-mm-diameter hollow electron gun was designed and built (Fig. 6.26c–e) [49, 50]. It is based on a tungsten dispenser cathode with a 9-mm-diameter hole bored through the axis of its convex surface. The peak current delivered by this gun is 1.1 A at 5 kV. The current density profile was measured on a test stand by recording the current through a pinhole in the collector while changing the position of the beam in small steps. The gun was installed in one of the TEL-2 in

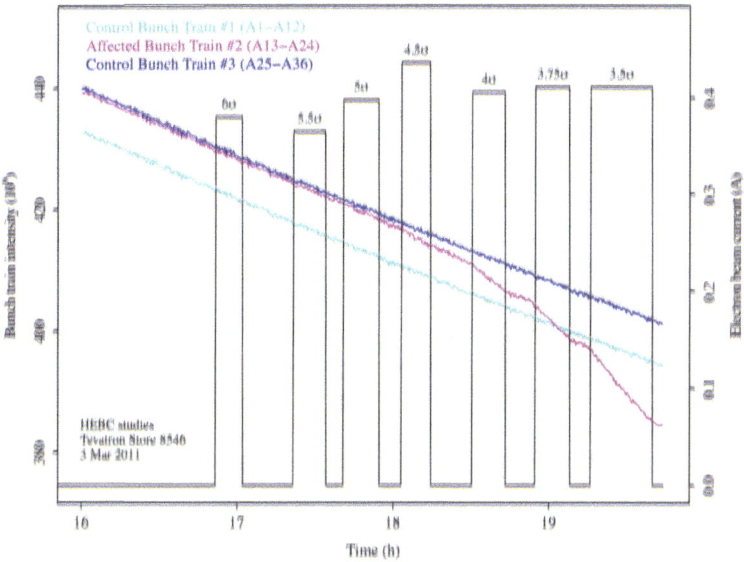

Fig. 6.27 Scraping effect of the hollow electron beam acting on one antiproton bunch train (*magenta*) at the end of a collider store. The intensities of the two control trains are shown in *cyan* and *blue*. The electron beam current (*black trace*) was turned on and off several times with different values of the hole radius, from $6\sigma_y$ to $3.5\sigma_y$ [27]

August 2010. The pulsed electron beam could be synchronized with practically any bunch or group of bunches.

The behavior of the device and the response of the circulating beams were measured for different beam currents, relative alignments, hole sizes, pulsing patterns, and collimator system configurations. Preliminary results were presented in [51, 52]. Here, we discuss a few representative experiments illustrating the main effects of the electron beam acting on antiproton bunches. Antiprotons were chosen for two main reasons: their smaller transverse emittances (achieved by stochastic and electron cooling in the Antiproton Source accelerators) made it possible to probe a wider range of confining fields and hole sizes; and the betatron phase advance between the electron lens and the absorbers was more favorable for antiproton collimation.

The particle removal rate was measured by comparing bunches affected by the electron lens with other control bunches. In the experiment described in Fig. 6.27, the electron lens was aligned and synchronized with the second antiproton bunch train, and then turned on and off several times at the end of a collider store. The electron beam current was about 0.4 A and the radius of the hole was varied between $6\sigma_y$ and $3.5\sigma_y$, with $\sigma_y = 0.57$ mm being the vertical rms beam size. The black trace is the electron-lens current. One can clearly see the smooth scraping effect. The corresponding removal rates are of a few percent per hour.

There is a concern related to adverse effects of hollow beam on the core of the circulating beam, because an asymmetry in the gun emission results in that the beam in the overlap region is not a perfect hollow cylinder. The asymmetry is amplified by evolution of electron beam distribution under space its charge, and by the bends in the transport system. The problem was approached from several points of view. First, one can see in Fig. 6.27 that no decrease in intensity was observed with large hole sizes, when the hollow beam was shadowed by the primary collimators. This implies that the circulating beam was not significantly affected by the hollow electron beam surrounding it, and that the effect on the beam intensity of residual fields near the axis was negligible. Second, we observed no difference in the emittance growth for the affected bunches. If there was an emittance growth produced by the electron beam, it was much smaller than that driven by other two main factors, namely IBS and beam–beam interactions. The effect of halo removal can also be observed by comparing beam scraping with the corresponding decrease in luminosity. The luminosity is proportional to the product of antiproton and proton bunch populations, and inversely proportional to the overlap area. If antiprotons are removed uniformly and the other factors are unchanged, luminosity should decrease by the same relative amount. If the hollow beam causes emittance growth or proton loss, luminosity should decrease even more. A smaller relative change in luminosity was observed, which is a clear indication of halo scraping. Also, the ratio between luminosity decay rates and intensity decay rates increased with decreasing hole size. Finally, one can attempt to directly measure the particle removal rate as a function of amplitude. This was done with collimator scans. A primary antiproton collimator was moved vertically in 50-μm steps towards the beam axis. All other collimators were retracted. The corresponding beam losses and decay rates were recorded. Particles were removed from the affected bunch train, but as soon as the primary collimator shadowed the electron beam, eliminating the halo at those amplitudes, the relative intensity decay rate of the affected bunch train went back to the value it had when the lens was off. Even with a hole size of $3.5\sigma_y$, the effects of residual fields on the core appeared to be negligible. The time evolution of losses during a collimator scan was also used to measure changes in diffusion rate as a function of amplitude, using an extended version of the technique presented in [26, 53].

Another observation was that the hollow electron lens mitigated the effects of beam jitter. In the Tevatron, beams oscillate coherently at low frequencies (from sub-Hz to a few Hz) with amplitudes of a few tens of microns, due to mechanical vibrations and ground motion. This causes periodic bursts of losses at aperture restrictions, with peaks exceeding a few times the average loss rate. When the collimators are moved inward, these loss spikes can cause quenches in the superconducting magnets or damage electronic components. In March 2011, to measure the loss spikes and the effects of the hollow electron beam, scintillator paddles were installed downstream of one of the antiproton secondary collimators (F48). These loss monitors could be gated to individual bunch trains. It was observed that losses from the two control trains were completely correlated, and that their frequency spectra showed strong peaks at 0.39 Hz and its harmonics (corresponding to the acceleration cycle of the Main Injector) and at 4.6 Hz (mechanical vibrations from the Central Helium Liquefier). The electron lens suppressed these peaks and eliminated correlations with the other trains. This can be interpreted as a reduction in the population of the beam tails. Note that it adversely affects the beam stability reducing the Landau damping for the affected bunches.

Losses generated by the electron lens were mostly deposited in the collimators, with small changes in the loss observed by the experiments. Alignment of the beams was done manually, with a setup time of about 15 min. Alignment is crucial for HEBC operation, and the procedures based on the electron-lens beam-position monitors were found to be reliable in spite of the different time structure of the electron and (anti)proton pulses. No instabilities or emittance growth were observed over the course of several hours at nominal antiproton intensities (10^{11} particles/bunch) and electron beam currents up to 1 A in confining fields above 1 T in the main solenoid. Most of the studies were done parasitically during regular collider stores.

Experiments at the Tevatron showed that the HEBC is a viable option for scraping high-power beams in storage rings and colliders. Its applicability to the LHC is under study. To make the device more versatile, larger cathodes and higher electron beam currents appear to be feasible, and experimental tests in this direction are planned.

6.5 Luminosity Evolution Model

6.5.1 Longitudinal Dynamics and Particle Loss from RF Bucket

While colliding at the top energy, the Tevatron proton and antiproton bunches lose their intensities as particles escape stable RF buckets. That occurs mostly due to the IBS and RF noise while infrequent occurrences of the longitudinal instabilities or trips of the RF power amplifiers can contribute large spills of particles to the uncaptured beam. Uncaptured beam particles are outside of the RF buckets, and therefore, move longitudinally relative to the main bunches, lose about 9 eV/turn due to the synchrotron radiation (SR), slowly spiral radially inward and die on the collimators, which determine the tightest aperture in the Tevatron during collisions. The typical time for a particle between its escape from the RF bucket and its reach of a collimator is about 20 min. The TEL operating in the abort cleaning regime (see Sect. 6.4.3) reduces the removal time of uncaptured beam from 20 min to about 2 min, thereby significantly reducing the particle population in the abort gap. That resulted in complete elimination of the quenches related to the uncaptured beam [31].

In the case of single harmonic RF, a particle phase trajectory in the longitudinal phase space (see Fig. 6.28) is described by the following equation [54]:

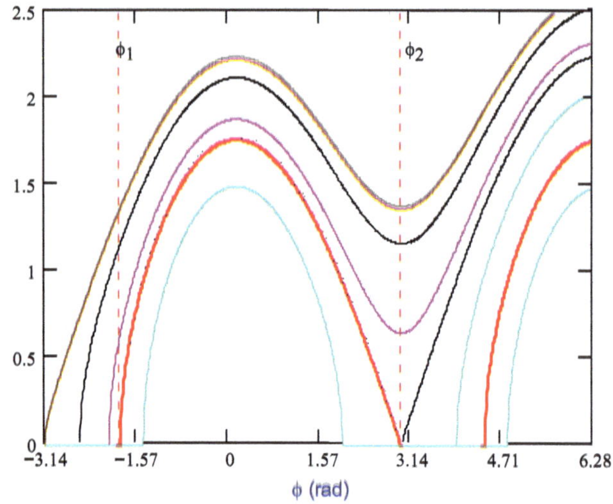

Fig. 6.28 Upper half of phase space trajectories in the vicinity of the separatrix (*red line*) for $\varphi_0 = 0.15$. Momentum spread (*vertical axis*) is presented in units of $(\eta h/Q_s)\ \Delta p/p_0$

$$\left(\frac{\Delta p}{p_0}\right)^2 = \frac{2Q_s^2}{\eta^2 h^2}\left(\cos\varphi - \cos\varphi_m + (\varphi - \phi_m)\sin\phi_0\right), \tag{6.125}$$

where $\Delta p/p_0$ is the relative particle momentum deviation, η is the slip factor, h is the harmonic number, Q_s is the synchrotron tune, ϕ is the RF phase, φ_0 is the accelerating phase, and φ_m determines the boundary of phase space trajectory. In the stationary state φ_0 is determined by particle energy loss due to synchrotron radiation eV_{SR}: $\sin\varphi_0 = V_{SR}/V_{RF}$. The SR radiation damping is neglected in Eq. (6.125) since the damping time is much longer than the store duration.

The outermost orbit, called the separatrix, determines the boundary of the RF bucket:

$$\left(\frac{\Delta p}{p_0}\right)^2 = \frac{2Q_s^2}{\eta^2 h^2}\left(\cos\varphi + \cos\varphi_0 + (\varphi - \pi + \varphi_0)\sin\phi_0\right). \tag{6.126}$$

If the equilibrium phase ϕ_0 is small, $\varphi_0 \ll 1$, the separatrix boundaries in the RF phase are:

$$\begin{aligned} \varphi_1 &\approx -\pi + \sqrt{4\pi\varphi_0} - \varphi_0, \\ \varphi_2 &= \pi - \varphi_0. \end{aligned} \tag{6.127}$$

As an example Fig. 6.28 presents the phase space trajectories for $\varphi_0 = 0.15$. The accelerating phase is much smaller for Tevatron at collisions: $\varphi_0 \approx 10^{-11}$, $\varphi_1 \approx -\pi + 10^{-5}$ and $\varphi_2 \approx \pi - 10^{-11}$. Thus, the Tevatron RF buckets are separated by a gap of ~10^{-5} rad. A particle with initial momentum above the RF bucket boundary is decelerated by energy loss due to SR and eventually passes through a gap between buckets to the lower momentum side where it is decelerated to the nearest apertures limiting the beam energy spread.

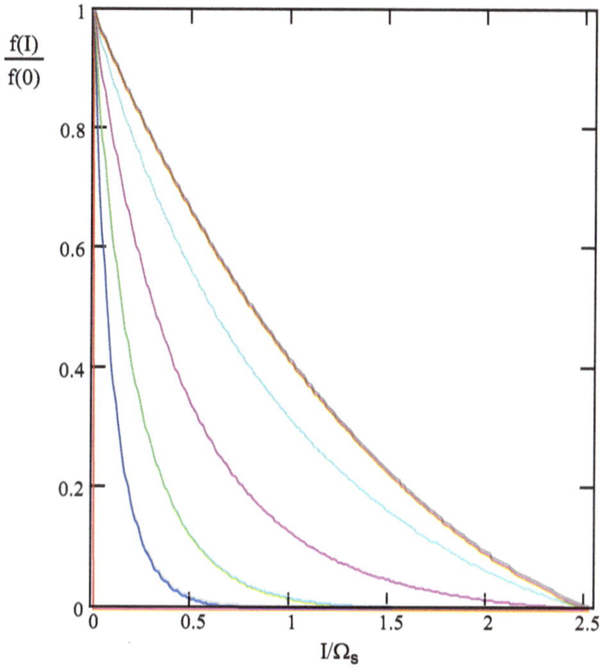

Fig. 6.29 Dependence of the distribution function on time for $Dt/\Omega_s^2 = 0, 0.125, 0.25, 0.5, 1$ and asymptotic at $t \to \infty$

6.5.2 Longitudinal Diffusion and Particle Loss

There are three major mechanisms creating the uncaptured beam. They are the diffusion due to amplitude and phase RF noises (see, e.g., [55]), multiple IBS and single IBS (Touschek effect) [56]. The RF bucket size at injection is 4.4 eV. Measurement showed that it is completely filled at the injection. The bucket size grows in the course of acceleration while the longitudinal emittance remains the same. At the acceleration end the bucket size reaches 11 eV s. Acceleration is sufficiently fast and therefore immediately after acceleration the bunch tails are negligible and the bunch occupies the same phase space as at injection. In the absence of tails the single IBS is the only mechanism for longitudinal particle loss. Shortly after acceleration the diffusion due to IBS and RF noise creates tails in the distribution function and results in additional beam loss, which significantly exceeds the loss due to single IBS (see, e.g., [57]). Therefore, in the following analysis we neglect the single IBS.

Figure 6.29 presents a numerical solution of diffusion equation (6.74) assuming (1) constant diffusion, $D(I) = D_0$, as a zero-order approximation, (2) the initial distribution being a δ-function, $f_0(I) = \delta(I)$, and (3) the boundary condition $f(I_b) = 0$ at the RF bucket boundary with $I_b = 8\Omega_s/\pi$. Figure 6.30 presents the corresponding beam intensity, rms momentum spread, and rms bunch length.

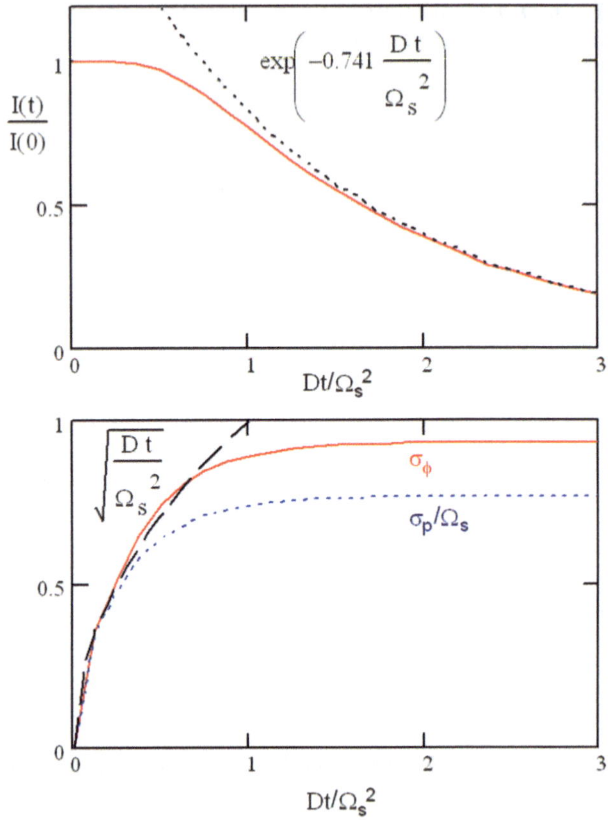

Fig. 6.30 Time dependence of beam intensity (*top*) and rms bunch length and momentum spread (*bottom*)

Initially, the bunch length and the momentum spread grow proportionally to $\sqrt{t}$ and the distribution function is close to the Gaussian distribution, $f(I,t) \propto (\Omega_s/Dt)\exp(-I\Omega_s/Dt)$. However when the bunch length becomes comparable to the bucket length, the non-quadratic behavior of the potential results in the bunch-length growing faster than the momentum spread. Finally, the distribution function and, consequently, the bunch length and momentum spread approach their asymptotic values, and the intensity decays exponentially as ~$\exp(-0.741\ Dt/\Omega_s^2)$.

The results of simulations yield the following approximate relationships between the bunch parameters:

$$\sigma_s \approx \Gamma_s \sigma_{\Delta p/p}\left(1+\frac{1}{4}\left(\frac{2\sigma_{\Delta p/p}}{\Delta P/P|_{\mathrm{sep}}}\right)^2+\frac{1}{6}\left(\frac{2\sigma_{\Delta p/p}}{\Delta P/P|_{\mathrm{sep}}}\right)^3\right),$$
$$\frac{1}{N}\frac{dN}{dt} \approx \frac{2.425(2\pi\sigma_s)^7}{\lambda_{\mathrm{RF}}{}^7+1.65(2\pi\sigma_s)^7}\left.\frac{d\left(\sigma_\varphi{}^2\right)}{dt}\right|_{\mathrm{LM}}, \tag{6.128}$$
$$\frac{d\left(\sigma_{\Delta p/p}{}^2\right)}{dt} \approx \left(1-\left(\frac{2\sigma_{\Delta p/p}}{0.765\Delta P/P|_{\mathrm{sep}}}\right)^5\right)\left.\frac{d\left(\sigma_{\Delta p/p}{}^2\right)}{dt}\right|_{\mathrm{LM}},$$

where $\Delta P/P|_{\mathrm{sep}}$ is the height of the RF bucket. The top equation accounts nonlinear dependence between the bunch length σ_s and the momentum spread $\sigma_{\Delta p/p}$ in the single harmonic RF bucket. The second and third equations account for appearing the particle loss from the bucket and the reduction in the momentum spread growth rate which occur when the bunch distribution achieves the bucket boundary.

6.5.3 Longitudinal Evolution Model

Numerous factors affect the Tevatron collider luminosity and its evolution in time. Each store is different and because of finite instrumentation accuracy it is practically impossible to state what was different or what came wrong for every particular store. Nevertheless the luminosity evolution is very similar for most of the stores. It is driven by some basic processes, which are not very sensitive to the details of distribution functions, and therefore the luminosity evolution can be described by comparatively simple parametric model [58] presented below. The model takes into account the major beam heating and particle loss mechanisms. They are (1) the emittance growth and the particle loss due to scattering on the residual gas, (2) the particle loss and the emittance growth due to scattering in IPs, (3) the transverse and longitudinal emittance growth due to IBS, (4) the bunch lengthening due to RF noise, and (5) the particle loss from the bucket due to heating of longitudinal degree of freedom. If the collider tunes are correctly chosen, the beam intensity is not too high, and the beams are well formed, then the beam–beam effects are not very important and the model describes the observed dynamics of beam parameters and the luminosity comparatively well. Detailed discussion of how the beam–beam effects and lattice nonlinearities interact with diffusion and how they can be incorporated into the model can be found in Chap. 8.

If aperture limitations are sufficiently large in comparison with the beam size ($A_{x,y} \geq 5\sigma_{x,y}$), then the multiple and single scattering on the residual gas atoms can be considered separately. In this case the single scattering causes the particle loss, while the multiple scattering causes the emittance growth.

The beam lifetime due to single scattering is described by the well-known formula [59]

Table 6.3 Gas composition used in the simulations

Gas	H_2	CO	N_2	C_2H_2	CH_4	CO_2	Ar
Pressure [nTorr]	1.05	0.18	0.09	0.075	0.015	0.09	0.15

$$\tau_{\rm scat}{}^{-1} == \frac{2\pi c r_{\rm p}{}^2}{\gamma^2\beta^3}\left(\sum_i Z_i(Z_i+1)\left(\frac{\overline{\beta_x n_i}}{\varepsilon_{mx}}+\frac{\overline{\beta_y n_i}}{\varepsilon_{my}}\right)\right)+\sum_i n_i\sigma_i c\beta, \tag{6.129}$$

where $\overline{\beta_{x,y}n_i} = \int \beta_{x,y}n_i ds/C$ are the average gas density weighted by beta-functions, $\varepsilon_{mx,my}$ are the horizontal and vertical acceptances, $r_{\rm p}$ is the proton classical radius, γ and β are the relativistic factors, the summing is performed over all residual gas species, and the averaging is performed over ring circumference. The first addend is related to the electromagnetic scattering and the second one to the strong interaction. Taking into account that the scattering angle due to strong interactions ($\theta \sim m_\pi c/p \approx 140$ μrad) significantly exceeds rms angles in the beam (~7 μrad) σ_i can be considered to be the total nuclear cross section with sufficiently good accuracy. At the collision energy of 980 GeV the beam lifetime is dominated by the strong interaction.

The emittance growth rate due to multiple scattering is closely related to the electromagnetic part of the single scattering lifetime and is determined by the following formula:

$$\frac{d\varepsilon_{x,y}}{dt} = \frac{2\pi c r_{\rm p}{}^2}{\gamma^2\beta^3}\left(\sum_i \overline{\beta_{x,y}n_i}Z_i(Z_i+1)L_{\rm c}\right), \tag{6.130}$$

where $L_{\rm c}$ is the Coulomb logarithm ($L_{\rm c} \approx 9$). Table 6.3 presents the gas composition used in the model. Overall pressure was scaled to fit the measurement results.

Similar to the gas scattering the scattering in the interaction point (IP) can be separated into the single scattering due to strong interaction and the emittance growth due to electromagnetic scattering. The total $\bar{p}p$ cross section consists of two parts: the inelastic cross section of 60 mbarn and the elastic cross section of 15 mbarn at 1 TeV energy. All particles scattered inelastically are lost immediately, while as shown in [59] about 40 % of elastically scattered particles remain in the beam (within 3σ). That happens because the beta-functions in the IP are small and, consequently, particle angles are large; so that the scattering angles are comparable to the particle angles (~100 μrad). The total cross section of particle loss obtained by fitting of the observations to the model is equal to 69 mbarn which coincides well with expectations.

The emittance growth due to electromagnetic scattering in one IP is equal to:

$$\frac{d\varepsilon_{x,y}}{dt} = \frac{4r_p^2 NL_{bb} f_0}{\gamma^2\beta^3\sqrt{(\varepsilon_{px}+\varepsilon_{py})(\varepsilon_{ax}+\varepsilon_{ay})}}. \tag{6.131}$$

Here ε_{px}, ε_{py}, ε_{ax}, and ε_{ay} are the emittances for proton and antiproton beams, f_0 is the revolution frequency, L_{bb} is the Coulomb logarithm ($L_{bb} \approx 20$), and N is the number of particles in the counter-rotating bunch. For two IPs and present Tevatron parameters it yields the antiproton emittance growth rate of about 0.0015 mm mrad/h. Although emittance growth rate is almost negligible in comparison with gas scattering the nuclear absorption in the IP is the main mechanism for antiproton loss during collisions.

Another important diffusion mechanism is determined by IBS. For the Tevatron collider parameters the longitudinal energy spread in the beam frame is significantly smaller than the transverse ones (e.g., the ratio of longitudinal to transverse particle velocities in the beam frame is about $v_{\|}/v_{\perp} \approx 0.02$ at collision energy, and $v_{\|}/v_{\perp} \approx 0.15$ at injection energy). In this case comparatively simple IBS formulas presented in Eq. (6.54) can be used. Tevatron has sufficiently smooth lattice and therefore IBS can be described with good accuracy in the smooth approximation. In this case Eq. (6.54) can be rewritten as follows [58]:

$$\begin{bmatrix} d\sigma_{\Delta p/p}{}^2/dt \\ d\varepsilon_x/dt \\ d\varepsilon_y/dt \end{bmatrix} = \frac{r_p^2 cNL_c \Xi_{\|}(\overline{\theta}_x,\overline{\theta}_y)}{4\sqrt{2}\gamma^3\beta^3\overline{\sigma}_x\overline{\sigma}_y\sigma_s\sqrt{\overline{\theta}_x^{\,2}+\overline{\theta}_y^{\,2}}} \begin{bmatrix} 1 \\ \overline{A}_x(1-\kappa) \\ \overline{A}_x\kappa \end{bmatrix}, \tag{6.132}$$

where

$$\begin{aligned} \overline{\sigma}_x &= \sqrt{\varepsilon_x\overline{\beta}_y + \overline{D}_x^{\,2}\sigma_{\Delta p/p}{}^2}\;, & \overline{\sigma}_y &= \sqrt{\varepsilon_y\overline{\beta}_y}\;, \\ \overline{\theta}_x &= \sqrt{\varepsilon_x/\overline{\beta}_x}\;, & \overline{\theta}_y &= \sqrt{\varepsilon_y/\overline{\beta}_y}\;. \end{aligned} \tag{6.133}$$

For Tevatron the averaged beta-functions, the dispersion and the horizontal motion invariant are: $\overline{\beta}_x = R/\nu_x = 49$ m, $\overline{\beta}_y = R/\nu_y = 49$ m, $\overline{D}_x = \int D_x ds/2\pi R = 2.84$ m, and $\overline{A}_x = \int A_x ds/2\pi R = 0.2$ m. To obtain Eq. (6.132) we neglected in variations of $\Psi(0,\theta_x,\theta_y)$ along the lattice in Eq. (6.54) and assumed that $\Psi(0,\theta_x,\theta_y) = 1$ and $\Psi_{x,y}(\theta_x,\theta_y) = 1$ (their accounting makes quite small correction), and we additionally introduced the coupling parameter κ which takes into account the redistribution of heating between horizontal and vertical degrees of freedom. An experimental value for κ is about 0.4. It is quite large and due to the strong coupling in Tevatron which additionally is amplified by the beam–beam interaction. To verify accuracy of the smooth approximation we performed averaging of Eq. (6.54) with Tevatron lattice parameters for the rms bunch length of 62 cm, zero coupling ($\kappa = 0$) and 1.6×10^{11} protons/bunch. It yielded the horizontal and longitudinal emittance growth lifetimes of

22.5 and 28.5 h, correspondingly. The use of smooth approximation yields 18.9 and 26.9 h. As one can see the difference is sufficiently small and therefore the smooth approximation has been used in the described below parametric model.

For Gaussian beams the luminosity of the collider is determined by the well-known formula:

$$L = \frac{f_0 N_b N_p N_a}{2\pi\beta^* \sqrt{(\varepsilon_{px} + \varepsilon_{ax})(\varepsilon_{py} + \varepsilon_{ay})}} H\left(\frac{\sqrt{\sigma_{sp}{}^2 + \sigma_{sp}{}^2}}{\sqrt{2}\beta^*}\right), \tag{6.134}$$

where N_b is the number of bunches, N_p and N_a are the number of protons and antiprotons per bunch, β^* is the beta-function in the interaction point (IP), and ε_{px}, ε_{py}, ε_{ax}, and ε_{ay} are the horizontal and vertical emittances for proton and antiproton beams. The hourglass factor $H(x)$ takes into account the finite value of the longitudinal bunch size. It is equal to:

$$H(x) = \frac{2}{\sqrt{\pi}} \int_0^\infty \frac{e^{-y^2}}{1 + x^2 y^2} dy \overset{x \le 3}{\rightarrow} \approx \frac{1}{\sqrt[3]{1 + 1.3x^2}}. \tag{6.135}$$

To describe the evolution of the luminosity we write a system of differential equations, which bounds up all basic parameters of the proton and antiproton beams:

$$\frac{d}{dt}\begin{bmatrix} \varepsilon_{px} \\ \varepsilon_{py} \\ \sigma_{pp}{}^2 \\ N_p \\ \varepsilon_{ax} \\ \varepsilon_{ay} \\ \sigma_{pa}{}^2 \\ N_a \end{bmatrix} = \begin{bmatrix} 2d\varepsilon_{px}/dt\big|_{BB} + d\varepsilon_{px}/dt\big|_{IBS} + d\varepsilon_{px}/dt\big|_{gas} \\ 2d\varepsilon_{py}/dt\big|_{BB} + d\varepsilon_{py}/dt\big|_{IBS} + d\varepsilon_{py}/dt\big|_{gas} \\ d\sigma_{pp}{}^2/dt\big|_{total} \\ -N_p\tau_{scat}{}^{-1} - dN_p/dt\big|_L - 2L\sigma_{p\bar{p}}/n_b \\ 2d\varepsilon_{ax}/dt\big|_{BB} + d\varepsilon_{ax}/dt\big|_{IBS} + d\varepsilon_{ax}/dt\big|_{gas} \\ 2d\varepsilon_{ay}/dt\big|_{BB} + d\varepsilon_{ay}/dt\big|_{IBS} + d\varepsilon_{ay}/dt\big|_{gas} \\ d\sigma_{pa}{}^2/dt\big|_{total} \\ -N_a\tau_{scat}{}^{-1} - dN_a/dt\big|_L - 2L\sigma_{p\bar{p}}/n_b \end{bmatrix}. \tag{6.136}$$

Here indices “p” and “a” denote protons and antiprotons, the derivatives $d\varepsilon/dt|_{BB}$ are the emittance growth rates due to scattering in the IP determined by Eq. (6.131) (factor of 2 takes into account 2 IPs), the derivatives $d\varepsilon/dt|_{IBS}$ are the emittance growth rates due to IBS determined by Eq. (6.132), the derivatives $d\varepsilon/dt|_{gas}$ are the emittance growth rates due to multiple scattering on the residual gas determined by Eq. (6.130), the derivatives $d\sigma^2/dt|_{total}$ are the momentum spread growth rates determined by Eq. (6.128), the derivatives $dN/dt|_L$ are the particle loss rate from bucket determined by Eq. (6.128), and the addends $2L\sigma_{p\bar{p}}/n_b$ determines particle loss in two collision points due to luminosity. We also took into account that the momentum spread growth rates used in Eq. (6.128) include contributions from the IBS and RF noise.

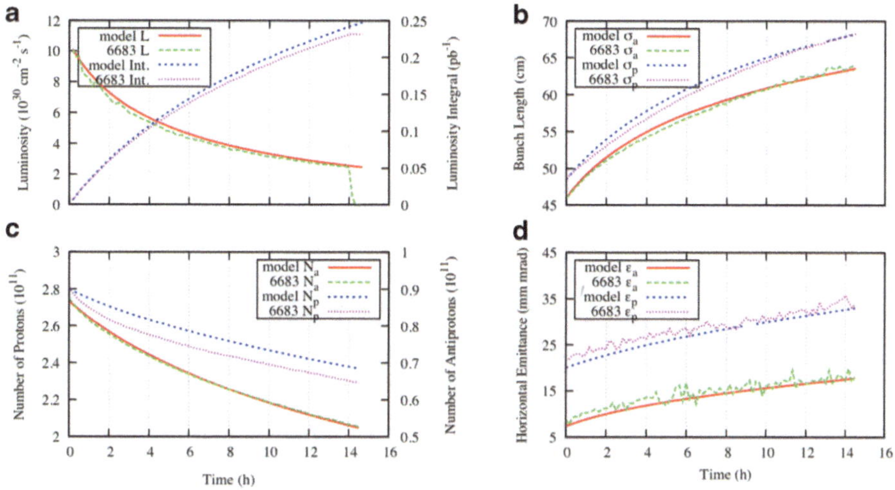

Fig. 6.31 Observed beam parameters in store 6683 compared to store analysis calculation (model). $L_0 = 3.5 \times 10^{32}$ cm^{-2} s^{-1}. (**a**) Single bunch luminosity and luminosity integral. (**b**) Intensity of proton bunch no. 6 and of antiproton bunch colliding with it (no. 13). (**c**) proton and antiproton rms bunch lengths. (**d**) proton and antiproton horizontal 95 % normalized bunch emittances

It is worth noting one more time that this model was able to predict evolution of the beam parameters in the case of weak beam–beam effects. When these effects are not small, it provides a reference for evaluation of their strength. We used that analysis on a store-by-store basis to monitor the machine performance in real time [60] because such calculations were very fast compared to a full numerical beam–beam simulation.

Figure 6.31 presents an example comparison of the evolution of beam parameters in an actual high luminosity store to the calculations. Note that there is no transverse emittance blow up in both beams, and the emittance growth is dominated by processes other than beam–beam interaction. The same is true for antiproton intensity and bunch length. The most pronounced difference between the observation and the model is seen in the proton intensity evolution. Beam–beam effects caused the proton lifetime degradation during the initial 2–3 h of the store until the proton beam–beam tune shift drops from 0.02 to 0.015. The corresponding loss of the luminosity integral was about 5 %.

The IBS model allows us to calculate the longitudinal beam loss rate in a typical Tevatron store (Fig. 6.32). The initial longitudinal loss rate is not equal to zero because of the Touschek effect. Later in the store, when more particles move closer to the boundaries of the RF buckets through diffusion processes, multiple IBS scattering starts to dominate over the single scattering effect. Note that for antiprotons, luminosity burning is the main loss contribution and the longitudinal loss due to IBS is much smaller than its total intensity loss rate. Normally about 40 % of antiprotons are "burnt" in the collisions due to elastic and inelastic interactions with protons.

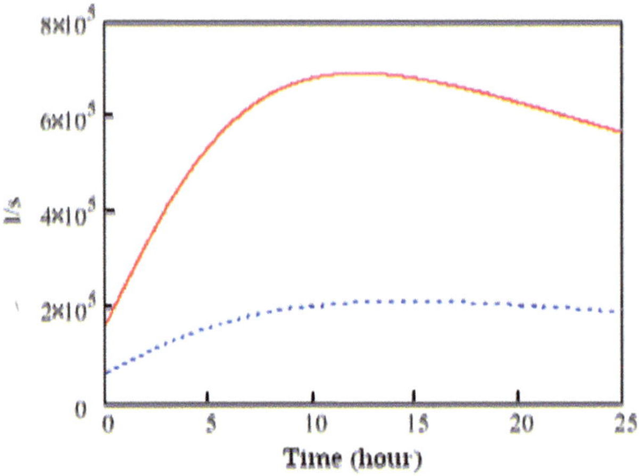

Fig. 6.32 Calculated longitudinal beam loss rate in unit of particles per second for a typical store using the IBS model, the *red curve* is for the proton bunch and the *blue dashed line* is for the antiproton bunch

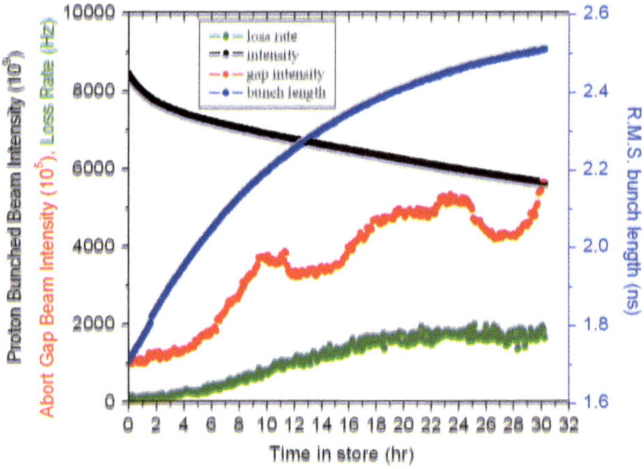

Fig. 6.33 Decay of proton bunch intensity (*black curve*) and growth of its length (*blue*) as well as abort gap loss rate at CDF detector (*green*) and proton abort gap intensity (*red*) in a typical HEP store (#5157, the TEL was on)

The calculated loss rate is in good qualitative agreement with the Tevatron observations. For example, Fig. 6.33 shows the evolution of the total proton bunched beam intensity, proton loss rate, proton rms bunch length, and the abort gap beam intensity during HEP store #5157. Bunch length and bunch intensity are reported from a wall current monitor (known as the "Sampled Bunch Display", described in Chap. 9). The loss rate is measured by gated scintillation counters on the CDF detector, which integrate over the time intervals corresponding to the abort gaps between the three proton bunch trains, while the simulation yields losses for the whole Tevatron storage ring.

References

1. E.M. Lifshitz, L.P. Pitaevskii, *Physical Kinetics*, vol. 10 (Pergamon Press, New York, 1981)
2. A. Piwinski, in *Proceedings of the 9th International Conference on High Energy Accelerators*, SLAC, Stanford, CA, USA, 1974, p. 405
3. J.D. Bjorken, S.K. Mtingwa, Part. Accel. **13**, 115 (1983)
4. L.D. Landau, Phys. Z. Sowjetunion **10**, 154 (1936)
5. M.N. Rosenbluth, W.M. MacDonald, D.L. Judd, Phys. Rev. **107**, 1 (1957)
6. S. Chandrasekhar, Rev. Mod. Phys. **15**, 1 (1943)
7. S. Nagaitsev, Phys. Rev. ST Accel. Beams **8**, 064403 (2005)
8. S. Ichimaru, M.N. Rosenbluth, Phys. Fluids **13**, 2778 (1970)
9. V. Lebedev et al., NIM A **391**, 176 (1997)
10. V. Parkhomchuk, *Physics of fast electron cooling*, in Proceedings of the Workshop on Electron Cooling and Related Applications, Karlsruhe, 1984
11. J.D. Jackson, *Classical Electrodynamics*, 3rd edn. (Wiley, New York, 1998)
12. C. Bernardini et al., Phys. Rev. Lett. **10**, 407 (1963)
13. J. Haissinski, LAL Orsay report 41-63 (1963)
14. V. Levedev, in *Proceedings HB-2004 Bensheim*, AIP Conference Proceedings, vol. 773, New York, 2005, p. 440
15. G. Dome, *Diffusion due to RF Noise*, in Proceedings of the CERN Accelerator School Advanced Accelerator Physics Course, Oxford, 1985, Preprint CERN 87-03 (April 1987), pp. 370–401
16. I.S. Gradshteyn, I.M. Ryzhik, *Tables of Integrals, Series and Products* (Academic, New York, 1980)
17. H. Hancock, *Theory of Elliptic Functions* (Dover Publications, New York, 1958)
18. V. Lebedev, L. Nicolas, A. Tollestrup, beams-doc-1155 (2004)
19. V. Shiltsev et al., Phys. Rev. ST Accel. Beams **8**, 101001 (2005)
20. V. Shiltsev, A. Tollestrup, Emittance growth mechanisms in the Tevatron beams. JINST **6**, P08001 (2011)
21. N.V. Mokhov, *Beam Collimation at Hadron Colliders*, in AIP Conference Proceedings, vol. 693, 2003, pp. 14–19
22. M.D. Church, A.I. Drozhdin, A. Legan, N.V. Mokhov, R.E. Reilly, *Tevatron Run-II Beam Collimation System*, in Proceedings of the 1999 IEEE Particle Accelerator Conference (IEEE Conference Proceedings, New York, 1999), pp. 56–58; Fermilab-Conf-99/059
23. I. Baishev, A. Drozhdin, N. Mokhov, STRUCT Program User's Reference Manual, SSCL-MAN-0034 (1994)
24. N.V. Mokhov, The MARS Code System User Guide, Version 13(95), Fermilab-FN-628 (1995); N.V. Mokhov et al., Fermilab-Conf-98/379 (1998); LANL Report LA-UR-98-5716 (1998); *nucl-th/9812038 v2 16 Dec 1998*; http://wwwap.fnal.gov/MARS/
25. L.Y. Nicolas, N.V. Mokhov, Impact of the A48 Collimator on the Tevatron BØ Dipoles, FERMILAB-TM-2214, June 2003
26. K.-H. Mess, M. Seidel, Nucl. Instrum. Methods A **351**, 279 (1994)
27. N. Mokhov, J. Annala, R. Carrigan, M. Church, A. Drozhdin, T. Johnson, R. Reilly, V. Shiltsev, G. Stancari, D. Still, A. Valishev, X.-L. Zhang, V. Zvoda, Tevatron beam halo collimation system: design, operational experience and new methods. JINST **6**, T08005 (2011)
28. J.M. Butler, D.S. Denisov, H.T. Diehl, A.I. Drozhdin, N.V. Mokhov, D.R. Wood, Reduction of TEVATRON and Main Ring induced backgrounds in the DØ detector. Fermilab FN-629 (1995)
29. A.I. Drozhdin, V.A. Lebedev, N.V. Mokhov et al., *Beam Loss and Backgrounds in the CDF and D0 Detectors due to Nuclear Elastic Beam-Gas Scattering*, in Proceedings of the 2003 IEEE Particle Accelerator Conference, Portland, 2003, pp. 1733–1735; Fermilab-FN-734
30. Tevatron Sequencer Software Help and Explanation. http://wwwbd.fnal.gov/webhelp_edit/pa1028/pa10281_1.html

31. X.-L. Zhang et al., Phys. Rev. ST Accel. Beams **11**, 051002 (2008)
32. E. Tsyganov, Fermilab preprint FNAL-TM-682 (1976)
33. V.M. Biryukov, Y.A. Chesnokov, V.I. Kotov, *Crystal Channeling and Its Application at High Energy Accelerators* (Springer, Berlin, 1997)
34. M. Maslov, N. Mokhov, L. Yazynin, Preprint SSCL-484 (1991)
35. R.A. Carrigan Jr. et al., Phys. Rev. ST Accel. Beams **1**, 022801 (1998)
36. R.A. Carrigan Jr. et al., Phys. Rev. ST Accel. Beams **5**, 043501 (2002)
37. R.A. Carrigan Jr. et al., Fermilab preprint FNAL-Conf-06-309 (2006)
38. N. Mokhov et al., in *Proceedings of the 2009 IEEE Particle Accelerator Conference,* Vancouver, 2009, p. 1836
39. V. Shiltsev et al., in *Proceedings of the International Particle Accelerator Conference*, Kyoto, 2010, p. 1243
40. W. Scandale et al., Phys. Rev. Lett. **101**, 234801 (2008)
41. R. Fliller et al., Phys. Rev. ST Accel. Beams **9**, 013501 (2006)
42. V. Biryukov, CATCH, 1.4 User's Guide. CERN SL/Note 93-74 AP (1993)
43. V. Shiltsev, in *Proceedings of LHC-LUMI-06 CARE-HHH-APD Workshop,* Valencia, 2006, CERN-2007-002 (2007), p. 92
44. V. Shiltsev, in *Proceedings of BEAM07 CARE-HHH-APD Workshop,* Geneva, 2007, CERN-2008-005 (2007), p. 46
45. D. Shatilov et al., in *Proceedings of the 2005 IEEE Particle Accelerator Conference,* Knoxville, 2005, p. 4138
46. G. Robert-Demolaize et al., in *Proceedings of the 2005 IEEE Particle Accelerator Conference,* Knoxville, 2005, p. 4084
47. V. Shiltsev et al., in *Proceedings of the European Particle Accelerator Conference,* Genoa, 2008, p. 292
48. J.C. Smith et al., in *Proceedings of the 2009 IEEE Particle Accelerator Conference,* Vancouver, 2009, WE6RFP031
49. G. Stancari et al., in *Proceedings of the International Particle Accelerator Conference,* Kyoto, 2010, p. 1698
50. G. Stancari et al., in *Proceedings of 14th Advanced Accelerator Concepts Workshop*, Annapolis, 2010, AIP Conference Proceedings 1299, 638
51. G. Stancari et al., in *Proceedings of the 2011 IEEE Particle Accelerator Conference,* New York, 2011, MOP147
52. G. Stancari et al., Phys. Rev. Lett. **107**, 084802 (2011)
53. M. Seidel, Ph.D. Thesis, Hamburg University, DESY 94-103 (1994)
54. D. Edwards, M. Syphers, *Introduction to the Physics of High Energy Accelerators* (Wiley, New York, 1993)
55. J. Steimel et al., in *Proceedings of the 2003 IEEE Particle Accelerator Conference,* Portland, 2003, p. 48
56. A. Wrulich, Single-Beam Lifetime, in *CERN Accelerator School, 5th General Accelerator Physics Course*, CERN Preprint 94-001 (1994)
57. V. Lebedev, in *Proceedings of the 33rd ICFA Advanced Beam Dynamics Workshop on High Brightness Hadron Beam* (Bensheim, 2004), AIP Conference Proceedings 773, in I. Hoffmann (ed) Comprehensive theoretical analysis of the longitudinal diffusion due to RF noise can also be found in S. Krinsky and J.M. Wang. Part. Accel. **12** (1982), p. 107
58. V. Lebedev, *Beam Physics at Tevatron Complex*, in *Proceedings of the 2003 IEEE Particle Accelerator Conference*, Portland, 2003, p. 29
59. N.V. Mokhov, V.I. Balbekov, Beam and luminosity lifetime, in *Handbook of Accelerator Physics and Engineering*, ed. by A.W. Chao, M. Tigner, 2nd edn. (World Scientific, Singapore, 2002), p. 218
60. A. Valishev, Tevatron store analysis package, Software code supported in 2006–2011, http://www-bd.fnal.gov/SDAViewersServlets/valishev_sa_catalog2.html

Chapter 7
Antiproton Production and Cooling

V. Lebedev, R. Pasquinelli, L. Prost, and A. Shemyakin

The progress in the antiproton production and cooling has been absolutely essential for the success of the Collider Run II. Improvements of the Tevatron optics and operation resulted in a gradual increase in the fraction of antiprotons lost in the proton–antiproton collisions in the interaction points. However, by the middle of 2004, it achieved its maximum of about 30–40 % (see Fig. 7.1) determined mainly by the intra-beam scattering (IBS) and the beam–beam effects (see Chap. 8). Since that time, it stayed basically unchanged through the end of the Run II. Further progress in the luminosity could not be achieved without an increase in the antiproton production. Figure 7.2 presents the weekly antiproton production in the course of Run II. One can see that starting from the beginning of 2005, the rate of antiproton production grew significantly reflecting an increased priority for antiproton production.

The major improvements came from an increase of number of protons on the antiproton production target from 5×10^{12} to 8×10^{12}/pulse (2004–2006), an increase of Debuncher acceptance from 20 to 35 mm mrad (2006), improvements of stochastic cooling in the Debuncher and Accumulator (2005–2008), commissioning of the Recycler as a third antiproton storage ring in the complex (2004), introducing electron cooling into Recycler operation (2005), and reducing time and improving efficiency for Accumulator-to-Recycler transfers (2006–2008). All this work resulted in a peak stacking rate of about 30×10^{10}/h with the best weekly averaged stacking rate of 24.5×10^{10}/h.

In this chapter, we consider all stages of the antiproton production and cooling and major limitations of the stacking rate. If it is not pointed out explicitly, all beam and machine parameters used in this section are referred to the Run II end.

V. Lebedev (✉) • R. Pasquinelli • L. Prost • A. Shemyakin
Fermi National Accelerator Laboratory, PO Box 500, Batavia, IL 60510, USA
e-mail: val@fnal.gov

V. Lebedev and V. Shiltsev (eds.), *Accelerator Physics at the Tevatron Collider*, Particle Acceleration and Detection, DOI 10.1007/978-1-4939-0885-1_7,

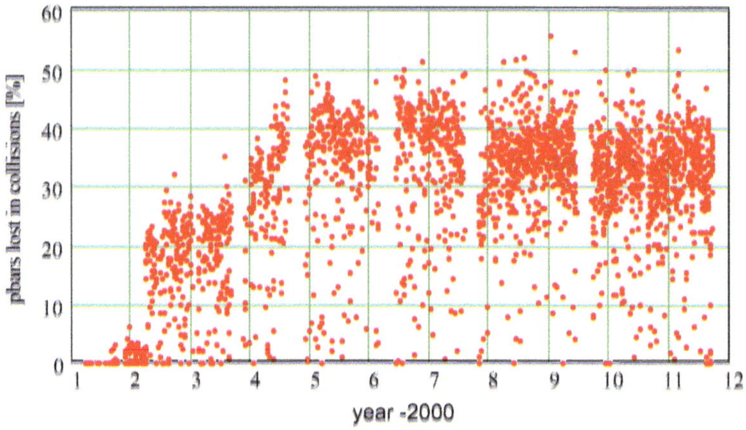

Fig. 7.1 Percentage of antiprotons lost in the proton–antiproton collisions in two interaction points to the store end; the cross section of the loss of 69 mb was determined from the luminosity evolution model

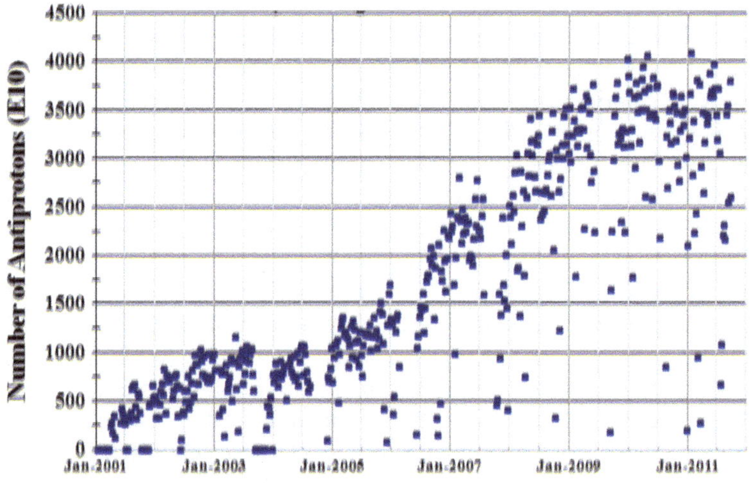

Fig. 7.2 Weekly antiproton production in the course of Run II

7.1 Production and Collections of Antiprotons

Antiprotons are produced by directing the 120 GeV proton beam to the antiproton production target every 2.2 s. The beam usually consists of 81 bunches with total intensity of about 8×10^{12}. Before extraction from the Main Injector, the bunches are shortened with bunch rotation performed by a voltage jump of the main RF system operating at 53.1 MHz. This process narrows the bunches in time at the expense of increasing the momentum spread ($\Delta p/p$). The $\Delta p/p$ of the antiprotons is

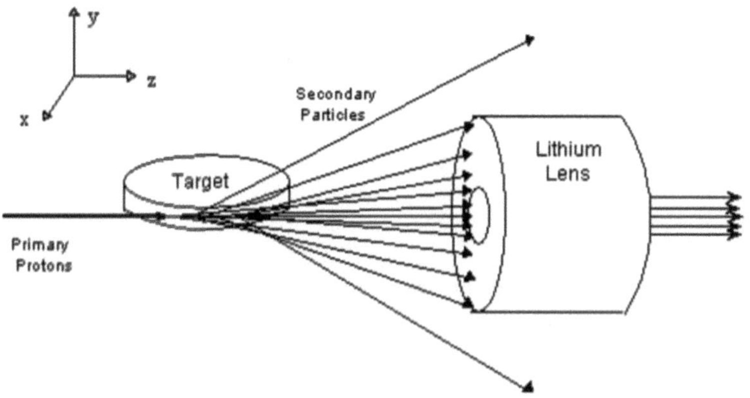

Fig. 7.3 The schematic of the antiproton production target and lithium lens

minimally affected by the $\Delta p/p$ of the protons hitting the target. By narrowing the bunches prior to extracting them from the Main Injector, the longitudinal phase space density of the antiprotons is increased, resulting in a smaller $\Delta p/p$ after a bunch rotation performed in the Debuncher. Bunch rotation reduces the antiproton momentum spread so that the antiprotons would be within the stochastic cooling momentum acceptance. To increase the transverse phase density of antiprotons, the proton beam is also focused into a tight spot with rms size of about 200 μm.

Figure 7.3 presents a schematic of the target. The target, surrounded by air, is attached to the bottom of a shield/motion control module. There were a few modifications of the target design. In its latest incarnation, the target consists of an Inconel 600 cylinder, 11.43 cm in diameter and 12.7 cm in height with an internal, cylindrical, air-cooled, copper heat exchanger. There are several motion control systems connected to the target. The target length can be adjusted by moving the target in the x direction. To reduce the average energy deposition and the peak temperature, the target is rotated with a period of 45 s. The principal target damage mechanism is surface oxidation due to high-temperature beam entry/exit points. Target rotation reduces the peak surface temperature, and a beryllium cover 6 mm thick used in the latest design excludes air from the target cylinder surface. The second damage mechanism is related to disruption of the crystalline structure leading to swelling of the target cylinder and eventual reduction of thermal conductivity between the air-cooled copper heat exchanger and the Inconel cylinder. With the loss of cooling, the beryllium cover breaks due to excessive hoop stress. The resulting damage is due to the snowballing effect of air contact with the cylinder, drastically reduced heat removal capability, high cylinder surface temperature, and finally rapid oxidation. The target was moved vertically by about 1 mm after each 2e17 protons, or about once per day during steady operations, in order to minimize localized swelling effects. At least several passes through the 75 mm operating range was possible extending the expected target lifetime to 6–12 months.

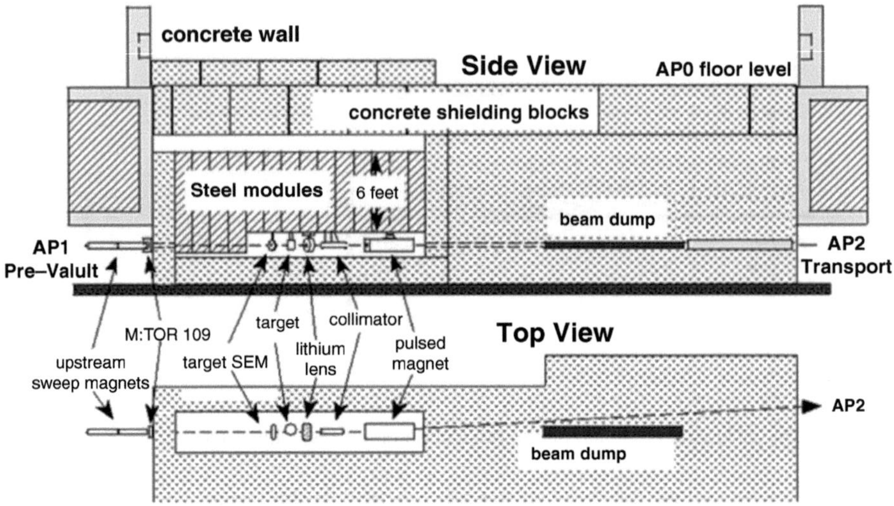

Fig. 7.4 The target vault layout; SEM stands for the secondary emission monitor; AP1 and AP2 are the names of transport lines from the Main Injector to the target and from target to the Debuncher, consequently

The angular spread of antiprotons coming out of the target is reduced by the lithium lens. Due to its short focusing length (~30 cm), the chromaticity of focusing is minimal resulting in high efficiency of beam transport to the Debuncher for antiprotons with large momentum spread ($\approx\pm 2.5$ %). Achieving the same transport efficiency with usual quadrupoles would not be possible. The lens has been designed to operate in conditions of extremely high radiation. Figure 7.4 presents the target vault layout.

After the lithium lens the proton and antiproton beams are separated by a 3° bending dipole, a single-turn, radiation-hard pulsed magnet, operating in extremely high radiation environment and bending antiprotons by about 3°. A collimator positioned upstream of the pulsed magnet intercepts the shower of secondary particles coming from the target and collection lens/transformer assembly. The protons are directed to the beam dump and the antiprotons to the transport line called AP-2. The line begins with a quadrupole triplet (actually built from four quads) and a downstream bending magnet. The triplet together with the dipoles makes this part of beam transport achromatic and matches the target beta-functions to the values of downstream line. The transport line with total length of about 290 m brings the antiproton beam to the Debuncher. An achromatic bend of 36.5° located in its middle determines the central momentum and momentum spread of antiprotons to be injected to the Debuncher. The momentum spread can be reduced by collimators located in the bend center where the dispersion achieves its maximum value.

The total proton beam power is about 70 kW. Only about 1 kW of it is deposited in the target. The remainder of the beam power is distributed among the collection

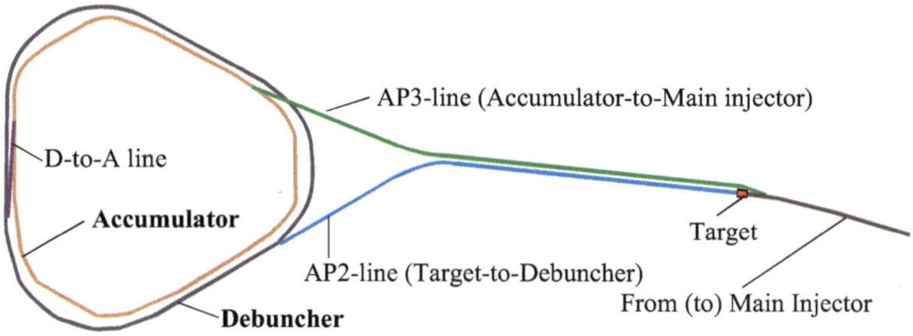

Fig. 7.5 Layout of Antiproton Source rings and transport lines

lens/transformer assembly (6 kW), the pulsed magnet collimator (7 kW), the pulsed magnet (7 kW), and the beam dump (25 kW). The remaining beam power is absorbed by the surrounding steel modules and beam dump shielding steel. Comparatively small power deposited in the target allows using air-cooling instead of water-cooling. That significantly simplifies its design and installation in a high radiation environment. The target, lithium lens, and 3° dipole are located at atmospheric pressure. Beryllium windows are used at the end of the AP1 line and at the beginning of the AP2 line to let the proton beam out of the vacuum chamber and to let the antiproton beam back into the vacuum chamber.

Antiprotons coming out of the target are accompanied by other negatively charged particles having the same momentum. Through the AP-2 line pions are the dominant species of the beam. Although about half of them decay in the beam line, they are still more populous than other species by about two orders of magnitude. Mainly the beam injected to the Debuncher consists of pions, muons, electrons, and antiprotons. Electrons and muons move faster than antiprotons, and therefore their motion is weakly affected by the Debuncher RF system. As a result the electrons are decelerated due to synchrotron radiation and hit the walls in less than 15 turns. Muons experience negligible deceleration and stay on orbit, but they decay with decay time of about 180 turns.

General layout of the Antiproton Source rings and transport beam lines is presented in Fig. 7.5. The AP-2 line brings the antiprotons to the Debuncher. After debunching and being stochastically cooled, they are sent to the Accumulator. The Accumulator stacks and cools antiprotons for ~40 min and then sends them to the Recycler via AP3 line.

The main Debuncher parameters are shown in Table 7.1. The Debuncher 2.2 s cycle begins from the injection of 81 antiproton bunches which inherit the longitudinal structure of MI protons. The Debuncher RF operates at 90th harmonic and is phase-locked to the MI injector RF signal at the time of injection. The RF system has its maximum voltage of 5 MV at the injection time. In about 50 μs, the RF drive signal is reversed to rapidly decrease the RF voltage at the end of bunch rotation which altogether takes about 100 μs. At the end of bunch rotation, the voltage is about 100 kV, and, then, it is lowered over about 12 ms to debunch bunches adiabatically. At the end of this process, about 95 % of antiprotons are within

Table 7.1 Main parameters of the Debuncher

Circumference (m)	505.3
Kinetic energy (GeV)	8
Betatron tune, ν_x/ν_y	9.77/9.78
Momentum compaction	0.017
Acceptance, $\varepsilon_x/\varepsilon_y$ (mm mrad)	35.3/34.6
Maximum relative momentum spread	±0.03
Peak RF voltage for bunch rotation (MV)	5
Harmonic number	90

Table 7.2 Main parameters of the Accumulator

Circumference (m)	474.1
Kinetic energy (GeV)	8
Betatron tune, ν_x/ν_y	6.684/8.680
Momentum compaction	0.0248
Acceptance, $\varepsilon_x/\varepsilon_y$ (mm mrad)	9/9
Maximum momentum acceptance	±0.0125

±0.5 % momentum spread—the acceptance of the longitudinal stochastic cooling. Then, the stochastic cooling in three degrees of freedom is switched on. At the end of the cooling, all antiprotons are assembled in one long bunch created by the barrier bucket RF which is on all the time but has little effect before the bunch is cooled. It creates a 200 ns gap for the extraction of antiprotons to the Accumulator.

The main Accumulator parameters are shown in Table 7.2. Every 2.2 s the Accumulator receives precooled antiprotons from the Debuncher. The injection kicker is located at high-dispersion region where stored and injected beams are separated. The beam arrives at the injection orbit at an energy that is approximately 140 MeV higher than that of the circulating core beam. During 100 ms after injection, the injected beam is adiabatically bunched and RF displaced by 53.1 MHz RF system (84th harmonic) to the deposition orbit where it is adiabatically debunched. This orbit is located approximately at the center of aperture. From this point, the beam falls under the action of the stochastic cooling force of the stacktail system which pushes it toward the main core beam located 60 MeV below the central orbit energy. A three-plane cooling of the main core beam is performed by core stochastic cooling systems. At the end of stacking cycle, the central part of the core is RF displaced to the injection/extraction orbit and, then, extracted to the Recycler. Normally the fourth harmonic system is used for this RF displacement. Consequently, four bunches are extracted. Usually the extraction consists of two or three consecutive extraction cycles to limit the longitudinal emittance of the extracted beam.

It was not clear at the Run II beginning what should be the optimal path for upgrades of the already existing infrastructure; in particular, how much stronger the lithium lens should be, does it have an optimal length, and how does the stacking rate depend on the Debuncher acceptance? There was a belief that a development of the liquid lithium lens [1] with much higher focusing strength was the most promising avenue. However, the study showed that only modest increase of the lens strength is required, but the machine acceptance increase is crucial for the stacking rate increase. The following sections describe the work which was carried out to answer these questions as well as what had to be done to increase the acceptances of the Antiproton Source rings and transfer lines and to increase the efficiency of stochastic cooling systems.

7.1.1 Estimate of the Antiproton Yield and Choice of Target Material

Figure 7.6 presents the dependence of antiproton production invariant cross section at zero angle, $(\gamma/\sigma_I)d^3\sigma/dp^3|_{\theta=0}$, on the kinetic energy of antiprotons for a thin nickel target bombarded by 120 GeV proton beam, where γ is the relativistic factor of antiproton and σ_I is the cross section of inelastic nuclear interaction. With this definition, the cross section weakly depends on the target material, and therefore Fig. 7.6 presents a good approximation for other materials as well. The distribution over transverse momentum has a weak dependence on the longitudinal momentum and is close to the Gaussian distribution with the rms transverse momentum of $p_{\perp a} \approx 0.9\sqrt{m_\pi m_p}$.

In the first approximation, the optimal target length is determined by balancing of antiproton production and absorption and is close to the nuclear interaction length. The acceptance of Antiproton Source is smaller than the rms emittance of the antiprotons coming out of the target. In this case, a rough estimate of the antiproton yield into the momentum spread $\pm\Delta p/p$ is

$$\kappa_a \equiv \frac{N_a}{N_p} \approx \left(\frac{1}{\sigma_I}\frac{d^3\sigma}{dp^3}\right)_{\theta=0} \frac{2\pi p_{\perp a}{}^2 \Delta p}{e} \overset{\gamma\gg 1}{\longrightarrow} \left(\frac{\gamma}{\sigma_I}\frac{d^3\sigma}{dp^3}\right)_{\theta=0} \frac{2\pi mc p_{\perp a}{}^2}{e}\frac{\Delta p}{p}, \tag{7.1}$$

where we took into account that maximum yield is achieved at the target length equal to the absorption length (here $e = 2.71\ldots$). The maximum transverse momentum of antiprotons, $p_{\perp a}$, is determined by the strength of the lithium lens. In the first approximation, it is proportional to the product of lens's length and its surface magnetic field. Both parameters do not depend on the beam energy. Consequently, $p_{\perp a}$ does not depend on the energy, and the number of antiprotons is proportional to the invariant cross section of Fig. 7.6. As one can see, the optimal energy for antiproton collection is close to 8 GeV—the energy chosen for the Antiproton Source operation.

The antiproton emittance is proportional to the target length, $\varepsilon \approx L\theta_{\max}{}^2/2$. Consequently, high-Z targets having smaller nuclear interaction length are preferable. However, the energy deposition per unit volume grows fast with Z, and at the Run II beam power, it exceeds the critical power density beyond which the target is destroyed. After a number of trials, nickel has been chosen as the best material for the target because it can sustain an unusually large energy deposition of up to

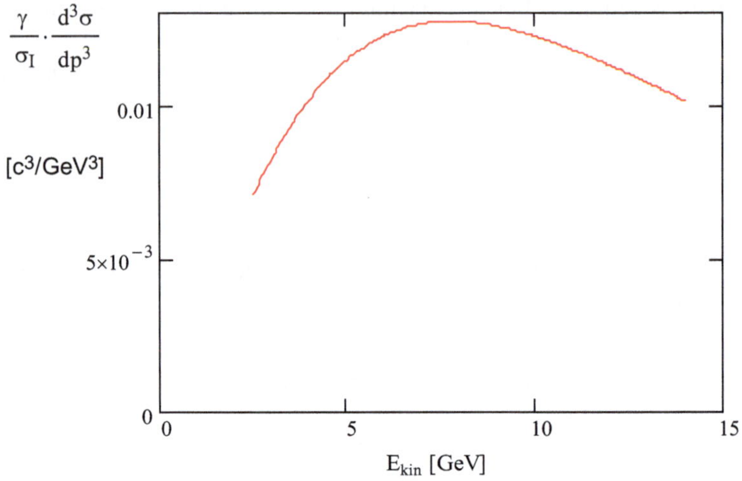

Fig. 7.6 Dependence of the invariant antiproton production cross section at zero angle on the kinetic energy of antiprotons for thin nickel target bombarded by 120 GeV beam (courtesy of S. Striganov)

1,000 J/g. Later it was found that a nickel alloy, Inconel 600, is an even better choice.

7.1.2 Optimization of Production and Collection of Antiprotons

The simulations of antiproton production in a nickel target were performed with MARS code [2] which takes into account all details of particles production and their interaction in the material. Figure 7.7 presents the simulations of total yield of antiprotons produced by 120 GeV proton beam into momentum acceptance of ± 2.25 % around 8 GeV kinetic energy as function of the target length. One can see that the total yield continues to grow to the target length equal to the inelastic scattering length of 15 cm (for nickel), but, as will be seen later, the phase density achieves its maximum at a smaller target length.

Figure 7.8 shows coordinates of antiprotons in the phase space produced by the proton beam with rms beam size of 100 μm in the 8 cm-long target. Particle x-coordinates were translated to the longitudinal coordinate at which the second-order moment $\langle x\theta_x \rangle$ is equal to zero. In the absence of antiproton scattering in the target, this coordinate (waist position) would be in the center of the target. Scattering shifts it downstream. For 8 cm target, this shift is equal to ~2.1 mm. Scattering leaves transverse distribution approximately Gaussian but increases the antiproton angular spread. For 8 cm target, an increase is ~20 %. The ellipse on the plot presents the

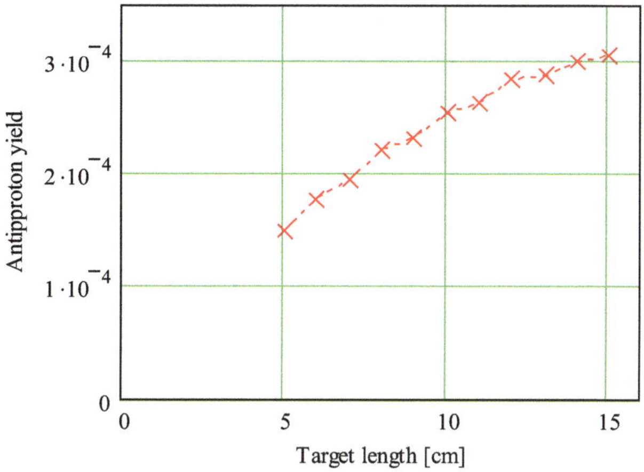

Fig. 7.7 Dependence of total antiproton yield (simulated with MARS) on the length of nickel target for proton beam of 120 GeV; momentum acceptance is ±2.25 %

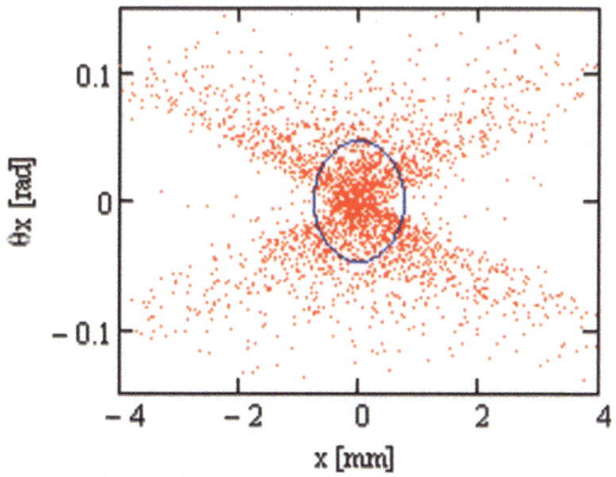

Fig. 7.8 Projections of coordinates of antiprotons to the x–θ_x-phase space simulated with MARS code for 120 GeV proton beam with rms beam size of 100 μm. The target length is 8 cm and the mean kinetic energy of antiprotons is 8 GeV. The circle inscribes phase space with $\varepsilon = 35$ mm mrad and $\beta^* = 1.5$ cm

boundary of phase space with acceptance $\varepsilon = 35$ mm mrad and beta-function $\beta^* = 1.5$ cm. The butterfly shape of the Fig. 7.8 distribution reflects the phase space transformation in the course of antiproton motion from a position s to the target center located at $s_0 = 2.1$ mm $\left((x, \theta_x)_s \rightarrow (x + \theta_x(s - s_0), \theta_x)_{s=s_0}\right)$.

An optimal target-to-Debuncher beam transport should deliver the maximum number of antiprotons to the limited Debuncher acceptance and momentum spread. Finding an optimal beta-function at the waist position, β^*, is the first step in the optics design. Figure 7.9 shows the antiproton yield as the function of β^* for different machine acceptances. The horizontal and vertical acceptances are considered to be equal; and the momentum acceptance is equal[1] ±2.25 %. Thus, the phase space of the accepted antiprotons is determined by the following equations:

[1] This value takes into account the finite value of beam betatron size and therefore is smaller than ±2.5 % quoted above (Sect. 7.1) and representing the momentum acceptance of particles with zero betatron amplitudes.

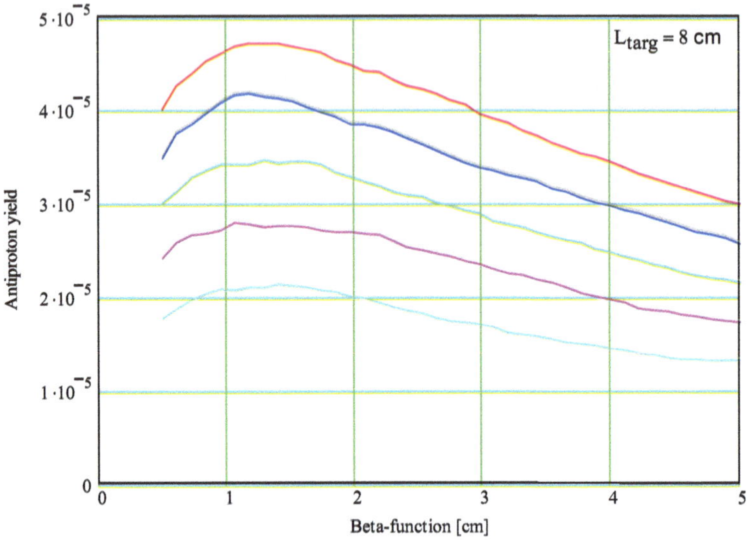

Fig. 7.9 Simulated dependence of antiproton yield into ±2.25 % momentum spread on β^* for the beam acceptances of 15, 20, 25, 30, and 35 mm mrad (corresponding curves follow from low to high values); other parameters are the same as in Fig. 7.8

$$\left(\frac{x_i^2}{\beta^*} + x_i'^2\beta^*\right) + \left(\frac{y_i^2}{\beta^*} + y_i'^2\beta^*\right) \le \varepsilon \quad \text{and} \quad \left|\frac{\Delta p_i}{p_0}\right| \le 0.0225.$$

For target lengths in the range of 5–12 cm, the maximum yield is achieved at the optimal beta-function. Its value practically does not depend on the acceptance and is equal to about 1/6 of the target length. Figure 7.10 shows a dependence of the antiproton yield on the ring acceptance. The yield weakly depends on the target length in the range of 6–10 cm and is proportional to the ring acceptance for acceptances below 50 mm mrad. However, longer target lengths are preferable because they require larger β^* and, consequently, weaker lithium lens focusing.

Thus, to achieve a large antiproton yield, one needs to have quite small beta-functions in the target center and to capture large momentum spread and acceptance. That requires a short distance focusing immediately after the target. Actually, a dependence of near target focusing on the momentum, $\Delta F/F = \Delta p/p$, excites the chromaticity of beam envelopes so that

$$\left.\frac{\Delta\beta}{\beta}\right|_{\max} \approx \frac{\beta_L}{F}\frac{\Delta p}{p} \xrightarrow{\beta_L \approx F^2/\beta^*} \frac{F}{\beta^*}\frac{\Delta p}{p}, \tag{7.2}$$

where $\beta_L \approx L_f^2/\beta^*$ is the average beta-function at the lens and $L_f \approx F$ is the distance between the target center and the lens center. For $\beta^* = 2$ cm and $\Delta p/p = 0.02$, a requirement to have $(\Delta\beta/\beta)_{\max} < 0.5$ yields $F < 50$ cm. This focusing distance

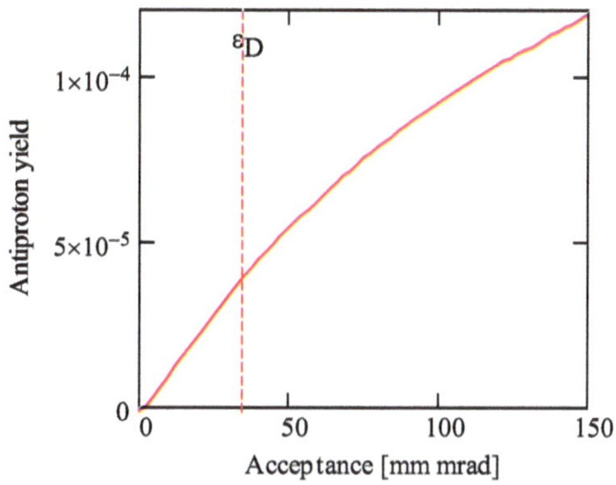

Fig. 7.10 Simulated dependence of the antiproton yield on the acceptance for 9.45 cm inconel target, $\beta^* = 1.6$ cm, $\varepsilon = \varepsilon_x = \varepsilon_y$, $\Delta p/p_{max} = \pm 0.0225$. Note that for unlimited transverse acceptance, the yield of antiprotons is equal to 2.3×10^{-4}; *vertical line* marks the Debuncher acceptance of 35 mm mrad

cannot be achieved with normal quadrupoles for the required acceptance. That leaves the lithium lens as the only practical choice for the first focusing element in the antiproton transport.

Unlike quadrupoles, the lithium lens has a material on the beam pass which results in additional scattering and absorption of antiprotons which reduce the yield. The effect of scattering is amplified by large value of beta-function in the lens (~3 m). The percentage of antiprotons which do not experience strong interactions with lens material and survive is about $\kappa_{lens} = \exp(-s_{Li}/L_{Li} - s_{Be}/L_{Be}) \approx 0.82$, where $s_{Li} = 15.5$ cm and $s_{Be} = 1.2$ cm are the total lengths of lithium and beryllium crossed by the beam and $L_{Li} = 97.7$ cm and $L_{Be} = 29.9$ cm are the nuclear collision lengths for lithium and beryllium, correspondingly. Multiple scattering in the lens can be estimated as $\sqrt{\overline{\delta\theta^2}} = \sqrt{s_{Li}/X_{Li} + L_{Be}/X_{Be}}(13.6\,\text{MeV})/\beta Pc = 0.635\,\text{mrad}$, where $X_{Li} = 155$ cm and $X_{Be} = 35.3$ cm are the radiation lengths for lithium and beryllium. For $\sqrt{\overline{\delta\theta^2}} \ll \sqrt{\overline{\theta^2}}$ the scattering in the lens causes the relative emittance growth equal to $\Delta\varepsilon/\varepsilon \approx \overline{\delta\theta^2}/\left(2\overline{\theta^2}\right)$. For the Debuncher acceptance of 35 mm mrad, the multiple scattering yields $\Delta\varepsilon/\varepsilon \approx 0.016$ and is less significant than the direct particle loss due to nuclear interaction.

Figure 7.11 presents the dependence of antiproton yield on the lithium lens strength. For every lens gradient, the distance between the lens and the target was adjusted to achieve the maximum yield. At the Run II beginning, the Debuncher acceptance was close to 20 mm mrad and the lens gradient was about 55 kG/cm. As one can see from Fig. 7.11, an increase of the lithium lens focusing would not result in a significant yield increase without an increase of the acceptance. Therefore, the acceptance increase was considered as a high-priority item in the plan of Antiproton Source upgrade. Simulations also verified that the length and radius of the lens were close to the optimal values. A correction of Debuncher orbit and optics in January of 2006 resulted in the Debuncher acceptance of about 35 mm mrad. An improved lens design yielded a lens gradient increase from 55 to 67 kG/cm. These and a few

Fig. 7.11 Simulated dependence of antiproton yield on the lithium lens gradient; rms proton beam size at the target is 130 μm, and target length is 8 cm

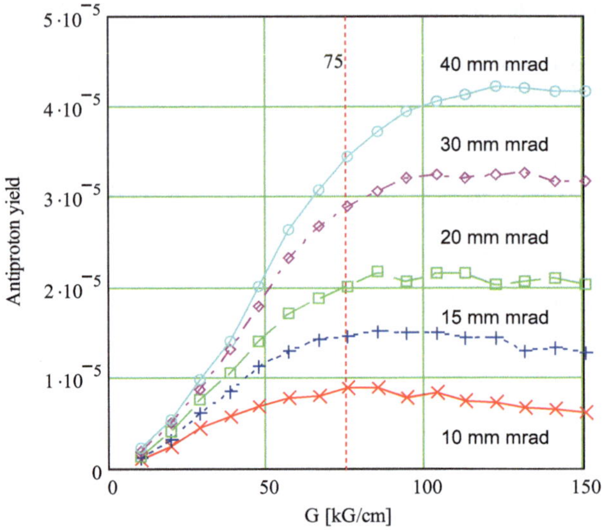

other less important improvements resulted in the antiproton yield[2] of about 3×10^{-5}. Detailed simulations of antiproton production which include their transport to the Debuncher and the tracking in Debuncher with realistic apertures resulted in the antiproton yield of 3.1×10^{-5} in a good agreement with experimentally measured yield. The rms proton beam size of 220 μm was used in the simulations.

Figure 7.12 presents the dependence of antiproton yield on the rms proton beam size for the different acceptances and the lens gradient of 75 kG/cm. One can see that the antiproton yield begins to decrease for sizes above 100 μm and this decrease is faster for smaller acceptances. The apertures and locations of target focusing quadrupoles allowed reduction of the rms spot size down to about 120–150 μm. However, for 8×10^{12} protons per pulse, a reduction of beam size below ~200 μm resulted in fast degradation of the target. It was driven by too large energy deposition in the target, resulting in instant melting of irradiated material and subsequent shock waves. A continuous operation with reduced beam size was tested with the previous target design which did not have the protecting beryllium cover. Such operation resulted in a horizontal "icicle" growing from the front face of the lithium lens due to little droplets of target material knocked out at each pulse. It resulted in a reduction of the antiproton yield due to additional scattering of antiprotons on the "icicle."

[2] This value of the yield was measured by DCCT (direct current transformer). Additional ~5 % loss occurs in the course of beam debunching and stochastic cooling in Debuncher.

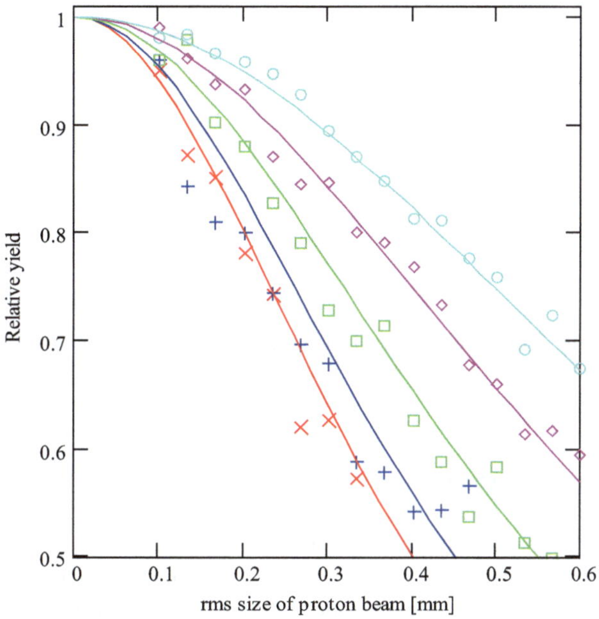

Fig. 7.12 Dependence of simulated relative antiproton yield on the rms size of proton beam for acceptances of 10, 15, 20, 30, and 40 mm mrad (curves follow from *bottom* to *top*); target length, 8 cm

7.1.3 Lithium Lens

The first lithium lens for the Antiproton Source was fabricated in Novosibirsk Budker Institute of Nuclear Physics in 1983 [3, 4]. From the beginning of the Antiproton Source, operation to the first years of the Collider Run II Fermilab continued to build lenses using similar design and technology [5]. The work on lens improvement was started at the Run II beginning and was carried out during most of the Run II time. This work resulted in a new design based on the diffusion bonding of lens elements. Figure 7.13 presents a cross section of this lens which inherited major ideas and parameters of the original design but allowed ~20 % increase of lens gradient. Similar to the original design, lithium is contained inside titanium cylinder with beryllium endcaps. It is loaded under pressure to prevent its separation from walls of titanium cylinder due to magnetic pressure. The titanium cylinder is cooled by water. The lens is inserted inside a high current 8:1 transformer to match it to the pulsed power supply. Pulses are formed by a discharge of a 4.5 mF capacitor charged to 2,300 V. The lens peak current of 437 kA creates effective lens gradient of 67 kG/cm at the time of beam crossing. The lens lifetime strongly depends on its gradient. At 67 kG/cm the lens lifetime is about 10^7 pulses (6–12 months). The maximum gradient for lens operation is 75.7 kG/cm with the capacitor charged to 2,600 V. Although it increased the number of antiprotons injected to Debuncher, its effect on the stacking rate was insignificant (<1 %) due to limited throughput of stacktail system (see details in the following sections). Therefore, at

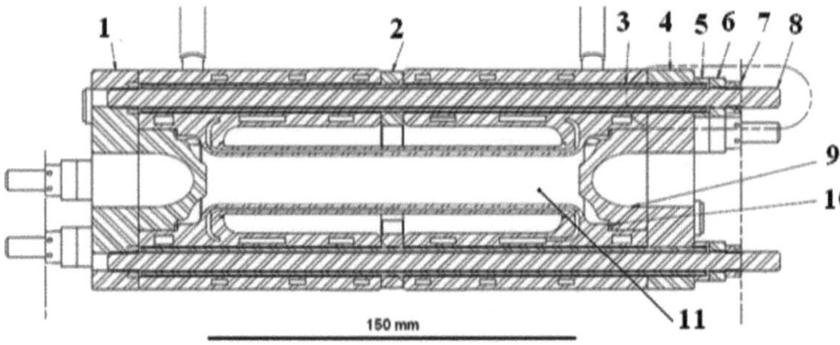

Fig. 7.13 Lithium lens: *1*, downstream window flange; *2*, body spacer; *3*, stud ceramic tube; *4*, upstream window flange; *5*, ceramic washer; *6*, stud washer; *7*, spiralock round nut; *8*, stud insulated; *9*, beryllium window; *10*, window gasket; *11*, lithium

normal operations the lower gradient was used which significantly extended the lifetime of the lens.

For the old lens design, one of the major limitations of the lens lifetime comes from fatigue in the material of titanium cylinder created by magnetic pressure, which stresses the cylinder longitudinally. For the new design, the major limitation of the lens lifetime was determined by longevity of electrical insulators in the eight-turn current transformer surrounding the lens. Several modifications to the lens transformer improved longevity during the course of Run II. The need for nearly all lens/transformer replacements came to be dominated by lens or transformer cooling water leaks, which led to failed electrical insulators.

A major nonlinearity in the lithium lens focusing is related to the skin effect. The lens current profile is close to a 350 μs-long half-period sinusoidal pulse (see Fig. 7.14). The skin depth at the characteristic frequency $1/(2 \times 350 \times 10^{-6}) \approx$ 1,400 Hz is 4.5 mm. That is twice smaller than the lens radius and implies that there is a significant delay in the penetration of magnetic field into the lens. Figure 7.15 shows results of calculations of magnetic field penetration into lithium cylinder with 1 cm radius. It was obtained by expanding the pulse into Fourier series, finding solution for harmonics, and performing inverse Fourier transform numerically. One can see that the maximum gradient is achieved at RF phases between 30 and 60°. Numerical comparison shows that for phases above 30°, the exact solution presented in Fig. 7.15 is very close to the solution for a single RF harmonic ($I(t) = I_0 e^{i\omega t}$):

$$B(r,t) = \frac{2I_0}{cr_0}\mathrm{Re}\left(\frac{\mathrm{ber}_1\left(\sqrt{2}r/\delta\right) + i\mathrm{bei}_1\left(\sqrt{2}r/\delta\right)}{\mathrm{ber}_1\left(\sqrt{2}r_0/\delta\right) + i\mathrm{bei}_1\left(\sqrt{2}r_0/\delta\right)}e^{i\omega t}\right), \tag{7.3}$$

where δ is the skin depth at frequency $f = 1/(2T)$, T is the pulse duration, I_0 is the current amplitude, r_0 is the radius of lithium cylinder, and ber(x) and bei(x) are the modified Bessel functions. Thus, Eq. (7.3) should be sufficient for most practical applications. Maximum linearity of the gradient is achieved at the phase of about

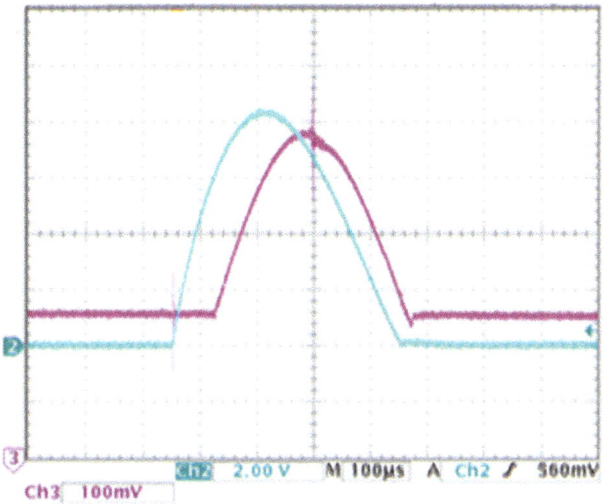

Fig. 7.14 Current waveforms for the lithium lens (*cyan*) and pulsed magnet (*magenta*) during antiproton stacking operations. The disturbance in the pulsed magnet waveform is due to beam crossing

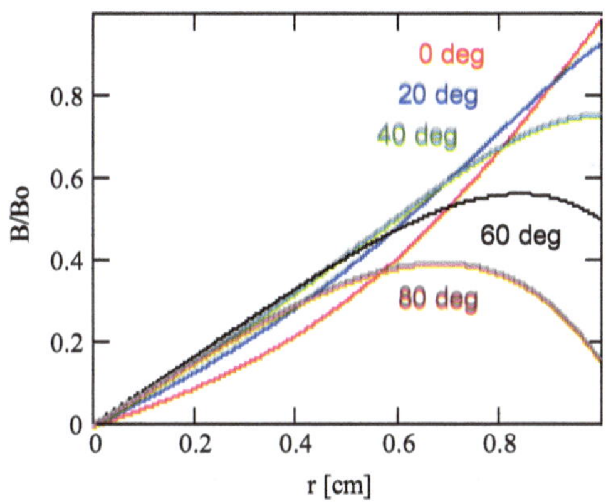

Fig. 7.15 Calculated dependence of lens magnetic field on radius for different times during half-period sinusoidal pulse of 360 µs long. Time is expressed through the RF phase so that the end and the beginning of the pulse correspond to ±90°

45°. At this time, the mean value of the gradient is ~80 % of the gradient calculated without the skin effect taken into account. Note that ~2 % loss of focusing is due to the current propagating in titanium cylinder holding lithium.

There are also other sources of lens nonlinearity. The first one is related to the nonuniform current distribution at the ends, and the second one to the temperature gradient resulting in a dependence of resistivity on radius. However, a contribution of these effects to the nonlinearity is much smaller than the contribution of the skin effect. Resistivity of titanium is 13 times the one of beryllium, and therefore it does not contribute much into the magnetic field penetration rate.

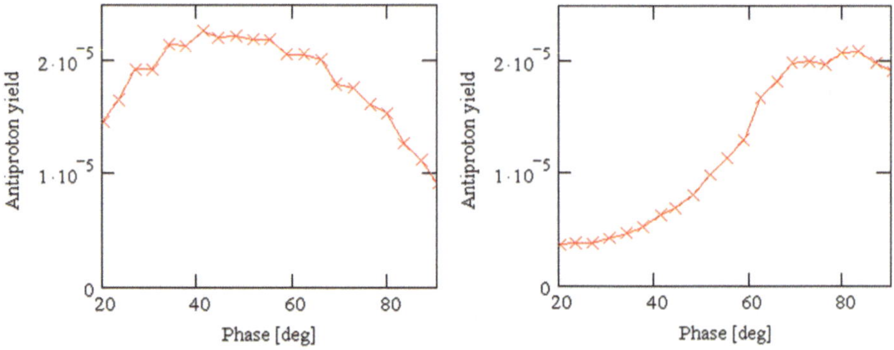

Fig. 7.16 Dependence of the antiproton yield on the arrival time for the pulse lengths of 360 μs (*left*) and 200 μs (*right*). Arrival time is expressed in the degrees of pulse RF phase. Acceptance is 20 mm mrad; the momentum spread is ±2.25 %

Simulations of the antiproton yield with lithium lens nonlinearity taken into account did not exhibit any significant yield drop in comparison with the linear lens simulations. Figure 7.16 shows that the yield is changing with change of the proton beam arrival time. One can see that the shortening of the lens pulse from 360 to 200 μs reduced the yield by only about 2 %, while nonlinearity, $B(r)/r$, grew from ±7 % to (+10 to −50)%. The effect is even smaller for larger acceptances. Figure 7.16 also demonstrates that due to stronger skin effect for shorter pulse, one needs to change the arrival time from 40 to 75°. It also requires an increase of the lens current by 1.4 times to compensate the gradient loss. Thus, a decrease of the lens power consumption due to shorter pulse is overcompensated by increased lens current, and the total power consumption is higher for shorter pulse. Similarly, the power consumption would grow for a pulse longer than 360 μs because in this case the lens current is not changed significantly and power grows proportionally to pulse length. Thus, the choice of 360 μs pulse length looks well optimized.

7.1.4 Optics Correction for Antiproton Source Transport Lines and Rings

The optics upgrades for Antiproton Source transfer lines and rings were aimed to maximize their acceptances. To prevent the emittance growth at transfers, a good optics match is required for the target-to-Debuncher and Debuncher-to-Accumulator transfer lines, called AP-II and D-to-A lines, correspondingly. Both transfer lines were measured with use of the differential orbits method (see Chap. 2). Optics of AP-2 line was corrected to increase its acceptance and to reduce chromaticity of its focusing. Figure 7.17 presents the beta-functions and dispersions through the line. The antiproton yield is relatively insensitive to the optics mismatches in AP-2 line because the phase space of antiprotons coming out of the

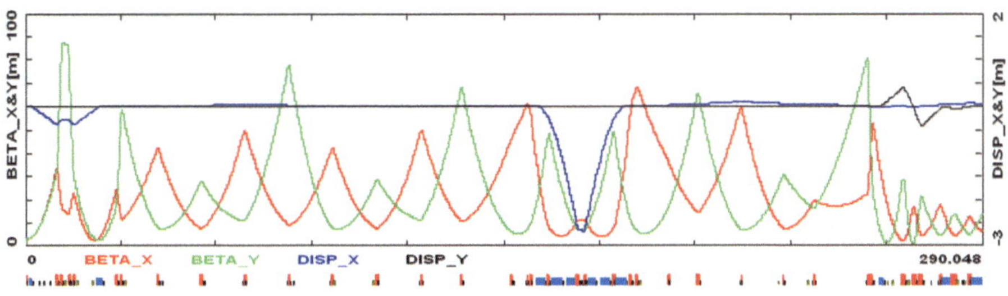

Fig. 7.17 Beta-functions and dispersions for AP-2 line

target significantly exceeds the line acceptance. Note that the quadrupoles, which match the vertical dispersion at the line end, are very strong and introduce large focusing chromaticity. However, it results in a little penalty for the stacking rate because the AP-2 line has larger acceptance than the Debuncher. The optics of the Debuncher-to- Accumulator beam line was found to be satisfactory and was left unchanged.

To minimize cost, many quadrupoles have been connected serially to a single power supply. When optics improvement work was started, we found that in many cases these connections are not optimal. However, if a satisfactory solution without a reconnection of quads could be found, it was preferred to the "perfect" solution.

7.1.5 Debuncher

The central step of the optics improvements in the Antiproton Source was maximizing the Debuncher acceptance, which necessitated detailed understanding of the Debuncher optics and aperture limitations [6]. Initially, the Debuncher upgrade plan anticipated an increase of physical aperture of eight stochastic cooling tanks (pickups and kickers of the frequency bands 3 and 4) which were major aperture limitation. After lattice measurements and their analysis, we found that a Debuncher optics modification would be an easier way to achieve the required acceptance. It also eliminated risks of deterioration of stochastic cooling systems. The method of the response matrix described in Chap. 2 was used to determine quadrupole errors and to build a model of machine optics. An estimate of the optics model accuracy yields that the beta-functions are known to about 5 %. It is mainly limited by the BPM resolution and the small number of steering elements in the machine. Achieving high accuracy of the optics model was essential to maximizing the Debuncher acceptance. Stochastic cooling tanks presented major aperture limitations. A new optics design was based on balancing beam envelopes in different stochastic cooling tanks. Figure 7.18 presents the beam envelopes for the Debuncher straight sections where the pickups are located. One can see that the beam size is minimized in pickups with small aperture that results in an envelope increase in the pickups with large aperture. Similar adjustments were done for the

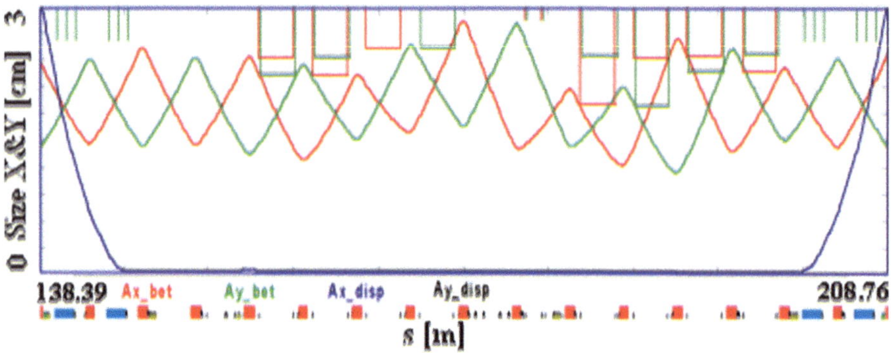

Fig. 7.18 Beam envelopes in the AP-10 straight section of the Debuncher for acceptances of 40.5 and 37.5 mm mrad for horizontal and vertical planes, correspondingly; *red line*, horizontal beam envelope; *green line*, vertical beam envelope; and *blue line*, the contribution to beam size due to relative momentum spread of 0.03. *Rectangles* present aperture limitations with color corresponding to the curve of the same plane

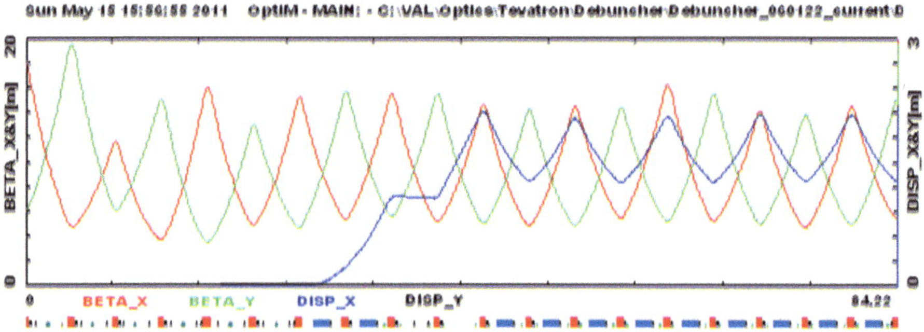

Fig. 7.19 Beta-functions and dispersions for a half of Debuncher super period (sector 50)

straight section where the kickers are located. To remove the last remaining aperture limitations, the injection region was rebuilt and the longitudinal Schottky monitor was removed in 2005.

The original Debuncher optics design had three periods and mirror symmetry in each of three sectors. The beta-functions in straight sections were periodic. The above-described adjustments broke perfect periodicity; however, it did not deteriorate the ring dynamic aperture. It was verified by both the multi-turn tracking and direct measurements of the Debuncher acceptance. The beta-functions and dispersions for a half of one of three Debuncher super periods are shown in Fig. 7.19. The optics of all three straight lines is slightly different. In addition to maximizing the machine acceptance, the betatron phase advance between pickups and kickers of transverse stochastic cooling was also corrected to be equal to $(90 + n \times 180)°$, where n is an integer.

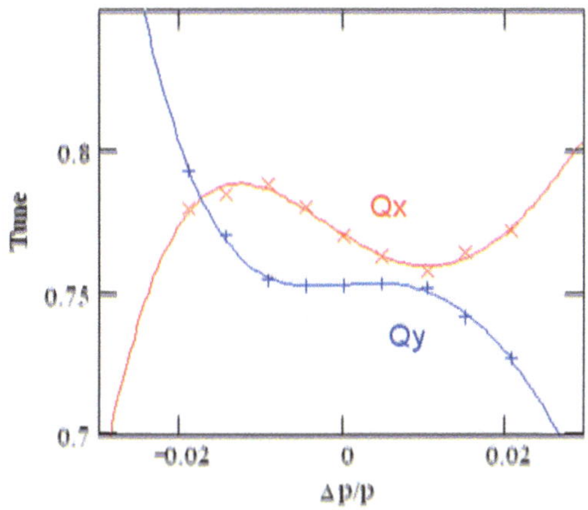

Fig. 7.20 Dependence of tunes on momentum computed using Debuncher optics model (*solid lines*) and measured (*crosses*)

Natural chromaticity is compensated by two families of sextupoles bracketing all arc quads. There is also a considerable nonlinearity introduced by the dipoles. The magnetic measurements of the Antiproton Source dipoles were performed early in the construction of the Antiproton Source and have not been totally reliable. In order to include the nonlinearities into the optics model, we adjusted nonlinear components of dipoles to match the model predictions to the measured tune dependence on the momentum and found that the sextupole and decapole components make major contributions ($B_2/B_0 = -3.6 \times 10^{-5}$ cm^{-2}, $B_4/B_0 = 1.6 \times 10^{-5}$ cm^{-4}). Figure 7.20 presents measured chromaticity and its comparison with the model. Sextupole correction has been applied to minimize tune variation in the central part of the aperture.

The free drift space in the straight sections is very tightly packed with stochastic cooling devices. In order to accommodate them, some of the dipole correctors were removed from the ring. There remain only ten horizontal and seven vertical correctors in the entire Debuncher. Supports for 35 Debuncher quadrupoles were made movable in the course of Run II (2003–2006). This solution was found to be extremely helpful for orbit correction in the case when only handful of correctors is present.

7.1.6 Accumulator

Similar to the Debuncher, the original Accumulator optics design had three periods and mirror symmetry in each of three sectors. There have been many incarnations of the ring optics in the course of Antiproton Source lifetime. The main objective for the present optics was an adjustment of the ring momentum compaction so that to obtain the slip factor optimal for the antiproton stacking [7]. As it will be shown below, the stacking rate is proportional to the slip factor as long as the longitudinal Schottky bands at the high-frequency end of the stacktail system band are not overlapped. An optimal slip factor was found to be close to 0.015. This value

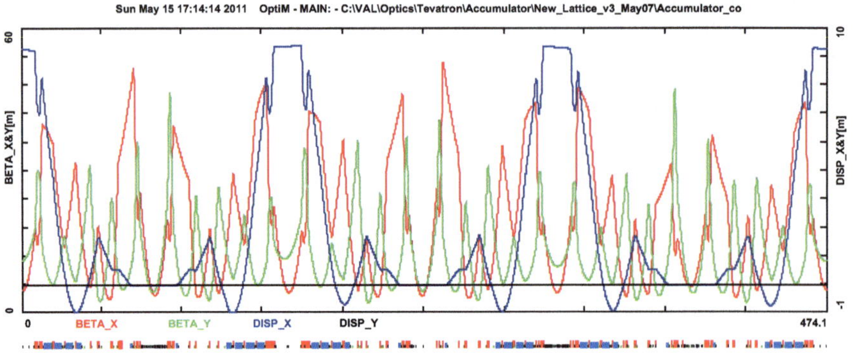

Fig. 7.21 Beta-functions and dispersion for the Accumulator

corresponds to momentum compaction of 0.026. In addition the correct phase advances between pickups and kickers of transverse stochastic cooling have to be maintained, and the aperture maximized for a known aperture limitations mainly presented by the stochastic cooling tanks. The stacktail pickups are located in a region with maximum dispersion. This dispersion was already optimized for the stacktail system, and therefore the momentum compaction correction was done by dispersion adjustments in the regions where the dispersion is negative. The results of the optimization are presented in Fig. 7.21. One can see significant deviations from the threefold symmetry. The same as for the Debuncher, the optics measurements were performed with the response matrix method. The obtained model predicts the beta-functions with accuracy of ~5 %. Special attention was paid to zeroing the dispersion functions in the straight sections to prevent the core heating by the stacktail system. The achieved dispersion leakage was less than 5 cm which corresponds to less than 0.5 % of the peak dispersion value. The optics design was based on the apertures of the elements taken from drawings. It resulted in the expected ring acceptance to be about 14–15 mm × mrad. However, only about 8–10π mm mrad was obtained. It was sufficient to support reliable machine operation. Due to the lack of available machine study time, this difference has not been well understood.

Although the model predicts the momentum compaction and, consequently, the slip factor with good accuracy, its independent measurement was important to build an accurate stacktail model and, then, to optimize the stacktail cooling system. Two independent methods were used. The first method was based on the direct measurement of γ-transition from the extrapolation of measurement of revolution frequency response to bend bus variations. In the second method, the slip factor was computed from the dependence of synchrotron frequency on the RF voltage. The measured slip factor at the central orbit was very close to the design value. However, the measurements showed strong dependence of slip factor on the momentum. The results are presented in Fig. 7.22. One can see that although the optics model qualitatively yields the same behavior, there are clear differences at the edges.

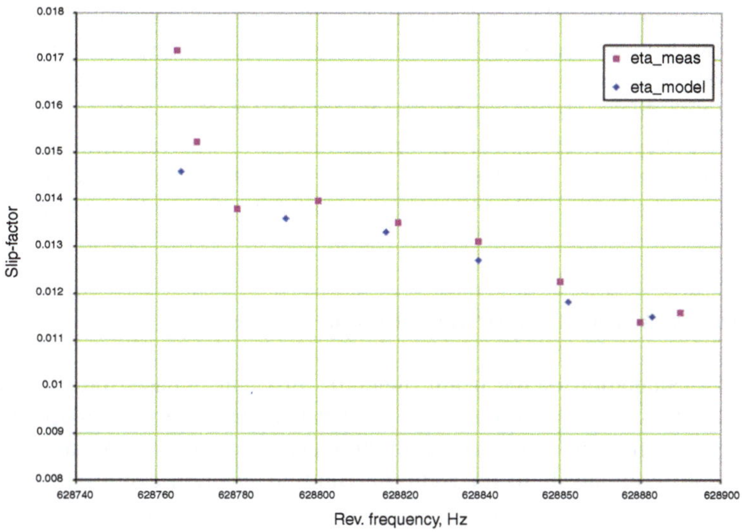

Fig. 7.22 Dependence of slip factor for Accumulator on the beam revolution frequency which is directly related to the relative momentum change

7.2 Stochastic Cooling of Antiprotons

Stochastic cooling technology at Fermilab has expanded considerably on the initial systems developed at CERN [8]. Cooling systems are utilized for increasing the phase space density of 8 GeV antiprotons in three Fermilab antiproton synchrotrons: Accumulator, Debuncher, and Recycler. A total of 25 independent systems are implemented, 21 in the Antiproton Source and 4 in the Recycler. The following sections present an introduction to the stochastic cooling theory used for optimization of the cooling systems, describe the technology choices for each segment of the cooling systems, and present cooling systems performance.

7.2.1 Stochastic Cooling: Theory

Reference [9] gives a good introduction to the subject of stochastic cooling. The model is based on ordinary differential equations allowing simple estimates of cooling rates, but it does not represent an accurate quantitative description of cooling process. More accurate approach based on partial differential equations has been developed in [10]. In the following sections, we consider an extension of this model for both transverse and longitudinal cooling. This model additionally takes into account the effect of Schottky band overlap and does not require a linear relationship between momentum offset and revolution frequency—the properties which were essential for modeling of some abovementioned systems. The results obtained with this model were essential for the choice of upgrade path, analysis of beam-based measurements, and the optimization of system performance. First, we will derive expressions for the beam dielectric functions and, then, proceed to a derivation of the Fokker–Planck equations describing the cooling process.

Beam Dielectric Function for Longitudinal Cooling

Usually, a calculation of the beam dielectric function is based on the azimuthal harmonics. This approach does not work if bands are close to overlap because in this case the amplitudes of the harmonics are changed within one revolution. Consequently, one needs to track the distribution function evolution within each turn. In this section, we limit ourselves to the case of a beam with sufficiently small intensity (or large momentum spread) so that the beam interaction with the vacuum chamber can be neglected. Then, the evolution of the longitudinal distribution function can be described by the following equation:

$$f(x, s_2, t) = f(x, s_1, t - T_{12}(x)), \tag{7.4}$$

where $x = (p - p_0)/p_0$ is the relative momentum deviation and $T_{12}(x)$ is the time of flight from point 1 to point 2. Equation (7.4) allows reducing the problem from one of finding the entire ring distribution function to one of finding the local distribution functions in the pickup and kicker. Figure 7.23 depicts the layout of a simplified cooling system. $f_1(x, t)$ is the distribution function immediately downstream of the kicker, $f_2(x, t)$ is the distribution function in the pickup, and $f_3(x, t)$ is the distribution function just upstream of the kicker. Since the particle momentum is changed only in the kicker and that the beam motion in the rest of the ring is described by Eq. (7.4), one can write the equations that relate these functions to one another:

$$\begin{aligned} f_2(x,t) &= f_1(x, t - T_1(x)), \\ f_3(x,t) &= f_2(x, t - T_2(x)), \\ f_1(x,t) &= f_3\left(x - \delta p(t)/p_0, t\right), \end{aligned} \tag{7.5}$$

where $\delta p(t)$ is the particle momentum change due to the kicker and $T_1(x)$, $T_2(x)$, and $T(x) = T_1(x) + T_2(x)$ are the kicker-to-pickup, pickup-to-kicker, and revolution times for a particle with momentum offset x, respectively. They are equal to

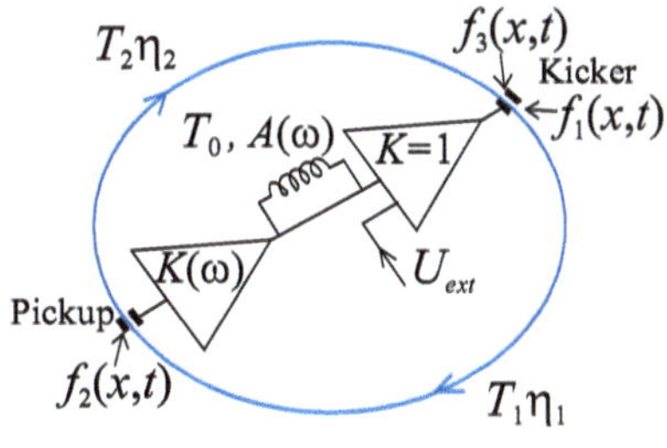

Fig. 7.23 Schematic of a longitudinal cooling system

$$
\begin{aligned}
T_1(x) &= \frac{T_{10}}{1-\eta_1 x-\eta_1^{(2)}x^2+\cdots} \approx T_{10}(1+\eta_1 x),\\
T_2(x) &= \frac{T_{20}}{1-\eta_2 x-\eta_2^{(2)}x^2+\cdots} \approx T_{20}(1+\eta_2 x),\\
T(x) &= \frac{T_0}{1-\eta x-\eta^{(2)}x^2+\cdots} \approx T_0(1+\eta x).
\end{aligned}
\tag{7.6}
$$

Here T_{10}, T_{20}, and $T_0=T_{10}+T_{20}$ are the kicker-to-pickup, pickup-to-kicker, and revolution times for the reference particle, respectively, $\eta=\alpha-1/\gamma^2$ is the ring slip factor, and η_1 and η_2 are the partial kicker-to-pickup and pickup-to-kicker slip factors so that $\eta T_0=\eta_1 T_1+\eta_2 T_2$. Expressing the distribution function as a sum of equilibrium value and perturbation, $f_k(x,t)=f_0(x)+\tilde{f}_k(x,t)$, $k=1,\ldots,3$ and leaving only the first-order term in the Taylor expansion of the third equation in Eq. (7.5), one obtains

$$
\begin{aligned}
\tilde{f}_2(x,t) &= \tilde{f}_1(x,t-T_1(x)),\\
\tilde{f}_3(x,t) &= \tilde{f}_2(x,t-T_2(x)),\\
\tilde{f}_1(x,t) &= \tilde{f}_3(x,t)-\frac{\delta p(t)}{p_0}\frac{df_0(x)}{dx}.
\end{aligned}
\tag{7.7}
$$

We will look for a solution in the form $\tilde{f}_k(x,t)=\tilde{f}_{k\omega}(x)e^{i\omega t}$ and $\delta p(t)=\delta p_\omega e^{i\omega t}$. Substituting these equations into Eq. (7.5), one obtains

$$
\begin{aligned}
\tilde{f}_{2\omega}(x) &= \tilde{f}_{1\omega}(x)\exp(-i\omega T_1(x)),\\
\tilde{f}_{3\omega}(x) &= \tilde{f}_{2\omega}(x)\exp(-i\omega T_2(x)),\\
\tilde{f}_{1\omega}(x) &= \tilde{f}_{3\omega}(x)-\frac{df_0(x)}{dx}\frac{\delta p_\omega}{p_0}.
\end{aligned}
\tag{7.8}
$$

Eliminating $\tilde{f}_{1\omega}(x)$ and $\tilde{f}_{3\omega}(x)$ from the above equations, one obtains

$$
\tilde{f}_{2\omega}(x)\exp(i\omega T_1(x)) = \tilde{f}_{2\omega}(x)\exp(-i\omega T_2(x)) - \frac{df_0(x)}{dx}\frac{\delta p_\omega}{p_0}. \tag{7.9}
$$

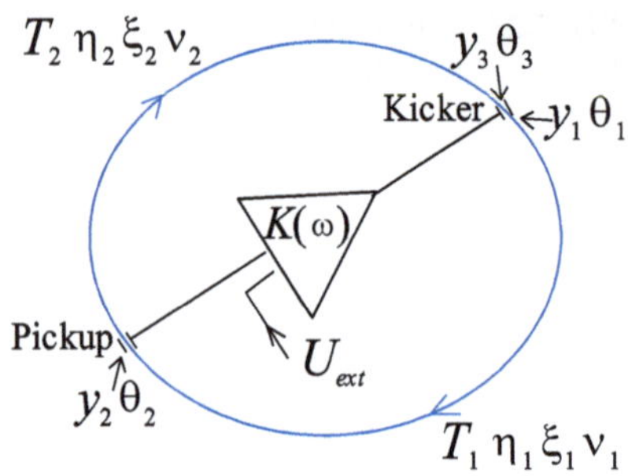

Fig. 7.24 Schematic of a transverse cooling system

Let the momentum kick be determined by the sum of the amplified pickup signal and an external harmonic perturbation (see Fig. 7.24) so that

$$\delta p_\omega / p_0 = \Delta p_{\text{ext}\omega}/p_0 + \int \mathrm{e}^{-i\omega T_{20}} \left[1 - A(\omega)\,\mathrm{e}^{-i\omega T_0}\right]\ G(x,\omega)\widetilde{f}_{2\omega}(x)\,dx. \qquad (7.10)$$

Here the term $\mathrm{e}^{-i\omega T_{20}}$ accounts for the signal propagation delay from the pickup to the kicker. It is equal to the pickup-to-kicker particle flight time for the reference particle, $\widetilde{f}_2(p, t - T_{20}) \to \widetilde{f}_{2\omega}(p)\mathrm{e}^{-i\omega T_{20}}$. The total system gain, $G(x,\omega)\left[1 - A(\omega)\,\mathrm{e}^{-i\omega T_0}\right]$, is expressed in a general form so that it can describe both Palmer and filter cooling. For Palmer cooling $A(\omega) = 0$ and the pickup momentum signal depends on the transverse displacement of the beam in a pickup located in non-zero dispersion. For filter cooling, the pickup signal does not depend on particle momentum, $G(x,\omega) \to G(\omega)$, and the cooling signal is formed by a notch filter. Thus, for filter cooling $A(\omega) \approx 1$, with a delay equal to the revolution time for the reference particle, T_0. Taking into account the distribution function normalization, $\int f_0(x)\ dx = 1$, and introducing the impedances of the pickup, Z_p, and kicker, Z_k, the pickup voltage is

$$U_{\text{pickup}_\omega} = I_0 \int Z_p(x,\omega)\, f_{2\omega}(x,\omega)\,dx, \qquad (7.11)$$

and the energy gain in the kicker is

$$\delta E_{\text{kicker}_\omega} = \mathrm{e}\frac{Z_k(\omega)}{Z_{\text{ampl}}} U_{\text{kicker}_\omega}. \qquad (7.12)$$

The system gain is then given by

$$G(x,\omega) = \frac{eI_0 Z_p(x,\omega) Z_k(\omega)}{\gamma\beta^2 mc^2 Z_{\text{ampl}}}\ K(\omega). \qquad (7.13)$$

Here I_0 is the beam current, $Z_{\text{ampl}} = 50\ \Omega$ is the impedance of the power amplifier, $K(\omega)$ is the total electronic amplification of the cooling system, c is the speed of light, e and m are the particle charge and mass, and β and γ are the relativistic factors.

Substitution of Eq. (7.10) into Eq. (7.9) yields

$$\widetilde{f}_{2\omega}(x)\left[e^{i\omega T_1(x)} - e^{-i\omega T_2(x)}\right] + \frac{df_0(x)}{dx}\left[\frac{\Delta p_{\text{ext}\omega}}{p_0} + e^{-i\omega T_{20}}\left[1 - A(\omega)e^{-i\omega T_0}\right]\int dx'\widetilde{f}_{2\omega}(x')G(x',\omega)\right] = 0. \tag{7.14}$$

Dividing both terms by $e^{i\omega T_1(x)} - e^{-i\omega T_2(x)}$, multiplying by $G(x,\omega)$, and integrating, one obtains

$$S_\omega + \frac{\Delta p_{\text{ext}\omega}}{p_0}\int \frac{df_0(x)}{dx}\frac{G(x,\omega)\,dx}{e^{i\omega T_1(x)} - e^{-i\omega T_2(x)}} + e^{-i\omega T_{20}}\left[1 - A(\omega)e^{-i\omega T_0}\right]S_\omega\int \frac{df_0(x)}{dx}\frac{G(x,\omega)\,dx}{e^{i\omega T_1(x)} - e^{-i\omega T_2(x)}} = 0, \tag{7.15}$$

where

$$S_\omega = \int dx'\widetilde{f}_{2\omega}(x')G(x',\omega). \tag{7.16}$$

Solving Eq. (7.15) for S_ω, one finally obtains the system response at the pickup location due to the external harmonic perturbation:

$$S_\omega = -\frac{1}{\varepsilon(\omega)}\frac{\Delta p_{\text{ext}\omega}}{p_0}\int_{\delta\to 0_+}\frac{df_0(x)}{dx}\frac{G(x',\omega)e^{i\omega T_2(x)}}{e^{i\omega T(x)} - (1-\delta)}dx, \tag{7.17}$$

where $\varepsilon(\omega)$ is the beam dielectric function given by

$$\varepsilon(\omega) = 1 + \left(1 - A(\omega)e^{-i\omega T_0}\right)\int_{\delta\to 0_+}\frac{df_0(x)}{dx}\frac{G(x,\omega)e^{i\omega(T_2(x)-T_{20})}}{e^{i\omega T(x)} - (1-\delta)}dx. \tag{7.18}$$

In the above equations, the rule for traversing the poles,[3] $\delta\to 0_+$, follows from the fact that for a complex Laplace transform, ω is shifted to the lower complex plane. Equation (7.18) shows that the dielectric functions for positive and negative frequencies are complex conjugates of one another: $\varepsilon(\omega) = \varepsilon(-\omega)^+$.

---

[3] Note that in Chap. 6 we look for harmonic solutions in the form of $\exp(-i\omega t)$. That changes signs for the rule to traverse the poles. All equations of this chapter can be converted to this notation by their complex conjugation.

To find the closed system response, we need to incorporate two additional factors into Eq. (7.17). They describe the pickup-to-kicker delay, $e^{-i\omega T_2}$, and the notch filter. Including these factors results in

$$S_{\omega_{\text{closed}}} = -\frac{1}{\varepsilon(\omega)}\frac{\Delta p_{\text{ext}\omega}}{p_0}\left(1 - A(\omega)\,e^{-i\omega T_0}\right)\int_{\delta\to 0_+}\frac{df_0(x)}{dx}\frac{G(x,\omega)e^{i\omega(T_2(x)-T_{20})x}}{e^{i\omega T(x)} - (1-\delta)}dx, \tag{7.19}$$

The response of the open-loop system can be obtained from Eq. (7.19) by setting $\varepsilon(\omega)=1$.

Far away from Schottky band overlap, the exponent in the denominator of Eq. (7.18) can be expanded near the revolution harmonic, $\omega = n\omega_0(1-\eta\ y)$, $\omega_0 = 2\pi/T_0$. That results in the generally accepted formula for the beam dielectric function at the nth revolution frequency harmonic of a particle with momentum offset y [10]:

$$\varepsilon_n(y) \approx 1 + \left(1 - A(n\omega_0)\,e^{2\pi i n\eta y}\right)\frac{1}{2\pi i n\eta}\int\limits_{\delta\to 0_+}\frac{df_0(x)}{dx}\frac{G(x,n\omega_0)}{x - y - i\delta\,\mathrm{sign}(n)}dx. \tag{7.20}$$

Here only the linear term in the expansion of revolution time in momentum has been included (see Eq. (7.6)). Note that although this expression is frequently used to describe signal suppression for filter cooling in a wide frequency range, it is only valid far away from band overlap. This allows the expansion of the last exponent in Eq. (7.20), yielding the following result for filter cooling with an ideal notch filter: $A(n\omega_0)=1$,

$$\varepsilon_n(y) \approx 1 - G(n\omega_0)\ y\int_{\delta\to 0_+}\frac{df_0(x)}{dx}\frac{1}{x - y - i\delta\,\mathrm{sign}(n)}dx, \tag{7.21}$$

where $G(0,n\omega_0)\equiv G(n\omega_0)$. Similarly, for the case of ideal Palmer cooling ($A(n\omega_0)=0$ and $G(x,n\omega_0)=G_x(n\omega_0)\ x$), one obtains

$$\begin{aligned}\varepsilon_n(y) &\approx 1 + \frac{1}{2\pi i\eta n}\int\limits_{\delta\to 0_+}\frac{df_0(x)}{dx}\frac{G_x(n\omega_0)x}{x - y - i\delta\,\mathrm{sign}(n)}dx\\ &= 1 + \frac{G_x(n\omega_0)}{2\pi i\eta n}y\int\limits_{\delta\to 0_+}\frac{df_0(x)}{dx}\frac{1}{x - y - i\delta\,\mathrm{sign}(n)}dx.\end{aligned} \tag{7.22}$$

Here in transformation from the top to bottom part in Eq. (7.22) we took advantage of the fact that $\int (df_0/dx)dx = 0$. Comparing Eqs. (7.21) and (7.22), one can see that there are similar dependencies of beam dielectric functions on momentum for both cooling methods.

Beam Dielectric Function for Transverse Cooling

For transverse cooling, as in the method used above for longitudinal cooling, the evolution of the beam distribution is considered at three points: (1) after the kicker, (2) in the pickup, and (3) before the kicker. The layout of the system is presented in Fig. 7.24. The beam dipole moment densities at each of three points are

$$d_k(t) = \frac{I_0}{c\beta}\int y_k(x,t)\, f_0(x)\, dx, \quad k = 1, 2, 3. \tag{7.23}$$

Here $f_0(x)$ is the longitudinal distribution function, and $y_k(x)$ is the average transverse beam displacement for particles with relative momentum deviations equal to x. Normalizing the beam displacements, $y_k(x)$, and angles, $\theta_k(x)$, by the beta functions so that $\widetilde{y}_k = y_k/\sqrt{\beta_k}$ and $\widetilde{\theta}_k = \theta_k\sqrt{\beta_k} + \alpha_k x_k/\sqrt{\beta_k}$, one can write a system of equations relating the beam displacements after and before the kicker:

$$\begin{aligned} \widetilde{y}_3(x,t) &= c(x)\widetilde{y}_1(x, t - T(x)) + s(x)\widetilde{\theta}_1(x, t - T(x)), \\ \widetilde{\theta}_3(x,t) &= -s(x)\widetilde{y}_1(x, t - T(x)) + c(x)\widetilde{\theta}_1(x, t - T(x)). \end{aligned} \tag{7.24}$$

Here $c(x) = \cos(2\pi\nu(x))$, $s(x) = \sin(2\pi\nu(x))$, $\nu(x) = \nu_0 + \xi x + \cdots$, ν_0 is the betatron tune, and ξ is the tune chromaticity. Passage through the kicker changes the beam angle but does not change the beam coordinate so that

$$\begin{aligned} \widetilde{y}_1(x,t) &= \widetilde{y}_3(x,t), \\ \widetilde{\theta}_1(x,t) &= \widetilde{\theta}_3(x,t) + \delta\widetilde{\theta}(t). \end{aligned} \tag{7.25}$$

We look for a solution in the form $\widetilde{y}_k(x,t) = \widetilde{y}_{k\omega}(x)\mathrm{e}^{i\omega t}$ and $\delta\theta(t) = \delta\theta_\omega \mathrm{e}^{i\omega t}$. Its substitution into Eqs. (7.24) and (7.25) yields

$$\begin{aligned} \widetilde{y}_{3\omega}(x) &= \left(c(x)\widetilde{y}_{1\omega}(x) + s(x)\widetilde{\theta}_{1\omega}(x)\right)\exp(-i\omega T(x)), \\ \widetilde{\theta}_{3\omega}(x) &= \left(-s(x)\widetilde{y}_{1\omega}(x) + c(x)\widetilde{\theta}_{1\omega}(x)\right)\exp(-i\omega T(x), \\ \widetilde{y}_{1\omega}(x) &= \widetilde{y}_{3\omega}(x), \\ \widetilde{\theta}_{1\omega}(x) &= \widetilde{\theta}_{3\omega}(x) + \delta\widetilde{\theta}_\omega. \end{aligned} \tag{7.26}$$

Solving these equations for $\widetilde{y}_{1\omega}(x)$ and $\widetilde{\theta}_{1\omega}(x)$ one obtains

$$\begin{aligned} \widetilde{\theta}_{1\omega}(x) &= -\frac{c(x) - \mathrm{e}^{i\omega T(x)}}{\mathrm{e}^{2i\omega T(x)} - 2c(x)\mathrm{e}^{i\omega T(x)} + 1}\mathrm{e}^{i\omega T(x)}\delta\widetilde{\theta}_\omega, \\ \widetilde{y}_{1\omega}(x) &= \frac{s(x)}{\mathrm{e}^{2i\omega T(x)} - 2c(x)\mathrm{e}^{i\omega T(x)} + 1}\mathrm{e}^{i\omega T(x)}\delta\widetilde{\theta}_\omega. \end{aligned} \tag{7.27}$$

Taking into account the relationship between coordinates and angles of points 1 and 2,

$$\begin{aligned}\widetilde{y}_2(x,t) &= c_1(x)\widetilde{y}_1(x,t-T_1(x)) + s_1(x)\widetilde{\theta}_1(x,t-T_1(x)),\\ \widetilde{\theta}_2(x,t) &= -s_1(x)\widetilde{y}_1(x,t-T_1(x)) + c_1(x)\widetilde{\theta}_1(x,t-T_1(x)),\end{aligned} \tag{7.28}$$

and transforming the time-dependent values to their Fourier harmonics, one obtains the following expression for the beam displacement in the pickup:

$$\widetilde{y}_{2\omega}(x) = \frac{s_2(x) + s_1(x)\mathrm{e}^{i\omega T(x)}}{\mathrm{e}^{2i\omega T(x)} - 2c(x)\mathrm{e}^{i\omega T(x)} + 1}\,\mathrm{e}^{i\omega T_2(x)}\delta\widetilde{\theta}_\omega. \tag{7.29}$$

Here $c_{1,2}(x) = \cos(2\pi\nu_{1,2}(x))$, $s_{1,2}(x) = \sin(2\pi\nu_{1,2}(x))$, $\nu_{1,2}(x) = \nu_{1,2} + \xi_{1,2}x + \cdots$, and $2\pi\nu_1$ and $2\pi\nu_2$ are the pickup-to-kicker and kicker-to-pickup betatron phase advances so that $\nu_1 + \nu_2 = \nu$, and ξ_1 and ξ_2 are the corresponding partial tune chromaticities so that $\xi_1 + \xi_2 = \xi$.

Similar to Eq. (7.10) the beam kick is determined by the sum of the amplified pickup signal and an external harmonic perturbation, $\Delta\widetilde{\theta}_{\mathrm{ext}\omega}$, so that

$$\delta\widetilde{\theta}_\omega = \int dx f_0(x)\widetilde{y}_{2\omega}(x) G_\perp(\omega)\mathrm{e}^{-i\omega T_{20}} + \Delta\widetilde{\theta}_{\mathrm{ext}\omega}. \tag{7.30}$$

We introduce the impedances of the pickup, $Z_{p\perp}$, and the kicker, $Z_{k\perp}$, so that the pickup voltage is given by

$$U_{\mathrm{pickup}_\omega} = I_0 Z_{p\perp}(\omega)\overline{y_\omega} = I_0 Z_{p\perp}(\omega)\int y_\omega(x) f_0(x)\,dx. \tag{7.31}$$

The change in transverse angle obtained by a particle in the kicker is

$$\delta\theta_{\mathrm{kicker}_\omega} = \frac{e}{mc^2\gamma\beta^2}\frac{Z_{k\perp}(\omega)}{Z_{\mathrm{ampl}}} U_{\mathrm{kicker}_\omega}. \tag{7.32}$$

This yields a system gain that is given by

$$G_\perp(\omega) = \frac{eI_0 Z_{p\perp}(\omega) Z_{k\perp}(\omega)}{\gamma\beta^2 mc^2 Z_{\mathrm{ampl}}}\sqrt{\beta_p\beta_k}K(\omega), \tag{7.33}$$

where β_p and β_k are the beta-functions in the pickup and kicker, respectively.

Substituting Eq. (7.30) into Eq. (7.29), one obtains

$$\widetilde{y}_{2\omega}(x) = \frac{\left(s_2(x) + s_1(x)\mathrm{e}^{2i\omega T(x)}\right)\mathrm{e}^{i\omega T_2(x)}}{\mathrm{e}^{2i\omega T(x)} - 2c(x)\mathrm{e}^{i\omega T(x)} + 1}\left(G_\perp(\omega)\mathrm{e}^{-i\omega T_{20}}\int dx f_0(x)\widetilde{y}_{2\omega}(x) + \Delta\widetilde{\theta}_{\mathrm{ext}\omega}\right). \tag{7.34}$$

Finding the system response to a harmonic transverse excitation is similar to that carried out for the longitudinal case. The result is

$$\overline{\widetilde{y}_{2\omega}} \equiv \int dx f_0(x)\widetilde{y}_{2\omega}(x) = \frac{\Delta\widetilde{\theta}_{\mathrm{ext}\omega}}{\varepsilon_\perp(\omega)}\int_{\delta\to 0_+} dx f_0(x)\frac{\left(s_2(x)+s_1(x)\mathrm{e}^{i\omega T(x)}\right)\mathrm{e}^{i\omega T_2(x)}}{\mathrm{e}^{2i\omega T(x)}-2c(x)\mathrm{e}^{i\omega T(x)}+1-\delta}, \tag{7.35}$$

where the beam dielectric function, $\varepsilon_\perp(\omega)$, is

$$\varepsilon_\perp(\omega)=1-\frac{G_\perp(\omega)}{2}\int_{\delta\to 0_+}\frac{\left[\mathrm{e}^{-i\omega T(x)}\sin(2\pi\nu_2(x))+\sin(2\pi\nu_1(x))\right]\mathrm{e}^{i\omega(T_2(x)-T_{20})}}{\cos(\omega T(x))-\cos(2\pi\nu(x))+i\delta\sin(\omega T(x))}f_0(x)dx. \tag{7.36}$$

The rule for traversing the poles is similar to that for Eqs. (7.17) and (7.18). As was for the case of longitudinal cooling, the dielectric functions for positive and negative frequencies are related to one another as complex conjugates: $\varepsilon_\perp(\omega)=\varepsilon_\perp(-\omega)^+$.

To find the closed system response, one needs to incorporate a factor describing the pickup-to-kicker delay, $\mathrm{e}^{-i\omega T_{20}}$, into Eq. (7.35). It yields

$$\overline{\widetilde{y}_{2\omega_{\mathrm{closed}}}}=\frac{\Delta\widetilde{\theta}_{\mathrm{ext}\omega}}{2\varepsilon_\perp(\omega)}\int_{\delta\to 0_+}\frac{\left[\mathrm{e}^{-i\omega T(x)}\sin(2\pi\nu_2(x))+\sin(2\pi\nu_1(x))\right]\mathrm{e}^{i\omega(T_2(x)-T_{20})}}{\cos(\omega T(x))-\cos(2\pi\nu(x))+i\delta\sin(\omega T(x))}f_0(x)dx. \tag{7.37}$$

The open-loop response of the system can be obtained from Eq. (7.37) by setting $\varepsilon_\perp(\omega)=1$.

Far away from Schottky band overlap, the cosines in the denominator of Eqs. (7.36) and (7.37) can be expended near a betatron sideband, $\delta\omega=\omega-(n\pm\nu)\omega_0$, $\omega_0=2\pi/T_0$, and we arrive at the standard formula for the transverse dielectric function [10]:

$$\varepsilon_{\perp n\pm}(\delta\omega)=1+\frac{G_\perp(\omega_{n\pm})}{4\pi\sin(2\pi\nu_0)}\int_{\delta\to 0_+}\frac{\sin(2\pi\nu_2)\mathrm{e}^{\mp 2\pi i\nu_0}+\sin(2\pi\nu_1)}{\pm\delta\omega/\omega_0-(\xi+\eta(\nu\pm n))x\mp i\delta}f_0(x)\,dx$$

$$\xrightarrow[\nu_0<<n]{2\pi\nu_2=\pi/2} 1+\frac{G_\perp(\omega_{n\pm})}{4\pi i}\int_{\delta\to 0_+}\frac{f_0(x)dx}{\delta\omega/\omega_0-(\eta n\pm\xi)x-i\delta},\quad \omega_{n\pm}=(n\pm\nu)\omega_0>0. \tag{7.38}$$

Fokker–Planck Equation for Longitudinal Cooling

It has been shown in [10] that if the number of particles simultaneously interacting with each other through the stochastic cooling system (number of particles in the sample) is sufficiently large, the evolution of the beam's longitudinal distribution is described by the Fokker–Planck equation:

$$\frac{\partial f_0(x,t)}{\partial t}+\frac{\partial}{\partial x}(F(x)f_0(x,t))=\frac{1}{2}\frac{\partial}{\partial x}\left(D(x)\frac{\partial f_0(x,t)}{\partial x}\right). \tag{7.39}$$

To find the drag force, $F(x)$, we consider the case where only one particle is present in the sample. The momentum change of that particle is determined by voltage excited in the kicker by this particle only. The kick value is determined by Eq. (7.10) where the system gain of Eq. (7.13) is determined by a single-particle current

$$G_1(x,\omega)\equiv\frac{G(x,\omega)}{N}=\frac{e^2Z_p(x,\omega)Z_k(\omega)}{T_0\gamma\beta^2mc^2Z_{\mathrm{ampl}}}K(\omega), \tag{7.40}$$

and N is the number of particles in the beam. The longitudinal distribution function in the pickup is

$$f_2(x,t)=\delta(x-x_0)\ T(x)\sum_n\delta(t-t_0-nT(x)), \tag{7.41}$$

where x_0 and $t_0+nT(x)$ are the momentum offset and time when the particle passes the kicker. The factor $T(x)$ is required to normalize the time-averaged distribution function, $\int\langle f_2(x,t)\rangle_t dx=1$. We also assume that the damping rate is much smaller than the revolution frequency. That allows one to ignore a revolution frequency change in the sum. The Fourier harmonic of the distribution function at frequency ω is

$$f_{2\omega}(x)=\delta(x-x_0)\ T(x)\sum_n\frac{\mathrm{e}^{-i\omega(t_0+nT(x))}}{2\pi}=\delta(x-x_0)\sum_n\mathrm{e}^{-i\omega t_0}\delta\left(\omega-\frac{2\pi}{T(x)}\right). \tag{7.42}$$

Substituting Eq. (7.42) into Eq. (7.10), integrating over momentum, and performing the inverse Fourier transform, one obtains the time dependence of the kicker strength:

$$\delta p(t)/p_0 = \sum_n e^{i\omega_n(x)(t-t_0-T_{20})}\left[1 - A(\omega_n(x))e^{-i\omega_n(x)_n T_0}\right] G(x,\omega_n(x)), \quad \omega_n(x) = \frac{2\pi n}{T(x)}. \tag{7.43}$$

The kick obtained by the particle is determined by the voltage on the kicker at the time the particle crosses the kicker:

$$\delta p/p_0 = \sum_n e^{i\omega_n(x)(T_2(x)-T_{20})}\left[1 - A(\omega_n(x))e^{-i\omega_n(x)T_0}\right] G(x,\omega_n(x)). \tag{7.44}$$

To find the cooling force, one needs to account for the screening of the particle signal by the other particles in the beam. If the number of particles in the sample is sufficiently large, this signal suppression is similar to the signal suppression of an external excitation expressed by Eq. (7.17). This results in

$$F(x) \equiv \frac{dx}{dt} = \frac{1}{T_0}\sum_{n=-\infty}^{\infty} \frac{G_1(x,\omega_n(x))}{\varepsilon(\omega_n(x))}\left(1 - A(\omega_n(x))e^{-i\omega_n(x)T_0}\right)e^{i\omega_n(x)(T_2(x)-T_{20})}, \tag{7.45}$$

where we neglected the dependence of revolution time on momentum in the denominator. Note that the terms for positive and negative n are complex conjugates; consequently $F(x)$ is a real function. Expanding the dependencies of the revolution and delay times on momentum using Eq. (7.6), one arrives to the cooling force presented in [10].

To determine the diffusion, we initially assume that it is produced by non-interacting particles. First, we relate $D(x)$ to the growth rate of momentum spread for an initial point-like momentum distribution, $f(x) = \delta(x - x_0)$. Multiplying Eq. (7.39) by $(x - x_0)^2$ and integrating over momentum, one obtains

$$\begin{aligned} \frac{d}{dt}\overline{(x-x_0)^2} &\equiv \int (x-x_0)^2\frac{\partial f}{\partial t}dx = -\int (x-x_0)^2\frac{\partial}{\partial x}(F(x)f)dx + \frac{1}{2}\int (x-x_0)^2\frac{\partial}{\partial x}\left(D(x)\frac{\partial f}{\partial x}\right) \\ &= 2\int (x-x_0)F(x)f(x)\,dx + \int f(x)\left(D(x) + (x-x_0)\frac{dD}{dx}\right)dx \xrightarrow{f(x)=\delta(x-x_0)} D(x). \end{aligned} \tag{7.46}$$

Second, we consider single-particle diffusion due to kicker noise. Let the particle energy change at turn n be equal to $\delta E_n = eU(nT(x))$, where $T(x)$ is the particle revolution period and $U(t)$ is the kicker voltage. The energy spread after N_r turns is

$$\overline{(\Delta E)^2} = \overline{\left(e\sum_{n=0}^{N_r-1} U(nT(x))\right)^2} = e^2\sum_{n=0}^{N_r-1}\sum_{m=0}^{N_r-1}\overline{U(nT(x))U(mT(x))} = e^2\sum_{n=0}^{N_r-1}\sum_{m=0}^{N_r-1}K_A((n-m)T(x))$$
$$= e^2\sum_{n=0}^{N_r-1}\sum_{m=0}^{N_r-1}\int_{-\infty}^{\infty} P_A(\omega)\mathrm{e}^{i\omega T(x)(n-m)}\,d\omega = e^2\int_{-\infty}^{\infty} P_A(\omega)\left|\frac{\mathrm{e}^{i\omega T(x)N_r}-1}{\mathrm{e}^{i\omega T(x)}-1}\right|^2 d\omega$$
$$\xrightarrow{N_r\gg 1} \frac{2\pi e^2 N_r}{T(x)}\sum_{n=-\infty}^{\infty} P_A\left(\frac{2\pi}{T(x)}n\right) \approx \frac{2\pi e^2 N_r}{T_0}\sum_{n=-\infty}^{\infty} P_A(\omega_n(x)). \tag{7.47}$$

where we have neglected the difference between $T(x)$ and T_0 in the denominator. $K_A(t)$ and $P_A(\omega)$ are the correlation function and the spectral density of the kicker voltage so that

$$K_A(t) = \int_{-\infty}^{\infty} P_A(\omega)\mathrm{e}^{i\omega t}\,d\omega. \tag{7.48}$$

Comparing Eqs. (7.46) with (7.47) and taking into account the relationship between the relative energy and momentum deviation, one obtains

$$D(x) = \frac{2\pi e^2}{{T_0}^2\left(\gamma\beta^2 mc^2\right)^2}\sum_{n=-\infty}^{\infty}\left|\frac{Z_k(\omega_n(x))}{Z_{\mathrm{ampl}}}\right|^2 P_U(\omega_n(x)), \tag{7.49}$$

where the relationship between the kicker voltage and the voltage of the power amplifier, $P_A(\omega) = |Z_k(\omega)/Z_{\mathrm{ampl}}|^2 P_U(\omega)$, has been taken into account.

The spectral density of the kicker noise consists of two contributions. The first is related to the noise of the electronics at the output of the power amplifier, $P_{U\mathrm{noise}}$, and the second is related to the particle noise. In the general case when Schottky bands can overlap the beam current shot noise for non-interacting particles can be presented as a sum over all Schottky bands:

$$P_I(\omega) = \frac{e^2 N}{2\pi T_0}\sum_{k=-\infty}^{\infty}\frac{f\left(x_f(\omega/k)\right)}{|k\eta(x(\omega/k))|}, \quad \eta(x) = -\frac{1}{\omega}\frac{d\omega}{dx}. \tag{7.50}$$

Here $\eta(x)$ is the momentum compaction for a given momentum, x, and the function $x_f(\omega)$ expresses the dependence of momentum on the particle revolution frequency. For frequencies in close proximity to the revolution frequency, this function is $x_f(\omega) \approx -(\omega - \omega_0)/(\omega_0\,\eta)$. In the absence of band overlap at each frequency there is only one non-zero addend in the sum. Taking into account the cooling system amplification and Schottky noise suppression due to particle interaction [10, 11], one obtains an expression for the diffusion coefficient:

$$D(x) = \frac{2\pi e^2}{{T_0}^2\left(\gamma\beta^2 mc^2\right)^2}\sum_{n=-\infty}^{\infty}\frac{1}{\left|\varepsilon(\omega_n(x))\right|^2}\left[\left|\frac{Z_k(\omega_n(x))}{Z_{\text{ampl}}}\right|^2 P_{U\text{noise}}(\omega_n(x))\right.$$

$$+\left|\frac{Z_p(x,\omega_n(x))Z_k(\omega_n(x))}{Z_{\text{ampl}}}K(\omega_n(x))\left(1-A(\omega_n)\mathrm{e}^{-i\omega_n(x)T_0}\right)\right|^2$$

$$\left.\times\frac{e^2 N}{2\pi T_0}\sum_{k=-\infty}^{\infty}\frac{1}{|k\eta(x)|}f\left(x_f\left(\frac{\omega_n(x)}{k}\right)\right)\right]. \tag{7.51}$$

After simplification we have

$$D(x)=\sum_{n=-\infty}^{\infty}\frac{1}{\left|\varepsilon(\omega_n(x))^2\right|}\left[\frac{2\pi e^2 P_{U\text{noise}}(\omega_n(x))}{{T_0}^2\left(\gamma\beta^2 mc^2\right)^2}\left|\frac{Z_k(\omega_n(x))}{Z_{\text{ampl}}}\right|^2\right.$$
$$\left.+\frac{N}{T_0}\left|G_1(x,\omega_n(x))\left(1-A(\omega_n(x))\mathrm{e}^{-i\omega_n(x)T_0}\right)\right|^2\sum_{k=-\infty}^{\infty}\frac{1}{|k\eta(x)|}f\left(x_f\left(\frac{\omega_n(x)}{k}\right)\right)\right]. \tag{7.52}$$

whereas in Eq. (7.43), $\omega_n(x)=2\pi n/T(x)\approx n\omega_0(1-\eta x)$. In the absence of band overlap, there is only one term with $k=n$ left in the sum over k:

$$D(x)=\sum_{n=-\infty}^{\infty}\frac{\frac{2\pi e^2 P_{U\text{noise}}(\omega_n(x))}{{T_0}^2\left(\gamma\beta^2 mc^2\right)^2}\left|\frac{Z_k(\omega_n(x))}{Z_{\text{ampl}}}\right|^2+\frac{N}{T_0}\left|G_1(x,\omega_n(x))\left(1-A(\omega_n(x))\mathrm{e}^{-i\omega_n(x)T_0}\right)\right|^2\frac{f(x)}{|n\eta(x)|}}{\left|\varepsilon(\omega_n(x))^2\right|}. \tag{7.53}$$

For the weak overlap case, two additional terms with $k=n\pm 1$ need to be included.

Fokker–Planck Equation for Transverse Cooling

The natural variables for a description of transverse cooling are the action-phase variables (I, ψ). We introduce the action so that

$$I=\frac{1}{2}\left(\beta_y\theta^2+2\alpha_y y\theta+\frac{1+{\alpha_y}^2}{\beta_y}y^2\right). \tag{7.54}$$

Here the same as above y and θ are the particle transverse coordinates, and β_y and α_y are the beta and alpha functions of the ring in the plane of cooling, and we assume that there is no horizontal–vertical coupling in the lattice. We also assume that the cooling force is linear in betatron amplitude, which, in practical terms, means that the electronics is not saturated and the pickup has a linear response across its aperture. These assumptions yield that an evolution of beam distribution can be described by the following equation [10]:

$$\frac{\partial f_\perp}{\partial t} + \lambda_\perp(x)\frac{\partial}{\partial I}(If_\perp) = D_{\perp/2}(x)\frac{\partial}{\partial I}\left(I\frac{\partial f_\perp}{\partial I}\right) \tag{7.55}$$

Here $f_\perp \equiv f_\perp(x, I, t)$ is the distribution function normalized so that $\int f_\perp(x, I, t)\, dI = f_0(x)$ and, as above, $\int f_0(x)dx = 1$, $\lambda_\perp(x)$ is the cooling rate, and $D_\perp(x)$ is the diffusion coefficient. $\lambda_\perp(x)$ and $D_\perp(x)$ do not depend on I because of system linearity in the transverse coordinate y.

In a manner analogous to longitudinal cooling, transverse cooling is created by particle self-interaction and is therefore not directly affected by band overlap but is still affected by screening. To find the cooling rate, we first consider single-particle damping. Introducing the ring transfer matrix, $\mathbf{M}$, and the partial kicker-to-pickup and pickup-to-kicker transfer matrices, $\mathbf{M}_1$, $\mathbf{M}_2$ such that $\mathbf{M} = \mathbf{M}_1\mathbf{M}_2$, and using the normalized transverse coordinates, we can write the total ring (pickup-to-pickup) transfer matrix in the following form:

$$\mathbf{M}_{\text{tot}} = \mathbf{M}_1(\mathbf{M}_2 + \mathbf{G}) = \mathbf{M} + \mathbf{M}_1\mathbf{G} = \begin{bmatrix} c(x) + G_{\perp 1}s_1(x) & s(x) \\ -s(x) + G_{\perp 1}c_1(x) & c(x) \end{bmatrix}. \tag{7.56}$$

Here we have made use of the fact that the angle change in the kicker is proportional to the particle displacement in the pickup:

$$\begin{bmatrix} \delta\tilde{x} \\ \delta\tilde{\theta} \end{bmatrix}_{\text{kicker}} = \begin{bmatrix} 0 \\ G_{\perp 1}x_{\text{pickup}} \end{bmatrix} = \begin{bmatrix} 0 & 0 \\ G_{\perp 1} & 0 \end{bmatrix}\begin{bmatrix} \tilde{x} \\ \tilde{\theta} \end{bmatrix}_{\text{pickup}} \equiv \mathbf{G}\begin{bmatrix} \tilde{x} \\ \tilde{\theta} \end{bmatrix}_{\text{pickup}}, \tag{7.57}$$

where $G_{\perp 1}(\omega) = G_\perp(\omega)/N$ is the single-particle system gain. The damping rate of the cooling system is determined by the eigenvalues of the total ring transfer matrix, $\mathbf{M}_{\text{tot}}$. These eigenvalues are

$$\Lambda_{1,2} = c(x) + G_{\perp 1}\frac{s_1(x)}{2} \pm \sqrt{\left(c(x) + G_{\perp 1}\frac{s_1(x)}{2}\right)^2 + (1 - G_{\perp 1}s_2(x))}. \tag{7.58}$$

Since the single-particle gain is small, $G_{\perp 1} \ll 1$, and only the linear term in the Taylor expansion of Eq. (7.58) is important, one obtains

$$\Lambda_{1,2} \approx e^{\pm 2\pi i\nu(x)}\left(1 + \frac{G_{\perp 1}}{2i}e^{-2\pi i\nu_2(x)}\right), \quad G_{\perp 1} \ll 1. \tag{7.59}$$

That yields the damping rates of both modes being equal:

$$\lambda_{1,2}(x) \equiv 2\ \ln|\Lambda_{1,2}| \approx -\mathrm{Re}\left(iG_{\perp 1}\mathrm{e}^{-2\pi i\nu_2(x)}\right), \quad G_{\perp 1} \ll 1. \tag{7.60}$$

Here the factor of two accounts for the difference between the damping decrements for the amplitude and the action. Expanding the particle signal in Fourier harmonics, summing their effect on the particle motion, and taking screening into account, one finally obtains [10]

$$\begin{aligned} \lambda_\perp(x) &= \frac{1}{T_0}\sum_{n=-\infty}^{\infty}\mathrm{Re}\left(\frac{G_{\perp 1}(\omega_{n\perp}(x))}{i\varepsilon_\perp(\omega_{n\perp}(x))}\mathrm{e}^{i\omega_n(T_2(x)-T_{20})-2\pi i\nu_2(x)}\right), \\ \omega_{n\perp}(x) &= \frac{2\pi n}{T(x)} - \nu(x) \approx \omega_0(n(1-\eta x)-(\nu+\xi x)). \end{aligned} \tag{7.61}$$

Here $\varepsilon_\perp(\omega)$ takes into account the screening of the particle field, and the term $\mathrm{e}^{i\omega_n(T_2(x)-T_{20})}$ takes into account the variation in particle arrival time at the kicker.

The diffusion coefficient for transverse cooling is obtained in a manner analogous to the determination of the diffusion coefficient for longitudinal cooling. It yields

$$D_\perp(x) = \frac{\pi\beta_k}{T_0{}^2}\sum_{n=-\infty}^{\infty}P_\theta(\omega_{n\perp}(x)), \tag{7.62}$$

where $P_\theta(\omega)$ is the spectral density of the angle kicks produced by the kicker. $P_\theta(\omega)$ consists of two contributions: the spectral density of amplifier noise, $P_{\perp U}(\omega)$, and the amplified shot noise of the beam. The shot noise of the beam at the pickup is

$$P_{\perp p}(\omega) = \frac{e^2|Z_{p\perp}(\omega)|^2\overline{y^2}}{2\pi T_0}N\sum_{m=-\infty}^{\infty}\frac{1}{|m\eta(x)+\nu'(x)|}f\left(x_f\left(\frac{\omega_{n\perp}(x)}{k}\right)\right), \tag{7.63}$$

where $\nu'(x) = d\nu(x)/dx$ is the local chromaticity. Substituting Eq. (7.63) into Eq. (7.62), taking into account particle screening, and using the definition of the single-particle gain, we obtain

$$\begin{aligned} D_\perp(x) = \sum_{n=-\infty}^{\infty}\frac{1}{|\varepsilon_\perp(\omega_{n\perp}(x))|^2}&\left(\frac{\pi\beta_k}{T_0{}^2}\left(\frac{e|Z_{k\perp}(\omega_{n\perp}(x))|}{mc^2\beta^2\gamma Z_{\mathrm{ampl}}}\right)^2 P_{\perp U}(\omega_{n\perp}(x))\right. \\ &\left. +|G_{\perp 1}(\omega_{n\perp}(x))|^2\frac{\overline{I(x)}N}{T_0}\sum_{m=-\infty}^{\infty}\frac{1}{|m\eta(x)+\nu'(x)|}f\left(x_f\left(\frac{\omega_{n\perp}(x)}{k}\right)\right)\right), \end{aligned} \tag{7.64}$$

where $\overline{I(x)} = \int f_{\perp}(x, I, t) I dI$ is the average action for a given momentum deviation x. In the absence of band overlap, there is only one term with $m = n$ left in the sum over m:

$$D_{\perp}(x) = \sum_{n=-\infty}^{\infty} \frac{\frac{\pi\beta_k}{T_0^2}\left(\frac{e|Z_{k\perp}(\omega_{n\perp}(x))|}{mc^2\beta^2\gamma Z_{\text{ampl}}}\right)^2 P_{\perp U}(\omega_{n\perp}(x)) + |G_{\perp 1}(\omega_{n\perp}(x))|^2 \frac{\overline{I(x)} N f(x)}{T_0|\nu'(x)+\eta(x)n|}}{|\varepsilon_{\perp}(\omega_{n\perp}(x))|^2}. \tag{7.65}$$

Using only the first nonzero terms in the Taylor expansion of $\omega_{n\perp}(x), \nu'(x)$ and $\eta(x)$ and using Eq. (7.38) for beam dielectric function, one obtains the result presented in [10]. Note that band overlap for transverse cooling occurs at significantly lower frequencies than for longitudinal cooling because there are two peaks per revolution band. They correspond to positive and negative betatron sidebands, $\omega_0(n \pm \nu)$, in the beam Schottky spectra. Even in the case of "optimal" tune $\nu = n/2 + 1/4$, the band overlap starts at twice smaller tune spread.

Cooling Rate Estimates

As one can see from the above sections, an accurate stochastic cooling description requires a solution of partial differential equation with cooling force and diffusion dependent on the particle distribution. It is close to impossible to find an analytical solution describing a distribution function evolution even for simplest cases. A numerical solution is quite complicated, and its accurate solution requires large volume of computations. However, it is the only reliable way allowing one to take into account important technical details of the system and to achieve reasonable accuracy for cooling dynamics. In this section, we will consider a simplified cooling description based on ordinary differential equations. We will assume the following:

- The gain is constant across the cooling band:

$$G(\omega) = \begin{cases} G_0, & f \in [f_{\min}, f_{\max}], \\ 0, & \text{otherwise} \end{cases}, \tag{7.66}$$

- The longitudinal particle distribution is Gaussian with rms relative momentum spread equal to σ_p.
- The system is far from the band overlap.
- $\eta(x)$ and $\nu'(x)$ do not change across momentum acceptance.
- The thermal noise is smaller than the particle Schottky noise. It is well justified for all Fermilab cooling systems in normal operating conditions.

Transverse Cooling

We start from the case of the transverse cooling. Multiplying Eq. (7.55) by I and integrating over I, one obtains

$$\frac{\partial \overline{I(x)}}{\partial t} - \lambda_{\perp}(x)\overline{I(x)} = D_{\perp 1/2}(x), \tag{7.67}$$

where $\overline{I(x)} \equiv \langle I(x,t)\rangle = \int If_{\perp}(x,I,t)dI \Big/ \int f_{\perp}(x,I,t)dI$ is the average action (rms emittance) for particles with relative momentum offset x. Equations (7.61) and (7.65) represent the damping rate and diffusion. Leaving only leading terms and substituting these equations into Eq. (7.67), one obtains

$$\frac{\partial \bar{I}}{\partial t} + \lambda_{\perp \text{eff}}(x)\bar{I} = 0,$$
$$\lambda_{\perp \text{eff}}(x) = \frac{1}{T_0}\sum_{n=-\infty}^{\infty} \text{Re}\left(\frac{-G_{\perp 1}(n\omega_0)}{\varepsilon_{\perp}(\omega_{n\perp}(x))} e^{in\eta x}\right) - \frac{Nf(x)}{2T_0|\eta|}\sum_{n=-\infty}^{\infty}\frac{1}{|n|}\left|\frac{G_{\perp 1}(n\omega_0)}{\varepsilon_{\perp}(\omega_{n\perp}(x))}\right|^2, \tag{7.68}$$

where we also assume 90° betatron phase advance between pickup and kicker and that the gain is smooth function of ω so that it does not change within few revolution bands. As one can see from Eqs. (7.67), each momentum bite would be cooled independently if $\lambda_{\perp}(x)$ and $D_{\perp}(x)$ would not depend on the distribution functions through $\varepsilon_{\perp}(\omega)$. Below we will assume that $\varepsilon_{\perp}(\omega) = 1$. An accuracy of such approximation will be estimated later. Assuming perfect phasing of the system, $\text{Im}(G_{\perp 1}(n\omega_0)) = 0$, $\text{Re}(G_{\perp 1}(n\omega_0)) = -G_0$, one obtains the effective cooling rate equal to

$$\begin{aligned}\lambda_{\perp \text{eff}}(x) &= \frac{2}{T_0}\sum_{n=n_{\min}}^{n_{\max}}\left(G_0 \cos(n\eta x) - G_0{}^2\frac{Nf(x)}{2n|\eta|}\right) \\ &= G_0\frac{2(n_{\max}-n_{\min})}{T_0}F_{\perp}(x) - G_0{}^2\frac{N}{T_0|\eta|}\ln\left(\frac{n_{\max}}{n_{\min}}\right)f(x).\end{aligned} \tag{7.69}$$

Here the factor of 2 in front of the sum accounts that both positive and negative frequencies contribute to the sum, $n_{\max} = T_0 f_{\max}$, $n_{\min} = T_0 f_{\min}$ are the harmonic numbers at the boundaries of rectangular cooling band, and the summations were performed with help of integration. This is well justified for a high-frequency cooling system $n_{\max}$, $n_{\min} \gg 1$, and $\eta_2 x \ll 1$. The form factor

$$F_{\perp}(x) = \frac{\sin(xn_{\max}\eta_2 T_2/T_0) - \sin(xn_{\max}\eta_2 T_2/T_0)}{x(n_{\max}-n_{\max})\eta_2 T_2/T_0}, \tag{7.70}$$

describes the cooling force reduction for a particle with momentum offset. It represents the so-called "bad mixing" which grows with an increase of partial slip factor between pickup and kicker, η_2, and reduces the cooling force for particles with large amplitudes. For correctly designed system, the "bad mixing" is sufficiently small for particles in the core; consequently, one can put that $F_\perp(x)=1$. Assuming that the average rms emittance, $\overline{I(x,t)} \equiv \overline{I}$, does not depend on longitudinal momentum and performing averaging over momentum, one obtains the average damping rate for cooling:

$$\lambda_{\perp a} = 2(f_{\max} - f_{\min})G_0 - \frac{N{G_0}^2}{2\sqrt{\pi}T_0\sigma_p|\eta|}\ln\left(\frac{f_{\max}}{f_{\min}}\right), \quad \sigma_p n_{\max}\eta_2 T_2/T_0 \ll 1. \quad (7.71)$$

The first term is related to the cooling. It is proportional to the gain G_0. The second addend is related to heating coming from the Schottky noise of other particles. It grows proportionally to ${G_0}^2$. The maximum damping rate is achieved at the optimal gain

$$G_{\mathrm{opt}\perp} = \frac{2\sqrt{\pi}\sigma_p|\eta|T_0(f_{\max} - f_{\min})}{N\ln(f_{\max}/f_{\min})}, \quad (7.72)$$

and is equal to

$$\lambda_{\mathrm{opt}\perp} = \frac{2\sqrt{\pi}\sigma_p|\eta|T_0(f_{\max} - f_{\min})^2}{N\ln(f_{\max}/f_{\min})}. \quad (7.73)$$

As one can see from Eq. (7.69), the beam core and the distribution edges in the momentum space are cooled with different cooling rates, and therefore, the transverse particle distribution does not stay Gaussian in the course of cooling, i.e., the above estimate is not self-consistent. Note also that Eqs. (7.71)–(7.73) are only applicable if the Schottky noise of particle is much larger than the thermal noise of electronics, i.e., the beam emittance is much larger than the equilibrium emittance determined by thermal noise only. In the case of high intensity beam, the IBS usually makes larger contribution to the diffusion than the thermal noise. The damping rate of Eq. (7.73) grows proportionally to the slip factor, η; however, this growth reduces when the Schottky bands start to overlap.[4] There are two peaks in one revolution band corresponding to the positive and negative betatron Schottky bands, $\omega_{n\pm} = \omega_0(n \pm \nu)$. The minimum band overlap happens for the betatron tunes equal to $(1+2m)/4$. Assuming $\pm 2\sigma$ width for each band, one obtains that the band overlap happens at $8\sigma_p n_{\max}\eta = 1$. That yields that the maximum damping rate is

[4] The damping achieves its maximum at near-complete band overlap. If the common mode signals are well suppressed, the cooling rate stays at this maximum with further increase of slip factor and band overlap. Achieving sufficient common mode suppression is not always possible. Therefore, most cooling systems operate without band overlap.

$$\lambda_{\max\perp} \approx \frac{\sqrt{\pi} f_{\max}(1 - f_{\min}/f_{\max})^2}{4N \ln(f_{\max}/f_{\min})} \xrightarrow{f_{\max}/f_{\min}=2} \approx 0.16 \frac{f_{\max}}{N}. \quad (7.74)$$

For tunes in vicinity of second-order resonance, $Q \approx (1+n)/2$, and sufficiently wide Schottky bands, the bands for positive and negative frequencies completely overlap, resulting in twice larger diffusion and reduction of optimal damping rate by a factor of 2. It can be compensated by an increase of slip factor, η, by a factor of 2 which is determined by a boundary of transverse and longitudinal bands overlap. Finally, it yields the same cooling rate as for $Q = (1+2m)/4$ described by Eq. (7.74).

Similar to the above calculation, one can prove that in the case of complete band overlap but still small "bad mixing" at all harmonics, the optimum gain is $G_{opt} = (2N)^{-1}$, and the maximum damping rate[5] is $\lambda_{\max\perp} = (f_{\max} - f_{\min})/(2N)$. However, this limit was not achieved yet in practical cooling systems.

Now let's consider the effect of signal suppression by the beam interaction through cooling system. Substituting the optimal gain of Eq. (7.72) into Eq. (7.38), we obtain

$$\begin{aligned} \varepsilon_{\perp n}(x) &= 1 + \frac{n_{\max} - n_{\min}}{2\sqrt{2}\,\ln(n_{\max}/n_{\min})n} \frac{1}{i\pi} \int_{\delta \to 0_+} \frac{e^{-y^2/(2\sigma_p{}^2)}\,dy}{x - y - i\delta} \\ &= 1 + \frac{n_{\max} - n_{\min}}{2\sqrt{2}\,\ln(n_{\max}/n_{\min})n} \Phi_D\left(\frac{x}{\sqrt{2}\sigma_p}\right), \quad n > 0, \end{aligned} \quad (7.75)$$

where $x = \delta\omega / |\eta n \omega_0|$ is the momentum offset corresponding to a frequency offset $\delta\omega$ from nth betatron sideband, $\omega_0(n \pm \nu)$, and the integral was expressed through the imaginary error function

$$\mathrm{erfi}(x) = \frac{2}{\sqrt{\pi}} \int_0^x e^{x^2} dx, \quad (7.76)$$

so that

$$\Phi_D(x) = \frac{1}{i\pi} \int_{\delta \to 0_+} \frac{e^{-y^2} dy}{x - y - i\delta} = e^{-x^2}(1 - i\,\mathrm{erfi}(x)). \quad (7.77)$$

A plot of function $\Phi_D(x)$ is presented in Fig. 7.25. Figure 7.26 presents $\varepsilon_{\perp n}(x)$ computed for optimal gain and a Schottky band at the low boundary of cooling band where the effect of particle interaction achieves its maximum. The corresponding stability diagram is presented in Fig. 7.27. Note that for perfectly phased system,

[5] Note that the damping rate definition has been introduced for damping beam emittance. The damping rate for the amplitude is twice smaller. Consequently, for the case of complete band overlap, the amplitude damping rate is $\lambda \approx (f_{\max} - f_{\min})/(4N)$.

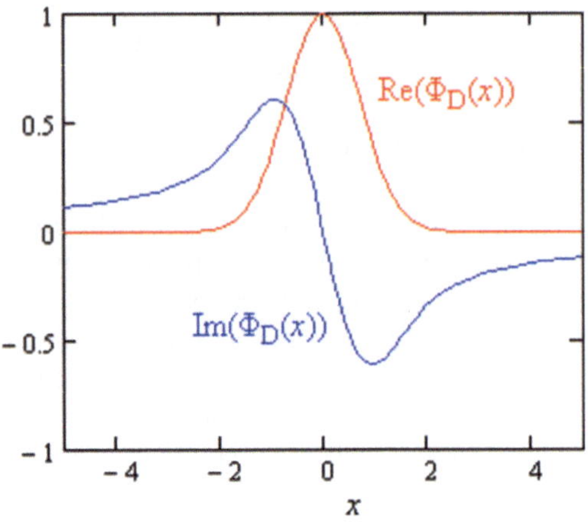

Fig. 7.25 Real and imaginary parts of function $\Phi_D(x)$

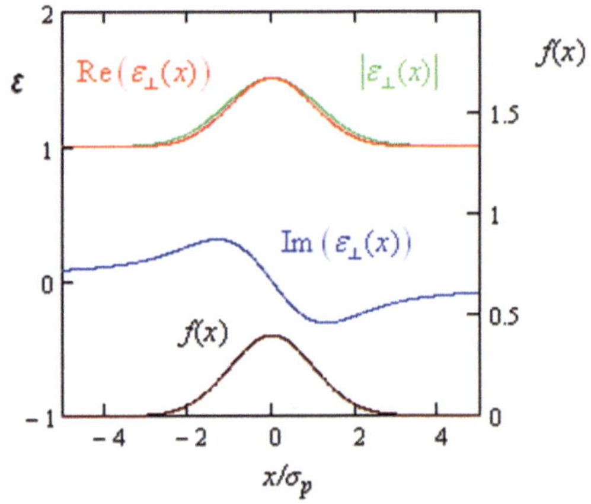

Fig. 7.26 Beam dielectric permeability computed with help of Eq. (2.72) at the low boundary of one octave frequency band, $f_{max}/f_{min} = 2$, and corresponding distribution function (*bottom curve*)

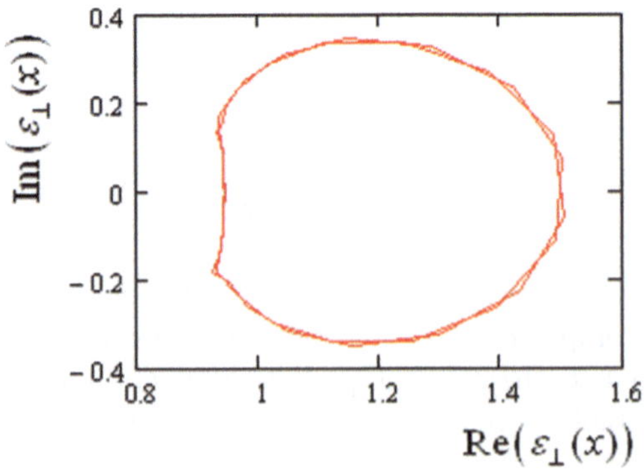

Fig. 7.27 The stability diagram for parameters of Fig. 7.26

the curve does not embrace the coordinate frame origin with gain increase and the system is stable as long as a single-particle gain is below 1, $G_1 < 1$. Note also that the imaginary part of $\varepsilon_{\perp n}(x)$ is the same for both lower and upper betatron sidebands; however, it changes its sign with change of slip factor. It was omitted in the above estimate to simplify it. If required the sign of imaginary part can be restored using Eq. (7.36).

The maximum of $\varepsilon_{\perp n}(x)$ is achieved at zero detuning, $x = 0$, where its imaginary part is equal to zero. For one octave bandwidth, $\varepsilon_{\perp n}(0)$ is equal to ≈ 1.51 at the low boundary of the band and decreases to ≈ 1.25 at the high boundary. In average it yields reduction of single-particle damping rate of ~1.4 times and reduction of diffusion of ~2 times. The reduction in the damping rate is mostly compensated by reduction in the diffusion. Actually, performing transformations similar to the one carried out above (see Eqs. (7.68)–(7.73)), one obtains the maximum damping rate to be

$$\lambda_{\perp \mathrm{opt}} \approx \frac{|\eta|}{2NT_0} \left\langle \left(\sum_{n=-\infty}^{\infty} \mathrm{Re}\left(\frac{G_{\perp 1}(n\omega_0)}{\varepsilon_{\perp n}(x)} e^{in\eta_2 x} \right) \right)^2 \Bigg/ \left(f(x) \sum_{n=-\infty}^{\infty} \frac{1}{|n|} \left| \frac{G_{\perp 1}(n\omega_0)}{\varepsilon_{\perp n}(x)} \right|^2 \right) \right\rangle_x . \tag{7.78}$$

As one can see, nominator and denominator are proportional to $(\mathrm{Re}\varepsilon_{\perp n})^{-2}$ and $|\varepsilon_{\perp n}|^{-2}$, correspondingly. It significantly reduces the effect of the signal suppression on the damping rate for a system with well-corrected gain.

In practical applications, the effect of signal suppression is used to set the optimal gain for cooling system. Equation (7.75) yields that at the optimal gain the signal suppression in the center of one octave band is ≈ 1.97 dB. Accurate accounting of signal suppression increases this value to ~3 dB. Note also that in vicinity of the optimal gain, the signal suppression reduces the dependence of average damping rate on the gain in comparison with the described above model which neglects the signal suppression.

As one can see from Eq. (7.75) for a system with constant gain across the band, the signal suppression is inversely proportional to the frequency and achieves its maximum at the low boundary of the band. Similarly the low-frequency contribution makes dominant contribution to the diffusion. Making the gain linearly growing with frequency,

$$G(\omega) = \begin{cases} G_0 T_0 \omega / 2\pi n_{\min} \;, & \omega/2\pi \in [f_{\min}, f_{\max}], \\ 0 \;, & \text{otherwise} \;, \end{cases} \tag{7.79}$$

results in that both the signal suppression and the ratio of cooling to diffusion stay constant across the band. Consequently, it maximizes the system performance for a system with given total bandwidth. Assigning $\varepsilon_{\perp n}(x) = 1$ and Gaussian distribution, one obtains from Eq. (7.78)

$$\lambda_{\perp\mathrm{opt}} \approx \frac{2\sqrt{\pi}|\eta|\sigma_p}{NT_0}\left(\sum_{n=n_{\min}}^{n_{\max}} \mathrm{Re}(G_{\perp 1}(n\omega_0))\right)^2 \Bigg/ \sum_{n=n_{\min}}^{n_{\max}} |G_{\perp 1}(n\omega_0)|^2/n. \tag{7.80}$$

Performing summation for a system with constant gain, one obtains the result of Eq. (7.73). A summation for the case of linearly growing gain of Eq. (7.79) yields

$$\lambda_{\perp\mathrm{opt}} \approx \frac{\sqrt{\pi}T_0|\eta|\sigma_p}{N}\left({f_{\max}}^2 - {f_{\min}}^2\right). \tag{7.81}$$

For a system with one octave band ($f_{\max}/f_{\min} = 2$), the linearly growing gain yields 4 % faster cooling than for the constant gain case. Such gain shaping was used to optimize performance of Accumulator and Recycler cooling systems in the course of Tevatron Run II.

Longitudinal Palmer Cooling

Following the approach presented above for transverse cooling, below we consider estimates for longitudinal cooling. We start from the case of longitudinal Palmer cooling where the gain can be factored so that $G_1(x, \omega_n(x)) = -G'(n\omega_0)x$ and $A(\omega) = 0$. Consequently, the cooling force and the diffusion described by Eqs. (7.45) and (7.53) can be rewritten in the following form:

$$\begin{aligned} F(x) &= -\frac{2x}{T_0}\sum_{n=0}^{\infty}\mathrm{Re}\left(G'(n\omega_0)e^{2\pi i n\eta_2 x}\right) \xrightarrow{G'(n\omega_0)=G' \text{ for } n\in[n_{\min},n_{\max}]} -\frac{2G'x}{T_0}(n_{\max} - n_{\min})\mathfrak{I}(x), \\ D(x) &= \frac{2Nf(x)x^2}{T_0|\eta|}\sum_{n=0}^{\infty}\frac{|G'(n\omega_0)|^2}{n} \xrightarrow{G'(n\omega_0)=G' \text{ for } n\in[n_{\min},n_{\max}]} \frac{2NG'^2 f(x)x^2}{T_0|\eta|}\ln\left(\frac{n_{\max}}{n_{\min}}\right). \end{aligned} \tag{7.82}$$

Here similar to the considered above case of transverse cooling, we assume that $\varepsilon(\omega) = 1$; the slip factor, $\eta(x)$, is constant ($\eta(x) = \eta$); and the thermal noise is negligible in comparison with the particle noise. The form factor $\mathfrak{I}(x)$ is the same as for the transverse cooling and is presented by Eq. (7.70). For the ideal rectangular band, the summing is straightforward and its result is presented in the right-hand side of Eqs. (7.82). To find the evolution of rms momentum spread, we multiply Eq. (7.39) by x^2 and integrate it over momentum deviation. Assuming that the particle distribution is Gaussian ($\overline{x^2} = {\sigma_p}^2$) and the "bad mixing" is sufficiently small, i.e., $\mathfrak{I}(x)=1$, one obtains

$$\frac{d\overline{x^2}}{dt} = -F'\overline{x^2} + D'\sqrt{\overline{x^2}}, \tag{7.83}$$

where

$$\begin{aligned} F' &= \frac{4G'}{T_0}(n_{\max} - n_{\min}), \\ D' &= \frac{3NG'^2}{4\sqrt{\pi}T_0|\eta|}\ln\left(\frac{n_{\max}}{n_{\min}}\right). \end{aligned} \tag{7.84}$$

That yields the initial damping rate, $\lambda_{\parallel} \equiv d\overline{x^2}/dt = F' - D'/\sigma_p$, and the equilibrium momentum spread, $\sigma_p = D'/F'$. Following the same procedure as for the transverse cooling, we obtain the optimal gain and the maximum damping rate:

$$\begin{aligned} G'_{\text{opt}} &= \frac{8\sqrt{\pi}\sigma_p|\eta|(n_{\max} - n_{\min})}{3N\ln(n_{\max}/n_{\min})}, \\ \lambda_{\text{opt}_{\parallel}} &= \frac{16\sqrt{\pi}\sigma_p|\eta|(n_{\max} - n_{\min})^2}{3NT_0\ln(n_{\max}/n_{\min})}. \end{aligned} \tag{7.85}$$

Comparing the above damping rate with Eq. (7.73) describing the damping rate for the transverse cooling, one can see that the only difference is in the numerical coefficients. The diffusion in the center of the distribution (small x values) is suppressed in the longitudinal cooling, resulting in the longitudinal damping being faster by 8/3 times.

The signal suppression due to beam self-interaction through the cooling system can be found from Eq. (7.22). Assuming Gaussian distribution and substituting the optimal gain of Eq. (7.85), one obtains

$$\begin{aligned} \varepsilon_{\parallel n}(x) &= 1 + \frac{2\sqrt{2}}{3\pi i {\sigma_p}^2 n}\frac{n_{\max} - n_{\min}}{\ln(n_{\max}/n_{\min})}\int\limits_{\delta\to 0_+}\frac{y^2 e^{-y^2/(2{\sigma_p}^2)}dy}{y - x - i\delta} \\ &= 1 + \frac{4\sqrt{2}}{3n}\frac{n_{\max} - n_{\min}}{\ln(n_{\max}/n_{\min})}\Phi_P\left(\frac{x}{\sqrt{2}\sigma_p}\right), \quad n > 0. \end{aligned} \tag{7.86}$$

Here $x = \delta\omega/|\eta n\omega_0|$ is the momentum offset corresponding to a frequency offset $\delta\omega$ from nth revolution frequency harmonic, and

$$\Phi_P(x) = \frac{1}{i\pi}\int\limits_{\delta\to 0_+}\frac{y^2 e^{-y^2}dy}{y - x - i\delta} = x^2 e^{-x^2}(1 + i\,\text{erfi}(x)) + \frac{x}{i\sqrt{\pi}}. \tag{7.87}$$

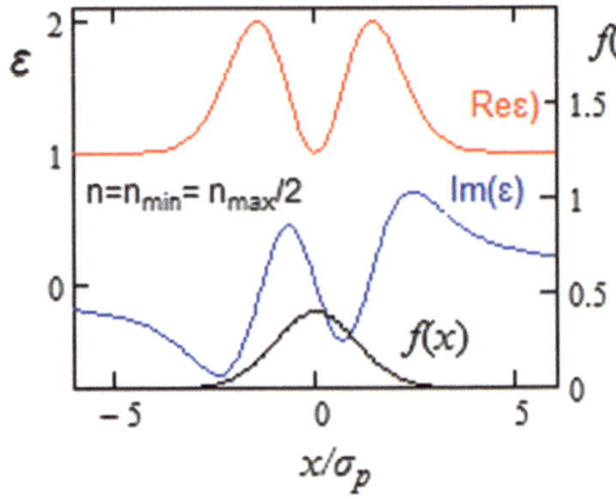

Fig. 7.28 Longitudinal beam dielectric function computed with help of Eq. (7.86) at the low boundary of one octave frequency band, $f_{max}/f_{min}=2$, and the corresponding longitudinal distribution function (*bottom curve*)

Figure 7.28 presents $\varepsilon_{\|n}(x)$ computed for the optimal gain and the Schottky band at the low boundary of cooling band where the effect of particle interaction achieves its maximum. In the Palmer cooling, the gain at zero momentum offset is equal to zero by design. Therefore, in difference to the transverse cooling, the signal suppression in the Palmer cooling is absent at zero momentum offset (compare Figs. 7.26 and 7.28). Similar to the case of transverse cooling, the imaginary part of $\varepsilon_{\|n}(x)$ changes its sign with slip factor sign change.

Longitudinal Filter Cooling

The longitudinal filter cooling requires the gain being pure imaginary, $G_1(x,\omega_n(x))=-iG(n\omega_0)$. Assuming $\varepsilon(\omega)=1$, constant slip factor $(\eta(x)=\eta)$, a rectangular band $(G_1(x,\omega_n(x))=-iG_0,\ n\in[n_{\min},n_{\max}])$, and perfect notch filter $(A(\omega)=1)$ and using Eq. (7.45), one obtains the cooling force

$$F(x)=\frac{2G_0}{T_0}\sum_{n=n_{\min}}^{n_{\max}}\mathrm{Im}\left(\left(1-e^{2\pi in\eta x}\right)e^{2\pi in\eta_2 x}\right)=-\frac{2G_0}{T_0}\Im_F(x). \tag{7.88}$$

For large enough harmonic numbers, the summation can be replaced by integration. That yields the following cooling force form factor:

$$\Im_F(x)=\frac{\cos\left(2\pi n_{\max}\eta_2 x\right)-\cos\left(2\pi n_{\min}\eta_2 x\right)}{2\pi\eta_2 x}-\frac{\cos\left(2\pi n_{\max}(\eta+\eta_2)x\right)-\cos\left(2\pi n_{\min}(\eta+\eta_2)x\right)}{2\pi(\eta+\eta_2)x}. \tag{7.89}$$

In difference to the Palmer cooling where the cooling range is determined by partial slip factor, the cooling range is now determined by the sum of the ring slip factor and partial slip factor. For $|\eta_2|<<|\eta|$ it can be roughly estimated by the equation $x_{\max}\approx 1/((\eta+1.3\eta_2)(n_{\max}+n_{\min}))$ describing the location of the first zero of Eq. (7.89). Consequently, the filter cooling cannot be used in a regime with band overlap.

Expending Eq. (7.88) in Tailor series over x and leaving only linear terms, one obtains

$$F(x) = -\frac{4\pi n\eta G_0}{T_0}x\sum_{n=n_{\min}}^{n_{\max}} n = -\frac{2\pi n\eta G_0}{T_0}\left({n_{\max}}^2 - {n_{\min}}^2\right)x. \tag{7.90}$$

As one can see, the gain at a given harmonic is proportional to the harmonic number. It was already discussed above that such dependence of gain on frequency is optimal for maximizing damping rate in a fixed frequency band. Considering the filter cooling at small amplitudes (i.e., in the linear part of the cooling force), assuming the Gaussian distribution, and following the procedure described in the above section, one obtains the optimal gain and the maximum cooling rate:

$$\begin{aligned} G_{0\text{opt}} &= \frac{4\sigma_p}{3\sqrt{\pi}N}, \\ \lambda_{\text{opt}F} &= \frac{8\sqrt{\pi}\sigma_p|\eta|}{3NT_0}(n_{\max} - n_{\min})^2. \end{aligned} \tag{7.91}$$

Using Eq. (7.21) one obtains the beam dielectric function at optimal gain for Gaussian distribution with sufficiently small momentum spread, $\sigma_p \ll 1/(\eta n_{\max})$:

$$\varepsilon_{\parallel}(x, n) = 1 + \frac{4\sqrt{2}}{3}\,\Phi_{\text{P}}\left(\frac{x}{\sqrt{2}\sigma_p}\right). \tag{7.92}$$

As one can see, the beam dielectric function in the filter cooling does not depend on the harmonic number, n. It follows from the independence of system gain on the frequency for the rectangular band, $G_1(x, \omega) = -\,iG_0$, which results in the effective gain being proportional to the harmonic number and independence of ε on n.

The independence of the signal suppression on the harmonic number allows one to make a simple estimate of the signal suppression effect on the damping rate. Similar to Eq. (7.78), one can obtain the damping rate for the filter cooling with the rectangular band and the signal suppression taken into account:

$$\begin{aligned} \lambda_{\text{opt}F} &= \frac{8\sqrt{\pi}\sigma_p|\eta|}{3NT_0}(n_{\max} - n_{\min})^2 F_\varepsilon, \\ F_\varepsilon &= \left(\frac{1}{{\sigma_p}^2}\int_{-\infty}^{\infty}\text{Re}\left(\frac{1}{\varepsilon_{\parallel}}\right)x^2 f(x)dx\right)^2 \Bigg/ \left(\frac{8\sqrt{\pi}}{3\sigma_p}\int_{-\infty}^{\infty}\left|\frac{x^3}{{\varepsilon_{\parallel}}^2}\right| f(x)\frac{df(x)}{dx}dx\right), \end{aligned} \tag{7.93}$$

which coincides with Eq. (7.91) for $\varepsilon_{\|} = 1$ (consequently $F_\varepsilon = 1$). Strictly speaking ε depends on the gain, G_0, which optimal value needs to be separately determined. For an estimate, we assume that the optimal gain is equal to the gain in Eq. (7.91). Computation of integrals for the Gaussian distribution with $\varepsilon_{\|}$ of Eq. (7.92) yields $F_\varepsilon \approx 1.037$, which justifies our negligence of ε in the above estimates and the assumption that accounting of signal suppression weakly affects the optimal gain.

Stacktail Cooling System

To maximize the antiproton accumulation rate, the stacktail cooling system was suggested by Simon van der Meer [8]. The idea is based on the exponential dependence of cooling force on the particle momentum. Newly injected antiprotons are coming first to the deposition orbit where the gain achieves its maximum. The resultant large cooling force allows fast clearing of the deposition orbit so that it would be prepared for next injection. The exponential reduction of the gain in the course of particle displacement from the deposition orbit to the core orbit yields an exponential increase of particle density, allowing an accumulation of large number of particles in a finite momentum acceptance of Accumulator ring.

Below we consider a simplified model representing an ideal stacktail system where the gain is parameterized as follows: $G(x,\omega) = G_\omega(\omega)\mathrm{e}^{-x/x_d}$. Then, Eqs. (7.39), (7.45), and (7.53) can be rewritten in the following form:

$$
\begin{aligned}
&\frac{\partial f(x,t)}{\partial t} + \frac{\partial}{\partial x}(J(x,t)) = 0, \\
&J(x,t) = F(x) f(x,t) - \frac{D(x,t)}{2}\frac{\partial f(x,t)}{\partial x}, \\
&F(x) = \frac{2}{T_0}\mathrm{e}^{-x/x_d}\sum_{n=0}^{\infty}\mathrm{Re}(G_1(n\omega_0)), \\
&D(x,t) = \frac{2}{T_0}\frac{f(x,t)}{|\eta|}\mathrm{e}^{-2x/x_d}\sum_{n=0}^{\infty}\frac{1}{n}|G_1(n\omega_0)|^2.
\end{aligned}
\tag{7.94}
$$

Here we renormalized the distribution function so that the particle number enclosed in the momentum range $[x_1, x_2]$ is equal to $\int_{x_1}^{x_2} f(x)dx$; similar to the above, we assume $\varepsilon = 1$; and we wrote the top equation in the form of the continuity equation. Actually, integrating it from x_1 to x_2, one obtains that the number of particles in this region is changed as the difference of ingoing and outgoing fluxes:

$$\frac{dN}{dt} = \int_{x_1}^{x_2} \frac{\partial}{\partial x}(J(x,t))dx = J(x_2,t) - J(x_1,t)$$

i.e., $J(x,t)$ represents the particle flux through momentum x.

Van der Meer solution, $f(x,t) = f_0 \exp(x/x_d)$, is a stationary solution of Eq. (7.94). The corresponding particle flux does not depend on x and is equal to

$$J = f_0 \frac{2}{T_0}\left(\sum_{n=0}^{\infty} \mathrm{Re}(G_1(n\omega_0)) - \frac{f_0}{2x_d|\eta|}\sum_{n=0}^{\infty}\frac{1}{n}|G_1(n\omega_0)|^2\right) \tag{7.95}$$

The flux achieves its maximum

$$J_{\max} = \frac{|\eta|x_d}{T_0}\left(\sum_{n=0}^{\infty}\mathrm{Re}(G_1(n\omega_0))\right)^2 \Big/ \sum_{n=0}^{\infty}\frac{1}{n}|G_1(n\omega_0)|^2, \tag{7.96}$$

for

$$f_0 = |\eta|x_d \sum_{n=0}^{\infty}\mathrm{Re}(G_1(n\omega_0)) \Big/ \sum_{n=0}^{\infty}\frac{1}{n}|G_1(n\omega_0)|^2. \tag{7.97}$$

Replacing summing by integration, we finally obtain

$$J_{\max} = |\eta|x_d T_0 W^2, \tag{7.98}$$

where W is the effective bandwidth of the system

$$W = \sqrt{\left(\frac{1}{2\pi}\int_0^{\infty}\mathrm{Re}(G_1(\omega))d\omega\right)^2 \Big/ \left(\int_0^{\infty}|G_1(\omega)|^2\frac{d\omega}{\omega}\right)}. \tag{7.99}$$

For the ideal system with rectangular band, one obtains

$$W = \frac{f_{\max} - f_{\min}}{\sqrt{\ln(f_{\max}/f_{\min})}}. \tag{7.100}$$

For the ideal system with linearly growing gain, one obtains

$$W = \sqrt{\frac{{f_{\max}}^2 - {f_{\min}}^2}{2}} \tag{7.101}$$

Comparing Eq. (7.98) and (7.99) with Eq. (7.80) and (7.85), one can see that the effective bandwidth of Eq. (7.99) characterizes both the damping rates of transverse and longitudinal Palmer cooling as well as the maximum flux of the stacktail system.

The theoretical results considered above have been extremely helpful in understanding and upgrading stochastic cooling systems during the course of Collider Run II. It has been particularly useful for upgrades of the Antiproton Source stacktail system which after the upgrades operates with nearly overlapped Schottky bands. It was achieved by a 15 % increase in the slip factor of the Antiproton Accumulator and an increase in gain at the high-frequency end of the system bandwidth. As long as the bands do not overlap, an accurate accounting of their interaction does not change the cooling rates significantly from the standard formulation. However, the effects of band overlap are important for interpreting stochastic cooling measurements. These results have been especially useful in the separation of transverse and longitudinal kicks from the stacktail system kickers. By design, the kickers should only make longitudinal kicks, but unfortunately also kick the beam transversely causing transverse emittance growth and, consequently, a lower antiproton stacking rate.

7.2.2 *Stochastic Cooling: Technology and Systems*

Stochastic cooling system technology at Fermilab has expanded considerably on the initial systems developed at CERN [8]. The cooling systems are utilized for increasing phase space density of 8 GeV antiprotons in three Fermilab antiproton synchrotrons: Accumulator, Debuncher, and Recycler. A total of 25 independent systems are implemented, 21 in the Antiproton Source and 4 in the Recycler; see Table 7.3. Figure 7.29 shows a typical schematic for a cooling system.

In the early commissioning stages, Debuncher antiproton yields amounted to a few 10^7 particles per Main Ring cycle. Near the end of Run II, some 2×10^8 antiprotons or approximately 20 μA of Debuncher beam current were produced on each 2.2 s Main Injector cycle. Total signal power at the pickup is typically in the picowatt range. In all but Accumulator core and Recycler cooling systems, cryogenically cooled pickups are required to achieve a positive signal-to-noise ratio. Stochastic cooling times range from about 2 s in the Debuncher to tens of minutes in the Accumulator core and Recycler systems. RF power levels to achieve these cooling times are kilowatts in the Debuncher and stacktail momentum cooling to tens of watts in other systems. A typical system gain can exceed 150 dB and must be carefully executed with good RF shielding for stability. The following sections will describe the technology choices for each segment of the cooling systems.

Pickups and Kickers

The pickup arrays are the devices that transform beam current to microwave signals that will be fed back at the appropriate gain and phase to the kickers, typically

Table 7.3 Locations and types of stochastic cooling systems

	Stochastic cooling systems at Fermilab 2011				
Machine	# systems	Frequency (GHz)	Type	Cooling time (s)	Max installed power watts
Debuncher	4	4–8	Momentum	2	6,400
	4	4–8	Horizontal	2	3,200
	4	4–8	Vertical	2	3,200
Accumulator	1	2–4	Stacktail momentum	1,200	6,400
	3	4–8	Core horizontal	1,200	15
	3	4–8	Core vertical	1,200	15
	1	2–4	Core momentum	1,200	400
	1	4–8	Core momentum	1,200	400
Recycler	1	0.5–1	Momentum	1,800	200
	1	1–2	Momentum	1,800	400
	1	2–4	Horizontal	1,800	200
	1	2–4	Vertical	1,800	200

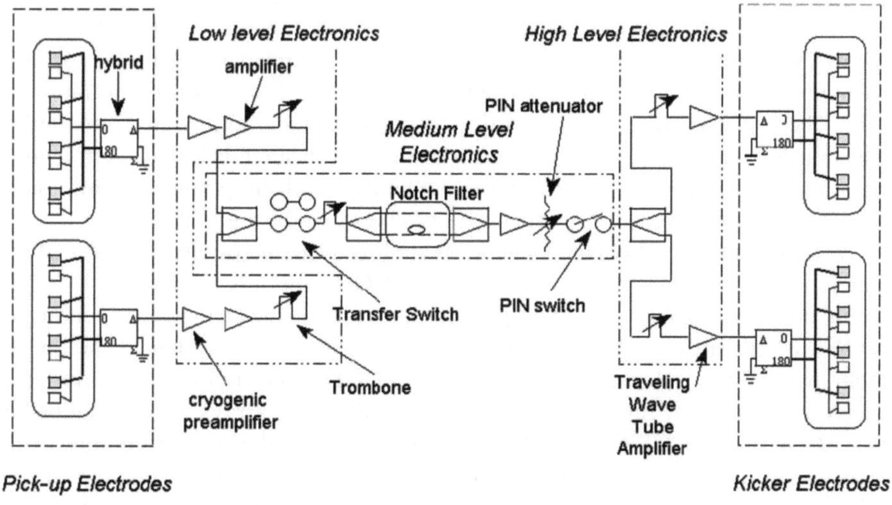

Fig. 7.29 Typical stochastic cooling block diagram

located downstream across the synchrotron. Stochastic cooling requires prompt signal feedback with no additional delay. Cutting a cord across the rings with the appropriate transmission medium is required to obtain appropriate insertion delays of the electronics. The pickup/kickers must be designed to provide the best possible transfer impedance from the beam while maintaining the desired bandwidth.

The first devices used in the Antiproton Source were three-dimensional stripline-type devices that were developed at Lawrence Berkeley National Lab (LBNL) in the early 1980s [13, 14]. Signals from each individual quarter wave loop are combined via a suspended stripline combiner. This combiner structure compensates for the

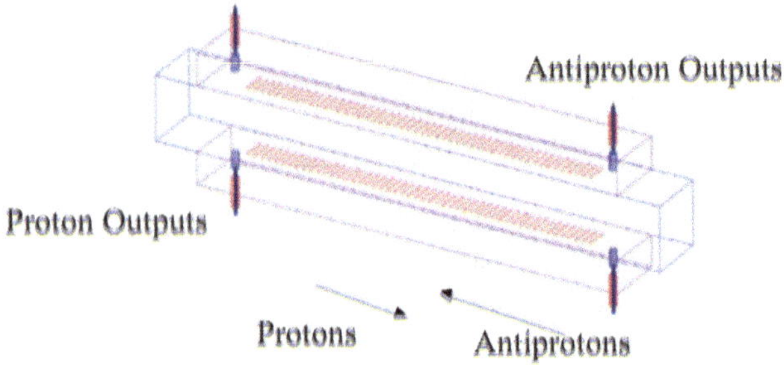

Fig. 7.30 Conceptual drawing of slotted waveguide array

propagation delays between loops with a precision of plus/minus two picoseconds. This same structure is used in the kickers to distribute the power to individual kicker loops. While these arrays proved worthy as the initial antennas, they were difficult and costly to manufacture. They also suffered from mechanical failures where a kicker loop antenna on several occasions would become unsoldered due to heat dissipation and fall into the beam aperture. The first stacktail system operated at 1–2 GHz. In 1999 an upgrade was implemented to increase the stacking rate necessitating an increase in bandwidth to 2–4 GHz. These frequencies were chosen based on available octave bandwidth catalog microwave components and traveling-wave tube (TWT) power amplifiers. New arrays were designed using microstrip and stripline techniques that proved to be more robust and less expensive to fabricate [15]. These planar structures take full advantage of integrating signal, combining and splitting on the same circuit board where the loops are fabricated. Termination resistors are the only component not an integral part of the printed circuit. Planar loops have been designed in frequency bands between 0.5 and 8 GHz.

A significant upgrade to the Debuncher cooling systems was completed in 2000. The initial Debuncher cooling operated at 2–4 GHz and this upgrade to 4–8 GHz would double the bandwidth. The signal-to-noise ratio in the Debuncher operating at 80 K was still sufficiently small that cooling was limited by system noise for the latter part of the 2.4-s cooling cycle. At 4–8 GHz, planar loops have reduced transverse sensitivity due to their physical size. Plunging arrays were developed at the ACOL ring at CERN to follow the beam profile during the cooling cycle [16]. Repetitive mechanical motion in a cryogenic environment was deemed a reliability risk for the Fermilab upgrade. A new technology of slotted waveguide pickups was developed which employed sandwiched waveguides with coupling slots [17]. Figure 7.30 depicts a typical array. The slotted waveguide technique is also used with three bands in the core transverse 4–8 GHz cooling systems. Schottky detector systems in the Tevatron, Recycler, and CERN LHC also use this pickup at 1.7 GHz and 4.8 GHz [12]. The sensitivity of slotted waveguide pickups is considerably higher than planar or stripline units while maintaining acceptable apertures (Fig. 7.31). The trade-off for higher sensitivity is reduced bandwidth.

Additional complications include the fact that most of the pickups operate at cryogenic temperatures and the kickers need to dissipate kilowatts of power while maintaining excellent beam vacuum. Considerable effort was expended on materials used inside the vacuum vessels. The planar loops that were developed for

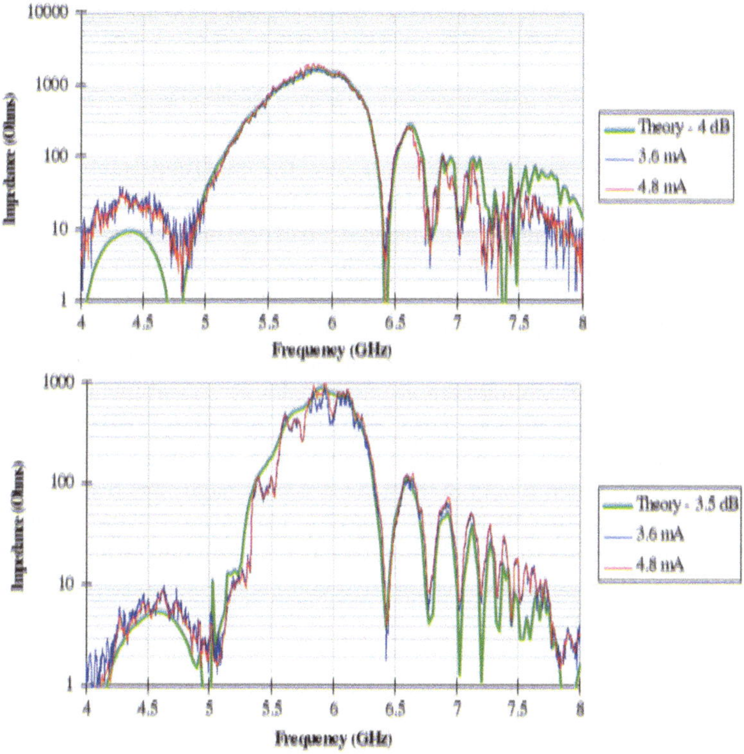

Fig. 7.31 Beam response for slotted waveguide pickup. *Top* difference mode, *bottom* sum mode. Comparison between theory and actual beam measurements in the Debuncher

stochastic cooling arrays are fabricated on Teflon circuit boards with substantial surface area. At CERN, the vacuum group has outlawed the extensive use of Teflon boards within the vacuum system, insisting on ceramics boards. While ceramics such as aluminum dioxide have excellent vacuum and microwave properties, they add considerable expense, mechanical fragility, and slower propagation velocities that are unattractive for cooling purposes. This was overcome at Fermilab with extensive use of sublimation pumping in the Accumulator/Recycler and meticulous cleanliness handling techniques. Stainless components and vacuum compatible materials such as Vespel are used where appropriate. The vacuum vessels are baked at 125 °C in situ for up to 4 days to improve degassing. Vacuum levels approaching 10^{-10} Torr are routine with beam lifetimes exceeding 500 h in both the Accumulator and Recycler.

Preamplifiers

The Debuncher ring has the most stringent cooling requirements with only 2.2 s to cool antiprotons in six-dimensional phase space. Signal-to-noise ratio in stochastic cooling systems is the critical parameter to achieving good cooling performance. Antiproton production yields are of order 20 ppm of incident protons on target. The net result is a circulating Debuncher beam current measured at 20 μA. There are two variables available to minimize the impact of low-level signals: temperature of the pickup array and noise performance of the first preamplifier.

The first operating Debuncher cooling systems consisted solely of transverse cooling with a bandwidth of 2–4 GHz. In an effort to minimize cost, the pickup temperature was chosen to be 80 K; based on liquid nitrogen refrigeration. The front-end noise performance is the sum of the equivalent noise temperatures (T_e) of the pickup array and the preamplifier. While the T_e of the array is the actual temperature of the termination resistor, the T_e of the amplifier is not the ambient temperature [18]. It is possible for the T_e of an amplifier to be below ambient. For the original Debuncher cooling systems, cryogenically cooled low noise amplifiers (LNA) were developed by Lawrence Berkeley Laboratory [19]. These discrete device GAsFET amplifiers achieved an equivalent noise temperature of 50 Kelvin (K) at an ambient temperature of 80 K in the 2–4 GHz band. The resulting front-end effective temperature of 130 K was achieved. The design of the amplifiers was based on the pioneering efforts of Weinreb et al. [20] utilizing custom bias for each the three stages of the amplifier. This added complexity a necessity to achieve the best noise performance.

A custom bias regulator was designed and located within a few feet of the amplifier. At cryogenic temperatures, silicon-based devices are not functional and must be outside the cryo-amplifier module.

Until the 1990s, cryogenically cooled microwave amplifiers were the sole domains of radio astronomy and accelerator physics applications. With the wide spread growth of cellular telephony, the use of cryogenic amplifiers with low noise performance could be employed by communications companies to increase the range of cellular repeater towers. Commercially available cryogenic amplifiers appeared in the market place. While a number of vendors indicated they could build such LNAs for Fermilab's stochastic cooling systems, only one, Miteq [21], proved viable. The Debuncher cooling upgrade was solidly based on reducing the front-end noise temperature from the original 130 to 10–40 K (amplifier noise effective temperature is frequency band dependent) [22]. The largest improvement would be from an upgraded cryogenic system operating at liquid Helium 4.5 K. Operating the amplifiers at this temperature and minimizing any insertion loss between pickup arrays and preamplifier resulted in the lowest possible noise performance. Miteq built a complete line of LNAs guaranteed to work at liquid helium, even though Miteq did not have the capability to test them below 80 K. These amplifiers have performed as advertised for 11 years without failure at 4.5 K.

The Debuncher upgrade consisted of 32 pickup arrays operating from 4 to 8 GHz in eight bands. Upon delivery, the Miteq LNAs were immediately tested at room temperature and at 80 K. The specified noise temperatures were achieved. A full-blown test of the pickup with amplifiers was then benchmarked at 4.5 K. It was during this test that the expected performance of liquid helium amplifiers was verified. Amplifier noise temperatures ranging from 5 to 30 K were achieved across the frequency span. An unanticipated setback occurred with bias applied to the amplifiers during the warm-up from liquid helium to room temperature, causing an unstable operating point resulting in half of the tested amplifiers self-destructing. Due to this phenomenon, a redundant interlock system was developed. While the amplifiers were known to be stable below 80 K, an arbitrary trip temperature of 40 K was chosen. Exceeding this temperature would clearly indicate a problem with the liquid helium cooling.

Recursive Notch Filters

Recursive notch filters are a basic building block of stochastic cooling systems. Figure 7.32 shows the configuration for a correlator-type notch filter, which is the equivalent of an analog two-tap finite impulse response (FIR) filter. Such a filter will provide deep (30 dB) notches at harmonics of the beam revolution frequency by means of the delay time difference between the two taps. A fundamental feature of FIR filters is a linear phase shift of 360° per revolution harmonic.

Notch filters are utilized in many of the momentum and transverse cooling systems in both Antiproton Source and Recycler synchrotrons. The recursive requirement is based on machine revolution frequency, providing gain shaping for particles of a specific momentum for momentum cooling. In transverse cooling, notch filters with repetition frequencies at harmonics of half the revolution frequency provide suppression of undesirable common mode (longitudinal) signals as well as reducing integrated thermal noise between revolution lines. The double notch spacing is required to maintain equal phase between upper and lower transverse sidebands. Reliable operation of stochastic cooling systems requires very high stability of the notch frequency spacing. Systems are stabilized to several parts per million of the revolution frequency. Combination of the two taps must be with 180° phase intercept to obtain integer harmonics of the revolution frequency.

A wide variety of technologies have been exploited in notch filter development: superconducting coaxial delay lines, bulk acoustic wave (BAW) delays, and microwave to fiber optic delay links. Each has advantages and disadvantages that will be discussed next.

The very first notch filters implemented at the start of the Antiproton Source operations spanned a bandwidth of 1–2 GHz and were based on superconducting coaxial delay lines [23–25]. The Accumulator has a revolution frequency of 628 kHz, corresponding to a revolution period of 1.6 μs. Furokawa (Japan) manufactured the superconducting coax used. Diameter of the coax is 0.085 in. with a solid niobium center conductor, lead plated outer copper conductor, and solid Teflon dielectric. Advantages are extremely low insertion loss, dominated by the loss tangent of the dielectric. Skin losses are virtually nonexistent. Low loss and high bandwidth of a coaxial line also affords excellent phase linearity and extremely low dispersion. These characteristics yielded deep notches with high dynamic range. Disadvantages include the need for a liquid helium system (costly to maintain and operate) and sensitivity to Dewar liquid He levels and pressures.

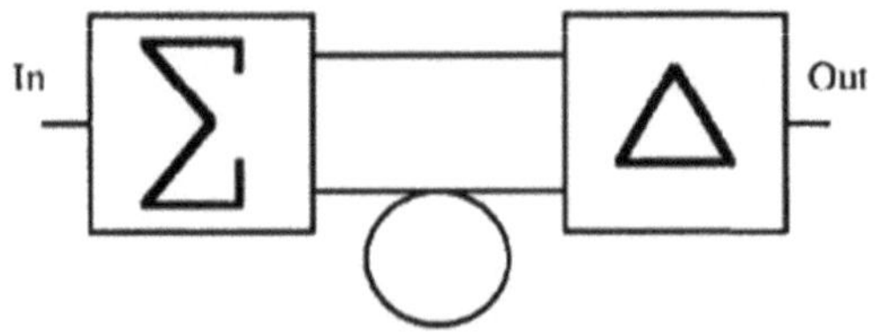

Fig. 7.32 Schematic block diagram of a two-tap correlator FIR notch filter. Wilkinson or microwave hybrid circuits realize the sum and difference function. This configuration provides notches at integer harmonics of the revolution frequency

The wide bandwidth of the coax allows for the injection of an out-of-band stable pilot reference frequency used for a phase-locked delay line feedback system keeping delays within specification. The superconducting notch filters were replaced by BAW filters after the first few years of operations to minimize operating cost and maintenance. To maximize stacktail bandwidth, the third stacktail filter reverted to superconducting delays in the last years of collider operations.

BAW devices were developed post-World War II for use as signal delays of microseconds required for radar systems. The BAW is a piezoelectric device that relies on the propagation of acoustic waves in the bulk of the crystal. Crystals can be fabricated from sapphire or quartz, with sapphire providing the best thermal stability coefficient. Resulting delays are of order 1 μs per centimeter of crystal. Electromagnetic waves are transformed to acoustic waves by means of antennas that are printed onto the piezo crystal. The impedance mismatch is significant, resulting in insertion losses of 30–40 dB. The mismatch also leads to a triple travel reflection which provides a reflected vector that has traveled through the delay three times and combines destructively with the incident delayed signal. Triple travel suppression is an important specification. Typically −20 dB suppression of triple travel wave results in acceptable performance of the notch filter. The high voltage-standing-wave ratio must be compensated by the use of circulators, which can restrict bandwidth. An integral amplitude and phase equalizer is also required to achieve the specified gain and phase flatness. Advantages include small size, low-cost operation (temperature-stabilized oven over cryogenic system), and good phase linearity. Disadvantages are limited choice of delays below 10 μs with octave microwave bandwidths, phase nonlinearity at band edges, triple travel signal, high insertion loss, and limited dynamic range (40–50 dB). BAW filters are used in the Accumulator stacktail and Debuncher transverse cooling systems [26].

Microwave to fiber optical delay links are the third technology used for recursive notch filters [27]. The first optical notch filters were developed for Debuncher transverse cooling systems [28]. Broadband microwave to fiber optic-based links became commercially available in the mid-1980s operating at 1,310 nm infrared wavelength. This is the zero dispersion wavelength in single-mode fiber. With the development of optical amplifiers via erbium doped fibers, 1,550 nm has become

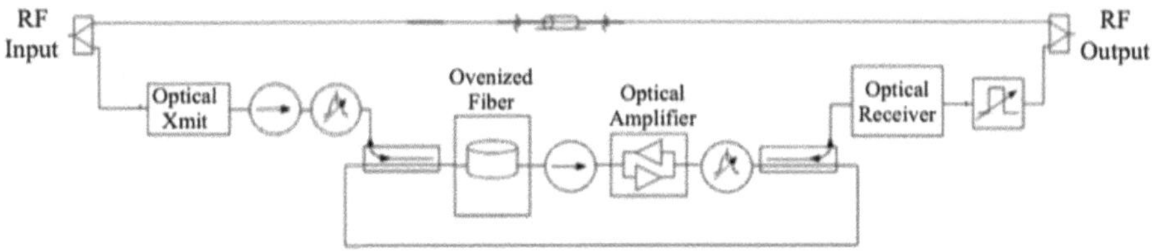

Fig. 7.33 Block diagram of optical IIR notch filter. Optical isolators and etalon band-pass filters required for stable operation as the optical amplifier is bidirectional

the industry standard for transmission links. This wavelength is the minimum insertion loss wavelength on single-mode fiber. By translating the microwave cooling signals to light carriers, signals could be transmitted with minimal gain variation and excellent phase and dispersion characteristics, perfect for long transmission trunks and notch filters.

Single-mode fibers moved quickly from experimental prototypes to the backbone of the telecommunications industry in the 1980s. Modern optical links span oceans with solely optical signals and optical repeaters located every 200 km. Sumitomo developed a very special temperature-stabilized single-mode fiber [29]. By adjusting the cladding material, the temperature coefficient of single-mode fiber can be made close to zero at room temperature. This characteristic was highly desirable for the application of the infinite impulse response filter (IIR) described below where signal delay needed to be kept to the most stringent tolerances. The Sumitomo fiber proved to be very expensive and has limited use for scientific applications where temperature stability is of the utmost importance.

In addition to FIR filters, with fully optical-to-optical amplification available, an IIR filter could be developed for bunched beam cooling in the Tevatron [30]. Figure 7.33 shows the basic block diagram. An IIR filter has the possibility of becoming unstable should loop gain exceed unity. An added complication is that fiber amplifier gain is a function of polarization of the optical propagation mode. Conventional optical amplifier gain can vary a few dB by merely flexing the fiber. A polarization-insensitive amplifier is essential for gain stability. The resulting transfer function of the IIR filter is shown in Fig. 7.34. Note the very steep notch and excellent phase flatness between Schottky bands. This was a requirement as Tevatron betatron sideband widths occupy a significant fraction of the band between revolution lines. While the filter worked as designed, it suffered from two very important shortcomings: poor noise performance and limited dynamic range. The optical amplifier is essentially a laser just below lasing threshold. The carrier inversion is a significant noise contributor. To be effective in the Tevatron bunched beam cooling, this filter needed to be placed immediately after the pickup with a minimum of 70 dB of dynamic range. The filter needed high signal-to-noise input and only achieved 40 dB dynamic range, a limitation of the optical link.

Debuncher momentum cooling consists of four systems within 4–8 GHz. The original upgrade included four BAW notch filters providing the momentum gain shaping function. Due to physical location in the tunnel, each filter was enclosed in its own oven. Each of these filters tended to wander within a specified tolerance, but

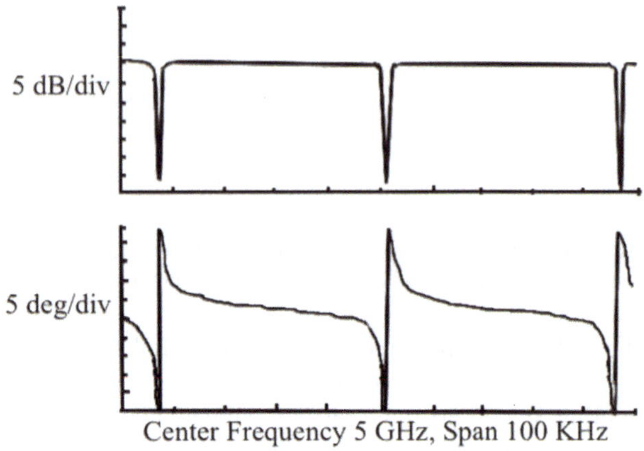

Fig. 7.34 Measured amplitude and phase response of IIR notch filter. Phase change within ±5 kHz of notch is 350°

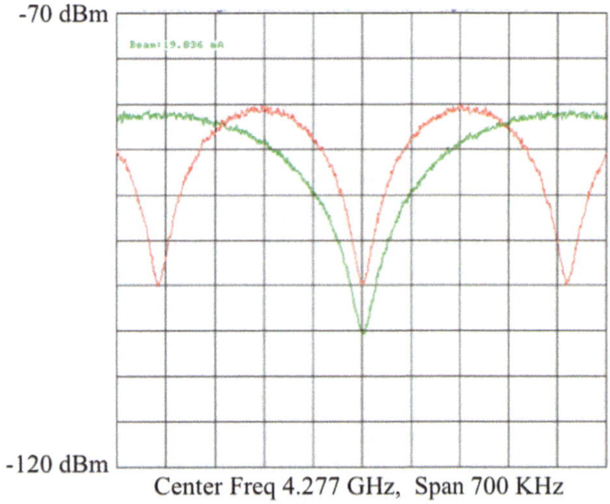

Fig. 7.35 Optical single/double notch filter gain transfer function zoomed in on a single revolution line. Optical switches insert double delay midway through the cooling cycle to increase the gain slope near the central momentum

the result was a widening of the beam momentum spread. This was overcome by making one optical notch filter covering the full 4–8 GHz band and locating it near the kickers where all four systems physically converged. The BAW devices could not accommodate the full 4–8 GHz bandwidth with the required linearity. Optical-based notch filters are a mature technology and used in the Recycler with delays of 11 μs. Passing all four systems through the same notch filter would guarantee all notches to be locked to the same revolution frequency. An added improvement would be to switch mid-cycle between a single delay to a double turn delay. The notch would have a steeper gain profile near the central orbit frequency and lower momentum spread results, Fig. 7.35.

Advantages of optical links are broad bandwidth, excellent phase linearity, almost ideal zero or 180° phase intercept, and delays well beyond a few microseconds. The largest drawback of optical delay lines is insertion loss of the link, limited dynamic range, sensitivity to radiation damage, and cost. The laser and photodiodes have impedances of a few ohms. With the need for broad bandwidth performance, resistive matching networks were initially used. The mismatch loss has been mitigated to some extent by the use of trans-impedance matching amplifiers to drive the laser and amplify the detected optical signal.

Signal Transmission

Stochastic cooling has a fundamental requirement that the feedback signal be applied to the beam promptly with less than one beam revolution period delay. The Antiproton Source is shaped in the form of a triangle (Fig. 7.36). Its optics is designed so that the short and long straight sections have high and zero dispersions, correspondingly. The velocity of wave propagation in the transmission media is significantly slower than beam velocity (coax cable ranges from 67 to 98 % *c*, fiber optics 67 % *c*). Short distance between pickup and kicker is required to make effective transverse cooling ("bad mixing" reduction). That yields short pickup-to-kicker delay time and forces all electronics including kicker power amplifiers to be placed in the tunnel for transverse cooling. Accumulator momentum cooling takes advantage of pickups located in high-dispersion ring sections and kickers in low dispersion. This fact allows the cooling feedback signal path to cut across the midsection of the ring, lending extra delay time for stacktail 2–4 GHz and core 2–4 GHz momentum cooling. These two systems have kicker TWT amplifiers located in surface buildings, facilitating their repair and maintenance. The large size of the Recycler Ring (3.3 km circumference) forces the signal path to be considerably longer than those of the Antiproton source, some 600 m. Here coax transmission would prove to have exceedingly high insertion loss with detrimental signal-to-noise degradation and high dispersion. A free space laser/microwave transmission link was implemented to span the chord between pickups and kickers. Figure 7.36 shows the locations of the cooling systems.

A variety of coaxial transmission lines are used for stochastic cooling signal transmission. Antiproton Source stochastic cooling bands range from 2 to 8 GHz. Each coax has a maximum transmission frequency above which waveguide modes are excited and dispersion becomes a serious issue. Insertion loss is also directly proportional to both the dimensions of the coax as well as the materials from which it is manufactured. The velocity of signal propagation is mostly controlled by the dielectric media between center and outer conductors. For stochastic cooling, the fastest possible propagation velocity is desirable. This rules out the use of solid dielectrics with propagation constants below 70 % of light velocity.

While air dielectric coax (98 % *c*) was a possibility and initially used on the stacktail betatron systems (these two systems were not very effective and eliminated when the stacktail was upgraded to 2–4 GHz), air coax has a significant disadvantage of transmission notches. This is due to evanescent waves that are launched from the periodically spaced center conductor dielectric supports. This is a collective effect that only materializes when transmission lengths approach hundreds of nanoseconds such as required for stochastic cooling. This well-

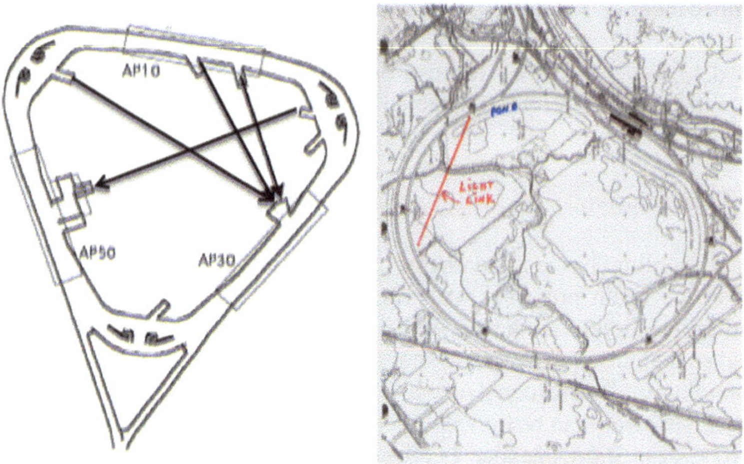

Fig. 7.36 Layout of Antiproton Source (*left*) and Recycler (*right*) stochastic cooling systems. *Chord lines* are location of stochastic cooling signal transmission across the rings

known phenomenon could be mitigated by random spacing of supports, but requires a custom manufacturing process. Diameters of rigid coax smaller than 7/8″ are not commercially available. Spiral air dielectric coax at ½″ diameter is available, but suffers from severe dispersion in the 4–8 GHz band. Foam dielectric coax, with a propagation constant of 88 % c, was chosen for most of the cooling systems in the Antiproton source. This type of coax has the best combination of high velocity, low insertion loss, and acceptable dispersion for lengths approaching 400 ns delays.

There is one application in the Accumulator where the 4–8 GHz core momentum cooling with a cord cutting one half of the ring can make use of slower fiber optic transmission. This special case was pursued because this cooling system has a full octave bandwidth. Note that the foam coax exhibits some 300° of phase roll-off over the octave with the required 400 ns transmission cable. A complicated equalizer was designed to compensate this dispersive effect, but did not yield optimal cooling performance [31]. Single-mode optical fiber has a velocity of propagation of 67 % c. The advantage of fiber is that the transmission link essentially is dispersion free, making full use of the octave bandwidth without the need for an equalizer. When used in a broadband application such as stochastic cooling, fiber optic links have a limited dynamic range of approximately 40 dB, which is adequate for a core cooling system.

For the Recycler, neither coax nor fiber is an option due to excessive insertion loss and dispersion (coax) or slow propagation velocity (fiber). Two transmission methods were pursued: over-moded waveguide at millimeter wavelengths and free space microwave-modulated laser beams. The over-moded waveguide utilizing TE01 transmission mode was extensively researched by Bell Labs in the early 1970s as a means of low loss long distance telephony links. A millimeter wave carrier frequency could be modulated with bandwidths of several GHz resulting in high propagation velocities with very low insertion loss and dispersion. In 1977, during the construction of National Radio Astronomy Observatory's (NRAO) Very Large Array (VLA) radio telescope in Socorro, New Mexico, this technique was adopted for linking the 27 dishes [32]. Fiber optics was not a mature technology at that time, and free space radio transmission was deemed incompatible due to interference with the telescopes. The over-moded waveguide technique was considered for the Recycler solely based on a visit to the VLA and learning that some 2 km of the waveguide could be made available to Fermilab. The spare circular waveguide was stored at NRAO's Greenbank, West Virginia facility. A cursory inspection of the Greenbank stock yielded only a 20 % salvageable quantity due to poor storage outdoors for two plus decades. This was insufficient to complete the Recycler project. Fabrication of new waveguide would prove to be prohibitively expensive and this option was dropped.

Microwave to optical fiber transmission links was a mature technology at the time of Recycler construction. These fiber optic-based links could be explored to substitute free space for fiber as the transmission media. A link was first installed in the Accumulator core transverse cooling as a prototype for the Recycler [33]. Three links were installed in the Recycler to support the four cooling systems utilizing fiber beam expanders, telescopes, optical positioning hardware, and a large cross section high bandwidth photodiode. One link was used for two cooling systems utilizing frequency division multiplexing. All of the optical hardware is housed in climate-controlled enclosures located at the ends of the buried transmission pipe. In an effort to minimize delays between the optical link and the tunnel electronics and ensure a stable environment not effected by daily temperature variations, a transmission steel pipe of 24 in. diameter is buried an average of 12 ft below the surface for the 1,850 ft chord across the ring. This diameter pipe was chosen as a trade-off between labor costs for positioning accuracy required for a smaller pipe over less stringent survey tolerances for a larger more costly pipe. A rough vacuum of a few milli-Torr is maintained to prevent distortion due to index of refraction changes in an air-filled pipe. Initial testing proved the vacuum was essential to the success of the link.

Power Amplifiers

A variety of power amplifiers are utilized in stochastic cooling systems based on solid-state and TWT technologies [34]. TWTs comprise the majority of RF power sources for stochastic cooling. Their wide band, high power, and radiation hardness are well suited to stochastic cooling. The TWTs utilized in all systems have a saturated broadband power level of 200 W and octave bandwidths covering 1–2, 2–4, and 4–8 GHz. The phase linearity and stability are carefully specified and maintained by a tightly regulated power supply for the helix voltage.

Due to smaller power solid-state power amplifiers are used in the core 4–8 GHz transverse cooling and Recycler 0.5–1 GHz momentum cooling. Solid state has the advantage of long life, no high voltage requirements, good gain flatness, and phase linearity. Core systems have amplifiers capable of delivering 5 W of power. The Recycler has 100 W units. Solid state has the disadvantage of being radiation sensitive.

Antiproton Source tunnel ambient temperature approaches 100 °F, which hampers convection cooling of solid-state amplifiers. A significant number of one-watt amplifiers in the Debuncher suffered from deteriorating gain due to high operating temperatures. The solution was to water-cool all these amplifiers with chilled water. Each unit is insulated to prevent condensation and has a separate remotely controlled voltage regulator that monitors voltage and current with trip indications for fault conditions. Due to very high power dissipation in TWTs, 2 kW, water-cooling is essential. For this application, 95 °F low conductivity water is used to minimize any condensation possibilities near the high voltage terminals. The filament voltage is regulated at the point of connection to the TWT. TWTs do not suffer from operation in a radiation environment, but the power supplies do. In all systems, power supplies are located remotely from the TWTs by tens to hundreds of feet. Careful power supply design must take into account the added cable capacitance to assure stable operation in these regulated supplies.

The power level required for stochastic cooling ranges from watts to kilowatts. When amplifiers approach saturated power levels, intermodulation distortion occurs. Odd order products can be translated back into the useful cooling band. Such “intermods” reduce cooling effectiveness. Most power amplifiers are operated 3–6 dB below saturated power to reduce this effect. The only exception are Debuncher cooling TWTs that are operated very close to saturated power as these cooling systems are power limited. Costs of microwave power at these frequencies approach tens to hundreds of dollars per watt, hence making best use of installed power a high priority.

Equalizers

Once all components of the cooling system are assembled, installed, and commissioned with beam, transfer functions that include the beam response are measured in the accelerator complex. Transfer coaxial switches included in every medium level stochastic cooling system allow direct calibrated vector measurement of the complete gain chain via a vector network analyzer and automated measurement software. The resulting transfer function serves as the starting point for equalizer/filter design.

Stochastic cooling performs most efficiently if the amplitude and phase characteristics can be tailored to optimum parameters as specified by system modeling [35]. As one might imagine, the concatenation of dozens of devices will inevitably vary from the ideal individual component measurements due to mismatches at various interconnections.

Equalizers are microwave circuits that when added to the feedback system provide linearization of phase, gain slope control, and frequency response limits. The technologies utilized for fabrication include microstrip and stripline

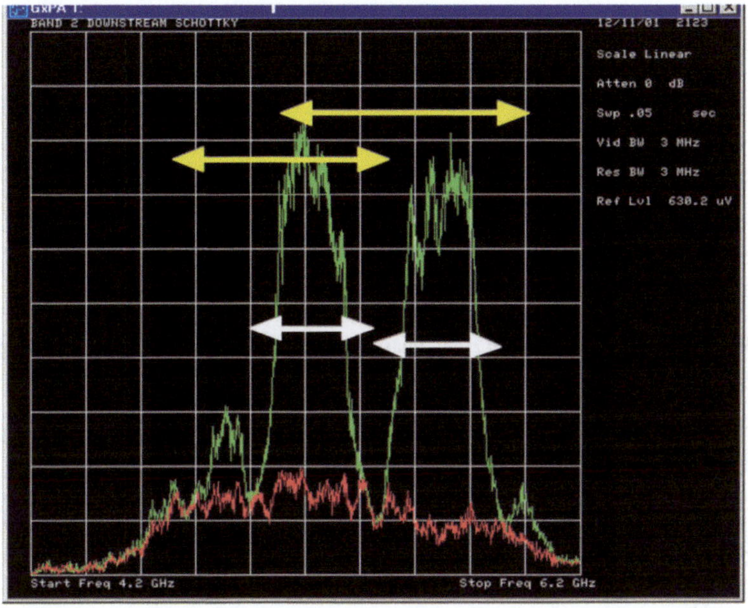

Fig. 7.37 Band 2 Debuncher momentum pickup spectrum linear scale reference level 630 μV, start 4.2 GHz stop 6.2 GHz. *Red*, no beam; *green*, with beam. *Yellow arrows* initial filter bandwidth, *white arrows* customized filter bandwidth. Thermal noise reduction observed is approximately a factor of 2

transmission lines. Resonators in the form of delay loops, transmission line stubs, hybrid circuits, coupled transmission lines, and circulators are also employed.

Simple equalizers have been designed to linearize individual components such as BAW delays or to band limit frequency response for rejection of unwanted signals. Precise band-pass filters were utilized in the Debuncher cooling upgrade to minimize the sub-band overlap [36]. Figure 7.37 shows the pickup spectrum from Debuncher Band 2. Red trace is no beam noise bandwidth, green trace with beam before the final filters were installed. The yellow arrows indicate the 3 dB bandwidth of the original installed filters, white arrows the bandwidth of customized filters. The addition of precise frequency centered band-pass filters also improved the rejection of thermal noise by approximately a factor of 2.

Some of the most complicated equalizers are those fabricated for the stacktail system [37]. Figure 7.38 depicts before and after beam transfer function measurements. Improvement in amplitude and phase response is evident, resulting in improved cooling performance.

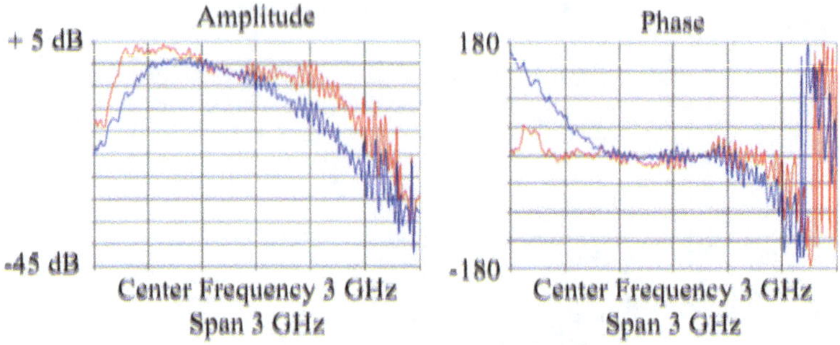

Fig. 7.38 *Left*, amplitude response; *right*, phase response of stacktail momentum cooling system. *Blue* before, *red* post-equalization where amplitude is enhanced, phase is flattened

7.2.3 Operational Optimization of the Stacktail System

An introduction of slip stacking in the Main Injector, improvements of antiproton production on the target, and an increase of Debuncher acceptance were complete at the beginning of 2006. It resulted in the antiproton flux coming to Debuncher of ~3.8×10^{11}/h (2.3×10^8 per cycle). This number was close to the expectations; however, the peak stacking rate of ~2×10^{11}/h achieved to this time was significantly lower. An estimate based on Eq. (7.96) showed that further improvement of stacking rate is impossible without an upgrade of the stacktail system which was the "bottleneck" in the staking process. Later we learned that all other stochastic cooling systems require performance improvements as well. Altogether there were 21 cooling systems in Debuncher and Accumulator and 4 in Recycler. Upgrades were carried out for all of them. Depending on the system, one or more of the following steps were applied: an increase of the effective system bandwidth with band equalizers, other modifications of existing hardware (notch filters, etc.), and optics corrections aimed on cooling improvements.

Design of equalizers correcting amplitude and phase of each system was based on the beam-based measurements of the system gain. The goal of band equalization was to minimize dependence of phase on the frequency and to make the effective gain linearly growing with frequency [38]. The sequence of upgrades carried out over 3 years (2006–2009) resulted in an increase of the peak stacking by 1.5 times to 3×10^{11}/h. This value is close to expectations based on the model predictions described below. Improvements of electron and stochastic cooling in Recycler and an introduction of the fast Accumulator-to-Debuncher transfers resulted in almost threefold increase of the stacking rate to Recycler. Improvements of the stochastic cooling theory [39] carried out shortly before 2006 and presented in Sect. 7.2.1 above have been extremely useful in choosing the upgrade path and resolving the problems encountered along the way. Below we describe an upgrade of the stacktail system initiated in 2006 which can be considered as the most profound and arguably the most complicated upgrade among upgrades of all the cooling systems.

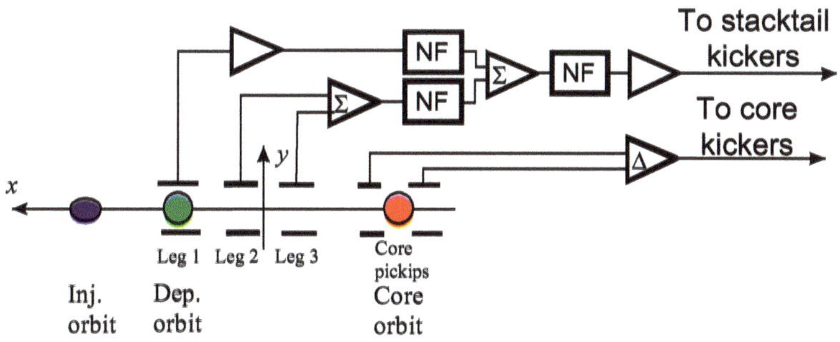

Fig. 7.39 Block diagram of the stacktail system ($x \equiv \Delta p/p$ corresponds to the radial direction). Signals from the top and bottom loops (plates) are summed

The schematic of the stacktail stochastic cooling system is presented in Fig. 7.39 [40]. The pickup electrodes are located in a straight line with large dispersion which makes particles being radially spread in accordance with their energy. Due to sufficiently large dispersion, the total radial size is significantly larger than the betatron size. The kickers are located in a straight line with zero dispersion on other side of the ring. The beam is injected to the injection orbit and then is RF displaced to the deposition orbit where it can be seen by stacktail pickups. The cooling signal is formed as a combination of signals of three pickup legs. Each leg signal is combined from signals of multiple pickups with appropriate delays. Leg 1 is located at the smallest particle density and has the highest sensitivity. Altogether it has 128 separate pickups combined from 8 pickup arrays. Half of them are above the beam plane and other half are below. Legs 2 and 3 have 48 and 16 pickups, correspondingly. The radial size of pickups is much smaller than the radial extension of the stack, making the signal dependent on a particle radial position. The width of the pickup plates, w, and half-gap between them, h, are chosen so that to form a desired exponential dependence of gain on the momentum (as required by Eq. (7.94)) for positions outside of pickups [41]:

$$g(x) = \frac{\operatorname{atan}\left(\sinh\left(\frac{\pi}{h}\left(x+\frac{w}{2}\right)\right)\right) - \operatorname{atan}\left(\sinh\left(\frac{\pi}{h}\left(x-\frac{w}{2}\right)\right)\right)}{2\operatorname{atan}(\sinh(\pi w/2h))} \xrightarrow{x \gg h, w} \frac{2\sinh(\pi w/2h)e^{-\pi x/h}}{\operatorname{atan}(\sinh(\pi w/2h))}. \tag{7.102}$$

Table 7.4 presents h and w parameters found from the beam measurements described below. It also presents parameters of core cooling pickups and locations of the pickup centers in radial direction. Normally the locations of injection, deposition, and core orbits were 8.52, 0.78, and −5.93 cm, correspondingly. Corresponding revolution frequencies are 628,749, 628,832, and 628,897 Hz where nonlinear dependence of revolution frequency on the orbit position is accounted (see Fig. 7.22). Parameter R presents relative gains of different pickups at their maximum sensitivities. Note that although pickups for all legs have the same width and gap, the effective width, w, for the leg 1 is larger. It is related to the

Table 7.4 Geometrical parameters of stacktail pickups and their radial positions obtained in the beam-based measurements

	w (cm)	h (cm)	r (cm)	Revolution frequency (Hz)	R
Leg 1	3	3.3	0.97	628,830	1
Leg 2	3	3	−0.20	628,840	0.34
Leg 3	3	3	−2	628,860	0.023
Core cooling 2–4 GHz	2	2.7	−3.45/−8.42	628,874/628,920	2×10^{-3}
Core cooling 4–8 GHz	0.76	3.2	−4.79/−7.18	628,886/628,908	1.6×10^{-3}

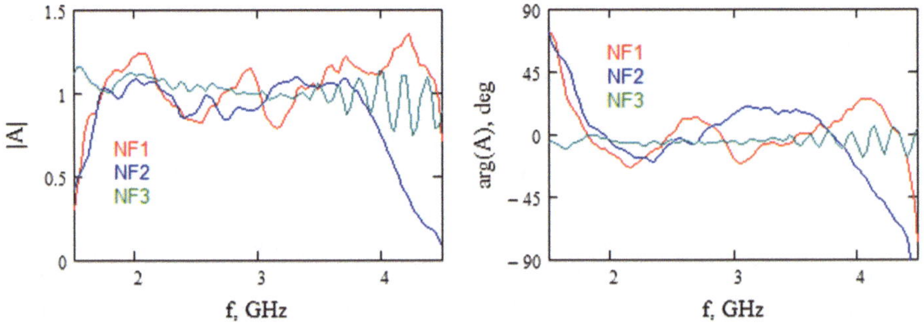

Fig. 7.40 Dependence of magnitudes and phases of notch filter coefficients, $A(\omega)$, on the frequency $f = \omega/2\pi$

long length of leg 3 pickup array and its imperfect alignment relative to the beam. Three notch filters (NF) are tuned to suppress effect of the stacktail signal on particles located in the beam core. Together with signal delays for each pickup, they also help to form the desired dependence of cooling force on the particle momentum. The frequency band of stacktail is 2–4 GHz. Two core momentum cooling systems (2–4 and 4–8 GHz) cool the beam longitudinally to the core.

The system gain in the stacktail model can be presented in the following form:

$$\begin{aligned} G(x,\omega) = G_0\left(1 - A_3(\omega)e^{i\omega T_3}\right) \times \Big[a_1 g_1(x) K_1(\omega,x) e^{i\omega\tau_1}\left(1 - A_1(\omega)e^{i\omega T_1}\right) \\ + \left(a_2 g_2(x) e^{i\omega\tau_2} K_2(\omega,x) + a_3 g_3(x) e^{i\omega\tau_3} K_3(\omega,x)\right)\left(1 - A_2(\omega)e^{i\omega T_2}\right)\Big] \\ + a_4 g_4(x) e^{i\omega\tau_4} K_4(\omega) + a_5 g_5(x) e^{i\omega\tau_5} K_5(\omega), \end{aligned} \tag{7.103}$$

where indices $i = 1, 2$, and 3 belong to the stacktail systems and $i = 4$ and 5 to the 2–4 GHz and 4–8 GHz core systems, correspondingly; T_i are the time delays in notch filters; τ_i are signal delays; and a_i are the gain coefficients. The complex functions $A_i(\omega)$ were separately measured by switching off and on the short and long legs of notch filters. Their plots are presented in Fig. 7.40. As one can see, the notch filters NF1 and NF2 are far away from being ideal. They are constructed using the BAW delays yielding bad performance on the band boundaries. The NF3 filter is based on a delay with superconducting (SC) coaxial cable and, as one can see, is significantly better. The form of Eq. (7.103) was chosen so that the functions $g_i(x)$ would describe dependence of the response function on the momentum for lower half of frequency band. The dependence of functions $K(\omega, x)$ on x corrects for the difference in the x response between high and low frequencies.

To measure the dependence on ω for complex functions $K_i(\omega, x)$, a small intensity cooled antiproton beam was RF displaced from the stack to a desired momentum, x_0, and then scraped in a high-dispersion region to obtain the longitudinal distribution close to the rectangular one. The width of distribution was much smaller than a typical stack width as well as a characteristic width of pickup response. In this case, the exponent in the denominator of Eq. (7.19) can be expended near a revolution harmonic, $\omega = n\omega_0 + \delta\omega$. Omitting the effect of notch filter (as in the measurements) and leaving only leading terms, one obtains for the open system response:

$$S_n(\delta\omega) = -\frac{\Delta p_{\text{ext}\omega}}{p_0} \frac{N}{2\pi i n \eta} \int_{\delta \to 0_+} \frac{df_0(x)}{dx} \frac{G_1(x, \omega)}{x + \delta\omega/(n\omega_0\eta) - i\delta} dx, \tag{7.104}$$

where $G_1(x, \omega)$ is the single-particle gain. Computation of integral for a rectangular distribution with the total width Δx is straightforward. Taking into account that $df_0/dx = (\delta(x + \Delta x/2 - x_0) - \delta(x - \Delta x/2 - x_0))/\Delta x$, one obtains that the response is

$$S_n(\delta\omega) = \frac{\Delta p_{\text{ext}\omega}}{p_0} \frac{2N\omega_0}{\pi i} \left(\frac{n\omega_0\eta\Delta x G_1(n\omega_0, x_0) - (dG_1/dx)\delta\omega}{(n\omega_0\eta\Delta x)^2 - 4\delta\omega^2} \right), \quad |\delta\omega| < |n\omega_0\eta\Delta x|, \tag{7.105}$$

where x_0 is the location of the distribution center and we expended the gain as following: $G(x, \omega) = G(x_0, \omega) + (x - x_0) dG/dx|_{x = x_0}$. As one can see, the response function magnitude in the distribution center ($\delta\omega = 0$) is inversely proportional to the harmonic number; and the response phase is shifted by 90° everywhere within distribution ($|\delta\omega| < |n\omega_0\eta\Delta x|$). This independence of the response phase on a frequency error represents significant advantage of a rectangular-like distribution over Gaussian-like.

Positions of the scrapers and width of the distribution were controlled with longitudinal Schottky monitor. The open-loop response was measured with a network analyzer programmed so that the measurements were carried out exactly at the beam revolution harmonics. A few hundred harmonics (usually, 401) uniformly distributed across the system band were used. Response of each leg was measured separately. In the course of such measurements, the notch filters and other pickup legs were switched off. The measurements were usually carried out at four radial positions (the same for all legs) distributed across the stacktail range. For each position, the excitation amplitude was adjusted so that, on one hand, it should be sufficiently large to reduce the effect of noise on measurements and, on the other hand, it should be sufficiently small to avoid changes of the distribution function in

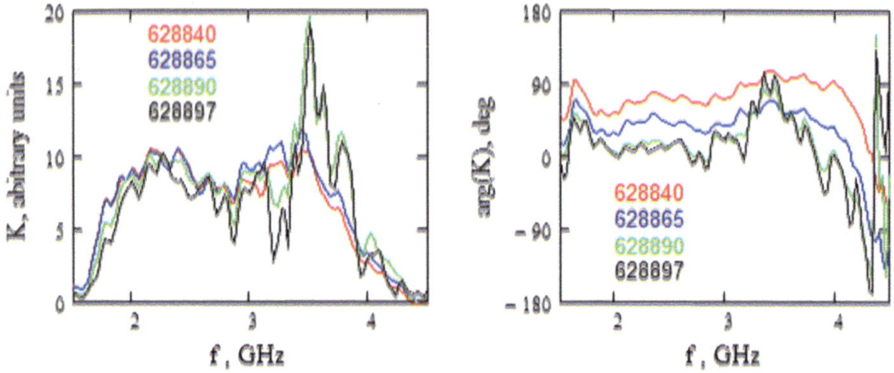

Fig. 7.41 Measurements of $K(\omega, x)$ on frequency, $f \equiv \omega/2\pi$, for leg 1 pickups acquired at four values of momentum which correspond to the following revolution frequencies: $f = 628{,}840$, 628,865, 628,890, and 628,897 Hz

the course of the measurements. The beam current was increased with moving orbit out of pickups so that to compensate for reduction of pickup sensitivity.

Results of the leg 1 measurements are presented in Fig. 7.41. The measured spectra were corrected so that they would be proportional to the gain function: i.e., their phases were shifted by 90° and magnitudes corrected proportionally to frequency in accordance with the system response of Eq. (7.105). The measurements acquired at different beam radial locations were scaled so that they would coincide at the lower half of the band. As one can see, the shapes of the responses are almost identical for frequencies of 628,840 and 628,865 Hz corresponding to locations near and within leg 1 pickups. For two other frequencies, the orbit is located far outside of the leg 1 pickup where the sensitivity is exponentially decreasing. However, the decay rates on the momentum, x, for low and high frequencies are different. In the upper half of the band, the frequency is approaching the cutoff frequency of the pickup structure, yielding a reduced decay rate. Consequently, the response at 3.5 GHz is about twice larger for revolution frequencies of 628,890 and 628,897 Hz. The phase deference presented in the right pane of Fig. 7.41 is related to the nonzero partial slip factor between pickup and kicker and coincides well with predictions of optics model.

A lack of control for the beam intensity and beam momentum width did not allow us measuring $g_i(x)$ with satisfactory accuracy using the described above method. Therefore, another type of measurements was used. It is based on measuring the stacktail Schottky noise in a narrow band (usually a single revolution frequency band at 2.4 GHz) for each pickup array and referencing it to the signal of 79 MHz Schottky monitor which is located at zero dispersion and therefore does not have a direct dependence of signal on the beam momentum. Relating all signals to the first harmonic of revolution frequency and computing a square root of signal ratio for a stacktail pickup and the Schottky monitor, one obtains a dependence of corresponding pickup sensitivity on the horizontal particle position within pickup (see Fig. 7.42). To acquire this data, all longitudinal cooling systems including the stacktail have to be stopped to avoid changes of the distribution during measurements. This was verified by acquiring longitudinal Schottky monitor signals at the beginning and at the end of measurements. Automation of the measurements allowed us to reduce time of data acquisition to about 10 min. Fitting the measurements to Eq. (7.102) resulted in pickup parameters presented in Table 7.4. The large dots in the bottom pane of Fig. 7.42 show results of described above transfer function measurements acquired with network analyzer at various revolution frequencies. Close agreement between the network analyzer and the Schottky measurements reassures us in the accuracy of both measurements. It has been crucially important for predictive power of the computer model. In the model, the functions $g_i(x)$ were computed using Eq. (7.102) with parameters presented in Table 7.4; the complex functions $K_i(\omega, x)$ were computed by interpolation of network analyzer measurements acquired at four revolution frequencies.

Let us introduce the complex cooling force:

$$F_C(x) = \frac{2}{T_0}\sum_{n=0}^{\infty} G(x, n\omega_0), \tag{7.106}$$

so that $F(x) = \mathrm{Re}(F_c(x))$ and $G(x, \omega)$ is determined by Eq. (7.103). To optimize the stacktail performance, we assume that the maximum particle flux which the stacktail can accept is determined by minimum value of the flux $J_{\max}(x)$ computed with Eq. (7.96), where

$$x_d \equiv x_d(x) = -\frac{G(x,\omega)}{dG(x,\omega)/dx}. \tag{7.107}$$

This assumption was later verified by solving the time-dependent Fokker–Planck equation (two top equations in (7.94)). The cooling force of the entire cooling system which includes three legs of stacktail and two core cooling system was computed. Parameters of the system (delays, gains, etc.) were adjusted in the computer model to maximize $\min(J_{\max}(x))$ on the way from the deposition orbit to the core which maximizes the staking rate. We also required approximately constant $x_d(x)$ in the central part of the stacktail and minimizing phase of the complex cooling force. For a given flux, the former maximizes the gain difference and, consequently, the particle density difference between the deposition and core orbits; and the latter reduces the stacktail power. Figure 7.43 presents the results of such optimization based on the measurements acquired after completion of all stacktail upgrades. One can see that the cooling force, $\mathrm{Re}(F_c)$, achieves its maximum at the deposition orbit, and then it exponentially decays in direction of the core orbit with minimum value of $p_0 x_d$ about 9 MeV and crosses zero at the core. The imaginary part of the cooling force is much smaller in the entire stacktail region (with exception of very small region around the point where the cooling force crosses zero). It would not be possible to achieve without notch filters, making

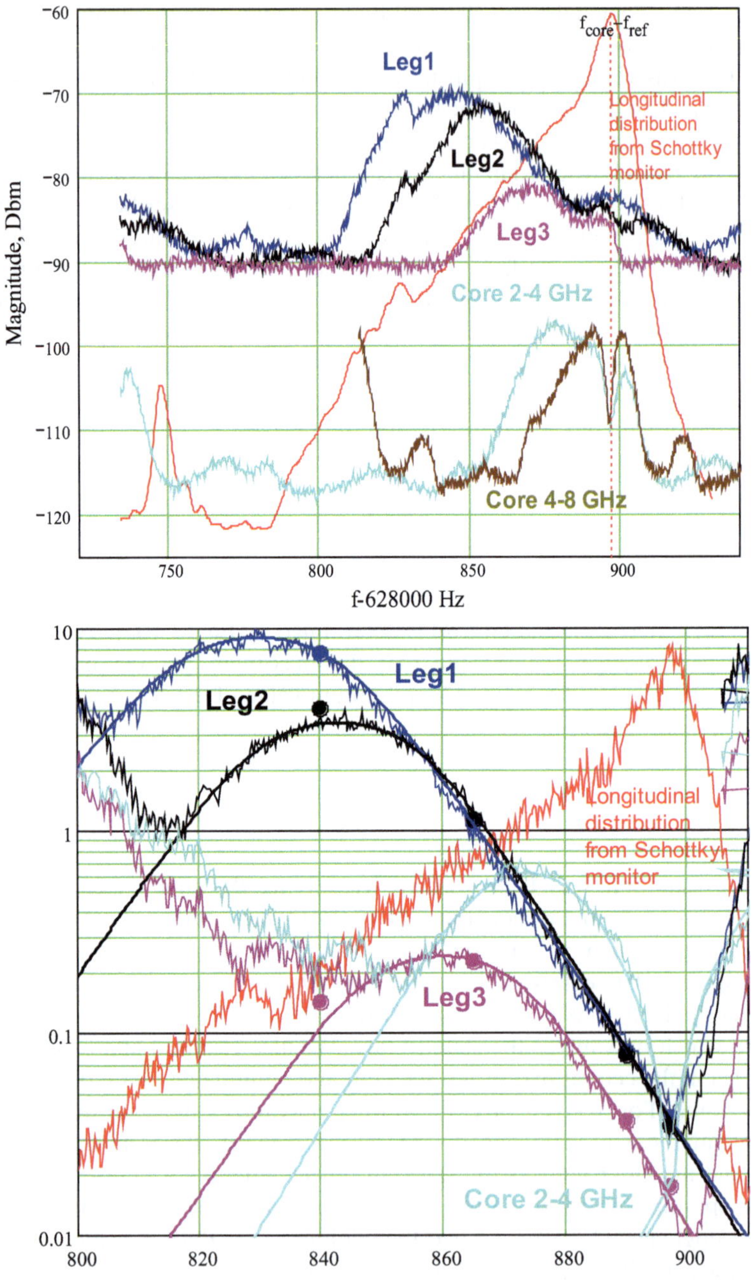

Fig. 7.42 *Top pane* shows the Schottky signals from all longitudinal cooling systems in the Accumulator at a stacking interruption. Leg 1, 2, 3 refer to the Schottky signal from each individual pickup of the stacktail system. Also included is the longitudinal distribution measured with 79 MHz Schottky monitor. *Bottom pane* shows the Schottky signals normalized to the longitudinal distribution. The *dots* represent the network analyzer transfer function measurements at various revolution frequencies. The close agreement indicates the transfer functions are truly representative of the system gain, which is most important for the computer model. In diffidence to the top pane the longitudinal distribution from Schottky monitor presented at the bottom pane is not smoothed

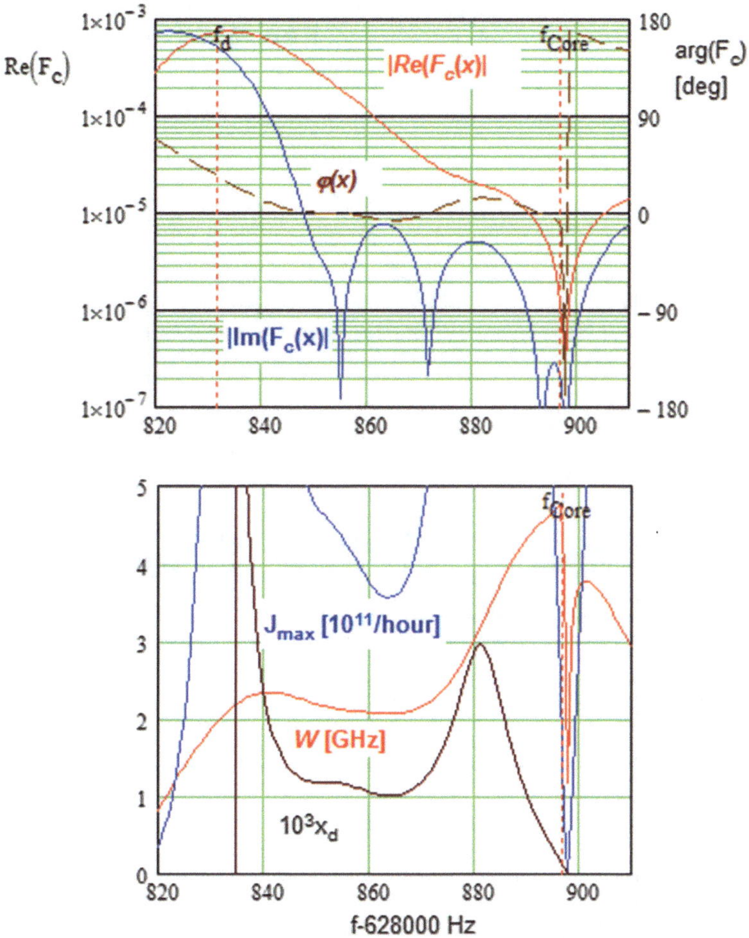

Fig. 7.43 Dependence of stacktail parameters on the revolution frequency; *top*: *red line*, cooling force; *brown line*, phase of the cooling force; *bottom*: *red line*, effective bandwidth; *brown line*, x_d; *black line*, maximum flux computed using Eq. (7.10)

negligible the core heating by the stacktail. The effective bandwidth has minor variations through the stacktail. It starts at ~2.4 GHz at the deposition orbit, slightly decreases, and then goes up to ~4 GHz at the core where the 4–8 GHz core cooling system dominates cooling force.

The static model predicts maximum stacking rate of ~35×10^{10}/h. A time-dependent solution of Fokker–Planck equation presented below yields approximately the same number. Experimentally, as can be seen in Fig. 7.44, we obtained a value slightly above 30×10^{10}/h.

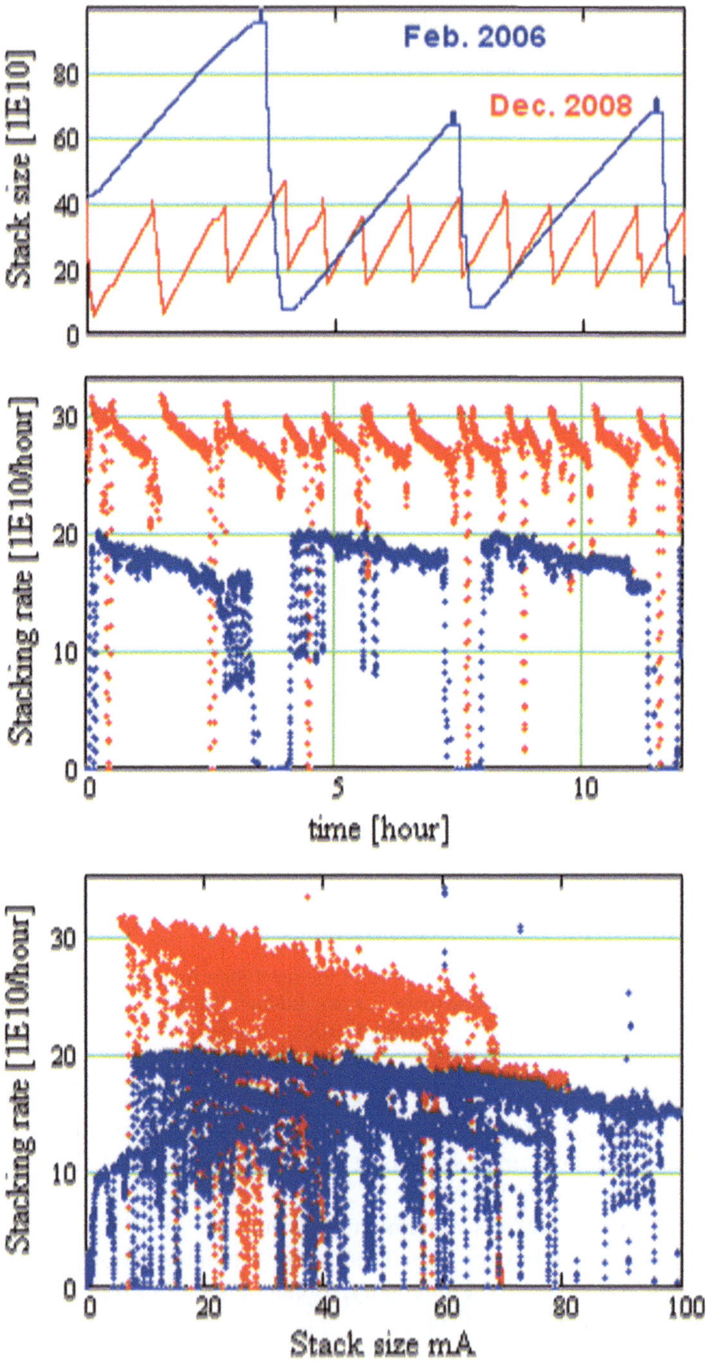

Fig. 7.44 Dependences of stack size on time (*top*) and corresponding stacking rate (*center*), and the stacking rate on the stack size (*bottom*) before (*blue dots*) and after (*red dots*) stacktail upgrades

Stacktail Equalizer

There were three major upgrades to the stacktail system: (1) the abovementioned replacement of the bulk acoustic-based notch filter by the one SC cable based, (2) optics upgrade aimed at an increase of slip factor described above, and (3) an installation of the equalizer correcting the phase and gain of the stacktail amplifier described below.

Left pane of Fig. 7.45 presents the leg 1 gain measured in October of 2006 before the installation of the equalizer. A delay was applied to the data so that the phase response would be flat in the center of the band. One can see two problems. The first one relates to large variation of the phase so that at frequencies below 2 GHz the system makes anti-damping (Re(K)<0) instead of damping resulting in reduced cooling power. The second one is related to the gain being peaked at 2.5 GHz, further narrowing the band. The right pane presents the similar measurements performed in August of 2009 after installation of the equalizer. One can see that the phase is almost perfectly corrected so that the real part and magnitude of the gain are almost undistinguishable, the band is significantly wider, and the gain is almost flat across the band. Actually, accounting for the entire system (where the major gain correction comes from the notch filters) makes the gain linearly growing with frequency as presented in Fig. 7.46. As it was pointed out above, such behavior maximizes the cooling rate. Note that the ripple visible on the data acquired after equalizer installation is real and is related to imperfections of the equalizer. The installation of stacktail equalizer (in Spring of 2007) yielded the effective bandwidth increase of 15 % from 1.75 to 2 GHz. As one can see from Eq. (7.98) corresponds to ~30 % increase in the stacking rate.

The behavior of the phase presented in the left pane of Fig. 7.45 is typical for a band-pass amplifier. It is related to causality and cannot be corrected without breaking it. Fortunately, causality is not directly present in the stochastic cooling, i.e., shortening the cable from pickup to kicker allows the signal to arrive earlier than the particle which radiated this signal in the pickup. This understanding directed us to the layout of the equalizer presented in Fig. 7.47. The equalizer consists of two parts: the phase part (left) correcting the phase and the amplitude part (right) correcting the amplitude. In the phase part, the signal was split into three paths for the prototype equalizer (presented in the picture) and into five paths for the final one. Each path has a resonator and delay line. Combining the signals together corrects the phase but cannot correct the magnitude to the desired level. The amplitude part does not affect the phase but corrects the amplitude. Its transfer function can be presented in the following form:

$$K_A(\omega) = (1 + 0.91\cos(\omega\tau))e^{-i\omega\tau}, \quad \tau = 195\ \text{ps}.$$

Signal splitting and combining are performed with commercial wideband splitters. Each resonator is implemented on a printed circuit board. Its frequency is determined by geometry and the quality factor by coupling. To compensate signal attenuation in the equalizer, an additional 20 dB wide band amplifier was added to

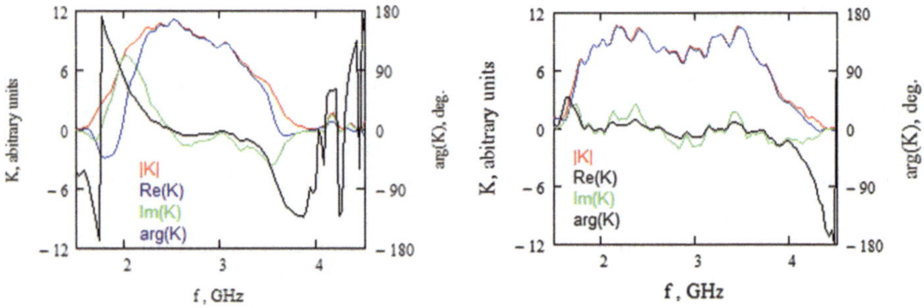

Fig. 7.45 Leg 1 gain measured before installation of stacktail equalizer (*left*) and after (*right*) at the revolution frequency of 628,840 Hz

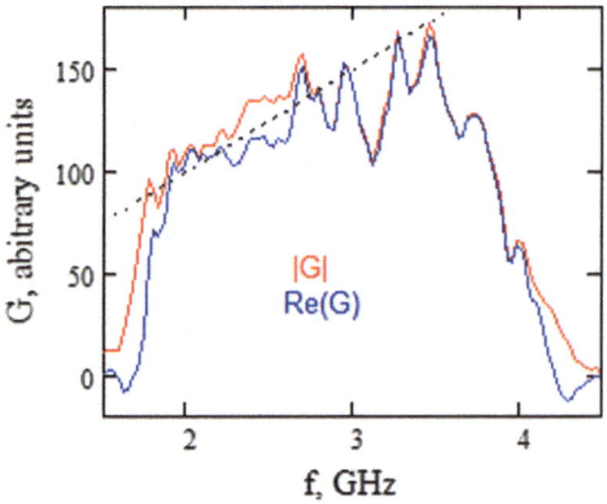

Fig. 7.46 The entire system gain reconstructed from measurements performed after installation of stacktail equalizer at the revolution frequency of 628,840 Hz. *Dashed line* shows linear dependence of gain on frequency

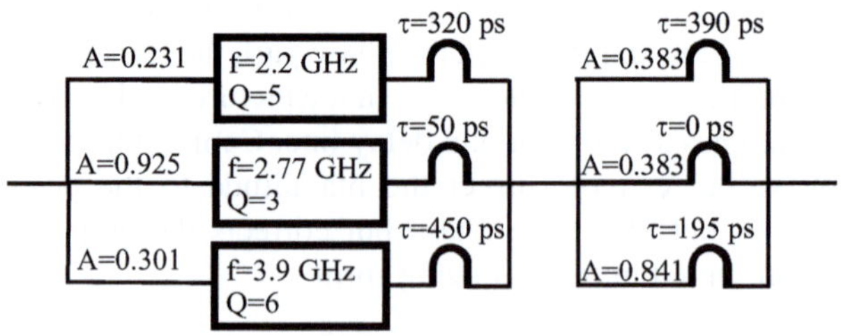

Fig. 7.47 Schematic of the stacktail prototype equalizer

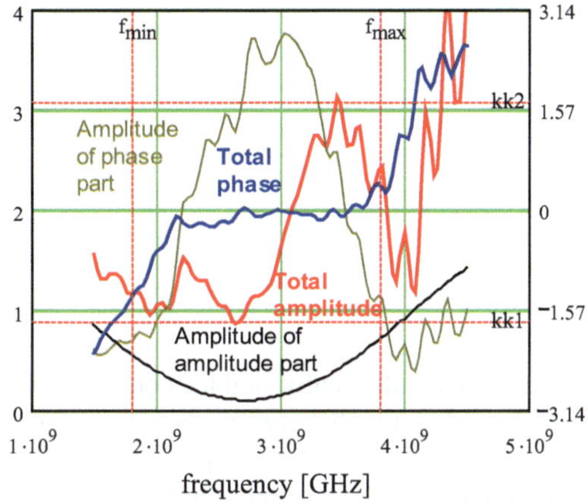

Fig. 7.48 Measurement results of the final stacktail equalizer

the system. Figure 7.48 presents results of measurements of the final stacktail equalizer. One can see that the phase behavior is opposite to the phase behavior of the system gain presented in the left pane of Fig. 7.45. The signal attenuation is maximized in the band center and minimized at its upper edge, resulting in a desired linear growth of the system gain with frequency.

Numerical Simulation of the Stacktail

The simulations are based on solving the Fokker–Planck equation (7.39) where all system parameters came from the beam-based measurements as described above. To be representative stacking simulations require computations for more than 1,000 cycles, resulting in a quite large number of time steps. The signal suppression is different at different harmonics, and its accurate accounting would increase the simulation time by orders of magnitude. Therefore, to accelerate the simulations, we neglected the particle interaction through the cooling system, i.e., kept $\varepsilon = 1$. That yielded a cooling force independent on the time and diffusion directly proportional to the distribution function and, thus, greatly simplified simulations. As it was shown above, such procedure does not affect cooling rates significantly if the gain is below or close to the optimal value. That allowed us to simulate a complete stacking cycle (~1 h) in less than 10 min on a single-processor computer. The beam dielectric function was calculated later to verify the system stability. Note that in the case of other cooling systems, where only one cooling cycle is computed, accurate accounting of signal suppression did not represent such a problem and was routinely used.

The following finite difference equation was used for approximation of the right-hand side of Fokker–Planck equation:

$$\frac{f_{n+1,k}-f_{n,k}}{\Delta t}=\frac{1}{2}\left(\frac{\partial f}{\partial t_{n+1,k}}+\frac{\partial f}{\partial t_{n,k}}\right),$$
$$\frac{\partial f}{\partial t_{n,k}}=\left(f_{n,k}\frac{F_{n,k+1}-F_{n,k-1}}{2\Delta x}+F_{n,k}\frac{f_{n,k+1}-f_{n,k-1}}{2\Delta x}\right)_n$$
$$+\left(f_{n,k+1}\frac{D_{n,k+1}+D_{n,k}}{4\Delta x^2}-f_{n,k}\frac{2D_{n,k}+D_{n,k+1}+D_{n,k-1}}{4\Delta x^2}+f_{n,k-1}\frac{D_{n,k}+D_{n,k-1}}{4\Delta x^2}\right)_n . \tag{7.108}$$

It is unconditionally stable, conserves the number of particles, and has better accuracy than the difference scheme presented in [42]. The variables in this equation are $\Delta x = x_{k+1} - x_k$, $\Delta t = t_{n+1} - t_n$, $f_k = f(x_k, t_n)$, $F_{n,k} = F(x_k, t_n)$, $D_{nk} = D(x_k, t_n)$, n enumerates steps in time, and k enumerates steps in x. Comparison of numerical simulations with measurements is presented in Fig. 7.49 [43]. One can see a step-by-step propagation of particles to the core. The particles which were not moved out of the deposition region (by stacktail) before the next pulse arrives are RF displaced in the direction of the injection orbit and lost for further stacking. One can see the buildup of such particles on the left from the deposition orbit. Experiments show that a higher stacking rate is achieved if the stacktail is on during RF displacement of the beam from the injection to the deposition orbit. The presence of RF strongly distorts the Schottky spectra, making the last spectrum of each stacking cycle unusable for analysis. In simulations newly arrived antiprotons replaced everything that was in the deposition region (no RF displacement), and therefore the peaks on the left are not present. These measurements allowed us to measure directly the absolute value of the system gain.

To estimate the effect of particle interaction on the computation results, ε was computed for a number of time steps and frequencies after a solution of the Fokker–Planck equation is obtained. Figure 7.50 presents the changes in ε within the time of cooling cycle "47" (starts ~100 s after stacking beginning). The beam is close to the stability boundary at the beginning of the cycle while a newly injected beam has narrow distribution; but at 100 ms, smoothening the beam distribution in the vicinity of the deposition orbit results in a reduction of particle interaction by more than a factor of 2. Consequently, as long as the stack size is small, the effect of particle interaction is also small during most of cycle time, resulting in relatively small ($\lesssim$10–20 %) overall correction for the stacking rate. Figure 7.50 also depicts that in the stacktail region (628,840–628,890 Hz), the $|\varepsilon - 1|$ does not exceed 0.25, i. e., the particle interaction has a small effect on the speed of the stack front propagation and, consequently, on the value of the gain deduced from the experimental data. With further staking, the particle number in the stack is growing, resulting in an increased interaction of core particle through cooling system. That forces reduction of the gain and, consequently, a reduction of stacking rate as presented in Fig. 7.44.

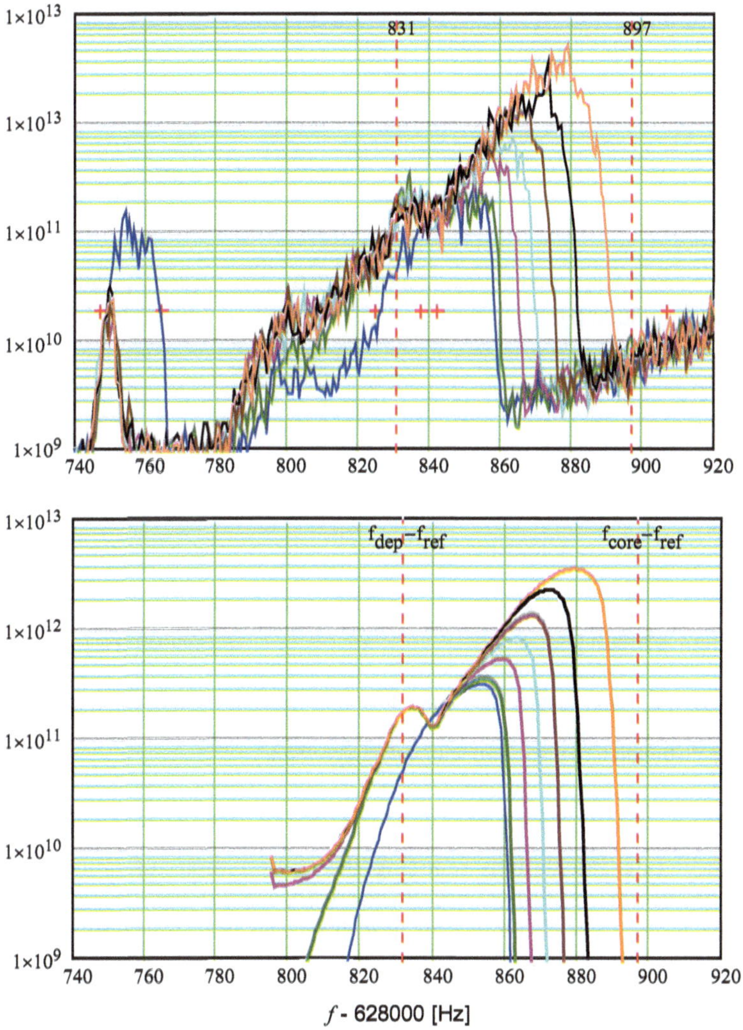

Fig. 7.49 Evolution of particle distribution over revolution frequency during the first 100 s of stacking; *top*, measurements; *bottom*, simulations. Curves are built at 1.54 s in cycle 2 and 0.22 s in cycles 3, 5, 8, 13, 23, and 47

Fast Computations of Beam Dielectric Function

The above-discussed computation of beam dielectric function was normally done by numerical integration. It is fast enough to verify the system stability but hardly can be used in simulations of Fokker–Planck equation if the signal suppression has to be correctly accounted. However, a proper procedure can

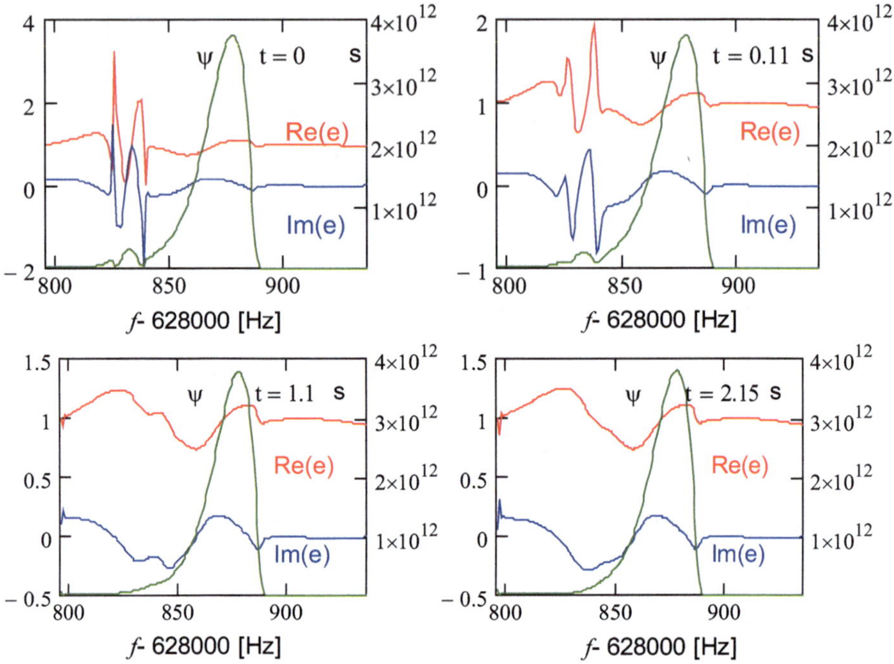

Fig. 7.50 The real (*red*) and imaginary (*blue*) parts of ε computed at different times in cycle 47 for stacking simulations presented in Fig. 7.49. Time marking the plots shows the time within cycle. The particle distribution (ψ) is shown with a *dark green line*

greatly accelerate calculations reducing the computations of multiple integrals to a matrix–vector multiplication. Below we consider how it can be done in application to the filter cooling. This algorithm was used for cooling simulations in the Debuncher.

To simplify formulas, we assume that the system is far from the band overlap. Taking also into account that the gain changes are negligible within one Schottky band, we can rewrite Eq. (7.20) in the following form:

$$\varepsilon_n(y) \approx 1 + \left(1 - A(n\omega_0)\,\mathrm{e}^{2\pi i n\eta y}\right)\frac{G(n\omega_0)}{2\pi i n\eta}\int_{\delta\to 0_+}\frac{df_0(x)/dx}{x - y - i\delta\,\mathrm{sign}(n\eta)}dx. \quad (7.109)$$

One can see that a calculation of the beam dielectric function requires a computation of the same integral for each harmonic. Replacing integration by summation, one obtains

$$\int_{\delta\to 0_+}\frac{df_0(x)/dx}{x - y_k - i\delta}dx \approx i\pi\left.\frac{df_0}{dx}\right|_{x=y_k} + \sum_j\int_{-\Delta x/2}^{\Delta x/2}\frac{f_j' + f_j'' x}{y_j + x - y_k}dx$$

$$= i\pi f_k' - \sum_j\left(f_j'\ln\left(\frac{p_k - p_j + \Delta p/2}{p_k - p_j - \Delta p/2}\right) + f_j''\left((p_k - p_j)\ln\left(\frac{p_k - p_j + \Delta p/2}{p_k - p_j - \Delta p/2}\right) - \Delta p\right)\right). \quad (7.110)$$

Replacing derivatives by their finite difference analogs ($f_j' = (f_{j+1} - f_{j-1})/(2\Delta x)$ and $f_j'' = (f_{j+1} + f_{j-1} - 2f_j)/\Delta x^2$), one obtains for the beam dielectric function

$$\varepsilon_n(y_k) = 1 + \left(1 - A(n\omega_0)\,\mathrm{e}^{2\pi i n\eta y_k}\right)\frac{G(n\omega_0)}{2\pi i n\eta}\frac{1}{\Delta x}\sum_m \Xi_{\mathrm{km}}\psi_m \quad (7.111)$$

Matrix Ξ_{km} is computed once at the beginning of calculations. Its elements are

$$\begin{aligned}\Xi_{km} &= a(k - m + 1) - a(k - m - 1) \\ &\quad + b(k - m + 1) + b(k - m - 1) - 2b(k - m), \\ a(k) &= \frac{1}{2}\ln\left(\frac{k + 1/2}{k - 1/2}\right), \quad b(k) = k\ln\left(\frac{k + 1/2}{k - 1/2}\right) - 1,\end{aligned} \quad (7.112)$$

where $\Delta x = x_{k+1} - x_k$. Note that the crossing of the pole, which introduces an imaginary part to the integral, comes from negative argument of the logarithm in functions $a(k)$ and $b(k)$ at $k = 0$.

In the course of Debuncher cooling simulations, the sum in Eq. (7.111) was computed once at every time step, resulting in fast computations of ε for multiple revolution frequency harmonics. Figure 7.51 compares results of these fast calculations to Eq. (7.92) for Gaussian beam. As one can see, good accuracy is achieved with the step size about half of the rms size.

Using similar procedure, it is straightforward to derive formulas for fast beam dielectric function computation in the general case. In the case of overlapping Schottky bands, matrices Ξ are different for different bands. To make time-

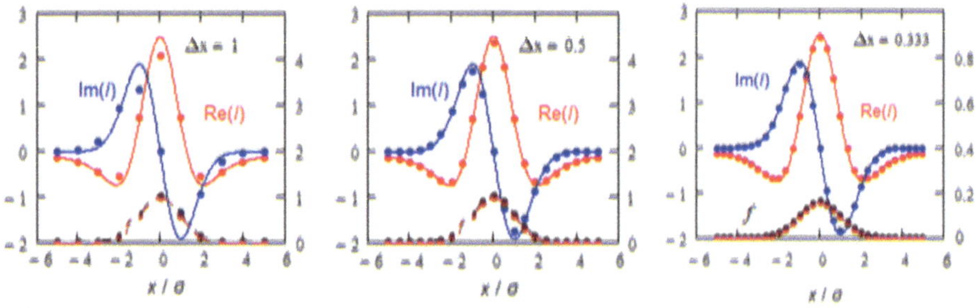

Fig. 7.51 Comparison of computation of Eq. (7.110) integral with $f(x) = e^{-x^2/2}$ using exact calculations (*solid lines*) and matrix approximation (*dots*)

effective computation, these matrices need to be precomputed and kept in computer memory.

Transverse Core Heating

Although the stacktail is a longitudinal cooling system and, on the first glance, should not excite the transverse motion, it actually does, affecting the motion of particles in the core and resulting in their transverse heating. The emittance growth is compensated by the core cooling systems. However, due to their insufficient cooling rates, the resulting emittance was not sufficiently small to make transfers to the Accumulator without particle loss or additional no-staking pause for transverse cooling.

Stacktail kickers have similar design and geometry to the stacktail pickups, and therefore, in accordance with the reciprocity theorem [10], the longitudinal kick and the pickup sensitivity depend similarly on the transverse coordinate. Expending Eq. (7.102) in Tailor series and taking into account that the pickup response satisfies to the Laplace's equation, one obtains the dependence of longitudinal kick on the particle transverse coordinates (X, Y):

$$U(X,Y) = U_0\left(1 + \kappa\frac{X^2 - Y^2}{2{a_{\text{eff}}}^2} + \cdots\right),$$
$$a_{\text{eff}} = \frac{h}{\pi}\cosh\left(\frac{\pi w}{2h}\right)\sqrt{\frac{\operatorname{atan}(\sinh(\pi w/(2h)))}{\sinh(\pi w/(2h))}}. \tag{7.113}$$

Here $\kappa = \pm 1$ with signs "+" and "−" assigned to the kickers rolled so that in the difference mode, they would be the horizontal or vertical kickers correspondingly, and the effective gap is $a_{\text{eff}} \approx 1.7$ cm for the Accumulator stacktail kickers ($h = w = 3$ cm, see Table 7.4). The Panofsky–Wenzel theorem [10] binds the transverse and longitudinal kicks—$\Delta\mathbf{p}_\perp = i v_0/\omega \nabla(\Delta p_\parallel/p)$. That yields

$$\frac{\delta p_x}{p} = \kappa \frac{i v_0}{\omega} \frac{\delta p}{p} \frac{x}{a_{\text{eff}}^2}, \quad \frac{\delta p_y}{p} = -\kappa \frac{i v_0}{\omega} \frac{\delta p}{p} \frac{y}{a_{\text{eff}}^2} \tag{7.114}$$

Below we consider the emittance growth in the horizontal plane. However, with exchange of index x by y, all formulas are directly applicable to the vertical plane as well.

The transverse kicks described above excite the emittance growth through three mechanisms. The first one is related to non-zero dispersion at the location of stacktail kickers. The second mechanism is related to offsets of kickers from the beam center, resulting in the transverse kicks proportional to the kicker offset. The emittance growth for both mechanisms is proportional to the beam Schottky noise on the betatron sidebands.[6] The third mechanism is associated with the parametric excitation of betatron motion, and, consequently, the emittance growth is excited by frequencies related to the sidebands of the double betatron frequency.

First, we consider the emittance growth related to the non-zero dispersion. In this case, a momentum change due to longitudinal kick, $\delta p/p$, in average excites the betatron motion with rms emittance increase:

$$\delta \varepsilon_x = \frac{1}{2} A_x \left(\frac{\delta p}{p} \right)^2, \tag{7.115}$$

where $A_x = (D_x{}^2 + (D'_x \beta_x + \alpha_x D_x)^2)/\beta_x$ is the dispersion invariant for x plane. Note that it is conserved in a straight line, i.e., it has the same value in all kickers. There is no quadrupoles in the straight section where all stacktail kickers are located. That makes a reference of all longitudinal kicks to a single point being straightforward. Consequently, the momentum change in Eq. (7.115) represents the total longitudinal kick of all kickers. Following the same procedure as in deriving Eq. (7.62), one obtains

$$\frac{d\varepsilon_x}{dt} = \frac{\pi A_x}{T_0{}^2} \sum_{n=-\infty}^{\infty} P_p(\omega_{n\perp}) \tag{7.116}$$

where $P_p(\omega)$ is the spectral density of momentum kicks excited by stacktail defined so that a single-turn momentum change is $\overline{(\delta p/p)^2} = \int P_p(\omega) d\omega$, $\omega_{n\perp} = \omega_{\text{core}}(n + \nu)$, ω_{core} is the revolution frequency in the core, and ν is the betatron tune.

[6] Here the same as above, we assume that the thermal noise of amplifiers is negligible in comparison to the particle noise. It is well justified for the stacktail in normal operating conditions.

Second, we consider the emittance growth related to transverse offsets of kickers. Let each kicker to make a horizontal kick $\theta_k \equiv \delta p_{xk}/p$. To find a single-pass emittance increase, we refer all kicks to the center of kicker straight section where the beta function achieves its minimum β^*. Then the total angular kick is $\sum \theta_k$ and the total displacement is $\sum S_k\theta_k$, where S_k is the distance between kth kicker and the location of β^*, and we took into account that there is no focusing in the kicker straight section. Averaging yields the emittance increase per turn to be equal to

$$\delta\varepsilon = \frac{1}{2}\left(\beta^*\left(\sum_{k=1}^{N_{\text{kick}}}\theta_k\right)^2 + \frac{1}{\beta^*}\left(\sum_{k=1}^{N_{\text{kick}}} S_k\theta_k\right)^2\right), \tag{7.117}$$

where N_{kick} is the number of kickers. Assuming that all kickers produce the same longitudinal kick $(\delta p/p)/N_{\text{kick}}$, using Eq. (7.114), and following the same procedure as in deriving Eq. (7.65), one obtains the emittance growth due to offset of kickers

$$\frac{d\varepsilon_x}{dt} = \frac{\pi}{2{T_0}^2}\sum_{n=-\infty}^{\infty} P_p(\omega_{n\perp})\left(\frac{v_0}{\omega_{n\perp}{a_{\text{eff}}}^2}\right)^2\left(\beta^*\left(\frac{1}{N_{\text{kick}}}\sum_{k=1}^{N_{\text{kick}}}\kappa_k x_k(\omega_{n\perp})\right)^2\right.$$

$$\left. + \frac{1}{\beta^*}\left(\frac{1}{N_{\text{kick}}}\sum_{k=1}^{N_{\text{kick}}} S_k\kappa_k x_k(\omega_{n\perp})\right)^2\right). \tag{7.118}$$

Here we also took into account that the electrical center of a kicker is frequency dependent in a practical system. Consequently, the kicker offsets are frequency dependent too.

Finally, we consider the emittance growth related to the parametric excitation. Each of eight stacktail kicker tanks has four kickers oriented in the same plane so that the higher-order modes could be damped. To minimize the parametric excitation, each next kicker tank is rolled to the orthogonal plane. In the first order, it cancels the parametric kick; however, in the second order, the perturbation is still significant.

Let x_0 and θ_0 be particle coordinates in front of the first kicker, and x_k and θ_k be particle coordinates immediately behind of the kth kicker. Then the iterative equations bounding up particle angles and coordinates are

$$\begin{aligned} \theta_{k+1} &= \theta_k + \Phi_{k+1}x_{k+1}, \\ x_{k+1} &= x_k + \theta_k(S_{k+1} - S_k), \end{aligned} \qquad k > 0, \tag{7.119}$$

where Φ_k is the focal strength of the kth kicker. In the first order, one obtains

$$\begin{aligned}
x_k^{(1)} &= x_0 + (S_k - S_1)\theta_0, \\
\theta_k^{(1)} &= \theta_0 + \sum_{j=1}^{k} \left(x_0 + \theta_0 (S_j - S_1)\right) \Phi_j, \\
x_k^{(2)} &= x_0 + \sum_{j=1}^{k} (S_j - S_{j-1}) \left(\theta_0 + \sum_{i=1}^{j} (x_0 + \theta_0 (S_i - S_1)) \Phi_i \right), \\
\theta_k^{(2)} &= \theta_0 + \sum_{j=1}^{k} \left(x_0 + \sum_{i=1}^{j} (S_j - S_i) \left(\theta_0 + \sum_{j=1}^{k} (x_0 + \theta_0 (S_i - S_1)) \Phi_i \right) \right) \Phi_j,
\end{aligned} \tag{7.120}$$

where the same as above the kicker coordinates, S_k, are referenced to the point where the beta-function achieves its minimum. Performing substitutions in Eq. (7.120) and relating the obtained transfer matrix, $\mathbf{M}_t$, to the location of beta-function minimum (to $S = 0$) after lengthy calculations, one obtains the transfer matrix describing instant particle coordinate jump at $S = 0$:

$$\begin{aligned}
\mathbf{M} &= \begin{bmatrix} 1 & -S_{N_{\text{kick}}} \\ 0 & 1 \end{bmatrix} \mathbf{M}_t \begin{bmatrix} 1 & S_1 \\ 0 & 1 \end{bmatrix} \\
&= \begin{bmatrix} 1 - \sum_{i=1}^{N_{\text{kick}}} \Phi_i S_i \left(1 + \sum_{j=1}^{i} \Phi_j (S_i - S_j) \right) & -\sum_{i=1}^{N_{\text{kick}}} \Phi_i S_i \left(S_i + \sum_{j=1}^{i} \Phi_j S_j (S_i - S_j) \right) \\ \sum_{i=1}^{N_{\text{kick}}} \Phi_i \left(1 + \sum_{j=1}^{i} \Phi_j (S_i - S_j) \right) & 1 + \sum_{i=1}^{N_{\text{kick}}} \Phi_i \left(S_i + \sum_{j=1}^{i} \Phi_j S_j (S_i - S_j) \right) \end{bmatrix}.
\end{aligned} \tag{7.121}$$

To find the emittance increase per turn, we consider a particle experiencing betatron oscillations with single-particle emittance (action) ε. Then its coordinates are

$$\mathbf{x} \equiv \begin{bmatrix} x \\ \theta \end{bmatrix} = \sqrt{2\varepsilon\beta^*} \begin{bmatrix} \cos\psi \\ -\sin\psi/\beta^* \end{bmatrix}. \tag{7.122}$$

After passing the "point-like" kicker, the new particle coordinate vector is $\mathbf{x}' = \mathbf{M}\mathbf{x}$, and the corresponding new emittance is

$$\varepsilon'(\psi) = \frac{1}{2} \left(\frac{x'^2}{\beta^*} + \beta^* \theta'^2 \right). \tag{7.123}$$

Expressing x' and θ' through elements of matrix $\mathbf{M}$ and vector $\mathbf{x}$ and performing averaging over phase ψ , one obtains

$$\varepsilon' = \frac{\varepsilon}{2}\left({M_{11}}^2 + {M_{22}}^2 + {M_{12}}^2 + {M_{21}}^2\right). \tag{7.124}$$

Substituting matrix elements of matrix in Eq. (7.121), we obtain the average emittance increase per turn:

$$\varepsilon' = \varepsilon\left(1 + \frac{1}{2}\sum_{i=1}^{N_{\text{kick}}}\sum_{j=1}^{N_{\text{kick}}}\Phi_i\Phi_j\left(\beta^{*2} + \frac{{S_i}^2{S_j}^2}{\beta^{*2}} + 2S_iS_j - \left(S_i - S_j\right)^2\right)\right). \tag{7.125}$$

Introducing the kick magnitude,

$$\Phi_q = \sum_{j=1}^{N_{\text{kick}}}|\Phi_i|, \tag{7.126}$$

and the effective β-function of parametric heating,

$$\beta_q = \frac{1}{\Phi_q}\sqrt{\sum_{i=1}^{N_{\text{kick}}}\sum_{j=1}^{N_{\text{kick}}}\Phi_i\Phi_j\left(\beta^{*2} + \frac{{S_i}^2{S_j}^2}{\beta^{*2}} + 2S_iS_j - \left(S_i - S_j\right)^2\right)}, \tag{7.127}$$

we obtain the emittance increase per turn:

$$\delta\varepsilon \equiv \varepsilon' - \varepsilon = \varepsilon\frac{{\beta_q}^2}{2}{\Phi_q}^2. \tag{7.128}$$

Following the same procedure as in deriving Eq. (7.62), one obtains the emittance growth due to parametric heating:

$$\left.\frac{d\varepsilon_x}{dt}\right|_{\text{param}} = \frac{\pi\varepsilon_x{\beta_q}^2}{{T_0}^2}\sum_{n=-\infty}^{\infty}P_\Phi\left(\omega_{nq}\right) = \frac{\pi\varepsilon_x{\beta_q}^2{\text{v}_0}^2}{{T_0}^2{a_{\text{eff}}}^2}\sum_{n=-\infty}^{\infty}\frac{P_p\left(\omega_{nq}\right)}{{\omega_{nq}}^2}, \tag{7.129}$$

where $\omega_{nq} = \omega_{\text{core}}(n + 2\nu)$ are the sideband frequencies of double betatron frequency and $P_\Phi(\omega)$ is the spectral density of Φ_q, and to obtain the second equality, we used its relationship to the spectral density of momentum kicks: $P_\Phi(\omega) = {\text{v}_0}^2\ P_p(\omega)/(\omega {a_{\text{eff}}}^2)^2$ (see Eq. (7.114)). It was also implied that all kickers are perfectly phased, i.e., their longitudinal kicks are added up with the same sign, while signs can be different for focusing. In other words, the total momentum kick is proportional to Φ_q defined in Eq. (7.126). Note also that in difference to the emittance growth rates due to kicker offsets and dispersion mismatch, which do not depend on the emittance and are driven by the betatron frequency sidebands, the parametric heating is proportional to the beam emittance and is driven by the sidebands of double betatron frequency.

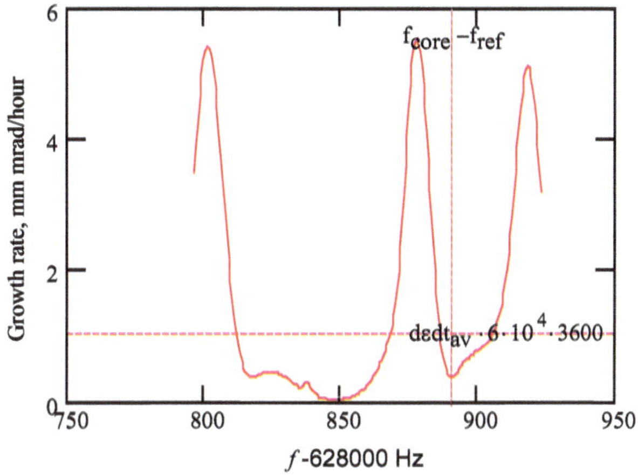

Fig. 7.52 Dependence of transverse heating rate on the revolution frequency before the Accumulator optics upgrade. *Horizontal line* marks the heating rate averaged over particle distribution

In normal operating conditions, the spectral density of stacktail momentum kicks used in Eqs. (7.116), (7.118) and (7.129) is mainly related to the beam Schottky noise which significantly exceeds the thermal noise of amplifiers. In this case, the spectral density can be expressed through the particle distribution and the system gain. Using Eqs. (7.12), (7.13) and (7.50) one obtains

$$P_p(\omega) = \frac{T_0}{2\pi}\sum_{k=-\infty}^{\infty}\frac{G_{ST}\left(x_f(\omega/k),\omega\right)f\left(x_f(\omega/k)\right)}{\left|k\eta\left(x_f(\omega/k)\right)\right|}, \quad \eta(x) = -\frac{1}{\omega}\frac{d\omega}{dx}, \tag{7.130}$$

where the stacktail gain $G_{ST}(x, \omega)$ is determined by Eq. (7.103) and the slip factor $\eta(x)$ and the function $x_f(\omega)$ are defined in Eq. (8.50) and the paragraph following it.

The stacktail system uses eight kicker tanks located close to each other in one straight section. Each of kicker tanks has four kickers oriented in the same plane. Initially one of these 32 kickers was used for the longitudinal 2–4 GHz core cooling other 31 for the stacktail. It became apparent that parametric heating was a problem. It was resolved after one of two kickers on each side of kicker straight section was switched off and another one was transferred to the core cooling system, so that only 30 kickers were used in the stacktail. That reduced the effective beta-function of the parametric heating, β_q, from 2.3 to 0.6 m, resulting in negligible parametric heating.

After the equalizer installation, we observed the strong transverse heating again. This time it was excited by a resonance in the kickers which became much more apparent with the increased bandwidth. The resonance occurs at 3.25 GHz and results in a resonant displacement of kicker electrical center with frequency. The amplitude of the displacement is ~2 mm and the quality factor is ~27. Figure 7.52 presents dependence of computed horizontal heating rate on the revolution frequency. Tables 7.5 and 7.6 present measured heating and cooling rates for the horizontal degree of freedom before the optics upgrade. Although we performed very accurate correction of both vertical and horizontal dispersions (to a few cm

Table 7.5 Heating and cooling rates during normal operation of the antiproton stochastic cooling systems

Heating mechanisms	mm mrad/h
IBS heating at 50 mA	~3
Stacktail heating	5–6
Noise of core systems	~2
Total heating=total cooling	~10

Table 7.6 Estimates of the stacktail system heating mechanisms

Heating mechanisms	mm mrad/h
Parametric heating	~0.25
Dispersion mismatch	~2.4
Kicker offset (res. at 3.25 GHz)	~1.2–2.2
Unaccounted (most probably due to geometric kicker offset)	~1.1

level), the contribution of the dispersion mismatch, as one can see from the table, is not negligible.

The optics upgrade increased the slip factor and resulted in the displacement of heating peaks (related to lower and upper betatron sidebands) so that the core became better centered between the peaks. That also reduced the heating. In addition it reduced IBS and improved the core cooling, resulting in acceptable values for transverse emittances.

7.3 Electron Cooling and Its Fundamentals

7.3.1 Introduction

Electron cooling is the method of increasing the phase density of "hot" heavy charge particles, ions, or antiprotons, through Coulomb interaction with a "cold" electron beam propagating with the same average speed. The method was proposed by G. Budker in 1967 [44], successfully tested in 1974 with low-energy protons [45], and later implemented at a dozen of storage rings (see, e.g., the review [46]) at nonrelativistic energies E_e<300 keV.

An electron cooler was envisioned as an important part of the Recycler Ring upgrade already in the Recycler Technical Design Report [47]. The main cooler parameters were chosen according to the then-stated Recycler goals: store antiprotons coming from the Accumulator, "recycle" antiprotons left over from Tevatron stores, and prepare bunches for Tevatron shots. Because of the longitudinal injection scheme of the Recycler, the main emphasis was made on longitudinal cooling. With a typical required cooling time of tens of minutes, the scheme without a strong magnetic field in the cooling section was shown to be satisfactory. Later changes, most notably the decision to do not use antiprotons from the Tevatron for "recycling" and the decreased emittances of bunches coming from the

Accumulator, relaxed the operational requirements for the electron cooling strength and allowed operation with a good safety factor. As soon as the issues with reliable recirculation of the electron beam had been resolved, for the first time, relativistic electron cooling was demonstrated [48] and within days was put into operation.

This chapter discusses formulae for nonmagnetized electron cooling, the setup and reasoning behind choosing the adopted cooler's scheme, and then goes through the main difficulties with the realization of the project: electron beam transport, stability of the beam recirculation, electron angles in the cooling section, and energy matching. The last section of this chapter describes the cooling force measurements and results.

For brevity, the Recycler Electron cooling system will be referred as REC.

7.3.2 Electron Cooling Formulae

A heavy charged particle moving in a free electron gas with a velocity distribution $f_e\left(\vec{v}_e\right)$ experiences a friction force that in a model of binary collisions can be written following [49]:

$$\vec{F}_b\left(\vec{V}_p\right) = -\frac{4\pi e^4 n_{eb}}{m_e}\eta \int L_C \frac{f_e\left(\vec{v}_e\right)}{\left(\vec{V}_p - \vec{v}_e\right)^2} \frac{\vec{V}_p - \vec{v}_e}{\left|\vec{V}_p - \vec{v}_e\right|} d^3 v_e, \tag{7.131}$$

where n_{eb} is the electron density in the beam rest frame, m_e the electron mass, $\vec{V}_p$ the velocity of the heavy particle, and $\eta = L_{cs}/C$ indicates the portion of the ring circumference C occupied by the cooling section of length L_{cs}. L_C is the Coulomb logarithm

$$L_C = \ln\left(\frac{\rho_{\max}}{\rho_{\min}}\right). \tag{7.132}$$

with the minimum and maximum impact parameters, $\rho_{\min}$ and $\rho_{\max}$, in the Coulomb logarithm defined as

$$\rho_{\min} = \frac{e^2}{m_e\left(\vec{V}_p - \vec{v}_e\right)^2}, \rho_{\max} = \min\left\{R_D, R_e, \left|\vec{V}_p - \vec{v}_e\right| \cdot \tau_f\right\}. \tag{7.133}$$

The maximum impact parameter can be determined by the electron beam radius R_e (typically the case in REC), by the Debye radius R_D, or by the relative displacement of particles during the flight time through the cooling section $\tau_f = (L_{cs}/\gamma\beta c)$, where γ and β are the relativistic factors of co-propagating particles in the lab frame.

Note that the presence of a strong longitudinal magnetic field ~1 kG, standard for low-energy coolers, makes cooling dynamics significantly more complicated (see, e.g., [46, 49]) and typically provides significantly stronger cooling. We did not find a way to numerically simulate this effect for the REC case, where electrons are immersed into a much weaker field of $B_{CS} = 105$ G but believe that Eq. (7.131) is a reasonable approximation. One may speculate that the magnetic field does not affect collisions with "small" impact parameters ρ

$$\rho \leq \rho_L \approx \frac{v_e \cdot m_e c}{eB_{CS}} \tag{7.134}$$

but enhances cooling at "large"

$$\rho > \rho_L \tag{7.135}$$

by "suppressing" the transverse electron velocities contribution. For typical REC parameters, $\rho_{\min} \approx 1 \times 10^{-7}$ cm, $\rho_L \approx 0.02$ cm, $\rho_{\max} \approx 0.2$ cm, and the value of the Coulomb logarithm is significantly larger for the region of Eq. (7.134), $\ln(\rho_L/\rho_{\min}) \approx 12$, than $\ln(\rho_{\max}/\rho_L) \approx 2.3$ for large impact parameters. An enhancement of the cooling force by the magnetic field for large impact parameters is unlikely to offset this difference because the typical value of the antiproton velocity (in the beam frame) is only a factor of 2–3 lower than the electrons'. Probably, a more convincing argument comes from measurements, where a strong dependence of cooling properties on the transverse electron angles was observed.

Assuming a Gaussian distribution for all velocity components

$$f\left(\vec{v}_e\right) = \frac{1}{(2\pi)^{3/2}\sigma_{ex}\sigma_{ey}\sigma_{ez}} \exp\left(-\frac{{v_{ex}}^2}{2\sigma_{ex}^2} - \frac{{v_{ey}}^2}{2\sigma_{ey}^2} - \frac{{v_{ez}}^2}{2\sigma_{ez}^2}\right) \tag{7.136}$$

and neglecting variations of the Coulomb logarithm in order to take it out from the integrand, Eq. (7.131) can be reduced to a single integral (so-called Binney formula, see, e.g., [50]). For example, the expression for the longitudinal component of the cooling force in the beam frame becomes

$$F_{bz}\left(V_{px}, V_{pz}, V_{pz}\right) = -\frac{4\Lambda}{\sqrt{\pi}} \cdot \int_0^\infty dt \frac{\exp\left(-\frac{t^2{V_{px}}^2}{1+2\cdot\sigma_{ex}^2 t^2} - \frac{t^2{V_{py}}^2}{1+2\cdot\sigma_{ey}^2 t^2} - \frac{t^2{V_{pz}}^2}{1+2\cdot\sigma_{ez}^2 t^2}\right)}{\sqrt{\left(1+2\cdot\sigma_{ex}^2 t^2\right)\left(1+2\cdot\sigma_{ey}^2 t^2\right)\left(1+2\cdot\sigma_{ez}^2 t^2\right)}} \cdot \frac{V_{pz} t^2}{1+2\cdot\sigma_{ez}^2 t^2}$$

$$\Lambda \equiv \frac{4\pi e^4 n_{eb}}{m_e} \eta L_C \tag{7.137}$$

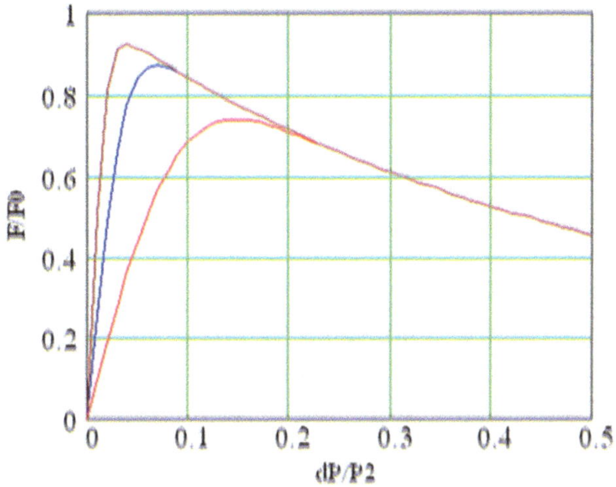

Fig. 7.53 Longitudinal force as a function of the momentum offset calculated with Eq. (7.138). *Vertical axis*, $F_{lz}(\Delta p_p)/F_0$, and *horizontal axis*, $\Delta p_p/p_2$. The curves show calculations for different ratios of electron velocities in the beam frame $p_1/p_2 \approx \sigma_{ez}/\sigma_{et}$: *red*, 10; *blue*, 25; *brown*, 50

This formula can be simplified further for the analysis of the longitudinal cooling force measurements. These measurements use a pencil antiproton beam with a large average momentum offset, so that the transverse antiproton velocity components can be neglected. Assuming also $\sigma_{ex}=\sigma_{ey}\equiv\sigma_{et}$, it is convenient to use Eq. (7.137) to fit the longitudinal cooling force as a function of momentum offset in the lab frame in the form of [51]

$$F_{lz}(\Delta p_p) = -F_0 \frac{2}{\sqrt{\pi}} \int_0^{\frac{\Delta p_p}{p_1}} \frac{e^{-u^2} u^2}{u^2 + \left(\frac{\Delta p_p}{p_2}\right)^2} du \tag{7.138}$$

with the parameters related to the lab-frame electron beam properties as follows:

$$\begin{aligned} p_1 &= \delta W_e \cdot \sqrt{2} \frac{M_p}{\beta m_e c} \\ p_2 &= \vartheta_t \cdot \sqrt{2} \gamma^2 \beta c M_p \\ F_0 &= \frac{\Lambda}{\sigma_{et}{}^2 - \sigma_{ez}{}^2} = \frac{n_{el}}{\vartheta_t{}^2} \cdot \frac{4\pi \cdot e^4 \eta \cdot L_c}{m_e c^2 \gamma^3 \beta^2}, \end{aligned} \tag{7.139}$$

where $\vartheta_t = \frac{1}{\gamma\beta c}\sqrt{\sigma_{et}{}^2 - \sigma_{ez}{}^2}$ is an effective angle, $\vartheta_e = \frac{\sigma_{et}}{\gamma\beta c}$ is a 1 day electron rms electron angle in the cooling section, $\delta W_e = p_e \sigma_{ez}$ is rms scatter of the electron energy, $n_{el} = \gamma\ n_{eb}$ is electron density in the lab frame, M_p is the proton mass, and p_e is the electron momentum. This parameterization (7.138) is convenient for fitting purposes because F_0 determines the maximum of the curve (the force maximum approaches F_0 at $\sigma_{et} \gg \sigma_{ez}$), the parameter $p_1$is approximately the position of the curve's maximum, and $p_2$characterizes the curve's drop rate after the maximum. Figure 7.53 presents results of calculations with Eq. (7.138).

Under the assumptions used for Eq. (7.137) and $\sigma_{ex} = \sigma_{ey} \equiv \sigma_{et}$, the cooling rates of a coasting antiproton beam with a Gaussian distribution

$$f_p\left(\vec{V}_p\right) = \frac{1}{(2\pi)^{3/2}\sigma_{px}\sigma_{py}\sigma_{pz}}\exp\left(-\frac{V_{px}^2}{2\sigma_{px}{}^2} - \frac{V_{py}^2}{2\sigma_{py}{}^2} - \frac{V_{pz}^2}{2\sigma_{pz}{}^2}\right) \tag{7.140}$$

with $\sigma_{px} = \sigma_{py}$ can be expressed through elementary functions [51]

$$\frac{d\sigma_{pz}}{dt} = -\sqrt{\frac{2}{\pi}}\frac{\Lambda \cdot \sigma_{pz}}{M_p\sqrt{\sigma_{ez}{}^2 + \sigma_{pz}{}^2}\left(\sigma_{ex}{}^2 + \sigma_{px}{}^2\right)} \cdot f_{\text{long}}(\alpha);$$

$$\frac{d\sigma_{px}}{dt} = -\sqrt{\frac{1}{2\pi}}\frac{\Lambda \cdot \sigma_{px}}{M_p\left(\sigma_{ex}{}^2 + \sigma_{px}{}^2\right)^{3/2}} \cdot f_{tr}(\alpha);$$

$$f_{\text{long}}(\alpha) = \begin{cases} \dfrac{1 - \sqrt{\dfrac{\alpha}{1-\alpha}} \cdot \operatorname{acos}(\sqrt{\alpha})}{1-\alpha}, & \alpha < 1 \\ \dfrac{1}{3}, \quad \alpha = 1 \\ \dfrac{\sqrt{\dfrac{\alpha}{\alpha - 1}} \cdot \operatorname{acosh}(\sqrt{\alpha}) - 1}{\alpha - 1}, & \alpha > 1 \end{cases}; \quad \alpha = \frac{\sigma_{ez}{}^2 + \sigma_{pz}{}^2}{\sigma_{ex}{}^2 + \sigma_{px}{}^2};$$

$$f_{tr}(\alpha) = \begin{cases} \dfrac{\operatorname{acos}\sqrt{\alpha} - \sqrt{\alpha(1-\alpha)}}{2(1-\alpha)^{3/2}}, & \alpha < 1 \\ \dfrac{1}{3}, \quad \alpha = 1 \\ \dfrac{\sqrt{\dfrac{\alpha}{\alpha - 1}} \cdot \operatorname{acosh}(\sqrt{\alpha}) - 1}{\alpha - 1}, & \alpha > 1 \end{cases}; \tag{7.141}$$

Note that for the transverse rate, the calculated single-pass force is decreased in Eq. (7.139) by a factor of 2 to take into account averaging over betatron oscillations. Functions appearing in Eq. (7.139) are shown in Fig. 7.54. For small values of the parameter $\alpha \ll 1, f_{tr}(\alpha) \approx \pi/4 - \sqrt{\alpha}/2$, $f_{\text{long}}(\alpha) \approx 1 - \pi/2 \cdot \sqrt{\alpha}$.

For a practical implementation, Eq. (7.141) gives only an upper estimate for the cooling rate. The most important flaw in the model is the assumption of the electron beam density and velocity distributions being constant across the area determined by $\rho_{\max}$. In the case of the Recycler cooler operational parameters, the rms radius of the antiproton beam is roughly the same as the radius of the area within the electron

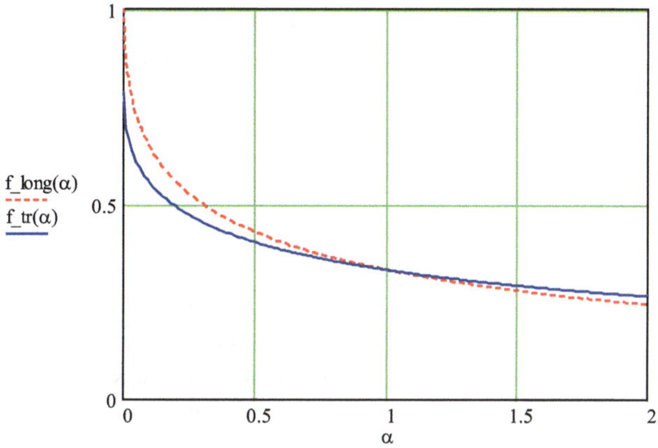

Fig. 7.54 Functions referenced in Eq. (7.141) (see text)

beam where electron cooling is effective. As a result, a large portion of antiprotons travels through the cooling section outside of the electron beam; hence the cooling rate decreases dramatically. With the tails of the antiproton distribution being typically fatter than Gaussian, this effect is even more pronounced. In addition, bringing the Coulomb logarithm out of the integral tends to slightly overestimate the cooling force for small relative velocities.

7.3.3 Cooler Setup: Electron Beam Design Parameters and Choice of the Scheme

Based on preliminary cooling scenarios and estimations of the cooling rates, the design parameters were specified in [61] (reproduced in Table 7.7).

Table 7.7 assumes a scheme with a DC electron beam, a longitudinal magnetic field at the cathode region and in the cooling section, and lumped focusing in the beam transport lines. Before finally choosing this scheme, other possibilities were considered as well.

By the time the Recycler Electron Cooling project started, several schemes for coolers with electron energy of several MeVs or higher had been investigated [52–54] (and the development of other schemes continues at present [55, 56]). However, none of these developments came close to demonstrating the electron beam parameters necessary for electron cooling in the Recycler.

In all previously existed coolers, a strong (~1 kG) longitudinal magnetic field was used to transport the electron beam and enhance the cooling force, but all of them worked at nonrelativistic energies $E_e < 300$ keV. Therefore, a straightforward extrapolation of the previous experience would be a version with a higher-voltage electrostatic accelerator and a continuous strong longitudinal magnetic field from the cathode to the collector. This was seriously discussed but eventually abandoned. On one hand, the Recycler did not require the benefits from cooling enhancement

Table 7.7 Electron cooling system design parameters

Parameter	Value	Units
Electrostatic accelerator		
Terminal voltage	4.3	MV
Electron beam current	0.5	A
Terminal voltage ripple (rms)	500	V
Cathode radius	2.5	mm
Magnetic field at the cathode	≤600	G
Cooling section		
Length	20	m
Solenoid field	≤150	G
Vacuum pressure	0.1	nTorr
Electron beam radius	6	mm
Electron beam divergence	≤80	μrad

by the strong field. On the other hand, the beam generation scheme tested for 1 MeV [53] was not easily scalable to 4 MeV and would have required significant R&D efforts (similar to what is being presently developed for the COSY cooler at Novosibirsk [56]). Combined with more expensive beam lines and cooling section as well as higher operating costs, it looked undoable within a realistically available budget and time.

The most affordable solution in terms of time and cost seemed to be the scheme of relativistic cooling proposed for the SSC MEB [57], which assumed focusing in the cooling section by lumped elements. Namely, the beam is generated without a longitudinal magnetic field, and then short solenoidal lenses periodically placed in the cooling section compensate the beam divergence caused by space charge and beam emittance. This scheme allows using an industrially manufactured electrostatic accelerator, and the cooling section is significantly cheaper than in the case with a strong magnetic field. This scheme was critically analyzed in [58]. The authors argued that the requirement of low transverse velocities in the cooling section results in a large value of the beta function, which makes the beam susceptible to perturbations. Specifically, a drift instability due to interaction of the beam with its wall image charges and an ion instability were pointed out as possible showstoppers. In addition to these arguments, cooling inside the lenses is ineffective because of the large azimuthal velocity of the electrons. The corresponding decrease of the effective cooler length does not allow a frequent placement of these lenses.

Note that the contradiction between small angles and strong focusing is very different in the scheme with continuous magnetic field, because an electron propagates along the field line corresponding to its origin. If the field magnitude is changed in the entire beam line, the value of angles stays the same.

The decision was made to combine the advantages of these two schemes by introducing a longitudinal magnetic field in the cooling section that is large enough to counteract possible perturbations but still low to allow beam transport outside of the cooling section with lumped elements and usage of a standard electrostatic accelerator with minimal modifications. An important requirement of this new scheme is the cathode immersion into a longitudinal field as well, so that the corresponding magnetic fluxes are matched [59].

Note that applicability of such scheme is critically dependent on the magnetic flux required in the cooling section. When a beam with no transverse velocities inside a solenoid exits into a free space, conservation of the canonical angular momentum results in a coherent angular rotation of the beam. In the paraxial ray approximation, it is equivalent to the appearance of an effective normalized emittance [59]

$$\varepsilon_{B,\mathrm{eff}} = \frac{e\Phi}{2\pi m_e c^2}, \tag{7.142}$$

where $\Phi = B_{CS} R_{CS}^2$ is the magnetic flux through the beam cross section in the solenoid. As in the case with a real emittance, the beam transport with lumped focusing is possible only if this emittance is low enough. For example, let's consider a transport channel for $\gamma = 10$ with a typical beam radius of ~1 cm and the beta function of ~1 m. If the required beam radius in a cooling section solenoid is $R_{CS} \sim 1$ cm, Eq. (7.142) limits the solenoid magnetic field to ~300G. To use a lumped focusing during acceleration (i.e., at lower γ), the magnetic flux should be decreased even further in comparison with this example by reducing both the beam size and the magnetic field strength in the cooling section down (in the REC case, to $R_{CS} = 2$–4 mm and $B_{CS} = 100$–200 G).

7.3.4 Electron Cooler Setup Description

A schematic of the electron cooler is shown in Fig. 7.55, and its elevation views are presented in Fig. 7.56.

Electrons are emitted by a thermionic cathode, accelerated inside an electrostatic accelerator, Pelletron [60], and transported through a beam "supply" line to the cooling section where they interact with antiprotons circulating in the Recycler. After separation of the beams by a 180° bend, electrons move through the "return" beam line out of the tunnel and then through a "transfer" line back to the Pelletron. Inside the Pelletron, the electron beam is decelerated in the second (deceleration) tube and is absorbed in a collector at the kinetic energy of 3.2 keV. While the final design of the gun and its performance have not been published, the main ideas for a low-halo gun are described in [98].

The vacuum chamber is pumped down by ion and titanium sublimation pumps. The typical diameter of the beam line vacuum chamber is 75 mm (the beam aperture is limited by the BPM's inner diameter of 47 mm).

When both main bending magnets under the Pelletron are turned off, the beam can be recirculated through a short beam line, denoted as U-bend in Fig. 7.56. This so-called U-bend mode was used for commissioning purposes. For instance, in this mode we were able to reach DC beam currents of up to 1.8 A at the nominal energy.

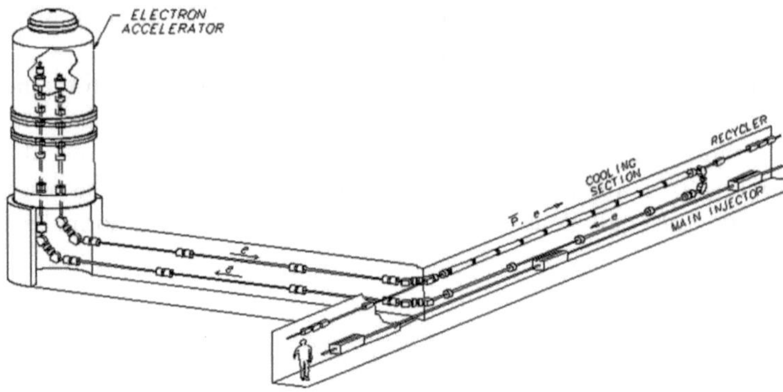

Fig. 7.55 Schematic of the electron cooler

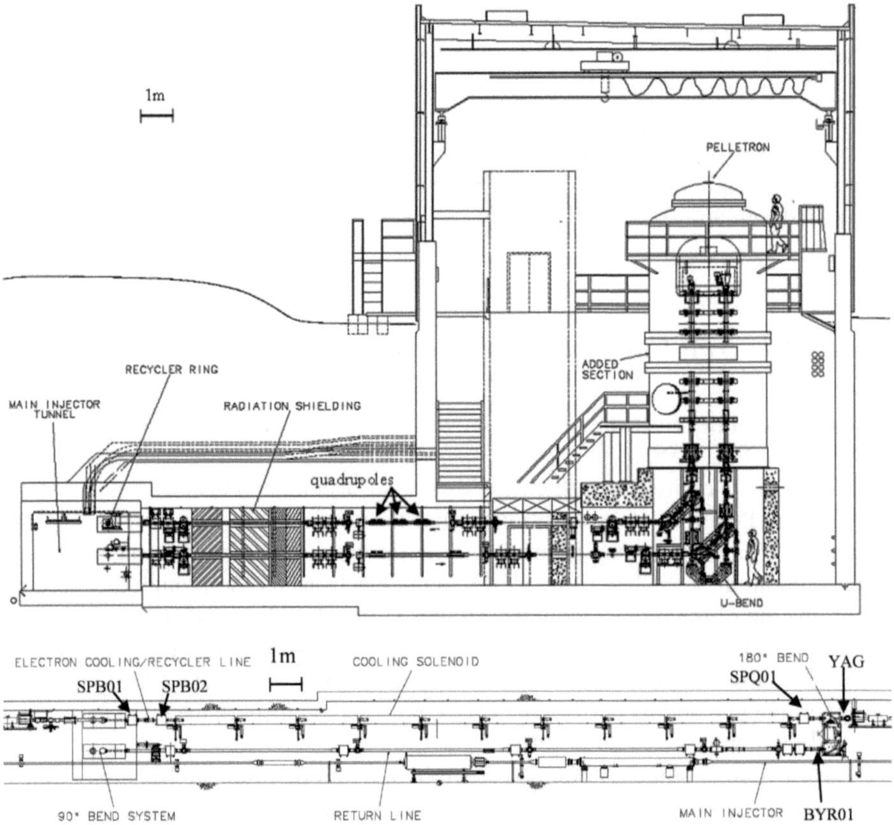

Fig. 7.56 Upper, elevation view showing the Pelletron and beam lines ("supply" and "transfer"). Lower, elevation view of the portion of the Main Injector tunnel containing the cooling section and the "return" line

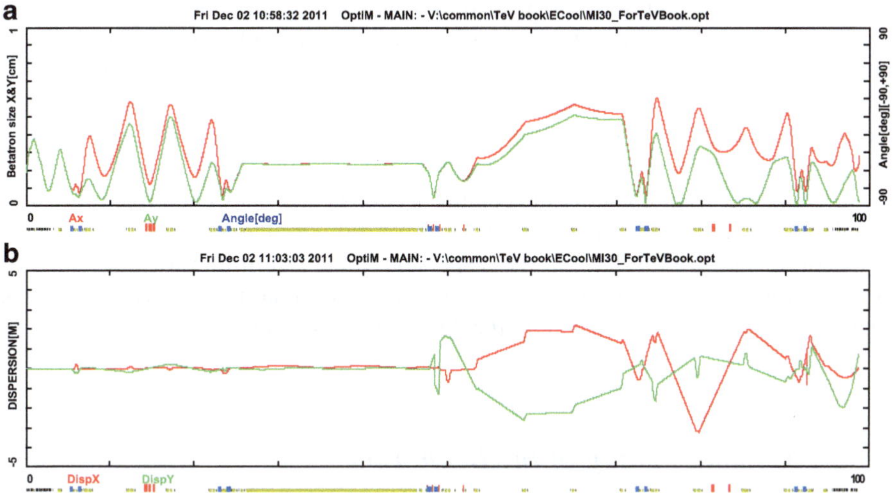

Fig. 7.57 The electron beam envelope (**a**) and dispersion (**b**) simulated with the OptiM code. Settings of optical elements are as they were in regular operation, except for minor adjustments to the correction quadrupoles in the supply line. Initial conditions are adjusted to have a parallel beam in the cooling section. $I_e = 0.1$ A

7.3.5 Main Points of Concern for the Design of the Cooler

The following issues have been identified as the most important for successful operation of the electron cooler:

- Electron beam transport
- Stability of the beam recirculation
- Electron angles in the cooling section
- Energy ripple, stability, and absolute calibration

These issues are discussed in detail in the following sections.

7.3.6 Electron Beam Transport: Design

The beam envelope generated by the OptiM code [62] is shown in Fig. 7.57a. The main features of the electron cooling line design are as follows:

- Magnetic flux through the beam cross section at the cathode is equal to the flux at the cooling section (for a negligible space charge).
- Ideally, the beam line between the Pelletron and the cooling section has a rotation-invariant map.
- Zero dispersion in the cooling section.
- Possible rotation invariance and zero dispersion in the return line.

The first three conditions are needed to obtain and maintain low electron angles in the cooling section, and the last one was considered to be useful for the beam transport inside the deceleration tube (but eventually was not used). The idea of the transport channel and its practical implementation in the cooler's prototype are described in detail in [63]. The final beam line differs from the prototype only by the length and the number of focusing elements.

Keeping the beam line after the cooling section nearly dispersion free was tested but found to be disadvantageous because of full discharges. For an optimum protection from full discharges, the dispersion in the return line was made significantly larger than in the deceleration tube, ensuring that a beam with a lowered energy is lost well before reaching the deceleration tube electrodes.

7.3.7 Commissioning of the Beam Line

The relatively low momentum of the electrons makes the beam sensitive to various residual or fringe magnetic fields, which complicated the initial beam transport. In addition, calibration of the lenses' strength and knowledge of the initial conditions in simulations was found to lack the accuracy needed to pass the beam through without significant tuning of focusing.

Commissioning of the beam line was made first with a pulse mode. The gun generates 2 μs, 1 Hz pulses that are analyzed by the beam line BPMs demodulating at 130 kHz [64]. When the beam hits the vacuum chamber somewhere close to a BPM, the secondary electron shower creates a large signal at the BPM sum (intensity) output. The intensity was recorded as a function of the current of an immediately upstream corrector, and the current was set to center the distribution. An indication of the loss-free beam passage through a BPM was independence of the intensity on the transverse position in some interval of corrector currents. Typically, this intensity was by several times lower than its peak with scraping.

The most complicated tuning was inside the deceleration tube, where there are no BPMs. The main indicators were the beam loss to the bottom tube electrode, the peaks in the anode power supply current, and the total drop of the terminal voltage during the pulse.

After passing through the entire beam line a pulsed beam at ~0.5 A with losses not resolvable in the pulsed mode (<10 mA), a several-mA DC beam went through as well. Further tuning toward higher DC beam currents was made with analyzing aperture restrictions in various locations by measuring the beam loss as a function of a transverse position.

The initial hope was to use the optics model for fine-tuning of the beam angles in the cooling section. The optics was measured by deflecting beam by dipole correctors, one at a time, or by changing the beam energy while recording BPM readings [65]. The resulting differential trajectories were analyzed with OptiM. Multiple sets of these measurements allowed correcting errors in electrical connections as well as adjusting calibrations of the optical elements.

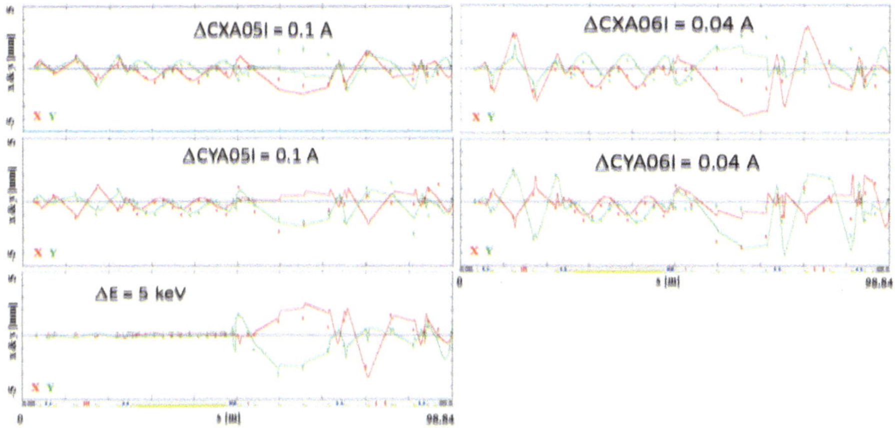

Fig. 7.58 Example of a standard optics measurement. Points are the measured differential trajectories, and the solid lines are fits by OptiM with fudge factors adjusted in previous measurements. Upper four plots show responses to kicks by two pairs (*X* and *Y*) of correctors (last pairs in the Pelletron and first pair outside, before the first 90° bend), and the lower plot is the result of an energy increase. All trajectories are differential, i.e., shown after subtraction of the unperturbed trajectory. *Red* and *green* curves represent *X* and *Y* projections, correspondingly. The error bars for the data points are the statistical errors of measurements

While the procedure permitted to correct the calibrations down to the level of several percents, more accurate results were not achieved. Sets of differential trajectory measurements made with different correctors or days apart disagreed well above the statistical BPM errors of ~10 μm. As a result, the agreement of the model with the measured trajectories was good only within relatively short portions of the beam line, but attempts to trace the entire cooler were not satisfactory. An example of such measurements and its comparison with the corresponding OptiM simulation is shown in Fig. 7.58. The differences between the data and the simulations seen on Fig. 7.58 were likely the result of mechanical drifts of the optical elements, insufficient stability of the power supplies, and nonlinearity of the focusing fields.

Another complication is the accuracy to which the initial conditions are known. They were derived from simulating the gun with the UltraSAM code [66] and then propagating the simulation through the acceleration tube with the BEAM code [67] to the tube exit, where the OptiM optics file begins. The simulations were tested against measurements of the beam profile with an optical transition radiation (OTR) monitor, which is mounted right below the first bend. While a good agreement was reported [68], resolution of the measurements might not be at the level needed to rely upon for fine-tuning.

Finally, simulations did not take into account the effect of secondary ion accumulation, which makes a significant contribution to focusing.

Not knowing the optics with good precision, tuning of focusing into the cooling section was made by using of a set of 11 scrapers with round openings [69]. With the electron beam off, one scraper is inserted into the beam path, such that when the beam is reestablished, it goes through the scraper hole. Then, the beam was moved by correctors in eight directions until ~10^{-5} portion of the beam was scraped each time, and the beam shape was determined from known calibration of the correctors. Repeating the procedure in all scrapers one by one gave the beating of the beam ellipse along the cooling section. While by design this diagnostic allows only to align the beam halo, which angles can differ significantly from the core [70], it was sufficiently good to demonstrate the first cooling and put the cooler into operation. All further tuning was made using cooling itself as an indicator of the electron beam angles.

7.3.8 Stability of the Beam Recirculation: The Energy Recovery Scheme and Beam Loss Limitations

To keep the dissipated energy low while using a MW-range DC beam in the cooling section, the cooler employs the energy recovery scheme. After acceleration and interaction with antiprotons, electrons return the energy to the terminal by decelerating in the second Pelletron tube down to the energy of ~3 keV at the collector surface and flow through the collector power supply toward the cathode to repeat the journey. At Fermilab this process is called "beam recirculation." A simplified electrical schematic is shown in Fig. 7.59.

This scheme puts stringent limitations on the beam loss. The most obvious reason is the low current provided by the Pelletron chains (up to ~400 μA), which is by several orders of magnitude lower than the beam current. However, a beam loss inside the Pelletron tubes is restricted even more. A loss comparable with the current flowing through the tube resistive divider (~40 μA) significantly redistributes the potential along the tube. The resulting change in the beam envelope causes even larger loses, and the beam recirculation is lost in a matter of milliseconds.

A stable long-term operation requires even much lower losses in the tubes, at the level of several μA. We interpret that as a result of a charge accumulation on the tube ceramic and following partial discharges in the acceleration gaps. These discharges occur all the time, with frequency dependent on the tube voltage gradient and amount of the beam loss. By itself, a discharge of a single gap cannot significantly change the overall voltage distribution. The structure of the Pelletron column contains large aluminum discs called separation boxes, which are connected every ~60 cm (2′) to both tube resistive divider as well as to a dedicated column resistive divider. When only 1 of 42 gaps contained between neighboring separation boxes is discharged, the effect on the voltage outside this portion of the tube is negligible. However, with some probability, plasma from such discharge can shorten also several neighboring gaps, while the capacitance between the separation boxes still holds the total voltage constant. If the unaffected portion of the tube is capable of holding the entire voltage, the gaps are charged up again, and the

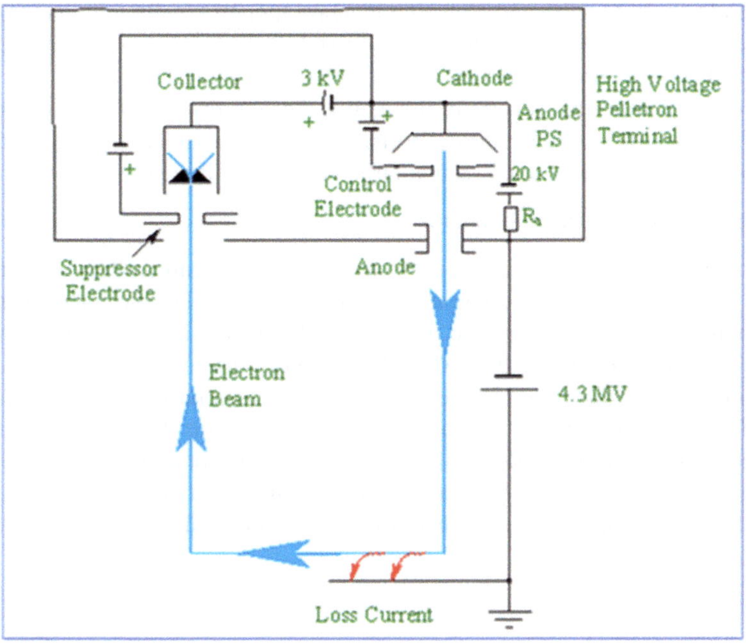

Fig. 7.59 Simplified schematic of beam recirculation

recirculation is not interrupted. Even if the envelope modification is large but is causing a beam loss only somewhere outside of the Pelletron, the protection system interrupts the beam and normal operation can be restored in a matter of seconds (so-called a beam trip). Otherwise, the entire tube shortens, and the Pelletron voltage drops to nearly zero.

7.3.9 Steps to Limit the Beam Losses to Acceleration Tubes

Because decreasing the beam loss to the tubes was recognized as the only way to provide long uninterrupted recirculation, several distinctive steps were made at the design and R&D phases.

- Electron gun with a negatively biased control electrode

Experience of MeV-energy electrostatic electron accelerators showed that one of the problems is to turn on the beam, because in absence of a strong longitudinal magnetic field, the beam may strongly diverge at low currents. It was alleviated by employing a gun with an electrode near the emitter that is negatively biased with respect to the cathode (the control electrode) [98]. In this gun, the emitting area is determined by the position of the equipotential surface at the cathode so that at a small current, only electrons from a narrow portion near the axis are accelerated.

As a result, the beam size and diversion increase monotonically with increasing the beam current from zero and then roughly stabilizes, so that a focusing channel optimized for the nominal current accepts lower currents as well.

An equally important feature of this gun is the suppression of electron emission from the side surface of the emitter. With a positively biased control electrode, such electrons are accelerated and create a large beam halo, increasing the beam loss.

- Effective electron beam collector

Another source of current loss is electrons from the beam collector. Part of this loss is secondary electrons escaping from the collector. Development of the collector with a transverse magnetic field [71, 72] allowed decreasing the portion of these electrons to less than 10^{-5}. Also, IBS increases the energy spread in the beam, and low-energy tails may be reflected while decelerating near the collector [73]. This effect was taken into account when designing the final version of the collector, because it limits the applicability of the traditional method of decreasing the secondary electron flow from a collector by creating a potential minimum near the collector entrance. Also, it determines the minimal value of the collector potential.

- Optimum beam tuning

We realized that tuning is optimum when the beam loss to the acceleration/deceleration tubes is minimized, which not necessarily coincides with the minimum of the overall beam loss. In particular for the deceleration tube, it meant passing toward the ground as many as possible of the electrons escaping from the collector, even at the cost of increasing the total beam loss. Typically, the changes of the tube resistive divider currents after turning the electron beam were used as indication of a beam loss to the tubes.

- Suppression of the electrons emitted from the control electrode

While the gun design suppresses the beam halo from electrons emitted from the cathode, particles emitted from the control electrode have a similar energy as the main beam and may propagate out of the gun. However, because the initial conditions for these particles are very different from those of the main beam, such particles are lost in the acceleration tube, deteriorating the long-term stability of the beam recirculation. This effect is difficult to model on a low-energy test bench, and all the significant stability improvements were made as results of experiments with the Pelletron.

First, the recirculation stability dramatically improved when the aperture in the gun anode was decreased to the minimum size allowed by the envelope of the primary beam [74]. With the smaller aperture, most of the electrons emitted from the control electrode are lost at the anode.

An important mechanism for generating electrons from the control electrode is through the impact of secondary ions. Without special precautions, the ions created by electrons in the beam line are captured by the beam space charge and can travel along the beam toward the acceleration tube, where they are accelerated and end up in the gun irradiating, in part, the control electrode. This mechanism was suppressed by creating a potential barrier at the entrance of both tubes with a positive voltage on the corresponding BPM plates, which increased the typical time between beam interruptions from dozens of minutes to hours.

Operationally, the recirculation stability was found to deteriorate when the pressure in the tubes was rising. To keep a typical interval between beam trips above several hours, the first ion gauges below the tubes had to be kept below ~0.3 nTorr. We attribute this to the same effect as in the previous paragraph, i.e., electrons generated by impacts of secondary ions, but in this case, by ions generated inside the tubes.

Finally, unwanted electrons were found to be emitted also from the inner surface of the negatively biased control electrode [74]. Following the hypothesis that the emission is related to the cathode material being sputtered onto the control electrode by ions striking the emitter (primarily during full discharges), we looked for a material of the control electrode that would suppress the emission from barium films. Replacing the originally copper electrode by tantalum and then by hafnium solved the problem.

7.3.10 Full Voltage Discharges

The most destructive events in operation are full discharges, when the Pelletron voltage drops to nearly zero in a matter of a microsecond. A significant portion of the acceleration tube is shorted out by plasma before the increased voltage gradient on other tube parts fires the protective spark gaps on the gas side. Currents flowing in vacuum modify the electrodes surfaces, decrease the electric strength, and create a large burst of pressure (to 0.01 mTorr), which may take hours to recover from. Also, much higher currents flowing on the gas side produce large electromagnetic waves, damaging the equipment inside and sometimes even outside of the Pelletron. Full discharges were common during the R&D and commissioning phases, and significant efforts were put to decrease their frequency.

Two factors were recognized to be of primary importance in avoiding full discharges: a high electric strength of the tubes and preventing the primary beam from reaching the tube electrodes.

- At the R&D phase, it became obvious that the electric strength of the tubes has to exceed significantly the nominal potential gradient. At that time, the total length of accelerating/decelerating tubes was ~3 m (10′), divided into five 2 f. sections. Each section was conditioned individually to ~1.1 MV (without any apparent discharges for many minutes) reaching occasionally the maximum of ~1.3 MV. This conditioning resulted in the entire tube being able to hold ~5 MV without beam, but at the same time operation with beam was stable only at 3.5 MV, ~0.6 times what one might expect from the conditioning of individual sections. A dedicated experiment showed that the factor of 0.6 remains roughly the same if a portion of the tubes is partially shortened. As a result, the decision was made to add one more section in the final Pelletron assembly. In accordance with the pattern just described, raising the total tube length to 3.6 m eventually allowed stable operation at 4.3 MV.

- Even if the tube electric strength is appropriate, a full discharge is very likely when the beam core touches the tube electrodes [75] because a large portion of the tube can be affected simultaneously. Apart from studies and tuning, such situation is possible in the case of a sudden increase of the beam loss anywhere along the beam line. Because the lost current discharges simultaneously the terminal and the cathode–anode effective capacitance, both the Pelletron and gun voltages drop down, changing the beam envelope and, at locations with high dispersion, the beam trajectory. Initially, such changes of the beam envelope in the acceleration tube were the leading cause of the full discharges. This was greatly suppressed by optimizing focusing such as to pass the beam through the acceleration tube for a range of gun parameters as wide as possible. The deceleration tube was protected by creating a large dispersion in the return line while minimizing it in the tube as shown in Fig. 7.57b.

Another critical element to fight the full discharges was the implementation of a fast protection system [76], closing the gun in about 1 μs after detecting a drop of the terminal voltage or a high beam loss (through radiation monitors).

With all these elements in place by the fall of 2005, operation at the beam current of 0.5 A was still producing one full discharge every several days on average. Switching to 0.1 A for normal operation decreased the frequency of the full discharges dramatically, to a few per year (excluding studies and cases with broken equipment).

7.3.11 Beam Trips

During regular operation, there are typically several beam trips per day. A trip starts with the protection system detecting either a drop of the terminal voltage or an elevated radiation near the beam line. Then an analog circuitry shuts the beam off by applying a large negative voltage to the gun control electrode, and a software loop turns off the anode voltage and disables the gun settings as an extra layer of protection. Following such an event, another software loop checks the Pelletron voltage and vacuum and, if they are within their tolerances, restores the electron beam in about 30 s after the beam trip.

The majority of the trips can be associated with one of the following reasons: false trigger due to Main Injector beam losses or from excessive noise from the electronics reading the terminal voltage, high beam loss near the entrance of the deceleration tube due to an electron beam trajectory alteration (e.g., caused by a large variation of the fringe fields of the Main Injector), and a partial discharge in one of the tubes. While reasons for some of the trips are unclear, the negligible impact of the interruptions on operation at this frequency did not justify additional efforts to understand them.

7.3.12 Electron Angles in the Cooling Section

The value of the electron angles in the cooling section dramatically affects the cooling properties. For typical operational parameters, the rms transverse electron velocity σ_{ex} (in the beam frame) significantly exceeds the antiproton velocity and the longitudinal electron velocity component. For this case, Eq. (7.141) predicts that the longitudinal and transverse cooling rates (time derivatives of the emittances) depend on the angles (or transverse velocities) as ${\sigma_{ex}}^{-2}$ and ${\sigma_{ex}}^{-3}$, correspondingly. Therefore, operation is very sensitive to variations of the angle value.

The origins of the angles can be roughly divided into four categories:

- Incoherent angles originated from the thermal electron velocities at the cathode
- Angles coming from an envelope mismatch
- Nonlinearities in the beam line
- Coherent dipole motion

The following sections describe these effects in detail. To give a sense of scale, let us note in advance that the estimated 1D effective electron angle was in the best cases ~100 μrad.

7.3.13 Thermal Angles

As it was noted above, envelope matching requires equality of magnetic fluxes through the beam at the cathode and in the cooling section (neglecting the space charge)

$$B_{\text{cath}}{R_{\text{cath}}}^2 = B_{CS}{R_{CS}}^2, \tag{7.143}$$

where R_{cath} and R_{CS} are the beam radii and B_{cath} and B_{CS} are the magnetic field strengths at the cathode and the cooling section, respectively. Ideal matching means that no additional coherent motion is excited, and the incoherent transverse momentum at the cathode with temperature T_{cath}, $p_{T\text{cath}} = \sqrt{m_e k_B T_{\text{cath}}}$ gives a rise of the transverse momentum in the cooling section

$$p_{T\_CS} = p_{T\text{cath}}\sqrt{\frac{B_{CS}}{B_{\text{cath}}}}. \tag{7.144}$$

The corresponding 1D electron angle is

$$\alpha_T = \frac{1}{\gamma\beta}\sqrt{\frac{k_B T_{cath}}{m_e c^2}\frac{B_{CS}}{B_{cath}}}. \tag{7.145}$$

For the cooler parameters, $T_{cath} = 1{,}050°C$, $B_{cath} = 86$ G, $B_{CS} = 105$ G, Eq. (7.145) gives a 1D thermal contribution of $\alpha_T = 57$ μrad.

7.3.14 Envelope Mismatch

An envelope mismatch (a focusing error) results in electron angles linearly growing with radius and the beam shape and size changing along the cooling section. Initial tuning was made by measuring the envelope with scrapers [69]. The beam envelope was found to be close to a round and was adjusted with two lenses right upstream of the cooling section. The residual 2D angle at the beam periphery was reported to be 0.22 mrad [70]. This was good enough to obtain electron cooling rates sufficient for cooling demonstration and early operation. Attempts to further adjust the lenses according to the cooling rates did not lead to any noticeable improvements.

However, there were accumulating indications of a deficiency to the procedure. The beam boundary was determined by scraping ~10^{-5} of the beam intensity. Therefore, this procedure was sensitive only to the beam halo, which properties may be very different from the core's [70]. From the very beginning, the measured beam size exceeded the prediction from Eq. (7.143) by a factor of 1.3, well above measurement errors. Later, comparison of the cooling force measured at the beam axis and cooling rates predicted by Eq. (7.141) showed a significant discrepancy that could be resolved only with the assumption of a large envelope mismatch [99]. Finally, direct imaging in a pulse mode with a scintillator clearly revealed that the beam core was elliptical [77], which would explain the relatively poor cooling rates. By adjusting quadrupoles upstream of the cooling section, the beam was properly matched in the pulse mode, but it did not improve cooling, likely because the residual ion background in the DC mode significantly modified focusing. There was no nondestructive transverse beam size diagnostics available to tune the beam envelope, and the accuracy to which the beam optics were measured did not allow using simulations for fine-tuning as well.

Eventually, the quadrupoles were tuned based on the cooling properties. The drag rate was measured for several values of a quadrupole current, the quadrupole was set to the value giving the highest force, the next quadrupole was varied, etc. This procedure noticeably improved the cooling properties at the beam current $I_e = 0.1$A [77]. Estimation made with OptiM showed that the applied changes to the quadrupole currents should correct the angles of a 0.1 A electron beam at the radius of 2 mm by ~0.3 mrad (2D). A crude estimation of what is left uncompensated can be made judging by the sensitivity of tuning of the quadrupole settings, and it is likely a half of this value. Assuming a linear dependence of this angle with the beam radius, the rms 1D contribution average over the beam cross section is estimated to be ~50 μrad.

An additional source of angles linear with the radial offset, which exists even for ideal matching, is the beam rotation caused by the electron space charge and longitudinal magnetic field B_{CS} in the cooling section. For $I_e = 0.1\text{A}$, electron beam radius $a_e = 2\text{mm}$, and $B_0 = 105\text{G}$, this drift angle at the beam periphery is

$$\alpha_{dr} = \frac{E_{sc}}{\beta\gamma^2 B_0} = \frac{2I_e}{\beta^2\gamma^2 B_{CS} a_e c} \approx 10\,\mu\text{rad}. \tag{7.146}$$

but its contribution to the sum is negligible.

7.3.15 *Nonlinear Perturbations*

Several effects may result in a nonlinear perturbation of the electron motion: space charge of the electron beam itself, higher-order field components in the focusing elements, and the electric field of the background ions.

The nonlinear angles caused by the beam space charge are likely a minor contributor to the cooling properties. At the standard operating anode voltage of 20 kV and the beam current of 0.5 A, for which the gun was optimized, the current distribution at the cathode is close to being flat (Fig. 7.60). Because the beam envelope is determined mainly by the effective emittance Eq. (7.142), the distribution remains nearly flat in the beam line; hence, the force due to space charge is mainly linear. The flat distribution was demonstrated by measurements in pulse mode [68]. For lower currents, the relative nonlinearity is increasing while the linear component diminishes. Estimations show that for the operational current $I_e = 0.1\text{A}$ and focusing tuned to minimize angles near the beam axis, the additional angle due to the current density profile at a radius of 1.8 mm is ~30 μrad, insignificant in comparison with other contributions to the total angle.

The next effect, nonlinearity in the focusing elements, became obvious during measurements of a pulsed electron beam imaged onto a YAG crystal installed right downstream of the cooling section. When the beam was tightly focused at the YAG, the image clearly showed higher-order perturbations (Fig. 7.61).

To determine the source of the perturbations, the beam positions in the BPMs were measured as functions of currents in various correctors, and the angle imposed by lenses to the beam centroid at different offsets was calculated and analyzed [78]. For an ideal, optically thin solenoidal lens, this angle α_{ideal} is increasing with the offset r as

$$\alpha_{\text{ideal}} = \frac{r}{F_{\text{ideal\_0}}}\left(1 + k_{\text{lens}} r^2\right) \tag{7.147}$$

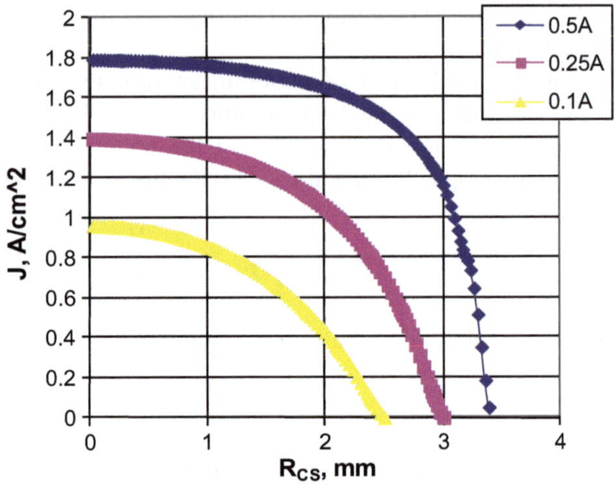

Fig. 7.60 Current density distribution in the cooling section. The current density distribution was simulated at the cathode with the code UltraSAM [67] and then adjusted according to Eq. (7.143). The anode voltage was 20 kV. The values of the beam current are indicated on the plot

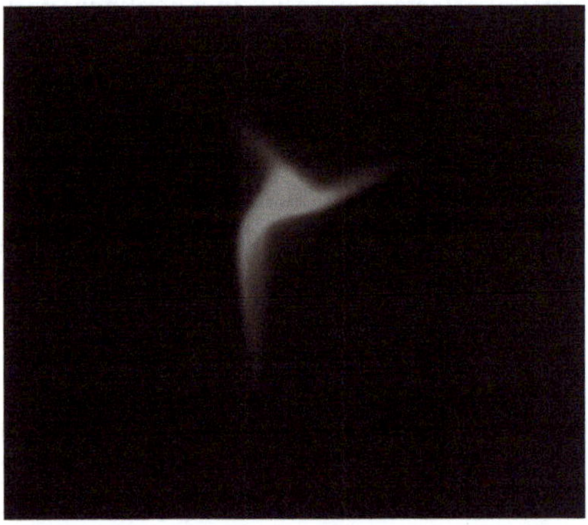

Fig. 7.61 Image of the beam tightly focused at the YAG

where $F_{\mathrm{ideal\_0}}$ is the focusing length for a paraxial trajectory and k_{lens} is coefficient determined by the lens geometry. While the angle dependence on the offset reconstructed from the measurement differed from the function of Eq. (7.147), one still can characterize the deviation from linearity in similar manner, calculating the coefficient k_{lens} for a radius typical for the measurements (~1 cm). For most of lenses, this coefficient was found to be within a factor of 2 from the expected for an ideal lens (k_{ideal}~3×10^{-3} cm^{-2}). However, in several cases these deviations were significantly higher. Most notably, one of the lens doublets in the supply line (i.e., between the first vertical and the first horizontal bends), called SPS03, showed the coefficient being ~6 times higher than k_{ideal}. For optimum steering (i.e., trajectory where nonlinearity is minimal) and typical beam sizes in this lens predicted by OptiM simulations for beam currents of 0.1 and 0.5 A (i.e., ~3 mm and ~5 mm radius, respectively), the additional electron angle in the cooling section at the beam periphery is estimated to be ~40 and ~100 μrad, correspondingly. While no complete simulations were made, taking into account aberrations in all lenses together with beam misalignments would likely double these numbers. Note, however, because of strong (approximately cubic) dependence of these perturbations on the offset, the beam core should not be dramatically affected. The estimation for the 1D rms angle for a well-aligned 0.1 A electron beam gives ~20 μrad. In operation, the beam trajectory was periodically realigned to keep the beam center within 1 mm from the lens axes.

The strongest nonlinear contribution comes from the residual gas ions. This effect is described separately in the next section.

7.3.16 Effect from the Ions Generated by Beam-Background Gas Interactions

Ions created in the beam line by the electron impact on the residual gas molecules can significantly modify its focusing properties. The initial kinetic energy of the secondary ions is close to thermal, and the electric field of the electron beam prevents ions from escaping radially starting from mA-range currents. With no ion clearing mechanisms, the ion density would increase until reaching the electrons (i.e., up to the neutralization factor of $\eta \sim 100$ %). At $\eta \sim 100$ % the focusing effect from ions is by a factor of $\gamma^2 \sim 100$ higher than defocusing from the beam space charge. Electron beam envelope simulations with the OptiM code predict that the electron beam space charge becomes important in the beam line at the beam current of $I_e \sim 0.1$ A. Therefore, for the operational range of 0.1–0.5 A, the effect of ions should be significant at $\eta \sim 1$ %, thus requiring effective ion clearing.

To address this requirement, all capacitive pickups monitoring the beam position in the cooler (BPMs) have a negative DC voltage offset on one of its plates, while the other plate is DC grounded. The resulting electric field prevents the creation of a potential minimum inside the pickup and removes ions in the vicinity of each BPM. To further estimate the process of ion accumulation, we assume the residual gas to be hydrogen at 0.3 nTorr. The calculated time of reaching $\eta \sim 1$ % is ~200 ms. It is much longer than the time for a thermal velocity H_2^+ ion to fly ~5 m between two neighboring BPMs, ~3 ms, and, therefore, clearing with the electric field in BPMs should be effective. However, significant size variations of both the electron beam and the vacuum pipe along the beam line create local potential minima that prevent ions from traveling to the clearing field in the BPMs. Also, solenoidal lenses providing focusing in the beam line are additional barriers for ions. Because the electric field inside the electron beam is primarily radial, the transverse component of the ion velocity is typically much higher than the longitudinal. Due to the conservation of the transverse adiabatic invariant, even the modest magnetic fields of the lenses ($\leq$600 G) can reflect the ions, further increasing a steady state concentration. The hope was that the focusing effect of the ion background would

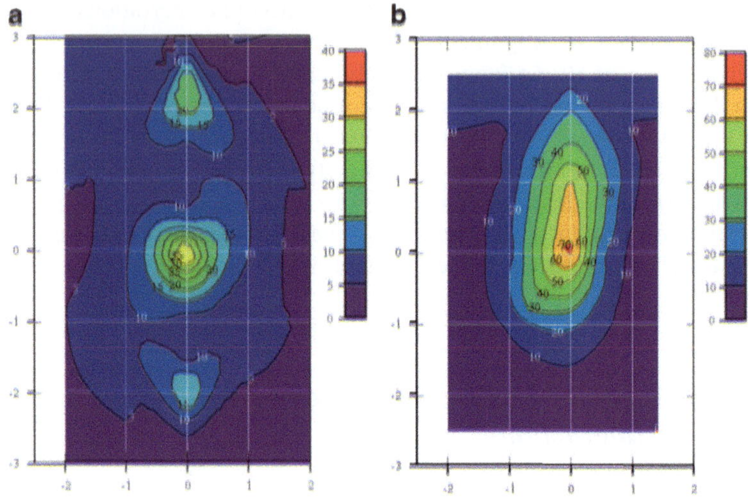

Fig. 7.62 Contour plot of drag rates across the electron beam without (**a**) and with (**b**) ion clearing by beam interruptions. Voltage jump of 2 kV, $I_e = 0.3$ A. In the mode with ion clearing, the interruption frequency was 100 Hz. Contour levels are in MeV/c/h. Horizontal and vertical axes are the corresponding beam displacements in mm. Note the difference by a factor of 2 in scales for the color bar

be mainly in a linear tune shift and, therefore, could be compensated by adjusting the lens settings.

Indeed, the cooling properties of the electron beam were found satisfactory for what became the standard of operation mode, at $I_e = 0.1$ A. However, the cooling efficiency (characterized by drag rate measurements) peaked at 0.1–0.2 A (i.e., curve June 13, 2006 in Fig. 7.72), while it is supposed to be monotonically increasing with I_e. Note that adjusting quadrupoles upstream of the cooling section significantly improved cooling at 0.1 A but did not noticeably change its performance at higher currents.

This insensitivity to focusing settings correlated with results of transverse scans of the drag rates at $I_e = 0.3$ A shown in Fig. 7.62a. A detailed discussion of the drag rate measurements follows in Sect. 7.3.26; here we assume that the drag rate is equal to the longitudinal cooling force average over a small-size (pencil) antiproton beam. The drag rates were measured in various positions of this pencil beam with respect to the electron beam (experimentally, the electron beam was moved). Because the cooling force is determined mainly by the local properties of the electron beam and the value of the voltage jump is chosen to measure the drag rate near the maximum of the force curve of Fig. 7.53, the force is roughly proportional to the local value of j_e/ϑ_e^2. The results of such measurements in 28 points are shown as a contour plot in Fig. 7.62a. With only three narrow areas providing significant drag rates, this profile corresponds to high-order focusing perturbations that cannot be corrected by adjusting solenoidal lenses and quadrupoles.

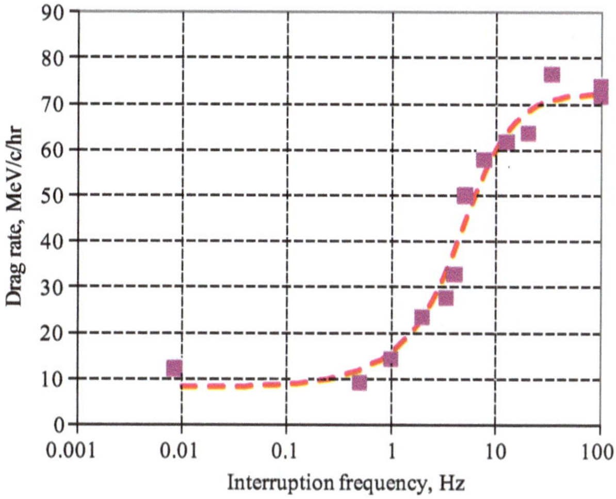

Fig. 7.63 Drag rate as a function of the interruption frequency f_{int} for $I_e = 0.3$ A and separation between beams of 1 mm. January 2, 2011. The interruption pulse was 2 μs. Focusing was optimized on axis at 20 Hz. The squares represent the data, and the line is the model

The only remedy to decrease the average ion concentration was found to be periodic interruptions of the electron beam. In the potential well created by electrons, ions gain the kinetic energy of up to 10 eV (at $I_e = 0.3$ A). Thus, if the electron beam is abruptly turned off, an H_2^+ ion reaches the vacuum pipe in 1–2 μs. The capability of interrupting the electron current for 1–30 μs with a frequency of up to 100 Hz was implemented in the electron gun modulator in 2009 [79]. While the clearing voltage applied to the BPMs was always on and certainly decreased the ion density, for brevity we will further refer to the operation with these interruptions as to the "mode with ion clearing."

Ion clearing significantly increased the area of the electron beam cross section with good cooling (Fig. 7.62b) as well as improved the drag rate measured on axis at higher electron currents (curve January 2, 2011). The latter is related to the finite transverse size of the "pencil" antiproton beam in the measurements.

Dependence of the drag rate measured at 1 mm offset on the interruption frequency is shown in Fig. 7.63. The results can be compared with the following model:

(a) The beam space charge outside of the Pelletron tubes is relatively small, so that the envelope electron angle in the cooling section changes linearly with variation of the beam current and the radius r, $\Delta\alpha = k_{sc} \cdot r \cdot \Delta I_e$. According to OptiM simulations, the coefficient $k_{sc} \approx 1$ rad/A/m.
(b) The effect of accumulated ions is similar to the one due to the beam own space charge,

$$\Delta\alpha_i = k_{sc} \cdot r \cdot \Delta I_e \cdot \delta \cdot \eta \cdot \gamma^2 \tag{7.148}$$

where $\delta < 1$ is a fitting coefficient representing the portion of the beam line where ions can be accumulated.

(c) Neutralization drops instantaneously to zero when the beam is interrupted, increases linearly with time until reaching an equilibrium at some value η_0, and stays constant afterward. This can be represented by the following function:

$$\eta(t) = \begin{cases} t/\tau_c, t \le \tau_0 \\ \eta_0, t > \tau_0 \end{cases} \quad \tau_c = \frac{1}{n_a \sigma_i \beta c}; \ \tau_0 = \eta_0 \tau_c \tag{7.149}$$

where n_a is the atom density and σ_i is the ionization cross section of hydrogen.

(d) The cooling force F_c changes with the additional envelope angle introduced by neutralization as

$$F_c = \frac{F_0}{1 + (\Delta\alpha_i/\alpha_0)^2}, \tag{7.150}$$

where α_0 and F_0 are the rms angle and drag force at optimum focusing.

(e) The measured drag rate F_d is the cooling force averaged over the period between interruptions (assuming that the pencil antiproton beam is sensitive mainly to the electron angles in the location of its center).

$$F_d = f_{\text{int}} \int_0^{1/f_{\text{int}}} F_c(t)\, dt \tag{7.151}$$

The model described by Eqs. (7.149)–(7.151) was compared with the measurements using the measured value of $F_0 = 73$ MeV/c/h, the rms angle estimated at the beam center $\alpha_0 \approx 0.1$mrad, and fitted parameters of $\delta = 0.5, \eta_0 = 0.02$. The result, shown as a dashed line in Fig. 7.63, follows the experimental data reasonably well.

For operation the most important result of ion clearing is an increase of the longitudinal cooling rate by a factor of 2 with otherwise similar parameters. Also, it was interesting to find out that the clearing with BPMs, according to the above comparison with the model, was quite effective, down to ~2 %.

7.3.17 Coherent Dipole Motion

A coherent dipole motion of the beam in the cooling section adds a corresponding transverse velocity to individual electrons, decreasing the cooling efficiency. This motion can be caused either by errors in the beam entrance position and angle or by dipole magnetic fields inside the cooling section.

The main sources of slow (longer than minutes) changes in the position of the beam coming toward the cooling section are mechanical drifts of focusing elements and drift of their power supplies. This contribution is suppressed by a software feedback loop, which adjusts currents of two pairs of dipole correctors immediately upstream of the cooling section according to position measurements in two pairs of BPMs inside the section. The measurements are synchronized with the Main Injector cycle and, therefore, are typically made every 2.2 s.

The original plan also foresaw the use of a faster (up to ~10 Hz) feedback loop to compensate the effect from the MI stray fields, but it was not implemented mainly because the corresponding oscillations were found to contribute very little into the angle budget. This was determined from an analysis of time-dependent readings of the cooling section BPMs [80]. For every time sampling point, deviations of the positions in 11 BPM pairs were fitted to a helical trajectory. The fit was found to be close to the raw signal, so the BPM noise contribution was small. Then, the beam position and angle at the cooling section entrance were reconstructed from each fit. The FFT of the resulting set showed the MI field contribution to be ~30 μrad rms. A similar procedure applied without MI ramps [81] showed that contributions from the Pelletron vibrations (mainly 20, 29.8, and 59.5 Hz lines [82]) and the power grid (60 Hz) accounted together for ~20 μrad rms (1D).

Because the cooling section has good magnetic shielding, external magnetic fields are decreased by a factor of >1,000 inside [83] and do not create any measurable perturbations. On the other hand, dipole magnetic fields created by imperfections in the cooling section magnetic field are significant and typically the main component of the total angle. After installation, the transverse magnetic fields were measured and compensated with dipole correctors to the level where the resulting rms angle was estimated to be ~50 μrad [84]. However, beam trajectories measured several months afterward indicated much larger values. Because studies at the R&D stage proved that the individual 2-m modules are rigid, it was concluded that the reason for the trajectory's drift is mechanical shift of the modules with respect to one another. Indeed, to produce a helix with a 1D rms angle of 100 μrad, it is enough for one of the supports of a solenoidal module to be shifted by ~0.1 mm, which is modest in comparison with millimeter-range tunnel drifts observed after construction of the electron cooler's building. Later, there were also hints that the cooling section deforms significantly (in this scale) during fast changes of the cooling section temperature, for example, as result of a cooling water system failure.

These drifts were compensated multiple times during operation of the cooler using the cooling section dipole correctors. Each of ten 2 m-long modules of the cooling section is equipped with eight pairs of 20 cm-long printed circuit dipole correctors (so-called main correctors), mounted outside of the solenoid winding along the entire length of the module, and two pairs of narrow correctors attached to the inner surface of the solenoids at each end of the module (so-called end correctors). A simultaneous change of all main correctors' currents in a module by the same amount creates a dipole field roughly equivalent to an inclined solenoid (with respect to its initial position); adjustment of an end corrector has an effect similar to a shift of the module end with respect to the neighboring module. Hence, the right combination of these two types of adjustments should be able to compensate the mechanical drifts. The difficulty is in determining the appropriate values of the corrector currents. Note that the simple alignment of the beam to the centers of

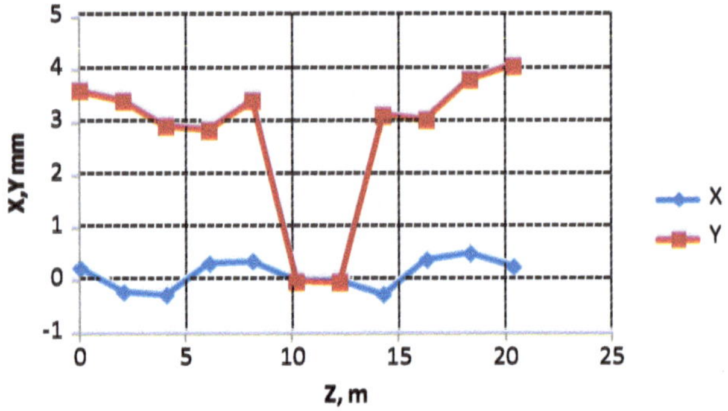

Fig. 7.64 Trajectory in the cooling section when measuring the drag rate in the module positioned between 10 and 12 m. Points show data from BPMs located at the beginning of each module

the cooling section BPMs with, for example, the end correctors only does not guarantee the straightness of the trajectory; in practice, such attempts made cooling only worse.

After testing several procedures, the following, based on the drag rate measurements, was eventually developed and implemented.

1. The antiproton beam trajectory is measured with the cooling section BPMs in order to "realign" them. Because of large antiproton momentum and effective magnetic shielding, this trajectory is a straight line. Therefore, deviations from the line reported by BPMs results from the mechanical offsets of the BPMs. These deviations are then subtracted in the software that calculates the beam positions. Thus, once implemented, the measured trajectory should again be a straight line within the measurement errors. Because before installation all BPMs were calibrated in both antiproton and electron modes (89 and 32 kHz, correspondingly) at a bench, the subtraction corrects the measurements in the electron mode as well.
2. A pencil antiproton beam is prepared for the dag rate measurements.
3. The electron beam is quickly shifted far from the antiprotons everywhere in the cooling section except for one module (see an example in Fig. 7.64), and the drag rate is measured. The electron trajectory is returned to the standard position to ensure the same initial conditions of the antiprotons in each measurement.
4. All (x or y) main correctors of the module are changed by the same value, and the step (3) is repeated. During the measurement, the electron beam is kept centered in the BPMs at both ends of the module.
5. After repeating step (4) for several values of the correctors current both in x and y, the values corresponding to the largest drag are determined. A typical example of the measurement result is shown in Fig. 7.65.
6. After measuring all modules, the changes of the main correctors providing the best drag rates are applied. Finally, the resulting trajectory is aligned to the centers of all BPMs with the corresponding end correctors.

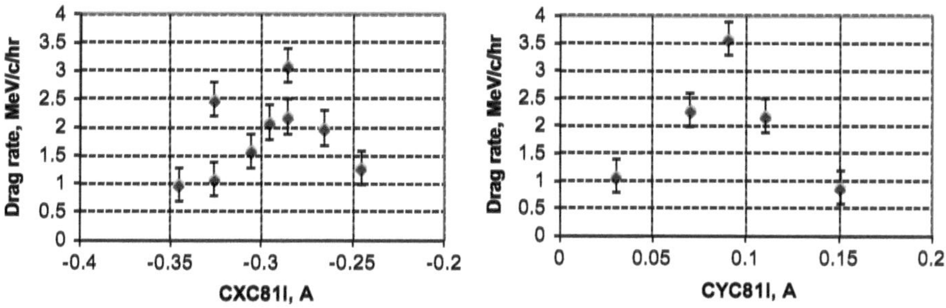

Fig. 7.65 Typical measurement during the cooling section field alignment. First, the drag rate of a single module is measured as a function of changes to all X main correctors in section #8 (*left plot*). The horizontal axis is the current of one of these correctors. Then, the X correctors are set to the "optimum" values, and the measurement is repeated with the Y main correctors. In this specific set, the currents of the X correctors were increased by 40 mA, i.e., CXC81I was changed from −0.326 A to −0.286 A, and the Y corrector currents were lowered by 20 mA, i.e., CYC81I was changed from 0.11 to 0.09A. The error bars show typical statistical errors of individual measurements

Because for a single module, the drag rate is low, the statistical noise is high, and the procedure allows correcting only large errors. Also, it is time-consuming (~20 h) and was repeated mainly after the complex's shutdowns, when deformations of the cooling section were maximal.

One can subtract the corrector settings before and after the procedure and feed them into a tracking program. The resulting differential trajectory typically yields a 1D rms angle ~ 100 μrad, indicating the scale of the possible improvement.

7.3.18 Summary of Electron Angles

Table 7.8 summarizes the contributions from the various effects discussed above to the total electron angle in the cooling section for the best tuning of each component. The values are obtained from averaging the angles over the transverse section of the beam, assuming a 2 mm radius and using the angle estimates for the beam current of 0.1 A. Obviously, different ways of averaging are relevant for different scenarios, so the table mainly shows the scale of the effects.

For the calculation of the total angle, individual contributions are summed in quadratures.

Table 7.8 Contributions to the total electron angle in the cooling section

Effect	Angle (μrad)	Method of evaluation
Thermal velocities	57	Calculated from the cathode temperature
Envelope mismatch	~50	Resolution of tuning + OptiM simulations
Dipole motion (above 0.1 Hz)	~35	Spectra of BPMs in the cooling section
Dipole motion caused by field imperfections	~50	Simulation of electron trajectory in the measured magnetic field
Nonlinearity of lenses	~20	Trajectory response measurements
Ion background	<10	Cooling measurements
Total	~100	Summed in quadratures

Shown values are 1D, rms

7.3.19 Electron Energy Spread, Stability, and Matching

The effective energy spread of the electrons in the cooling section is determined by intra-beam processes and by fluctuations of the Pelletron terminal voltage, with the contribution of the longitudinal thermal velocities at the cathode being negligible. A closely related effect, very important for operation, is a slow drift of the electron energy.

Core of the energy distribution is formed by multiple IBS and by the dissipation of density micro-fluctuations. For the case relevant for the Fermilab cooler, when the transverse temperature is much higher than longitudinal (in the beam frame) and the beam envelope is determined by conservation of the canonical momentum, the resulting rms energy spread was estimated in [73] to be ~90 eV.

7.3.20 Terminal Voltage Ripple

Because the typical cooling time is minutes, the effect of fluctuations of the electron energy on the cooling process caused by the terminal voltage ripple at Hz-range frequencies is heavily averaged and equivalent to an increase of the beam energy spread. The most accurate way to measure these fluctuations was found to be by analyzing the BPM readings in a high-dispersion region [82, 85]. Optical analyses of different frequency components showed measurable energy fluctuations only at frequencies in the 1–6 Hz range. Lower frequencies are effectively suppressed by the energy stabilization system, and the higher frequencies are eliminated by the large capacitance of the terminal shell (~300 pF). The typical rms value of the terminal voltage ripple is about 150 V.

The ripple comes primarily from the chain current fluctuations at the chain rotation frequency of 1.8 Hz and its harmonics [85] and somewhat depends on the chain condition and settings of the energy stabilization system. This system uses a generation volt meter (GVM) with a DC to ~10 Hz bandwidth as the main tool for measuring the electron energy. In addition, there are capacitive pickups mounted on

Table 7.9 Elements affecting the energy drift

	Sensitivity	Cause	Remedy
Variations of the building temperature	500 eV/K	Temperature sensitivity of the GVM preamplifier	Temperature of the pre-amplifier is stabilized to within ±0.5 K Software loop
Variations of the Pelletron temperature	400 eV/K	Distance between the terminal and GVM depends on temperature	The tank temperature is kept within ±0.2 K Software loop
Chain current drift or corona current from the terminal	100 eV/μA	Insufficient gain of the analog feedback loop	Software loop adjusting the chain current
SF6 pressure	~500 eV/psi	Effect of SF6 permittivity on GVM reading	Software loop

Table 7.10 Values of the rms angle, in mrad, from fitting the data in Fig. 7.74 with Eq. (7.131)

Helical angle (1D), α_h	Fitted rms angle, σ_{et}	$\sqrt{\sigma_{et}(0)^2 + \alpha_h{}^2}$
0	0.134	0.134
0.1	0.180	0.167
0.2	0.285	0.241

The last column compares the fitted angles with the simplest prediction

the Pelletron tank opposite to the terminal shell, but their circuitry was found ineffective and they were not used for HV regulation. More details on the performance of the energy regulation system can be found in [85].

7.3.21 Terminal Voltage Drift

Several mechanisms responsible for the energy drift and corresponding solutions are listed in Table 7.9.

In addition to stabilization of the temperatures, two software loops were implemented. One of them adjusts the chain current to eliminate the difference between setting and reading of the terminal voltage.

The second loop modifies the set point of the terminal voltage to keep constant the beam position in a high-dispersion region right after the 180° bend that follows the cooling section. Because the beam position in the low-dispersion (<10 cm) cooling section is stabilized and the field in the bend magnet is regulated with NMR sensors, a fixed position implies a constant energy. After 0.1 Hz filtering of the BPM signal, the residual beam motion and electronics noise limit the resolution to about 30 eV. It is enough to keep the energy at the right value. However, failures of the NMR system caused by high radiation in tunnel, tunnel temperature variations, and drifts of the trajectory in the cooling section change the calibration of the loop.

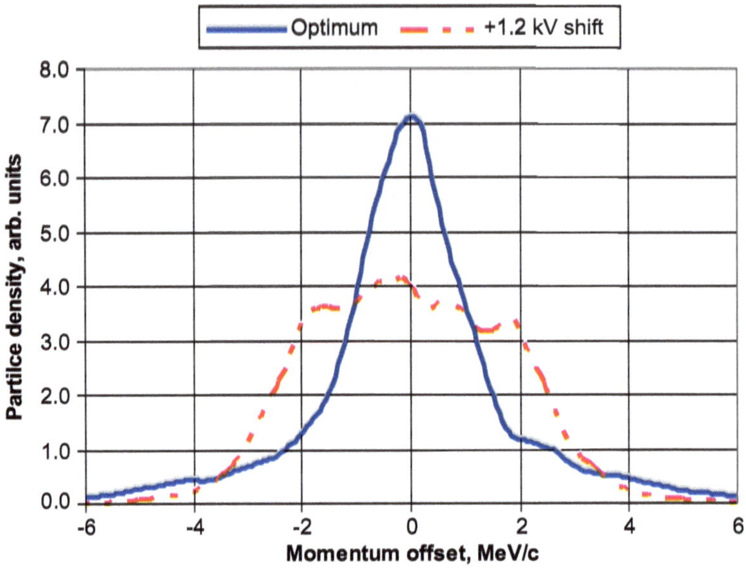

Fig. 7.66 Momentum distribution of the antiproton beam electron cooled at two different electron energies. The *solid blue line* shows data for optimum energy tuning, and the *dash-dot red line* represents data for cooling with the electron energy shifted by 1.2 keV. September 10, 2008. 105×10^{10} antiprotons, bucket length 5.8 μs; electron beam is on axis. Vertical scale is linear

The most sensitive (in operation) indication of an energy mismatch is the shape of the Schottky momentum distribution, which becomes flat near its maximum (Fig. 7.66). Approximately monthly, the calibration of the loop is adjusted by making the momentum distribution as peaky as possible.

7.3.22 *Initial Energy Matching*

One of the problems during cooler commissioning was the initial matching of the electron and antiproton energies. Because of slow cooling times, the cooling effects are indistinguishable if the energy error is larger than ~0.1 %. The absolute energy of antiprotons was known to better than 0.1 % from fix-target experiments in the Accumulator and cross-calibrations between machines. The electron energy, determined mainly by the terminal potential, was known initially with much worse accuracy. After assembling the Pelletron, the GVM calibration was verified with a 100 kV external power supply and a calibrated resistive divider. The next step for calibrating the electron energy was to measure the length of a Larmor spiral pitch in the cooling section [86] at the energy close to the nominal. The resulting change of the GVM calibration was 5.2 %. The precision of this measurement, determined by the calibration of the Hall probe used for the longitudinal field measurements and

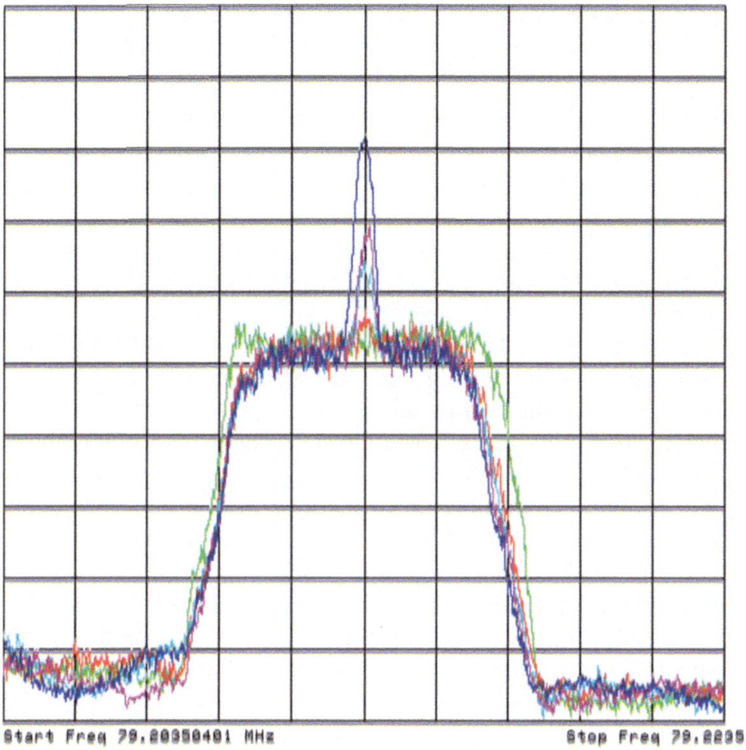

Fig. 7.67 Evolution of the antiproton distribution function after turning on the 0.5 A electron beam. The vertical axis represents the longitudinal Schottky power (arb. units, log scale), and the horizontal axis is the frequency which scale converted into momentum units is 25 MeV/*c*/div. The antiproton beam intensity was 5×10^{10}, and its transverse emittance was 2 μm rad. Traces were taken 15 min apart

by the beam position measurement errors, was estimated to be ~0.2 %. It was sufficiently low to observe the first interaction between beams using a specially developed procedure [87]. Antiprotons were smeared over the Recycler momentum aperture (~ ±0.4 %). Electron cooling creates a peak where the antiprotons and electrons energies match (Fig. 7.67). In the first such observation of the cooling force, the electron energy was found within 3 keV (~0.07 %) of its optimal value.

7.3.23 Cooling Force Measurements

For operation, the figures of merit to assess cooling are the speed at which the longitudinal and transverse emittances decrease (cooling rates) and the equilibrium emittance values. However, knowledge about details of the electron beam properties can only come from the drag rate measurements discussed in this section.

7.3.24 Procedure of a Drag Rate Measurement

1. A coasting antiproton beam is scraped to a low intensity. Empirically, the optimum intensity is determined to be ~1×10^{10} particles, large enough for the longitudinal Schottky signal to remain reasonable and low enough to avoid accumulation of ions in the absence of gap in the beam. Longitudinal stochastic cooling is turned off, but the transverse system stays on.
2. With the electron beam in a standard configuration ("on axis," when the antiproton and electron beam centers coincide), antiprotons are cooled to an equilibrium.
3. If necessary, the electron beam state is changed. For example, for studying the radial dependence of the cooling force, the electron beam trajectory is shifted parallel to the axis in the cooling section.
4. Immediately after, the electron energy is changed by shifting the terminal voltage set point by $\delta U_t = 0.5/10$ kV.
5. While the antiprotons are dragged to the new equilibrium, their longitudinal distribution is recorded every 15–17 s, and the average momentum $\overline{p_p}$ and the distribution rms width σ_p are calculated.
6. After 2 min, the electron energy is returned to its original value, as well as other parameters, and antiprotons are cooled again into the "standard" equilibrium.
7. The drag rate $\overline{p_p} \cdot$ is calculated from the linear fit of the $\overline{p_p}(t)$ data recorded during $\tau_{\text{meas}} = 2\,\text{min}$ with the offset electron energy.

The procedure is illustrated in Fig. 7.68. Note that in a standard measurement, the points are recorded more frequently and for a shorter period of time than in this illustration.

One of the obvious limitations of the procedure is a relatively long measurement time. It was chosen to have a reasonable number of points (5–7) for a linear fit, taking into account the significant statistical noise of the mean of the measured distribution. The notion that the drag rate is measured at the momentum offset equal to the initial value of $\overline{\Delta p_{p0}} = \frac{M_p}{m_e} e \cdot \delta U_t$ is self-consistent only if the measured rate $\dot{\overline{p}}$ is low enough so that the shift during the measurement is small in comparison with the initial offset:

$$\dot{\overline{p}} \cdot \tau_{\text{meas}} \ll \overline{\Delta p_{p0}} \tag{7.152}$$

For the most frequently used voltage jump of $\delta U_t = 2$ kV, Eq. (7.152) limits $\dot{\overline{p}} \ll 100\ \text{MeV}/c/\text{h}$. This inequality was typically satisfied at the early stages of the project but did not remain true with improving the cooling properties. Because we have not found a better way to measure the cooling force, we still present some data where Eq. (7.152) is not fulfilled. Therefore, results for small momentum offsets and large drag rates underestimate the value that would be measured if a fast measurement could be made (Figs. 7.69, 7.70, and 7.71).

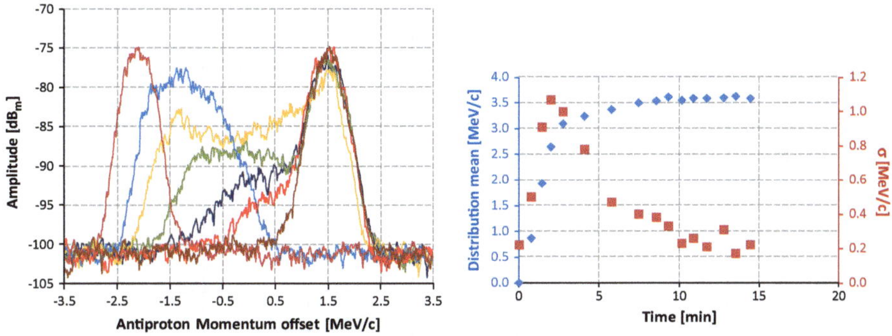

Fig. 7.68 *Left*, evolution of the antiproton momentum distribution recorded by a Schottky monitor after a 1.9 keV jump of the electron energy. $I_e = 0.5$A with ion clearing at 100 Hz. The time between the first and the last traces is 7 min. January 2, 2011. *Right*, corresponding evolution of the mean and rms values of the momentum distribution. The drag rate calculated with a linear fit to the first four points is 71 (MeV/c)/h (not a standard measurement)

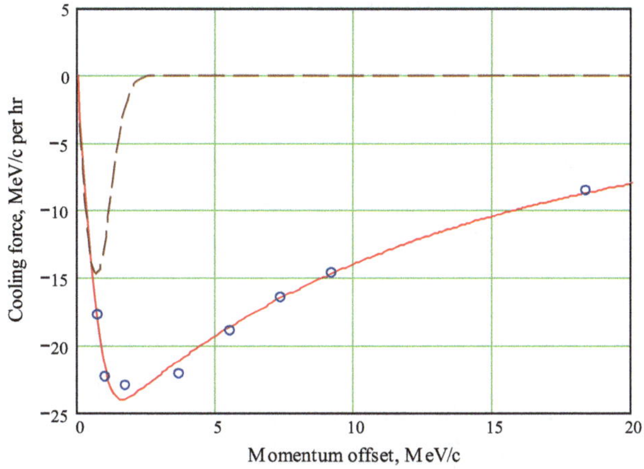

Fig. 7.69 Drag rates measured on February 6, 2006 (*blue circles*). $N_p = 3.5 \times 10^{10}$ antiprotons, transverse emittance (95 %, normalized, measured with a Schottky pickup) is $\varepsilon_{p,95\%n} \approx 1.5p$ mm mrad, initial rms momentum spread $\delta p_p = 0.2$ MeV/c, electron beam current $I_e = 0.1$ A. The *red solid line* represents Eq. (7.138) at $\delta W_e = 370$ eV, $\theta_e = 0.2$ mrad and $j_e = 0.94$ A/cm^2 (as predicted from the gun simulations), and $L_c = 14$. Contribution of the second derivative term is shown by the dashed brown line (for the presentation purpose, it is negated and shown for $\delta p_p = 1$ MeV/c)

7.3.25 Drag Rate and the Longitudinal Cooling Force

The drag rate represents the longitudinal cooling force averaged over all antiprotons. Because these measurements were always made with a pencil beam, one can expand the force near its value for a central particle. Assuming that the 6D

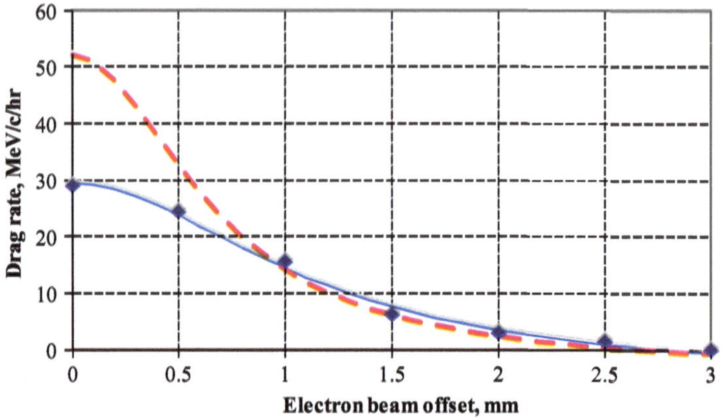

Fig. 7.70 Drag rate as a function of the electron beam offset. The points are results from July 3, 2007. The voltage jump was 2 kV, $I_e = 0.1$ A, $N_p = 4 \times 10^{10}$. During the measurement, the emittance measured with a flying wire was ~ 0.5 μm (95 %, normalized), which corresponds to $a_p \sim 0.5$ mm. The *red curve* is the reconstructed cooling force

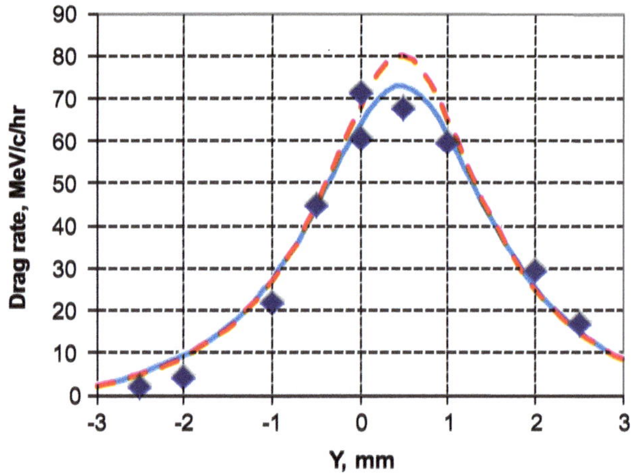

Fig. 7.71 Drag rate as a function of the electron beam offset. The points are results from December 31, 2010 (the same set as in Fig. 7.62b). The voltage jump was 2 kV, $I_e = 0.3$ A, $N_p = 1.3 \times 10^{10}$. During the measurement, the emittance measured with a flying wire was <0.3 μm (95 %, normalized), and the corresponding value of a_p was estimated to be ~0.25 mm. The blue curve is the best fit to the described model with $F_0 = 80$ MeV/c/h, $a = 4.3$ mm (from simulations), and $b = 1.2$ mm. Also, to better fit the data, the curve is shifted by 0.47 mm in comparison with the position reported by BPMs. The red curve shows for comparison the reconstructed cooling force

antiproton distribution is symmetrical and identical in horizontal and vertical directions, the instantaneous drag rate can be expressed as

$$\begin{aligned}\dot{\bar{p}} &= \overline{F_{lz}(\Delta p_p, x_p, y_p, p_{px}, p_{py})} \\ &\approx F_{lz} + \frac{\partial^2 F_{lz}}{\partial \Delta p_p{}^2} \cdot \frac{\overline{\Delta p_p{}^2}}{2} + \frac{\partial^2 F_{lz}}{\partial p_{px}{}^2} \cdot \overline{p_{px}^2} + \frac{\partial^2 F_{lz}}{\partial x_p{}^2} \cdot \overline{x_p^2} \\ &= F_{lz} + \frac{\partial^2 F_{lz}}{\partial \Delta p_p{}^2} \cdot \frac{\overline{\Delta p_p{}^2}}{2} + \left(\frac{\partial^2 F_{lz}}{\partial p_{px}{}^2} \cdot \frac{p_p^2}{\beta_{CS}} + \frac{\partial^2 F_{lz}}{\partial x_p{}^2} \cdot \beta_{CS} \right) \varepsilon_{px}\end{aligned} \tag{7.153}$$

where all functions on the right side are estimated for the central particle $(\overline{\Delta p_p}, 0, 0, 0, 0)$, $p_p = \gamma \beta M_p c$, $\beta_{CS} \approx 30$ m is the beta function in the cooling section, and $\varepsilon_{px}$is the antiproton transverse rms emittance. Note that both correction terms in Eq. (7.153) decrease the drag rate.

To interpret a drag rate as a cooling force experienced by the central particle, the antiproton beam needs to have a small rms momentum spread $\delta p_p = \sqrt{\overline{\Delta p_p{}^2}}$ and a small transverse emittance. To estimate the contribution related to the finite momentum spread, let's assume the dependence on the momentum offset is the same for of the drag rate and the cooling force and calculate the second derivative from the fit to measured data. In the case of the example shown in Fig. 7.69, this contribution is small (<10 %) if the momentum offset is >2 MeV/c at the typical case of $\delta p_p < 1$ MeV/c. The relative value of this correction does not change significantly for lower transverse electron velocities and remained small also for later measurements, when the electron beam quality had been optimized.

The relative contribution of the second correction term in Eq. (7.153) can be expressed as

$$\frac{1}{F_{lz}} \frac{\partial^2 F_{lz}}{\partial p_{px}{}^2} \cdot \overline{p_{px}^2} = f_{px\_contr} \cdot \left(\frac{\vartheta_{px}}{\vartheta_t} \right)^2 \tag{7.154}$$

where f_{px_contr} is a function of the electron velocities and the antiproton momentum offset and θ_{px} is the antiproton 1D angle in the cooling section. For all interesting parameters, estimations from Eq. (7.137) give $f_{px\_contr} = 1.5/2$, and in drag rate measurements the antiproton angles were an order of magnitude lower than the electrons'. Hence, this contribution is negligible.

The last term in Eq. (7.153) was found to be significantly more important and can be estimated from the dependence of the measured drag rate on the radial position with following assumptions:

1. The contribution of far collisions can be neglected, and the force is determined by local values of the electron current density j_e and rms angle at the position of the pencil antiproton beam as in Eq. (7.139), $F_{lz} \prec \frac{j_e}{\vartheta_t{}^2}$.

2. The electron rms angle is determined by its value at the beam center, α_0 and the envelope mismatch angle added in quadratures. The latter is linear with radius, and the total angle can be expressed as $\vartheta_t^2 = \alpha_0^2(1+(x^2/b^2))$ (similar to Eq. (7.150)).
3. The current density distribution is calculated from the gun current density simulations (see Fig. 7.60) and approximated by a parabola

$$j_e = j_0 \begin{cases} 1 - \dfrac{x^2}{a_e^2}, x \le a_e \\ 0, x > a_e \end{cases} \tag{7.155}$$

.

4. Transverse distribution in the antiproton beam is Gaussian.
5. The cooling force can be presented as a product of a component determined by the momentum offset $F_0(\Delta p_p)$ and a radial dependence:

$$F_{lz}(x) = F_0(\Delta p_p) \cdot \begin{cases} \dfrac{1 - \left(\frac{x}{a_e}\right)^2}{1 + \left(\frac{x}{b}\right)^2}, x \le a_e \\ 0, x > a_e \end{cases} \tag{7.156}$$

.

6. The drag rate at a radial offset is the cooling force of Eq. (7.156) integrated over a Gaussian antiproton radial distribution with the rms size a_p.

Comparison of the model with the measured drag rates is shown in Fig. 7.70 as the solid curve, which is calculated with $a_e = 2.9$ mm from the gun simulations, $a_p = 0.5$ mm estimated form the flying wire measurements, and the best fitting values of $F_0 = 53$ MeV/*c*/h and $b = 0.68$ mm. The cooling force in the center is higher than the measured drag rate by almost a factor of 2. Correspondingly, the relative contribution of the last term in Eq. (7.153)

$$\frac{1}{F_{lz}} \frac{\partial^2 F_{lz}}{\partial x_p^2} \cdot \overline{x_p^2} = 2\overline{x_p^2}\left(\frac{1}{a^2} + \frac{1}{b^2}\right) = -4.56 \cdot 0.5^2 = -1.14. \tag{7.157}$$

is large, and the expansion is not valid.

This effect was creating a large scatter of the drag rates related to the difficulty to accurately measure and control the transverse emittance. Eventually to decrease this scatter, the transverse stochastic cooling system was left on during the measurements; the antiproton beam was scraped down to the limit at which a reasonable resolution of the Schottky detector remained, $N_p \sim 1 \times 10^{10}$; and between the measurements the electron beam was set into the state of the strongest cooling. However, equally important was a decrease of the electron angles spread across the beam as it was discussed above. These measures allowed improving the reproducibility of the results, and the antiproton beam transverse emittance measured with the flying wire was <0.3 μm (the emittance measurements were unreliable below this level).

An example of the improved measurement is shown in Fig. 7.71. In this case, the measured drag rates were close to the longitudinal cooling force, and the estimation from Eq. (7.157) gives a relative contribution of 8 %.

7.3.26 Drag Rates and Electron Beam Parameters

With all the difficulties associated with the drag rate measurements, they still provided valuable estimations of the cooling force and, consequently, of the electron beam angles at the location of the antiproton beam. Figure 7.72 shows the dependence of the drag rate on the beam current recorded over the years. All measurements were made with the same voltage jump, $\delta U_t = 2$ kV, and on axis. Each curve was recorded at a fixed set of focusing parameters. If the electron angles stay unchanged, the cooling force should be changing proportionally to the local current density (shown by the dashed curve). Significant deviations of the drag rates from this trend was the strongest reason for measuring the transverse distributions of the cooling force (as in Fig. 7.62a) and eventually developing the ion clearing procedure described above. The results with ion clearing are presented in Fig. 7.72 by the curve labeled January 2, 2011. Note that most of the data points of this set do not satisfy Eq. (7.152) and the flatness of the curve at larger currents can easily be explained by the inadequacy of the procedure. We were not able to explain and correct the dramatic deterioration of cooling around February of 2011 and the corresponding decrease of the drag rate (curve March 7, 2011), partly because of lack of resources toward the end of the run.

Measurements of the drag rate as a function of the voltage jump, interpreted as the cooling force vs. momentum offset, give some information about the absolute value of the electron angles in the cooling section. An example of measurements is shown in Fig. 7.73. The data were fitted to Eq. (7.131) with the Coulomb logarithm under the integral (curve 1) and Gaussian electron velocity distributions. For these measurements, the estimated effect of the final size of the antiproton beam, discussed in the previous section, was small, ≤ 10 %, and was not taken into account.

The only fitting parameter was the rms spread of the electron transverse velocities, σ_{et}, while the longitudinal velocity spread is assumed to be determined by the measured voltage ripple, and the current density is estimated from the gun simulations and magnetic measurements. The first point was excluded from the fit because

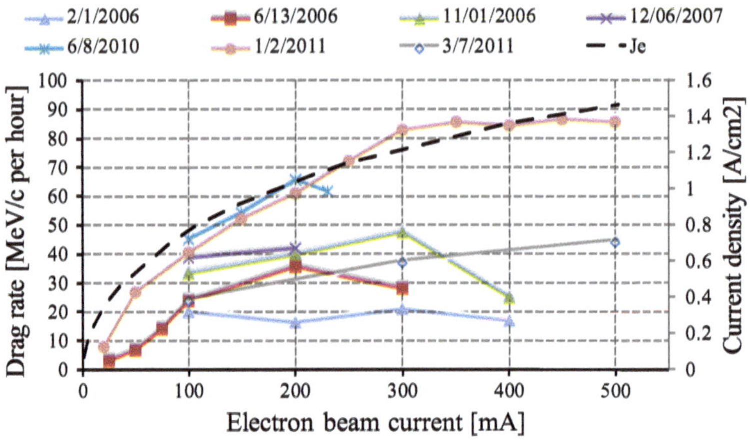

Fig. 7.72 Drag rate as a function of the beam current measured on axis at various dates with a 2 kV voltage jump. The current density calculated at the beam center (*dashed curve*) is shown for comparison

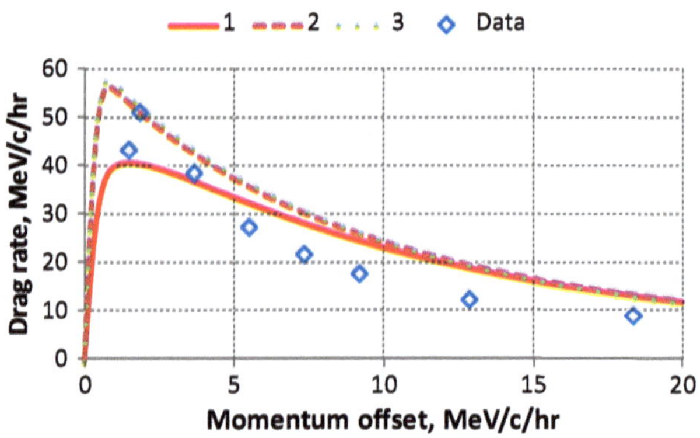

Fig. 7.73 Dependence of the drag rate on the momentum offset. The *diamonds* represent data measured on January 4, 2011 with $I_e = 0.1$ A "on axis." See the text for the curves description

it was far from satisfying inequality of Eq. (7.152). For comparison, curve 2 shows a calculation (for the same σ_{et}) using the simplified Eq. (7.138) with the Coulomb logarithm calculated for each momentum offset according to Eqs. (7.132)–(7.133), and curve 3 is the same but with a constant value of the logarithm estimated at the momentum offset of 10 MeV/*c*. The simplified formula gives a good approximation. The value of σ_{et} found in the fit, 0.13 mrad, is close to what has been found from other measurements in Table 7.5, 0.1 mrad.

If the electron angle is artificially increased, for instance, by mis-steering at the entrance of the cooling section, the drag rates decrease (Fig. 7.74), and the fitted

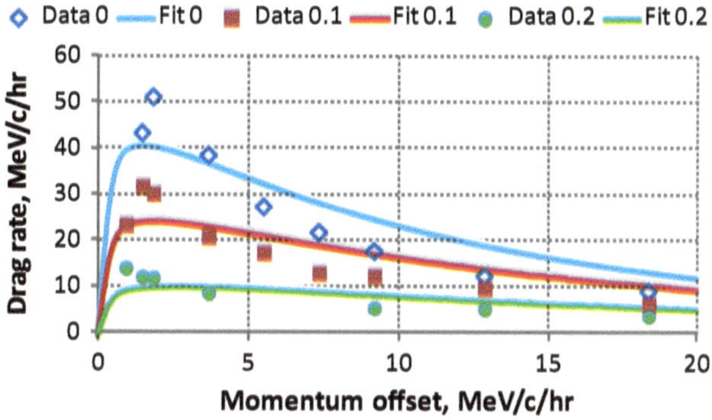

Fig. 7.74 Data and fits with Eq. (7.131) for a beam with helix-like trajectories. The *numbers* in the legend indicate the 1D values of the angle of the beam centroid in the cooling section in milliradian (verified with BPM measurements). The fitted angles are shown in Table 7.10

value of the angle increases in reasonable agreement with the sum in quadrature of the initial angle and the angle introduced by the trajectory perturbation (Table 7.9).

At the same time, the data points systematically deviate from the fitting curve by more than the statistical measurement errors (which were typically 1–3 MeV/c/h). It can be corrected by allowing one more fitting parameter. For example, the fit becomes almost perfect if the constant in front of the integral in Eq. (7.131) is artificially decreased by a factor of 2.5 and σ_{et} is set to 0.07 mrad. One may speculate that it is due to a large difference in dipole electron angles existing between modules in the cooling section; thus, only a portion of it works effectively. However, we did not see so dramatic effects experimentally. More likely reasons for the disagreement are the simplifications of the model itself.

One of them is the neglect of the longitudinal 105 G magnetic field. The field should modify the dynamics of binary collisions with large impact parameters (above the electron Larmor radius of ~0.2 mm), but we have not found a way to estimate this effect quantitatively. An indirect indication of the importance of the collisions with large impact parameters was seen in measurements with helix-like trajectories reported in [88], where the drag rates grew when decreasing the helix pitch at a constant helical angle.

Another possible cause of the disagreement is the assumption that the transverse inhomogeneity of the electron beam affects identically the force at all momentum offsets (Eq. (7.156)).

From these comparisons, we conclude that the model is capable of predicting the drag rates for weakly magnetized electron cooling to within ~50 %. Using the simplified formula of Eq. (7.138) makes analytical estimations and fitting much easier and without significant sacrifices in accuracy.

The Recycler Electron cooler was the first and, during its operation, the only relativistic ($\gamma > 2$) electron cooler. It significantly contributed to the success of Run II, resulting in an immediate increase of the integrated luminosity by ~25 % and removing a bottleneck in the antiproton production chain. An indication of the importance of electron cooling was a drop of the integrated luminosity rate by a factor of ~2.5 when the cooler was off for repairs.

Table 7.11 Main Recycler Ring parameters

	Units	Value
Circumference	m	3,320
Acceptance	π mm mrad	40.0
Fractional momentum aperture	%	±0.25
Maximum dispersion function	m	2.0
Average β_f	m	40.0
Average beam pipe radius	mm	23
Beam energy	GeV	8.89
Average beam relativistic β		0.998
Average beam relativistic γ		9.48
Maximum energy loss	GeV	0.089

In addition, technical solutions found during the project and better understanding of the peculiarities of the cooling process should help to build future coolers.

7.4 Cooling and Beam Manipulations in the Recycler

The Fermilab Recycler [47] is an 8 GeV ring using strontium ferrite permanent magnets, which provided additional storage for the accumulation and cooling of antiprotons. The main parameters are shown in Table 7.11.

The objectives of the Recycler were to accumulate and store antiprotons with high efficiency (e.g., low beam loss during injection/extraction, high lifetime), allow for fast and frequent transfers of antiprotons from the Accumulator, and provide bunches with low emittance to the Tevatron. In order to meet these goals, general machine improvements, operational optimizations, elaborate beam and RF manipulations, and appropriate cooling schemes had to be devised. Note that many other efforts, the installation of transverse dampers, for example (section instabilities), were indispensable to achieve the stated goals. Many are covered in other chapters of this book and will not be discussed here. Others go beyond the scope of this document.

Among general machine improvements, reducing the RR base pressure and changing the operating tunes had the most significant impact at the beginning of the Recycler commissioning, in particular on the (anti)proton beam lifetime. Operational optimizations included automation of various manipulations between different machines during transfers in and out of the Recycler and adjustments to the overall exploitation of the accelerator complex. Then, in the Recycler proper, various and complex RF manipulations were needed for injections, storing, and extractions of the antiprotons while limiting emittance dilution/growth and beam loss. Finally, the cooling systems' role was to mitigate heating mechanisms, including from RF manipulations, but also further reduce the beam phase space. More specifically, the missions for the cooling systems in the Recycler were to neutralize multiple Coulomb scattering (IBS and residual gas); neutralize the effects of heating due to the Main Injector ramps (i.e., stray magnetic fields); reduce the emittances, in particular longitudinally, of the stored antiprotons between transfers from the Accumulator; and reduce the phase space of the stored antiprotons in preparation for a Tevatron store. While it is natural to expect that maintaining low emittances would help preserve a good antiproton lifetime, some subtle effects, positive and negative, associated with cooling—stochastic and electron—will be discussed.

Most of the characteristics listed here are intertwined; however, beam cooling has the largest leverage. For instance, the level to which emittance dilution and other emittance growth mechanisms can be tolerated depends on the capabilities of the cooling system. In turn, it imposes limits to the frequency of transfers in and out of the Recycler. At the same time, the evolution of the lifetime must be taken into account imposing in return restrictions on the way cooling may be used.

Thus, this chapter will present (1) a quick overview of an accumulation cycle, (2) the main heating mechanisms found in the Recycler, (3) the beam cooling schemes developed and applied, (4) a discussion of the antiproton beam lifetime as it relates to the Recycler operation, and (5) the ultimate performance of the Recycler ring.

7.4.1 Stashing[7] Cycle

A *stashing* cycle is defined as the steps that take place from the time antiprotons are generated to the time they are extracted to the Tevatron. Antiprotons are first produced and stored in the Debuncher/Accumulator rings. Then, they are sent to the Recycler in a series of small transfers, where they are accumulated and cooled until extraction for collisions in the Tevatron. With stored beam in the Recycler, barrier potential wells are used to time separate the "cold" *stashed* beam from the "hot" injected beam or to prepare the beam destined for extraction. The transfers and accumulation of antiprotons in the Recycler continue until an optimum stash size is reached. Many factors (e.g., Tevatron luminosity lifetime, stacking rate in the Accumulator, Tevatron availability, cooling performance in the Recycler) determine this optimum with the aim of maximizing the integrated luminosity delivered to the experiments. In turn, the stash size before extraction to the Tevatron could be anywhere from 150×10^{10} to 500×10^{10} antiprotons, depending on the accelerator complex conditions and performance.

For Run I and the beginning of Run II before the Recycler started to be integrated into regular operation, all the antiprotons delivered to the Tevatron were directly extracted from the Accumulator ring. During the commissioning of the Recycler and cooling systems, in particular the electron cooler, antiprotons directed to the

[7] To distinguish the antiprotons accumulated in the Accumulator ring from those accumulated in the Recycler ring, the term "stash" was adopted to describe the antiprotons stored in the Recycler (hence "stack" refers to antiprotons accumulated in the Accumulator).

Tevatron were coming from *both* the Accumulator and the Recycler (so-called mixed mode or combination shots). Eventually, once electron cooling became operational, all available antiprotons were accumulated in the Recycler and extracted from the Recycler to the Tevatron. It should be noted that at first, "Recycler only" shots did not result in an increase of the overall number of antiprotons available for collision. However, it delivered much brighter bunches than the Accumulator could and reduced the duration of the stores because antiprotons were produced faster in the Debuncher/Accumulator. At the end of Run II, the optimum procedure to maximize the *stashing* rate was to accumulate small stacks of ~25×10^{10} in the Accumulator and transfer them to the Recycler every 40–50 min. In these conditions, a typical store duration was ~15 h.

7.4.2 Heating Mechanisms

Heating mechanisms can be divided into three categories: heating due to the Main Injector ramp, heating from beam manipulations in the Recycler (i.e., nonadiabatic RF manipulations), and heating due to beam interactions with its environment (i.e., beam-background gas interactions) or itself (i.e., `IBS`). In order to reduce the burden put on beam cooling to a minimum, many efforts were directed toward limiting heating from external sources.

7.4.3 Main Injector Ramp Compensation: Shielding and Orbit Corrections

While it is not unusual to place multiple accelerators in a common enclosure to save on civil construction costs, this often complicates operations, especially if accelerators are affecting each other. At Fermilab, the influence of the Main Injector (MI) on the Recycler, placed in a common tunnel, was initially found to be unacceptable for a reliable operation of the Recycler.

1. Shielding [89]

Stray magnetic fields originate from both the quadrupole and the dipole bus (supply and return) excitations during the MI ramp (8–120/150 GeV). At the top of the ramp, the current in the quad and dipole busses are 3.5 and 9.2 kA, correspondingly. Both busses contribute about equally to the transverse magnetic field at the Recycler beam location. Figure 7.75 shows the measured transverse magnetic field in the Recycler beam pipe without a magnetic shield during the MI ramp. It was estimated that these fields can move the closed orbit of the 8.9 GeV/*c* Recycler beam by as much as 30 mm. The original magnetic shielding (one layer of Carpenter High Perm "49" alloy—"alloy 49" and one layer of silicone steel alloy

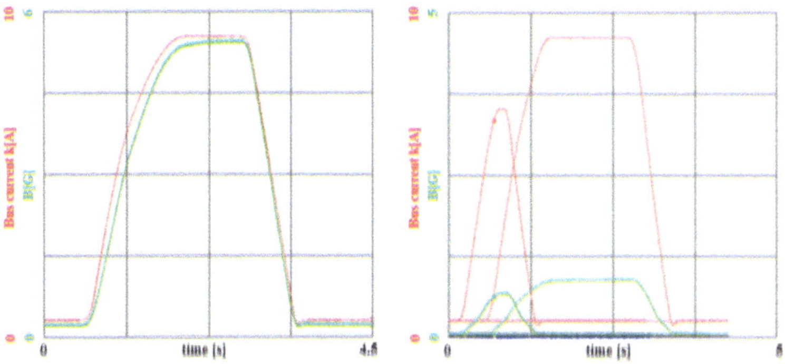

Fig. 7.75 Dipole bus current (*red*, 2.5 kA/div) and transverse magnetic field (*green*, 1.5 G/div) as a function of time during the 150-GeV MI ramp. The quad bus was also energized. *Left*, without shielding; *right*, after shielding (first implementation). On the *right* plot, the additional traces are a 120 GeV/*c* MI ramp and corresponding transverse magnetic field measurement

AS-0 separated by a fiberglass cloth) reduced the transverse magnetic field on average by a factor of 6 (Fig. 7.75). The measured rms Recycler closed orbit excursion during the MI ramp with this shielding was 5 mm in both planes. This was determined to be inadequate.

Consequently, several materials and wrapping techniques were tested on the bench (using large diameter Helmholtz coil producing fields of up to 10 G) with the goal of reducing stray magnetic fields by another factor of 10 or more. The final and most effective configuration is as follows:

- One layer of "alloy 49" with one layer of AD-MU-80 (from Ad-Vance Magnetics, Inc.) wound directly on top of it
- 1/8″ gap (fiberglass cloth)
- One layer of silicone steel alloy wrapped *along the direction of the spool roll*

It resulted in a reduction of stray magnetic fields by a factor of 30 with respect to the field measurement without magnetic field shown on Fig. 7.75. The rms closed orbit excursions for both planes were measured to be about 1 mm.

From this experience, a couple of useful considerations revealed themselves: first, the attenuation coefficients for multiple layers of shielding material (if separated by a small gap) add rather than multiply; second, it is important to keep in mind that some materials can be anisotropic and the roll direction can affect the shielding performance.

It should be noted that in addition to magnetically shielding the Recycler, the MI busses themselves are shielded where possible (i.e., not interfering with beam line components), although merely with steel covers.

2. Active ramp compensation [90]

While the improved shielding made the orbit excursions of the antiproton beam manageable, an active compensation scheme was devised. With the powered

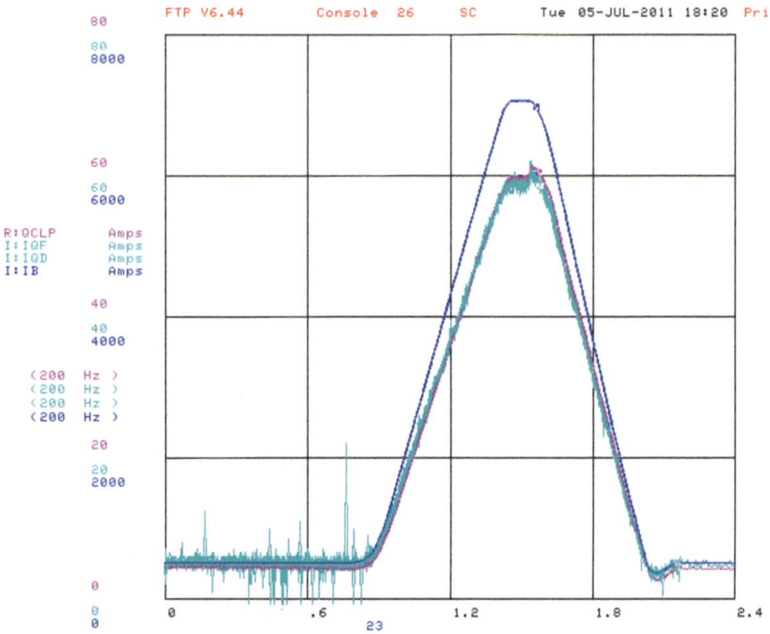

Fig. 7.76 Dipole bus current (2 kA/div), quadrupole bus current difference (20 A/div), and QCL current (20 A/div) as a function of time during the 120-GeV MI ramp

correction elements, a base orbit is defined with the MI energized at its 8 GeV level. Then, when the MI busses currents (dipole and quadrupole) increase during each ramp, hence the transverse magnetic field experienced by the Recycler, the dipoles correctors' currents ramp accordingly such that the antiproton beam remains on the previously defined 8 GeV orbit. However, because the bus current for the defocusing quadrupoles and the focusing quadrupoles differ by ~60/80 A at flattop (120/150 GeV), the net field induced by the quadrupoles busses needs to be corrected *globally* around the ring. The orbit correction with dipole correctors, which is local by definition, is not designed to address this issue. In addition, the current output range of these correctors is not adequate. Another level of compensation is therefore needed. This is achieved with the so-called Quad Compensation Loop (QCL), which is a single wire wound around the MI in between the interleaved focusing and defocusing quadrupole busses and cancels (on average) the difference in current. Reciprocally, this cannot take into account local variations (e.g., geometry of the beam line components), hence the need for the orbit correction with the Recycler correctors. Evidently, this current loop also ramps accordingly with the MI cycle. This is illustrated in Fig. 7.76.

Note that the QCL current settings are not calculated based on the MI reference bus current for each ramp but by actually measuring the current difference between the two busses. The quads, skew quads, and sextupoles are not ramped.

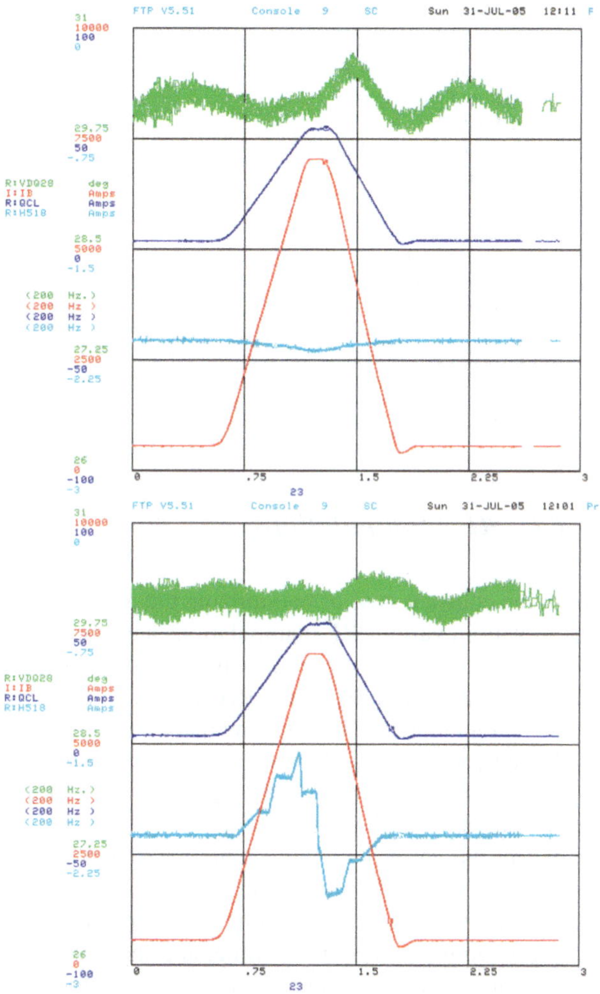

Fig. 7.77 Phase error during the 120 GeV MI ramp (2.4 s period). *Left*, without path length compensation; *right*, after implementing the path length compensation; *red*, dipole bend bus current (2.5 kA/div); *blue*, QCL current (50 A/div); cyan, current output from one of the correctors within the 5-bump (0.75 A/div); *green*, phase detector (1.25°/div). Data taken with 6×10^{10} antiprotons in a linear bucket (as opposed to barriers for normal operation)

In addition to the direct effect of the stray magnetic field from the MI busses, the antiproton beam is subject to an energy variation due to the net vertical magnetic flux that goes through its orbit. A mere 10 V per turn changes the Recycler beam energy by 1 MeV during the MI ramp. Qualitatively, it had been observed that the horizontal closed orbit moves toward the inside of the ring in the areas of high dispersion. As a result, the beam trajectory *breathes* synchronously with the MI ramp. In turn, the beam path length varies and a phase error develops with respect to the RF frequency. While not too detrimental from the point of view of the beam orbit (or the beam lifetime), if not corrected, it would be a strong source of longitudinal heating. Thus, this phase error is compensated with a local 5-bump which varies the path length of the beam accordingly to the energy gain/loss due to the MI ramp. With this scheme, the phase error was reduced to less than 0.5° (Fig. 7.77).

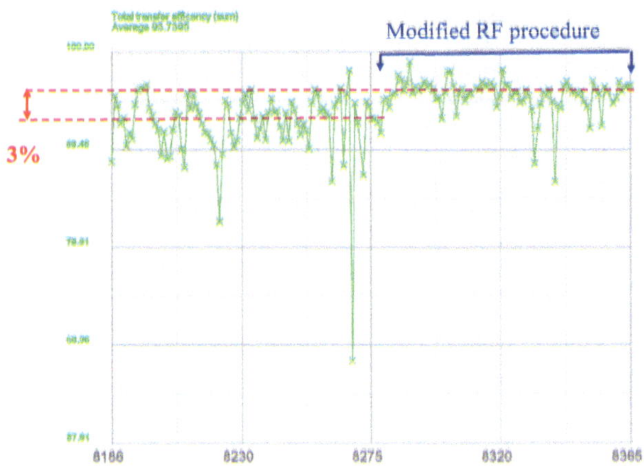

Fig. 7.78 Transfer efficiency as a function of the transfer number. The RF change was applied starting from transfer #8276. Note that the benefits from the RF modification may not be apparent at all times because of the accelerator chain conditions and destructive studies taking place concurrently

7.4.4 RF Manipulations

With the constant goal of maximizing the number of antiprotons to be delivered to the Tevatron, various RF manipulations have been developed for injection of fresh antiprotons and their merging with the stored beam as well as for the extraction process. One peculiarity of the Recycler is its use of a RF barrier system for all its beam manipulations. These RF manipulations are described in detail in Chap. 4, for both transfers into the Recycler and extraction to the Tevatron. Here, we will merely highlight those with the biggest impact on the overall beam loss and, to a lesser extent, beam heating.

While the antiprotons are accumulated and stored in the Recycler, the bunch occupies a little bit more than half the circumference of the ring. This is a compromise between maintaining the beam in a configuration favorable for (stochastic) cooling and beam lifetime, and the necessary space for injections/extractions. The stochastic system is more efficient for long bunches, when the particle density is the lowest. The beam lifetime typically improves in the long bunch configuration. When new beam is injected or the stored beam extracted, one of the most important feature of the RF manipulations described in Sect. 4.2 is the formation of the so-called anti-bucket.

In the injection case, the creation of an anti-bucket effectively *deepens* the potential well in the region of the stored beam. Consequently, high momentum particles are trapped in this potential well, hence clearing up the injection region. As a result only those particles not confined by the RF bucket, so-called DC particles, are lost when the injection kicker fires. In fact, the injection losses decreased significantly when it was realized that the anti-bucket was grown too rapidly for adiabatic capture. By merely slowing down this RF waveform formation, the transfer efficiency increased by ~3 % (Fig. 7.78).

In addition the modified RF manipulation significantly reduced the sensitivity of the transfer efficiency on the number of stored antiprotons. Typically, as the stash size increases, the population of the high momentum tails increases as well. In turn, with the slower growth of the anti-bucket, hence the better capture of the high momentum particles, the number of DC particles remains quasi-constant during the entire accumulation cycle.

In the extraction case, the anti-bucket effectively *raises* the potential of the bunches ready for extraction with respect to the mined bunches. Similar to the injection case, the presence of this higher potential reduces the number of DC particles present in the extraction region when the kicker fires, therefore reducing the amount of beam lost. However, the trailing edge of the extraction kicker pulse is long enough such as to seldom kick the particles contained in the high momentum RF bucket and the last couple of mined bunches. More than beam loss, here the kicker tail is a heating source for the affected bunches. To limit this effect, after each extraction, the ensemble of the remaining mined bunches was moved away from the extraction region, hence from the kicker pulse.

Before the advent of electron cooling, the longitudinal phase space of the stored antiprotons could be significantly larger than the phase space available in the mined buckets. Therefore, the momentum mining efficiency would be determined by the ultimate longitudinal emittance that could be achieved before growing the mined buckets. Thus, the amount of un-captured beam isolated in the high momentum RF bucket could be significant (up to ~20 % of the total stash), somewhat increasing the beam loss due to the kicker during extraction. Because the electron cooling rate is sufficiently high to reduce the longitudinal emittance of the "cold" beam to a level where virtually all the antiprotons are contained within the mined buckets—typically, less than 2 % of the stash remains in the Recycler after extraction—the high momentum bucket was deemed unnecessary and taken out of the RF manipulations leading to extraction. While this did not reduce the extraction beam loss further, it is believed that it increased the threshold limit to a resistive wall instability which the Recycler is susceptible to at large number of antiprotons and which results in a sudden, large beam loss.

7.4.5 Intra-beam Scattering

Small-angle multiple IBS is the dominant heating mechanism, which determines the equilibrium emittance achievable in the Recycler. The longitudinal IBS heating in particular has been of concern [47] because of a small longitudinal emittance required for Tevatron collisions. Figure 7.79 shows the calculated longitudinal IBS diffusion rate using the IBS theory and formalism from [91] (see also Sect. 6.1.2); the measured Recycler lattice functions have been used. One can notice from Fig. 7.79 (solid line) that longitudinal heating vanishes for a certain rms momentum spread, δp_0, and IBS has then a cooling effect above this momentum spread. For comparison, also shown in Fig. 7.79 is the heating (cooling) rate calculated in a smooth, round-beam, zero-dispersion approximation using the value of the average

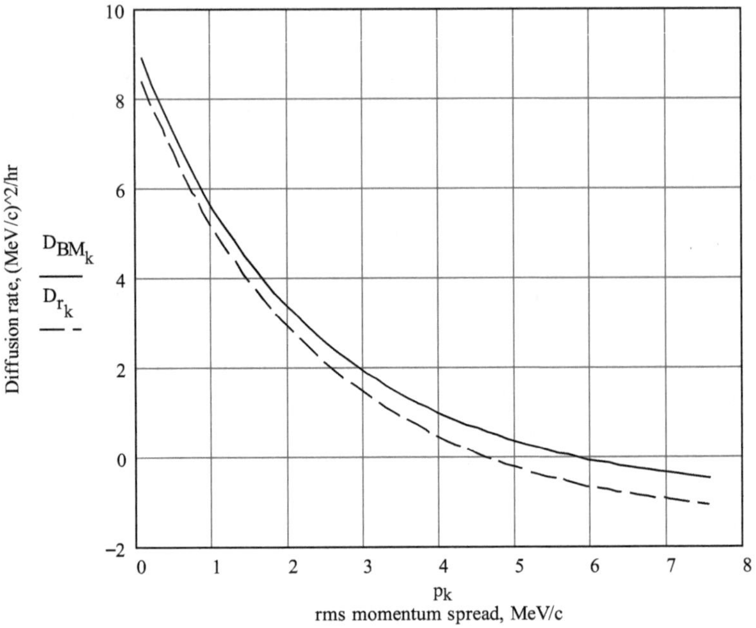

Fig. 7.79 The calculated longitudinal IBS diffusion rate as a function of the rms beam momentum spread in the Recycler for a coasting beam of 100×10^{10} antiprotons with a constant 5π mm mrad emittance. The solid line is modeled from [91]; the *dashed line* is the round-beam, zero-dispersion, smooth approximation

β-function, $\beta_{ave} = 30$ m and which shows the same characteristics. It has been pointed out previously [47] that one can exploit this feature of the longitudinal diffusion rate to minimize the longitudinal emittance growth by compressing the bunch length, and thus increasing the momentum spread, until the diffusion due to IBS vanishes or becomes negative (IBS cooling).

Figure 7.80 shows the value of the rms momentum spread, δp_0, for which the diffusion rate vanishes, as a function of the transverse beam emittance.

When stochastic cooling was the only cooling system available, “IBS cooling” was exploited in the Recycler to improve the performance of antiproton accumulation. Right after an injection, typical transverse emittances in the Recycler are 5–7π mm mrad; from the above theory, the point at which longitudinal heating vanishes for this range of transverse emittances—$\delta p_0 = 6$–7 MeV/*c*—cannot be reached in the Recycler without losing particles at the momentum aperture, hence deteriorating the beam lifetime significantly. However, a practical solution was adopted and proved to be beneficial. After each injection, the bunch was expanded so as to occupy the entire ring circumference (still contained within two RF pulses). While this state is not favorable for longitudinal IBS cooling, it does make the stochastic cooling system more efficient, in particular in the transverse direction. Once the transverse emittances reached 3–4π mm mrad, Figure 7.80 shows that IBS

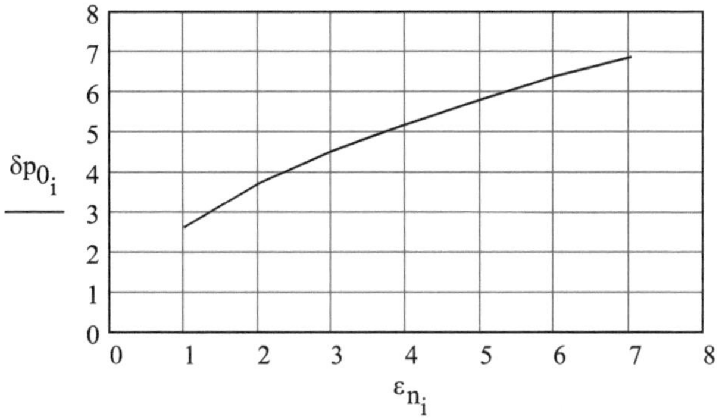

Fig. 7.80 Value of the rms momentum spread in MeV/c, for which the longitudinal IBS heating vanishes, as a function of beam transverse emittance (n, 95 %) in π mm mrad

cooling would occur if $\delta p_0 = 4$–5 MeV/c. This range of the momentum spreads is closer to what can be tolerated in the Recycler without beam loss. Thus, at this point, the bunch was regularly shortened (or *squeezed*) in order to maintain the momentum spread as high as possible, around 4 MeV/c, in a region where longitudinal IBS cooling could take place. Note that this technique is limited because the transverse diffusion rate due to IBS increases with increasing longitudinal charge density; hence, longitudinal IBS cooling effectively stops and the beam is in an equilibrium state, i.e., heating and stochastic cooling cancel one another. One should also note that the longitudinal stochastic cooling system remains on, also providing cooling in the longitudinal direction.

This technique allowed accumulating more antiprotons for a given longitudinal emittance. This is illustrated on Fig. 7.81 where the longitudinal emittance before mining is plotted against the number of antiprotons stored in the Recycler. In the figure, the green and blue horizontal dash lines represent the maximum and goal longitudinal emittances, respectively. The maximum value is dictated by the RF voltage available in the Recycler.

The data show the Recycler cooling performance before (blue diamonds) and after (orange triangles) changing the cooling procedure. Without IBS cooling, the minimum longitudinal emittance requirement for nine mined bunch extractions to the TeV could not be met for the number of antiprotons more than 60×10^{10}. With IBS cooling, it would allow stashes of up to 180×10^{10} antiprotons. The target longitudinal emittance could be achieved for up to 55×10^{10} antiprotons without IBS cooling and 130×10^{10} with IBS cooling. It should be noted that the limits shown on Fig. 7.81 assume that nine mined bunches are formed and extracted. In other words, if a lower number of mined bunches would be extracted, the longitudinal emittance target would decrease proportionally, which was often the case during mixed-mode operation. The IBS cooling procedure was abandoned when electron cooling became operational.

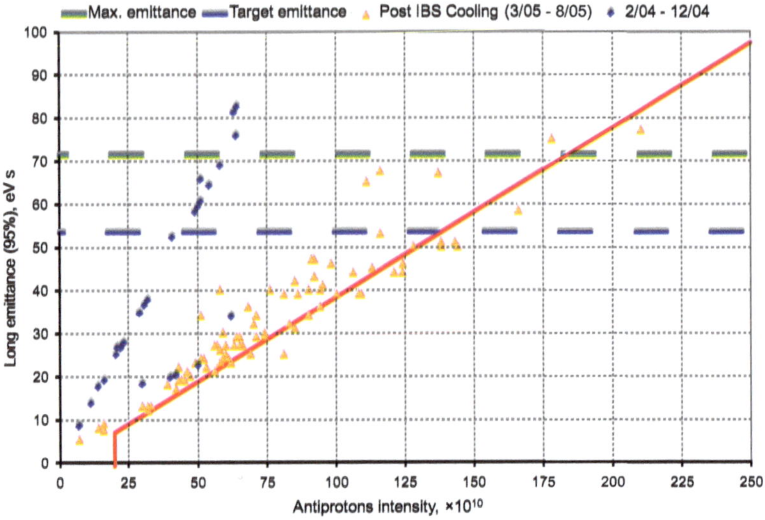

Fig. 7.81 Antiprotons bunch longitudinal emittance before mining. *Blue diamond*, before IBS cooling; *orange triangle*, taking advantage of IBS cooling; *green line*, longitudinal emittance corresponding to the maximum phase space available (2 eV s per 2.5 MHz mini-bunch); *blue line*, target emittance for 1.5 eV s per 2.5 MHz mini-bunch (goal). The *red line* is not a fit but represents an estimate of what could *reasonably* be achieved with the stochastic cooling system and the procedure taking advantage of IBS cooling based on early data. It served as a reference for operation

7.4.6 Beam-Background Gas Interactions

It is well known that elastic collisions between the beam particles and residual gas molecules cause emittance growth (and eventually beam loss) and that ultrahigh vacuum is of paramount importance for any storage ring to be viable. During its early commissioning, the beams, protons, or antiprotons circulating in the Recycler demonstrated poor lifetime (tens of hours, [92]) and a transverse emittance growth of the order of 5π mm mrad/h [93], which was unacceptable for its implementation into operation. After the number of ion pumps was doubled and the entire ring baked to >110 °C to remove the water content, the vacuum level and composition became adequate (currently $<3 \times 10^{-10}$ Torr, except at some localized areas such as the stochastic cooling pickup/kicker tanks where it can reach 10^{-8} Torr). Then, for a coasting beam with a very low number of antiprotons (~1×10^{10}) in order to limit IBS (discussed in the previous paragraph), a typical transverse emittance growth would be of the order of 0.5π mm mrad/h, one order of magnitude lower than initially measured. This is fairly close to the theoretical value of ~0.3π mm mrad/h obtained from Eq. (6.55) which can be reduced to:

$$\frac{d\varepsilon_{(\perp)n}}{dt} = 0.29 \times \beta_{\text{ave [m]}} \frac{P_{\text{[Torr]}}}{\beta^2 \gamma} \quad \left[\frac{\text{m} \cdot \text{rad}}{\text{s}}\right] \tag{7.158}$$

with $\beta_{\text{ave}} = 30$ m, the average beta function of the Recycler ring, $\gamma \approx 10$ and $P = 10^{-10}$ Torr, and in which the residual gas is pure nitrogen (N_2).

Similarly, as expected, the antiproton beam lifetime dramatically increased.

7.4.7 Beam Cooling After Electron Cooling Commissioning

The implementation of two cooling systems, stochastic and electron, was envisioned from the beginning of the Recycler Ring design [47]. Details of the stochastic cooling system and of the electron cooling system are described above, along with their intrinsic cooling characteristics. In this section, their practical implementations and operational performances are discussed. Because electron cooling is much more efficient than stochastic cooling, more emphasis will be put on the electron cooling system. Nevertheless, stochastic cooling is an integral part of the cooling procedure. Before electron cooling was operational, stochastic cooling was optimized accordingly to the requirements of the IBS cooling procedure detailed above. When electron cooling came into operation, the role of the stochastic cooling system somewhat changed. It became the primary system that could actively improve the beam lifetime. This will be discussed more extensively in the following section.

7.4.8 Procedure

As it was pointed out in the Introduction, the Recycler has mainly two goals: store antiprotons efficiently and prepare the beam for extraction to the Tevatron. In principle, electron cooling merely requires having the ion beam (i.e., the beam that is being cooled) and the electron beam overlap concentrically, with the electron beam transverse size larger than the ion beam's. In other words, the central orbit of the antiprotons in the cooling section and the electron beam centroid trajectory coincide. However, during accumulation, it was quickly realized that this configuration, which yields the maximum cooling rate, was neither beneficial nor required. In fact, and most importantly, the lifetime deteriorates rapidly in this mode of operation. Similar to low-energy coolers—see, e.g., [94]—the electron cooling induces additional diffusion.

The most practical solution to alleviate this issue was to displace the electron beam trajectory with respect to the antiproton beam orbit (a horizontal or vertical shift). To first approximation, the center of the electron beam is where the current density is the highest and the beam angles the smallest, hence the cooling force the

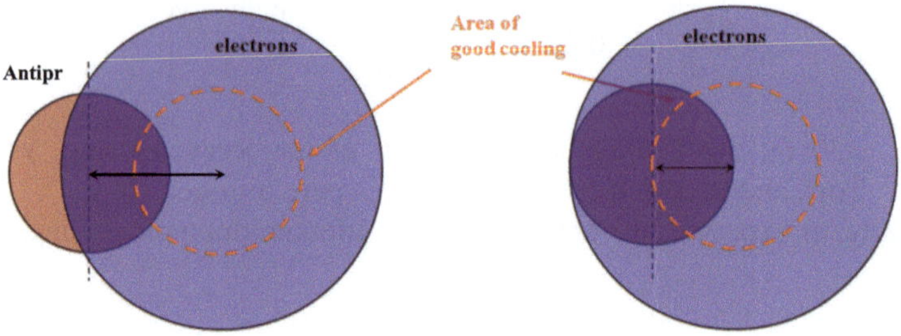

Fig. 7.82 Illustration of the cooling rate adjustment by a horizontal shift. The "area of good cooling" represents the transverse section of the electron beam where the current density and the beam angle are the most favorable for cooling. The *arrows* indicate the shift between the two beam centers

largest. Therefore, when the beam is offset, the majority of the antiprotons do not experience the maximum cooling force. On the other hand, with the beam offset, large betatron particles are being cooled more effectively, bringing in the "tails" of the beam distribution. This is particularly true when the transverse emittance of the ion beam is relatively large and the ion beam transverse dimension is actually larger than the electron beam cross section. This procedure can be regarded as "painting" and, in fact, is almost equivalent to the "hollow beam" concept for low-energy coolers. An illustrative cartoon is shown below (Fig. 7.82). Here, the "area of good cooling," or the lack thereof, is determined by various factors such as the beam current density uniformity or the electron transverse angle distribution.

Thus, although the cooling rate is not maximal, it provides sufficient cooling to permit efficient transfers from the Accumulator, and the lifetime does not deteriorate as much (see following section). When the stored beam needs to be prepared for extraction, so-called final cooling, the goal for cooling is slightly different than for accumulation. During "final cooling," reaching as small emittances as possible (without inducing an instability) takes priority over lifetime preservation (although the latter was not ignored; see RF manipulations section at the beginning of the chapter). Therefore, the electron beam offset is adjusted so as to obtain the appropriate cooling for reaching a predetermined target longitudinal emittance before mining; typically, it requires the electron beam to be brought closer to the antiproton beam central orbit. Once the longitudinal emittance target is attained, mining can occur and the electron beam brought to its final position, often on axis (i.e., antiproton central orbit) or very close to it in order to reduce the 6D emittance further. The electron beam offset positions both before mining and during extraction depend on the cooling performance. Then, during the whole extraction process, the beam remains at the offset thus determined.

Even with the electron beam being the primary cooling system, stochastic cooling remained needed. After extractions to the collider, the antiprotons left behind occupy a large phase space volume not favorable to electron cooling, in particular in the transverse directions. In addition, because stacking continues during extraction and the resulting stack may be large, the emittances of the beam injected into the Recycler may also be relatively large. For low intensity, large emittance beams, stochastic cooling is much more efficient than electron cooling: many of the particles with large transverse action experience the friction force from the part of the electron beam transverse distribution where cooling is the poorest. Thus, stochastic cooling provides some sort of precooling such that emittances enter a domain where electron cooling becomes dominant again. Then, mostly for large stashes, stochastic cooling plays a fundamental role in the lifetime management strategy.

7.4.9 Cooling Rates

While “drag rate” measurements were carried out to study and optimize the electron beam characteristics for cooling, the most relevant measurement to assess the efficiency of cooling during operation is what is referred to as a cooling rate in this text. Figure 7.83 gives a typical example of such a measurement.

In order to make comparisons possible between various electron beam tunes, all measurements were done in nearly similar beam conditions and followed a simple standard procedure:

1. The antiproton beam, confined by rectangular RF barriers, is first cooled with the stochastic cooling system only. This allows the antiprotons distribution to become nearly Gaussian and reduce IBS effects.
2. The initial rms momentum spread, δp, is 3–4 MeV/c. The bunch can be adjusted to reach this target.
3. The stochastic cooling system is turned off and the antiproton beam is let diffuse for 15 min.
4. The electron beam is turned on and cools the antiprotons for 15 min. The electron beam is “on axis.”

Then, the cooling rate is merely the difference between the time derivatives of the momentum spread/transverse emittances before and after turning on the electron beam. For the transverse direction, both the diffusion and cooling are fitted with straight lines. For the longitudinal direction, the data are fitted to an exponential decay curve. For the data shown on Fig. 7.83, transverse cooling rate (averaged over both directions) for the Schottky data is -2.4π mm mrad/h and for the flying wire data -5.6π mm mrad/h. For the momentum spread, the measured cooling rate is -7.3 MeV/c/h. The relatively large difference between the transverse cooling rate obtained from the Schottky data and the flying wire data is attributed to the respective sensitivity of these devices and the cooling properties of the electron beam. The Schottky pickup is more sensitive to tail particles than the flying wires. So, qualitatively speaking, the weight of particles with large betatron oscillations in the Schottky detector measurement is larger than for the flying wire, which principally captures the behavior of the core of the antiproton beam. Since electron cooling is less efficient for particles belonging to the transverse tails of the

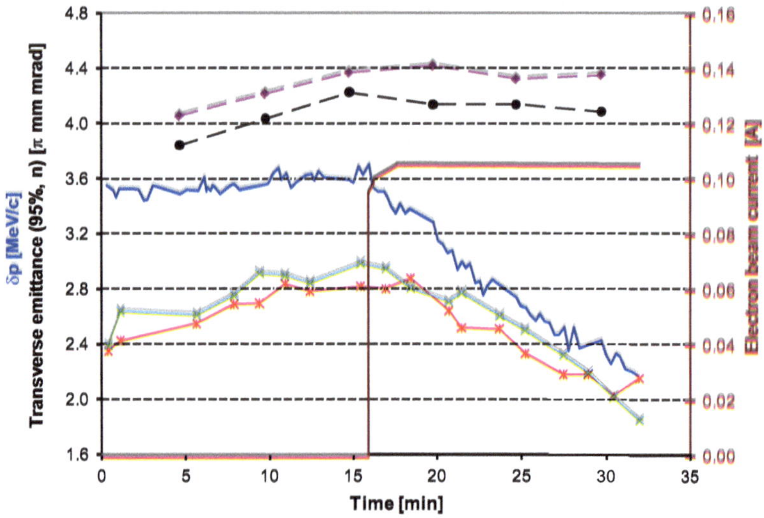

Fig. 7.83 Evolution of δp (*blue line*) and of the transverse emittances (95 %, normalized) during a cooling rate measurement. The *brown line* indicates when the electron beam was turned on and set to 100 mA. $N_p = 36 \times 10^{10}$, bunch length = 5.4 μs. For the transverse emittances, both the measurements obtained from a Schottky pickup (*violet diamonds* and *black circles*) and from flying wires (*green crosses* and *red stars*) are displayed

antiprotons distribution, one should expect the measured cooling rate to be less from Schottky data than from flying wire data.

Note that while the cooling rate measurements were carried out mostly to assess/check the electron cooling performance, attempts were made to characterize the stochastic cooling system as well. Issues encountered for the measurements of the stochastic cooling systems were that rates are quite low, thus easily subject to changes in the machine conditions, and the settings difficult to repeat from one measurement to the next. In addition, as mentioned previously, with electron cooling, the stochastic cooling rates are in effect irrelevant during normal operation.

Figure 7.84 summarizes most of the electron cooling rate measurements made between 2006 and 2010. Over that duration, the cooling rate for a given transverse emittance significantly increased due to several key improvements to the electron beam quality and are highlighted on the plot: improvement of the electron beam envelope through quadrupole magnets upstream of the cooling section, beam-based alignment of the magnetic field in the cooling section, and, for high beam current (>100–150 mA), ion clearing (in the electron beam). The arrows indicate the potential cooling rate increase, resulting from each of these beam optimization steps. Note, for instance, that the maximum cooling rate measured for a 3π mm mrad transverse emittance antiproton beam (from flying wire data) is almost twice the cooling rate obtained from the measurement shown on Fig. 7.84 −13.0 MeV/*c*/h vs. −7.3 MeV/*c*/h.

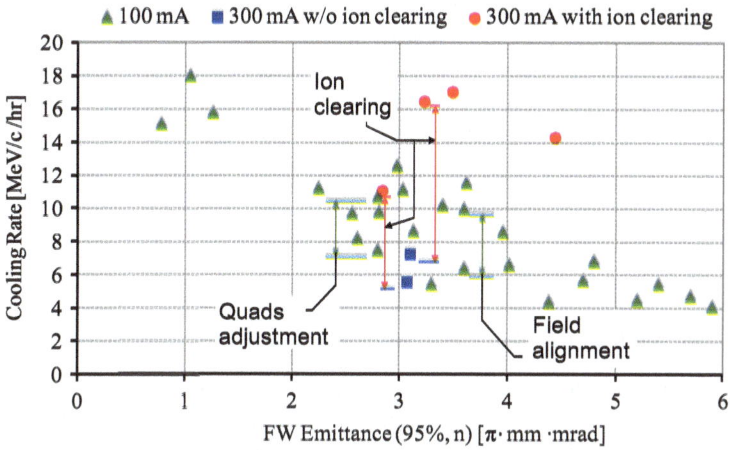

Fig. 7.84 Longitudinal cooling rate in 2006–2010. The arrows connect data points measured successively over the same day to show the potential cooling rate increase for several improvements brought to the electron beam tuning

Also visible on the plot is fairly strong dependence of the cooling rate on the transverse emittance of the antiproton beam. More details on how these adjustments improved the electron beam quality, hence the cooling rate, are discussed in Sect. 7.3.

7.4.10 Antiprotons Lifetime

In the Recycler, where antiprotons are typically accumulated for ~15 h, preserving the antiproton beam lifetime is crucial. The lifetime depends on the residual pressure, beam parameters, and tuning of the cooling systems. Below we will consider mainly the case of normal operation, i.e., with the cooling systems.

7.4.11 Vacuum Quality

At the initial stages of the Recycler commissioning, the lifetime was determined by interaction of the beam with the residual gas. During the September 2003 shutdown, the entire ring was baked and the pressure improved to ~0.1 nTorr, dominated by hydrogen [95]. Therefore, the lifetime calculations made in [92] (for the case of "minimum pressure") are applicable to the current state of the Recycler vacuum. Assuming that the cooling systems are capable of suppressing multiple Coulomb scattering, the main vacuum-dependent loss mechanisms (and the lifetime calculated in [92]) are single Coulomb scattering (~1,500 h), inelastic scattering (~3,300 h), and nuclear scattering (~7,000 h). Their compounded effect results in the estimation of the best life time achievable in the Recycler of ~900 h, which agrees, within measurement errors, with the best measured value of ~1,000 h.

Note that in the absence of cooling, multiple Coulomb scattering becomes dominant. The typical lifetime observed in the time of stochastic cooling failures, ~100 h, is in a reasonable agreement with the estimation in [92] (~130 h) as well.

7.4.12 Beam Parameters

In operation, the beam is characterized by the following parameters:

1. Transverse emittances measured by Schottky detectors. Typically, horizontal and vertical emittances are the same within the accuracy of the monitors' calibrations, so their average value is used as a single beam parameter.
2. Transverse emittances measured with flying wire detectors, mostly employed during injections and extractions. Their sensitivity to transverse tails is limited compared to the Schottky detectors, and the measurements obtained from the flying wires are deemed to be adequate for the core of the beam only.
3. Momentum spread measured by a Schottky detector and the longitudinal emittance calculated from the momentum spread as measured and the bunch length.
4. *Beam longitudinal distribution*. The values derived from the distribution and saved in the data logger are the length of the bunch, the maximum linear density measured by a resistive wall monitor, and the so-called DC beam, a value equal to the difference between the number of particles measured by a DC current transformer (DCCT) and by the signal from a toroid, and which is an approximate measure of the amount of antiprotons not captured by the RF potential well.
5. *Operating tunes*

   We have not found a single parameter from this list or a combination of these parameters that would uniquely determine the lifetime and were not able to come up with a cohesive model either. It is, in part, related to the fact that the lifetime is determined by the behavior of far transverse tails, which are not measured effectively. Below are some relevant empirical observations.

1. *Large transverse emittance*

The lifetime drops dramatically if the beam emittance is large. It was observed, for example, after beam instabilities or after injections into a nearly empty Recycler from a large antiproton stack. As a rule of thumb, the lifetime (calculated over 10 min) drops below 100 h if the Schottky emittance (95 %, normalized) is above 10 μm. Note that the measured acceptance of the Recycler (normalized) is 60/40 μm (H/V).

2. *Injections*

Immediately after each injection into the Recycler, the antiproton beam lifetime is poor. It is likely related to the incoming antiprotons higher emittances (the core

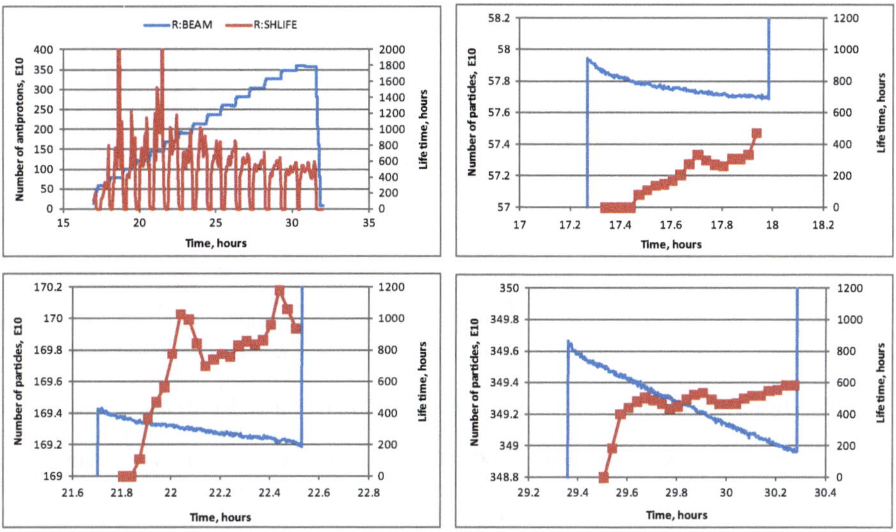

Fig. 7.85 Number of antiprotons and lifetime (10-min running average). October 10–11, 2010. The average lifetime for the cycle is 427 h. The *left upper plot* shows the entire cycle, and others present three intervals between transfers in more detail (in the same scale)

95 %, normalized emittance measured with flying wires is 4–6 μm, while it is ≲3 μm for the accumulated *stash*) and the transverse tails created by injection errors. Figure 7.85 shows the contributions of these effects on the lifetime for three intervals between transfers at different stages of accumulation: immediately after the Tevatron shot, in the middle of the cycle, and before the last injection (left to right and top to bottom). For all cases, the beam loss before the lifetime reaches an approximate equilibrium is nearly the same and on the order of 0.2×10^{10} antiprotons. Since this value is independent of the stash size, it is reasonable to assume that the lost particles primarily come from the freshly injected beam. Each transfer delivers ~20×10^{10} antiprotons to the Recycler in three "parcels" separated by ~3 min, and additional ~0.3×10^{10} antiprotons are lost between parcels arrivals. Therefore, the total lost portion of the injected beam is ~2 %.

3. *Linear density*

In operation, the beam is expanded to its maximum length when there are significant delays with transfers. It came mainly from practical experience of a noticeable increase of the life time for longer bunches. While it is difficult to make a conclusive experiment, the data shown on Fig. 7.86 seems to favor the linear density (Fig. 7.86a) rather than the transverse emittance (Fig. 7.86b) or 6D phase density as the beam parameter most likely to correlate with the value of the beam lifetime. One may speculate that it is related to peculiarities of the stochastic cooling system performance.

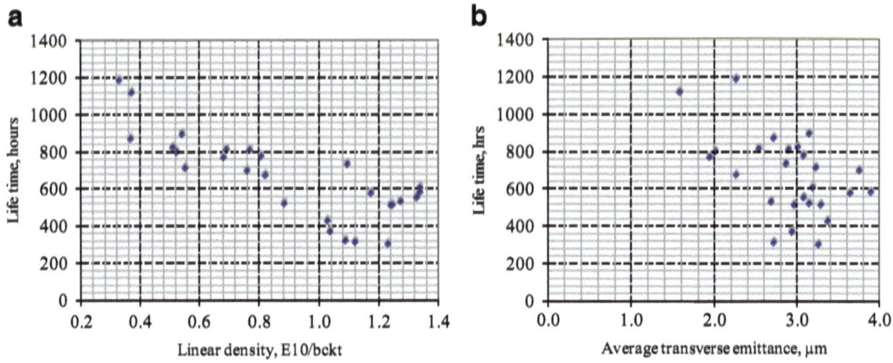

Fig. 7.86 Beam lifetime in a steady state as a function of (**a**) the linear density and (**b**) the average Schottky emittance (n, 95 %). The lifetime was recorded after at least 1 h without transfers or beam manipulations. December 2008–January 2009. The linear density is calculated as a ratio of the total number of antiprotons, $\times 10^{10}$, to the length of the RF barriers gap, in units of 53-MHz buckets—the Recycler perimeter in this unit is 588 buckets

4. *High momentum particles*

Tails of the momentum distribution can be lost if the particles reach the momentum aperture of the machine. The total measured aperture was ~ 120 MeV/c but, depending on tuning, was sometimes asymmetric with respect to the operational momentum, so that the particle loss was determined by the closest boundary with a minimum value of ~35 MeV/c. This value was still significantly larger than the typical rms momentum spread of the antiprotons, 2–4 MeV/c. Even during mishaps with longitudinal manipulations when the rms momentum spread had increased to up to 8 MeV/c, the lifetime was not affected (~500 h) for low-intensity stacks. In similar cases with large stacks, the more prominent effect on the lifetime was indirect and related to the resulting transverse emittance growth. We interpret this effect as being caused by a large number of particles not captured by the standard RF barriers of ~18 MeV/c height (DC particles). Traveling outside of the RF bucket, they "shortened out" the ion clearing gap, allowing ion accumulation, emittance growth, and eventually degradation of the life time.

5. *Tunes*

In regular operation, the Recycler tunes were ~0.414/0.418 in 2005–2006 and ~0.462/0.466 in 2006–2011. Dedicated scans with low-intensity beam did not show any lifetime difference between these two tune regions. While no special measurements have been made at the largest operational intensities, a direct influence of the tunes on the lifetime was likely below the scatter related to other factors.

7.4.13 Cooling Systems

Two Recycler cooling systems, electron and stochastic, contribute very differently to the antiproton lifetime. A simplified description of operational experience is that, properly tuned, stochastic cooling improves the lifetime, while applying of strong electron cooling degrades the lifetime. While we do not have a quantitative model, qualitatively this difference is understandable. The stochastic cooling system cools effectively the far tail particles, which determine the lifetime. In contrast, only antiprotons with small betatron amplitudes, comparable with the electron beam size of a few millimeters, are cooled by electrons. Any diffusion caused by the electron beam is compensated by its cooling effect for antiprotons in the core but drives away the particles in the transverse tails. In addition, electron cooling typically increases the density on and around the axis of the cooled beam, which may create additional diffusion in itself. A dense core with long tails accompanied by the combination of a shrinking emittance with worsening life time is a very reproducible feature of strong electron cooling.

Note that a deterioration of the lifetime in the presence of an electron beam has been observed at low-energy coolers (so-called electron heating [94]), and different explanations have been proposed ([96, 97]). It is difficult to accurately compare these observations to the Recycler's long lifetimes. Often, there are delays of tens of minutes between changes in the operational conditions and the actual corresponding response of the lifetime. However, in contrast to "electron heating," no strong dependence of the lifetime on the density of the either beam was found.

Two operational recipes were developed:

1. Properly tuned stochastic cooling systems is critically important for the lifetime at any tested operational intensities (up to 6×10^{10} antiprotons). In several occasions, with a broken (and turned off to avoid heating) stochastic cooling system, the lifetime would not improve beyond ~100 h for all the electron beam configurations which were attempted.
2. Stronger electron cooling results (sometimes with a delay) in a more pronounced deterioration of the lifetime. And it deteriorates the most when the beams are completely aligned, i.e., propagate along the same axis. Therefore, the optimum operation mode during accumulation is to keep the electron beam in the cooling section at a parallel offset with respect to the antiproton axis. The value of the offset approximately equal to the electron beam radius (1.5–2.5 mm) is chosen to provide just enough cooling between the injections and depends on how well the electron cooler is tuned.

7.4.14 Operational Results

To judge the efficiency of preserving antiprotons in the Recycler Ring related to the finite lifetime, an average lifetime was calculated as follows:

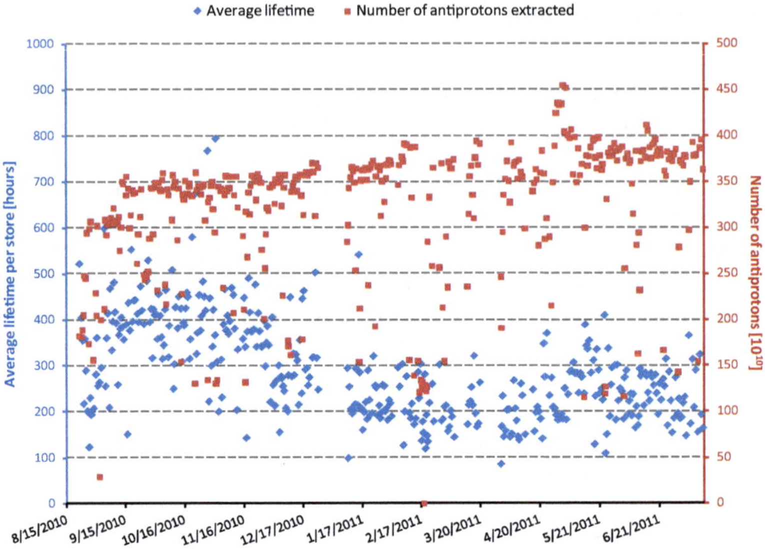

Fig. 7.87 Number of antiprotons extracted to the Tevatron (*red square*) and corresponding average lifetime (*blue diamond*) over the last year of operation (August 15, 2010–July 15, 2011)

$$LT = \frac{\sum N_i \Delta T_i}{\sum \Delta N_i}, \tag{7.159}$$

where ΔT_i is the time interval i between two transfers into or out of the Recycler, N_i is the average number of antiprotons in the Recycler during this interval, ΔN_i is the total beam loss (calculated from the DCCT signal) over the same interval, and the summation encompasses all intervals between two successive Tevatron accelerations. The value of the average lifetime varied significantly over time and from one storage cycle to another. An example is shown in Fig. 7.87.

One component of the scatter comes from the dependence of the lifetime on the linear density as described previously. With the beam length being constant in operation (5.8 μs during accumulation), the lifetime LT decreases as the maximum number of antiprotons in the cycle increases. Another important factor is the rate of transfers and quality of the incoming beam. Long delays with transfers decrease the relative contribution of having a poor life time right after a transfer (as in Fig. 7.85), and the large tails of the incoming bunches have a negative effect that may last for many hours.

However, when these effects are excluded either by the choice of the accumulation cycles to compare to or by averaging over many cycles, variations of the calculated lifetime remain large. In several occasions, the reasons for a worse lifetime were found to be due either vacuum leaks, a malfunction of the stochastic cooling system, or improvements made to the strength of electron cooling (corrected by applying a larger beam offset). On the other hand, we were not able to associate (and correct) the transition of the lifetime from ~400 to ~250 h presented in Fig. 7.87 to any of the phenomena described in this section. For a typical cycle duration of ~15 h, this change corresponds to loosing an additional ~1 % of the antiprotons sent to the Tevatron.

7.4.15 Antiproton Beam Parameters for Tevatron Shots

The characteristics of the bunches extracted from the Recycler correspond to both requirements from the Tevatron and the optimization of the accelerator complex operation. Table 7.12 lists the main beam parameters achieved in the Recycler as well as those used routinely during normal operation. Because the cooling scenario changed significantly with respect to the one initially envisioned, a comparison with goals expressed during the design phase, or even later on, is not relevant. So, while the "best achieved" column in the table shows the capabilities of the Recycler, the normal operation values are the ones that ultimately lead to the best performance of the accelerator complex in terms of integrated luminosity delivered to the experiments. For instance, while it was shown that the Recycler could store 600×10^{10} antiprotons with acceptable lifetime—actually one of the original goals—it would not have fitted within the operational scenario when taking into consideration the implications for the other rings, including the Tevatron and consequently for the integrated luminosity.

Other than limitations resulting from the optimization of the accelerator complex running conditions, several parameters could not be obtained concurrently purely because of the Recycler properties. For instance, a large *stash* with low emittances could lead to a resistive wall instability, which threshold depends on the phase density of the antiproton beam. This instability was experienced a few times during extraction, when cooling is the strongest. Also, as discussed above, a lot of attention was brought to the beam lifetime. In turn, it imposed practical limits on the phase space volume the most desirable for efficient accumulation. In addition, there are beam parameters which somewhat depend on one another, living only a small margin to work with. This is particularly true for the case of smaller stashes for which very small emittances and large lifetimes were obtained, while larger stashes had almost inevitably larger emittances and shorter lifetimes.

Therefore, the "normal operation" values in Table 7.12 are the result of multiple compromises and were routinely achieved. However, varying accelerator conditions required a flexibility that the Recycler could also provide. A good example is the "partial mining" procedure (Sect. 4.2), which permitted to extract only the number of antiprotons adequate for certain atypical situations. In these cases, the beam parameters in the Recycler were adjusted accordingly.

Table 7.12 Main antiproton beam parameters achieved in the Recycler during Run II

	Units	Best achieved	Normal operation
Number of antiprotons	10^{10}	600	360–390
$\varepsilon_{(s)n,95\%}$ (before mining)	eV s	48/55[a]	60–75
$\varepsilon_{(s)n,95\%}$ (average of the thirty-six 2.5 MHz bunches) in MI at 8 GeV	eV s	0.8/1.1[a]	1.1–1.3
$\varepsilon_{(\perp)n,95\%}$ (Schottky), stored beam	π mm mrad	N/A	3–4
Transfer efficiency (1 week average)	%	N/A	94–95
Extraction efficiency	%	99	97–98
Equilibrium lifetime	h	>1,000	N/A[b]
Time between transfers	min	N/A	40

[a]Strong dependence with the number of antiprotons, especially at the low end. The number of antiprotons for the emittances quoted in the table is ~160/~375 × 10^{10} particles, respectively
[b]During normal operation, there are no real equilibrium conditions

7.4.16 Final Performance

All the discussions presented in the preceding sections mostly focused on single processes and procedures, their implementation and performance, but mainly independently from one another. Ultimately, the Recycler performance is characterized by its ability to store antiprotons efficiently and deliver adequate beam parameters to the Tevatron.

In order to quantify the efficiency of the Recycler as a repository of antiprotons overall, a storage efficiency can be defined as

$$\frac{N_{\bar{p}}\quad_{\text{extracted to MI}} - N'_{\bar{p}}\quad_{\text{left in RR}}}{N_{\bar{p}}\quad_{\text{transferred from MI to RR}} + N'_{\bar{p}}\quad_{\text{left from previous extraction}}}, \tag{7.160}$$

where $N\bar{p}$ is the number of antiprotons measured by the Main Injector DCCT and $N\bar{p}'$ is the number of antiprotons measured by the Recycler DCCT. Thus calculated, the storage efficiency includes injection and extraction efficiencies from and to the Main Injector, losses due to the antiprotons lifetime and accidental losses (e.g., correctors' power supply trip, vacuum burst, and instability). On Fig. 7.88, the curves show the total number of antiprotons in the Recycler (blue diamonds) and losses (pink squares). For the beam intensity, the data points represent the number of antiprotons before and after injections/extractions and not the straight evolution over time. The losses are also discrete data and the zigzag pattern comes from the fact that the successive data points represent two different loss mechanisms, with one evidently more prone to result in losses than the other. The set of data points with low value is the loss during injections/extractions. The set of data with high value is the loss between injections, hence effectively due to the antiproton beam lifetime.

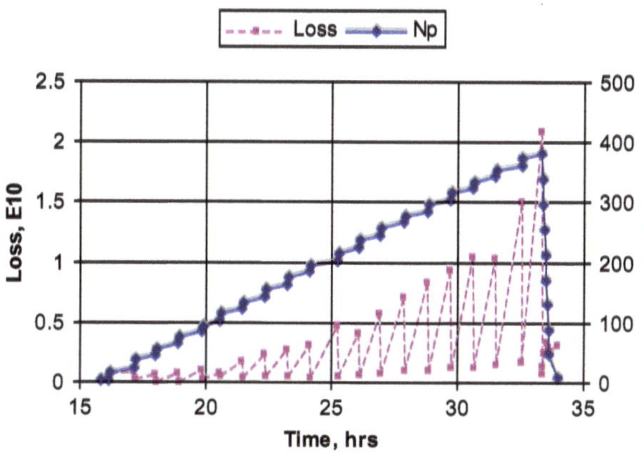

Fig. 7.88 Antiproton beam intensity and losses over a stashing cycle. *Blue diamonds* are the numbers of antiprotons in the Recycler before and after injections/extractions. The *pink squares* represent beam losses during injections/extractions (lower value data points) and between injections (higher value data points)

The data shown on the figure is fairly representative of a typical stashing cycle, where there is no accidental loss or operational issue. In this example, the total beam loss is 15.7×10^{10} antiprotons and the storage efficiency from Eq. (7.160) 92.9 %. Out of the 7.1 % of beam which is lost, 3.9 % is due to injection and extraction inefficiencies. The other 3.2 % come from the antiprotons lifetime.

Figures 7.89 and 7.90 depict the antiprotons lifetime throughout a stashing cycle from the first injection after loading the Tevatron to the last extraction to the Tevatron for another store (Fig. 7.89) and the corresponding emittances while accumulating the antiprotons (Fig. 7.90). Several features can be highlighted. The lifetime of the remaining beam after extraction is low and emittances are high. This is a direct consequence of the extraction process. In addition, because stacking continues while the Tevatron is being loaded with protons and antiprotons, the number of antiprotons transferred during the first set of transfers from the Accumulator is relatively large. The beam quality of large stacks being poorer than for smaller stacks, these first transfers also contribute to the poor lifetime and large emittances in the Recycler. However, as cooling progresses, in particular thanks to the stochastic cooling system, which as explained previously is more efficient than electron cooling for large emittances, the lifetime reaches >1,000 h after a couple more transfers. Concurrently, the emittances decrease, which also contribute to a better lifetime. Then, while emittances remain quite constant, and well within the acceptance of the Recycler, the lifetime decreases as a function of the antiprotons intensity. Nevertheless, the lifetime does not drop below 200 h until the last moments preceding mining and extraction, when strong cooling is applied—the zero values and the large fluctuations are consequences of the DCCT; as the lifetime degrades, the "noise" from the DCCT becomes negligible with respect to the particles loss, hence leading to a more satisfactory calculation of the lifetime.

As various systems were being built and/or improved, the accumulation scheme would evolve accordingly in order to maximize the integrated luminosity delivered to the detectors. The stashing profiles of Figs. 7.89 and 7.90 represent the operational optimum for maximizing the integrated luminosity at the end of Run II.

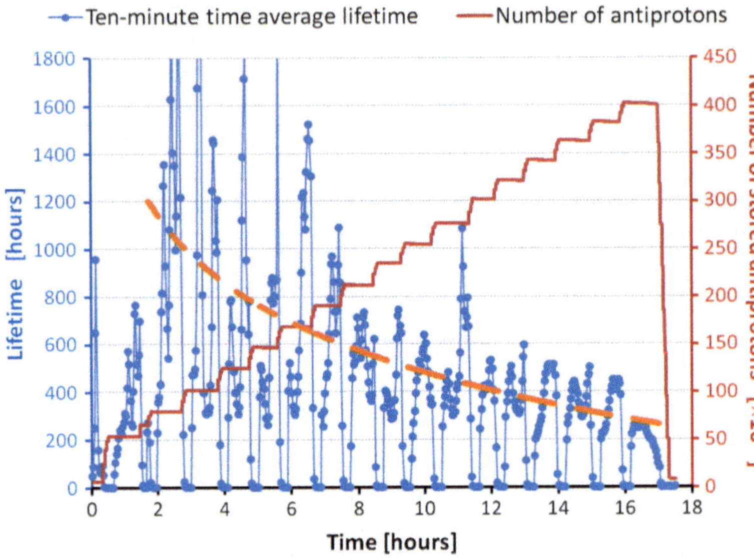

Fig. 7.89 Number of antiprotons and lifetime during stashing between two Tevatron stores. The *dashed orange* line is a fit to the lifetime data (excluding data from the first 2 h and zeros, which correspond to times of transfers into the Recycler and for which the lifetime is not calculable)

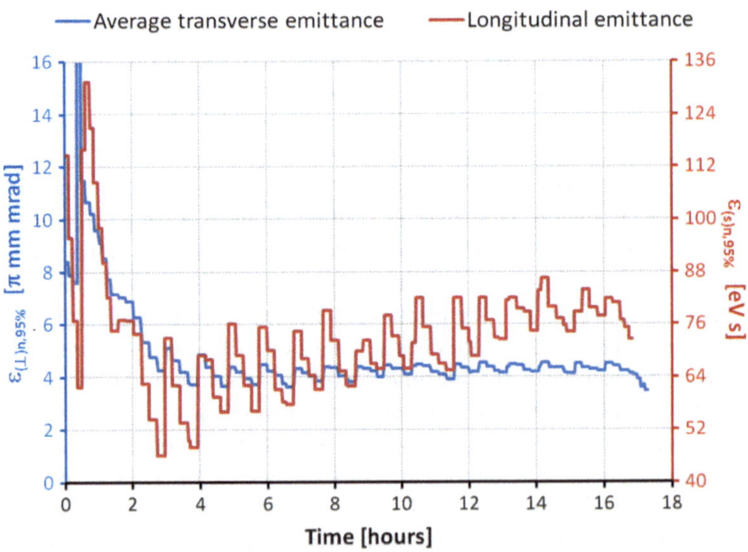

Fig. 7.90 Emittances (*blue transverse*, *red longitudinal*) evolution for the same period as in Fig. 7.89

In short, the rationale behind the stashing cycle that was adopted (and this is true at any point during commissioning of the Recycler) can be expressed in the following manner: Increasing the number of antiprotons that can be used for collisions starts with increasing the stacking rate in the Debuncher/Accumulator. Since the stacking rate decreases as the antiproton stack increases, emptying the Accumulator into the Recycler regularly allows maintaining a relatively small stack, hence improving the stacking rate on average. However, this also implies that losses in the transfer lines during extractions/injections must remain at a level where this procedure results in a net gain. Note that it is also true for extractions to the Tevatron. In turn, the Recycler Ring has to be able to maintain a high/good lifetime such that losses between Tevatron stores do not offset the gain in the stacking rate. Or, in other words, the Recycler had to and did maintain the right balance between cooling efficiency and storage efficiency until the end of Run II.

References

1. B. Bayanov et al., Liquid lithium lens with high magnetic fields. in *Proceedings of the 1999 I.E. PAC* (New York, 1999), pp. 3086–3088
2. N.V. Mokhov, The MARS Code System User Guide, Version 13(95), Fermilab-FN-628 (1995); N.V. Mokhov et al., Fermilab-Conf-98/379 (1998); LANL Report LA-UR-98-5716 (1998); *nucl-th/9812038 v2 16 Dec 1998*; http://wwwap.fnal.gov/MARS/
3. B.F. Bayanov et al., A lithium lens for axially symmetric focusing of high energy particles beams. NIM **190**, 9–14 (1981)
4. T.A. Vsevolozskaya, The optimization and efficiency of antiproton production within a fixed acceptance. NIM **190**, 479–486 (1981)
5. G. Dugan, P-bar Production and collection at the FNAL antiproton source. in *Proceedings of the 13th International Conference of High Energy Accelerators*, vol 2 (Novosibirsk, 1986), pp. 264–271
6. V. Nagaslaev, K. Gollwitzer, V. Lebedev, A. Valishev (Fermilab), V. Sajaev (Argonne). Measurement and optimization of the lattice functions in the Debuncher ring at Fermilab. in *FERMILAB-CONF-06-200-AD* (2006)
7. V.P. Nagaslaev, V.A. Lebedev, S.J. Werkema (Fermilab), Lattice optimization for the stochastic cooling in the Accumulator ring at Fermilab. in *FERMILAB-CONF-07-712-AD* (2007)
8. S. van der Meer, Rev. Mod. Phys. **57**, 689 (1985)
9. D. Möhl, Stochastic cooling, in *CERN Accelerator School, Fifth Advanced Accelerator Physics Course*, ed. by S. Turner (CERN, Geneva, 1995), pp. 587–671
10. J. Bisognano, C. Leemann, Stochastic cooling, in *Summer School on High Energy Particle Accelerators, AIP Conference Proceedings 87*, ed. by R.A. Carrigan et al. (American Institute of Physics, Melville, NY, 1982), pp. 584–655
11. V.V. Parkhomchuk, D.V. Pestrikov, Sov. Phys. Tech. Phys. **25**(7), 818 (1980)
12. A. Jansson et al., in *Proceedings of 2004 European PAC* (Lucerne, Switzerland), pp. 2777–2779
13. F. Voelker et al., in *Proceedings of 1983 PAC IEEE Conference* (Santa Fe, NM), pp. 2262–2263 http://www.jacow.org/
14. G. Lambertson et al., in *Proceedings of 1985 PAC IEEE Conference* (Vancouver, BC, Canada), pp. 2168–2170
15. D. McGinnis, Theory and design of microwave planar electrodes for stochastic cooling of particle neams. Microw. Opt. Tech. Lett. **4**(11), 439–443 (1991)

16. B. Autin et al., in *Proceedings of 1987 PAC IEEE Conference* (Washington, DC), pp. 1549–1551
17. D. McGinnis, Slotted waveguide slow-wave stochastic cooling arrays. in *Proceedings of 1999 PAC IEEE Conference* (New York), pp. 1713–1715
18. D. Pozar, *Microwave Engineering*, 3rd edn. (Wiley, New York, 2005), pp. 489–493
19. B. Leskovar, Low-noise cryogenically cooled broad-band microwave preamplifiers. in *172nd Meeting of the Low Temperature Electronics Symposium* (October 1987, Honolulu, HI)
20. S. Weinreb et al., FET's and HEMT's at cryogenic temperature, their properties and use in low-noise amplifiers. IEEE Trans. Microw Theory Tech **36**(3), 552–560 (1988)
21. http://www.miteq.com/products/amplifiers/cryogenic-amplifiers.php
22. R.J. Pasquinelli, Noise performance of the Debuncher stochastic cooling systems. Antiproton Source Note 661, March 2001, internal document
23. Y. Hoshiko, K1 superconducting communication transmission. in *Proceedings Fifth International Cryogenics Engineering Conference* (ICEC5, Tokyo, 1975), p. 282
24. R.J. Pasquinelli, Superconducting delay line for stochastic cooling filters, in *Proceedings of 1983 PAC IEEE Conference* (Santa Fe, NM), pp. 3360–3362
25. R.J. Pasquinelli, Superconducting notch filters for the Fermilab antiproton source. in *Proceedings of 12th International Conference on High Energy Accelerators* (Fermilab, Batavia, IL, August 1983), pp. 584–586
26. R.J. Pasquinelli, Bulk acoustic wave (BAW) devices for stochastic cooling notch filters. in *IEEE Nuclear Science* (May 1991)
27. R.J. Pasquinelli, Fiber optic links for instrumentation. in *Conference Proceedings of the Accelerator Instrumentation Workshop* (October 1990)
28. R.J. Pasquinelli, Optical notch filters for Fermilab Debuncher Betatron stochastic cooling. in *Proceedings of 1989 PAC IEEE Conference* (Chicago, IL), pp. 694–696
29. T. Kakuta, S. Tanaka, LCP coated optical fiber with zero thermal coefficient of transmission delay time. in *Proceedings of 36th International Wire and Cable Symposium*, vol 1 (Sumitomo Electric Industries, Ltd., Yokohama, Japan, 1987), pp. 234–240
30. R.J. Pasquinelli, Electro-optical technology applied to accelerator beam measurement and control. in *Proceedings of 1993 PAC IEEE Conference* (Washington, DC), pp. 2081–2085
31. D. McGinnis, J. Marriner, Design of 4–8 GHz stochastic cooling equalizers for Fermilab Accumulator. in *Proceedings of 1991 PAC IEEE Conference* (San Francisco, CA), pp. 1392–1394
32. S. Weinreb et al., NRAO, waveguide system for a very large antenna array. Microwave J **20**(3), 49–52 (1977)
33. R.J. Pasquinelli, Wide band free space transmission link utilizing a modulated infrared laser, TUA12. in *Proceedings of 1999 PAC IEEE Conference* (New York, NY), pp. 1094–1096
34. A.S. Gilmour Jr., *Microwave Tubes* (Artech House, Norwood, MA, 1986)
35. V. Lebedev, Improvements to the stacktail and Debuncher momentum cooling systems. in *Proceedings of COOL 09* (Lanzhou, China, 2009), MOA1MCCO02
36. R.J. Pasquinelli, Debuncher momentum cooling systems signal to noise measurements. Pbar note 667, Internal Antiproton Source Document (December 2001)
37. D. Sun et al., New equalizers for antiproton stochastic cooling at Fermilab. in *Proceedings of COOL-2007* (Bad Kreuznach, Germany, 2007), THAP16
38. V. Lebedev et al., in *Proceedings of COOL-2007* (Bad Kreuznach, Germany, 2007), p. 202
39. V. Lebedev, in *Proceedings of COOL-2007* (Bad Kreuznach, Germany, 2007), p. 39
40. R. Pasquinelli et al., in *Proceedings of COOL-2007* (Bad Kreuznach, Germany, 2007), p. 198
41. Tevatron Design Report, FERMILAB-DESIGN-1984-01 (1984)
42. I. Nesmiyan, F. Nolden, in *Proceedings of EPAC-2004* (Lucern, Switzerland), p. 2170
43. V. Lebedev et al., in *Proceedings of COOL-2009* (Lanzhou, China, 2009), p. 27
44. G.I. Budker, Sov. Atomic Energy **22**, 346 (1967)
45. G.I. Budker et al., IEEE Trans. Nucl. Sci. **NS-22**, 2093 (1975)
46. I.N. Meshkov, Phys. Part. Nucl. **25**(6), 631 (1994)

47. Fermilab Recycler Ring Technical Design Report, FERMILAB-TM-1991, 1996
48. S. Nagaitsev et al., Phys. Rev. Lett. **96**, 044801 (2006)
49. Y.S. Derbenev, A.N. Skrinsky, Particle Accelerators **8**, 1 (1977)
50. I.N. Meshkov et al., Physics guide of BETACOOL code, v.1.1, BNL note C-A/AP#262, p. 17
51. A. Shemyakin, L.R. Prost, in *Proceedings of APAC'07* (Indore, India, January 29–February 2, 2007), TUPMA094, pp. 238–240; also in A. Shemyakin, FERMILAB-TM-2374-AD (2007)
52. J.R. Adney et al., in *Proceedings of 1989 I.E. PAC* (20–23 March 1989, Chicago, IL), p. 348
53. M.E. Veis et al., in *Proceedings of EPAC'88* (Rome, 7–11 June 1988), p. 1361
54. G.P. Jackson, Modified Betatron approach to electron cooling. in *Proceedings of MEEC'97* (Novosibirsk, February 26–28, 1997)
55. I. Ben-Zvi, in *Proceedings of EPAC'06* (Edinburgh, UK, June 26–30, 2006), TUZBPA01
56. J. Dietrich et al., in *Proceedings of COOL'11* (September 12–16, 2011, Alushta, Ukraine), MOIO05
57. D.R. Anderson et al., in *Proceedings of EPAC'92* (Berlin, 1992), pp. 836–838
58. N. Dikansky et al., Electron beam focusing system. in *FERMILAB-TM-1998* (1996)
59. A. Burov, S. Nagaitsev, A. Shemyakin, Y. Derbenev, Optical principles of beam transport for relativistic electron cooling. Phys. Rev. **3**, 094002 (2000)
60. Pelletrons are manufactured by the National Electrostatics Corporation, www.pelletron.com
61. J.A. MacLachlan, A. Burov, A.C. Crawford, T. Kroc, S. Nagaitsev, C. Schmidt, A. Sharapa, A. Shemyakin, A. Warner, Prospectus for an electron cooling system for the Recycler. in *FERMILAB-TM-2061* (1998)
62. OptiM - V. Lebedev, A. Bogacz, Betatron motion with coupling. in *JLAB-ACC-99-19* (1999)
63. A. Burov, T. Kroc, V. Lebedev, S. Nagaitsev, A. Shemyakin, A. Warner, S. Seletskiy, Optics of the Fermilab electron cooler. in *Proceedings of APAC'04* (Gyeongju, Korea, March 22–26, 2004), pp. 647–649
64. B. Chase, private communication
65. The program for data recording was written by V. Lebedev and V. Nagaslaev
66. A.V. Ivanov, M.A. Tiunov, ULTRASAM—2D code for simulation of electron guns with ultra high precision. in *Proceedings of European, PAC'02* (Paris, 2002), p. 1634
67. M.A. Tiunov, BEAM – 2D-code package for simulation of high perveance beam dynamics in long systems. in *Proceedings of International Symposium* "SPACE CHARGE EFFECTS IN FORMATION OF INTENSE LOW ENERGY BEAMS". (JINR, Dubna, Russia, February 15–17, 1999)
68. A. Warner, A. Burov, K. Carlson, G. Kazakevich, S. Nagaitsev, L. Prost, M. Sutherland, M. Tiunov, in *AIP Conference Proceedings*, vol 821 (2006), pp. 380–385
69. T.K. Kroc, A.V. Burov, T.B. Bolshakov, A. Shemyakin, in *Proceedings of 2005 I.E. PAC* (Knoxville, USA, May 16–20, 2005), p. 3801
70. A. Burov, G. Kazakevich, T. Kroc, V. Lebedev, S. Nagaitsev, L. Prost, S. Pruss, A. Shemyakin, M. Sutherland, M. Tiunov, A. Warner, in *Proceedings of COOL'05* (Galena, USA, September 19–23, 2005), p. 139
71. A.Shemyakin, in *Proceedings of European PAC* (Vienna, Austria, 26–30 June 2000), pp. 1268–1270
72. L.R. Prost, A. Shemyakin, in *Proceedings of 2005 I.E. PAC* (Knoxville, USA, May 16–20, 2005), p. 2387
73. A. Burov, I. Gusachenko, S. Nagaitsev, A. Shemyakin, in *FERMILAB-CONF-05-462-AD* (2005)
74. A. Shemyakin, A. Burov, K. Carlson, V. Dudnikov, B. Kramper, T. Kroc, J. Leibfritz, M. McGee, S. Nagaitsev, G. Saewert, C.W. Schmidt, A. Warner, S. Seletsky, V. Tupikov, in *Proceedings of COOL'03* (Lake Yamanaka, Japan, May 19–23, 2003)
75. L. Prost, A. Shemyakin, in *Proceedings of COOL'05* (Galena, USA, September 19–23, 2005), p. 391

76. A. Warner, L. Carmichael, K. Carlson, J. Crisp, R. Goodwin, L. Prost, G. Saewert, A. Shemyakin, in *Proceedings of DIPAC-09 European Workshop on Beam Diagnostics and Instrumentation for Particle Accelerators* (Bazel, Switzerland, 2009), p. 222
77. A. Shemyakin, A. Burov, K. Carlson, L.R. Prost, M. Sutherland, A. Warner, in *Proceedings of 2009 I.E. PAC* (Vancouver, Canada, May 4–8, 2009), TU6PFP076
78. A. Burov, private communication (2007), available at http://beamdocs.fnal.gov/AD-public/DocDB/ShowDocument?docid=2966
79. A. Shemyakin, L. Prost, G. Saewert, in *Proceedings of 2010 Int'l PAC* (Kyoto, Japan, May 23–28, 2010), MOPD075
80. The measurements were made by P. Joireman, and the analysis was performed by A. Burov
81. A. Burov, private communication (2007), available at http://beamdocs.fnal.gov/AD-public/DocDB/ShowDocument?docid=3099
82. G. Kazakevich, A. Burov, C. Boffo, P. Joireman, G. Saewert, C.W. Schmidt, A. Shemyakin, in *FERMILAB-TM-2319-AD* (2005)
83. A.C. Crawford, S. Nagaitsev, A. Shemyakin, S. Seletsky, V. Tupikov, Fermilab preprint TM-2224 (2003)
84. V. Tupikov, G. Kazakevich, T.K. Kroc, S. Nagaitsev, L. Prost, A. Shemyakin, C.W. Shmidt, M. Sutherland, A. Warner, in *Proceedings of COOL'05* (Galena, USA, September 19–23, 2005), p. 375
85. A. Shemyakin, K. Carlson, L.R. Prost, G. Saewert, Fermilab preprint CONF-08-425-AD (2008)
86. S.M. Seletskiy, A. Shemyakin, in *Proceedings of 2005 I.E. PAC* (Knoxville, USA, May 16–20, 2005), p. 3638
87. S. Nagaitsev et al., in *Proceedings of COOL'05* (Galena, USA, September 19–23, 2005), p. 39
88. A. Khilkevich, L.R. Prost, A.V. Shemyakin, in *Proceedings of 2011 I.E. PAC* (New York, USA, March 28–April 1, 2011), WEP228
89. S. Nagaitsev, C. Gattuso, S. Pruss, J. Volk, Experience with magnetic shielding of a large scale accelerator. in *Proceedings of IEEE 2001 Particle Accelerator Conference* (Chicago, IL, USA, June 18–22, 2001), RPPH062
90. P. Lebrun, Recycler orbit length stabilization via the 5-bump at location 518–526. (unpublished)
91. J.D. Bjorken, S. Mtingwa, Intrabeam scattering. Particle Accelerators **13**, 115–143 (1983)
92. K. Gounder et al., Recycler ring beam life time. in *Proceedings of 2001 Particle Accelerator Conference* (Chicago, IL, USA, June 18–22, 2001), RPPH055
93. S. Mishra, Status of the Fermilab Recycler ring. in *Proceedings of European Particle Accelerator Conference* (Paris, France, June 3–7, 2002), MOPLE035
94. L. Hermansson, D. Reistad, Experiences of operating CELCIUS with hydrogen pellet target. Nucl. Inst. Meth. Phys. Res. A **441**, 140 (2000)
95. M. Hu, D. Broemmelsiek, A. Burov, J. Marriner, S. Nagaitsev, K.Y. Ng, Beam-based measurement at the Fermilab Recycler ring. in *FERMILAB-FN-0758-AD* (2004)
96. A. Burov, Electron drift instability in storage rings with electron cooling. Nucl. Inst. Meth. Phys. Res. A **441**, 23 (2000)
97. V.V. Parkhomchuk, New insights in the theory of electron cooling. Nucl. Inst. Meth. Phys. Res. A **441**, 9 (2000)
98. A. Sharapa, A. Shemyakin, S. Nagaitsev, Nucl. Instr. Meth. A **417**, 177 (1998)
99. L.R. Prost, A. Burov, K. Carlson, A. Shemyakin, M. Sutherland, A.Warner, in *Proceedings of COOL'07* (Bad Kreuznach, Germany, September 10–14, 2007), MOA2I04

Chapter 8
Beam–Beam Effects

V. Shiltsev and A. Valishev

8.1 Beam–Beam Effects in Tevatron: Introduction

Beam–beam effects became a subject of study as soon as there were colliders beginning with the first *e+e−* collider AdA in Frascati that started operating in 1962 and the first *p–p* collider ISR at CERN that started operating in 1971. Over the years many different issues related to electromagnetic interactions of colliding beams have emerged. In the Tevatron collider, the beam–beam problems take place in the context of beam losses and emittance growth due to long-range and head-on interactions. A comparative review of beam–beam performance of a number of hadron colliders [1] shows that the beam parameters operationally achieved in the Tevatron correspond to record high incoherent tune shift due to collisions (the figure of merit of beam–beam interaction):

$$\xi = N_{\mathrm{IP}} \frac{N_{\mathrm{p}} r_{\mathrm{p}}}{4\pi\varepsilon} \approx 0.025 - 0.030, \tag{8.1}$$

where r_{p} denotes the classical proton radius, N_{p} and ε are the opposite bunch intensity and emittance, correspondingly. Remarkably, the Tevatron working points (vertical and horizontal tunes) lie above the half integer between the fifth- and seventh-order resonances (between $3/5 = 0.6$ and $4/7 = 0.571$) and the beam–beam tune spread fully covers the tune area.

During the Collider Run II, beam losses during injection, ramp and squeeze phases were mostly caused by beam–beam effects. Figure 8.1 from [2] shows that

V. Shiltsev (✉) • A. Valishev
Fermi National Accelerator Laboratory, Accelerator Physics Center, PO Box 500, MS 221, Batavia, IL 60510, USA
e-mail: shiltsev@fnal.gov

V. Lebedev and V. Shiltsev (eds.), *Accelerator Physics at the Tevatron Collider*, Particle Acceleration and Detection, DOI 10.1007/978-1-4939-0885-1_8,

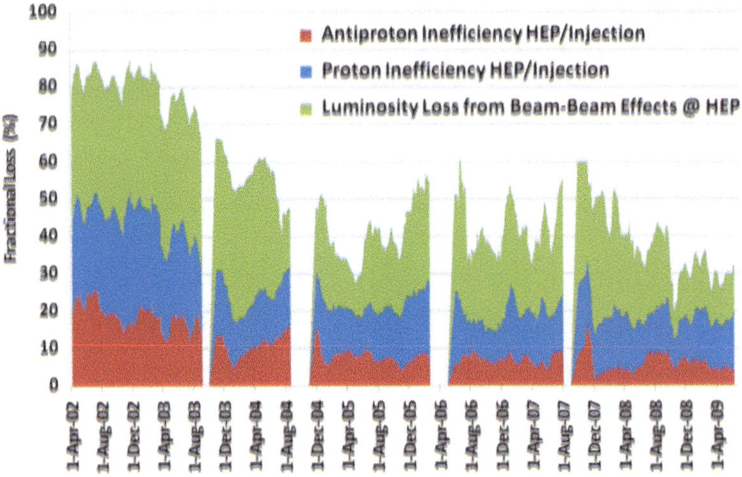

Fig. 8.1 Evolution of beam losses in 2002–2009. *Red* shows fractional loss of antiprotons between injection into the Tevatron and start of collisions, next (*blue*) one is for loss of protons, *green*—fractional reduction of the luminosity integral caused by beam–beam effects in collisions [2]

early in Run II, combined beam losses only in the Tevatron (the last accelerator out of total 7 in the accelerator chain) claimed significantly more than half of the integrated luminosity. Due to various improvements, losses have been reduced significantly down to some 20–30 % in 2008–2009, paving the road to a many-fold increase of the luminosity. In "proton-only" or "antiproton-only" stores, the losses do not exceed 2–3 %. So, the remaining 8–10 % proton loss and 2–3 % antiproton loss are caused by beam–beam effects, as well as some 5–10 % reduction of the luminosity lifetime in collision. Note that the proton inefficiency is higher than the antiproton one, despite the factor of 3–5 higher proton intensity. That is explained in the following chapters by significantly smaller antiproton emittances.

Beam–beam interactions differ between the injection and collision stages. The helical orbits should provide sufficient separation between the proton and antiproton beams in order to reduce detrimental beam–beam effects, e.g., tune shifts, coupling, and high-order resonance driving terms. Each bunch experiences 72 long-range interactions per revolution at injection, but at collision there are 70 long-range interactions and two head-on collisions per bunch at the CDF and D0 detectors—see Fig. 8.2. In total, there are 138 locations around the ring where beam–beam interactions occur. The sequence of 72 interactions out of the 138 possible ones differs for each bunch; hence the effects vary from bunch to bunch. The locations of these interactions and the beam separations change from injection to collision because of the antiproton cogging (relative timing between antiprotons and protons).

Initially, there were six separator groups (three horizontal and three vertical) in the arcs between the two main interaction points, *B*0 (CDF) and *D*0. During collisions, these separators form closed orbit 3-bumps in each plane. However,

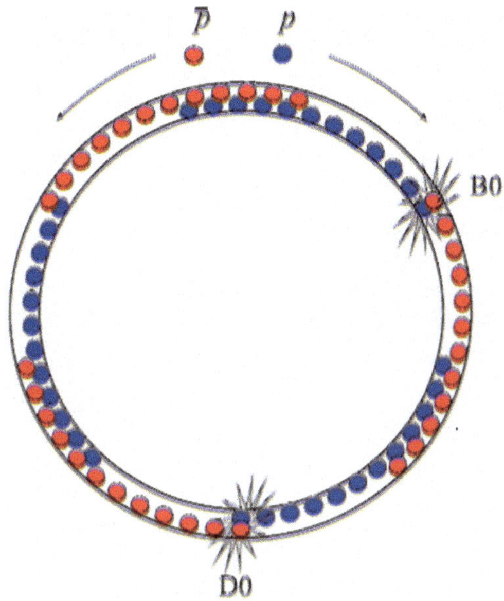

Fig. 8.2 Schematic of proton (*blue*) and antiproton (*red*) bunches in the Tevatron and the two head-on collision locations *B*0 and *D*0

the condition of orbit closure prevented running the separators at maximum voltages with exception for horizontal separators in the short arc from *B*0 to *D*0. This limited separation at the nearest parasitic crossings 59 m away from the main IPs aggravating the long-range beam–beam interaction. To increase separation at these parasitic crossings three additional separators were installed as to create closed orbit 4-bumps both in horizontal and vertical planes in the long arc (from *D*0 to *B*0) and in the vertical plane in the short arc.

There is more flexibility in the helix design for the preceding stages: injection, ramp, and squeeze. There are still some difficulties at these stages, including:

1. Irregularities in betatron phase advance over the straight sections, especially *A*0.
2. Aperture restrictions (physical as well as dynamic) that limit the helix amplitude at injection and at the beginning of the ramp.
3. The maximum separator voltage gradient of 48 kV/cm (limited by separator spark rate) leads to a faster drop in separation, $d \sim 1/E$, than in the beam size, $\sigma \sim 1/E^{1/2}$, during the second part of the ramp above the energy of $E = 500$ GeV.
4. The polarity reversal of the horizontal separation during the squeeze (to satisfy needs of HEP experiments) that leads to a short partial collapse of the helix.

A simple figure of merit is helpful when comparing different helix designs. The conventional choice is the *minimum* value of the so-called *radial separation*, *S*, over all possible parasitic interaction crossing points in units of the RMS betatron beam sizes $\sigma_{x,y\beta}$:

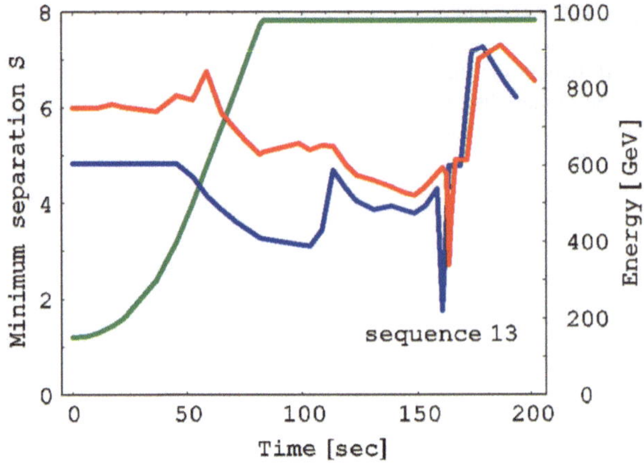

Fig. 8.3 Minimum radial separation, Eq. (8.2), on ramp and during the low-beta squeeze. The *green line* represents the beam energy on the ramp $E(t)$. The *blue* and *red lines* represent $S(t)$ for the helix configurations used ca. January 2002 and August 2004, respectively [7]

$$S = \sqrt{\left(\Delta x/\sigma_{x\beta}\right)^2 + \left(\Delta y/\sigma_{y\beta}\right)^2}. \tag{8.2}$$

The separation is normalized to a fixed reference emittance of 2.5 π mm mrad. Our experience has shown that less than 5–6σ separation causes unsatisfactory losses. Figure 8.3 shows the minimum radial separation S during the ramp and squeeze with the initial helix design (blue, ca. January 2002) and an improved helix (red, ca. August 2004). The long-range interactions contribute a tune spread of about [3]:

$$\Delta Q \approx \sum_{\text{parasitic encounters}} \frac{2\xi}{S^2} \approx 0.008, \tag{8.3}$$

as well as several units of chromaticity [4]. In the end of Run II operations, both species had about the same beam–beam tune shifts and are effectively in the strong–strong regime. That is because of much smaller antiproton emittances which were available due to electron cooling of antiprotons in the Recycler, starting in 2005. Consequently the antiprotons effectively experience only the linear part of the head-on beam–beam force and do not suffer much from it. Since 2006, antiproton losses due to beam–beam interactions during stores have been small, provided the tunes are well controlled. Protons on the other hand have tunes closer to twelfth-order resonances and are transversely larger than the antiprotons. Consequently during head-on collisions, they experience the nonlinear beam–beam force enhanced by chromatic effects and suffer beam loss and emittance growth. A review of beam–beam observations in Run I can be found in [5, 6].

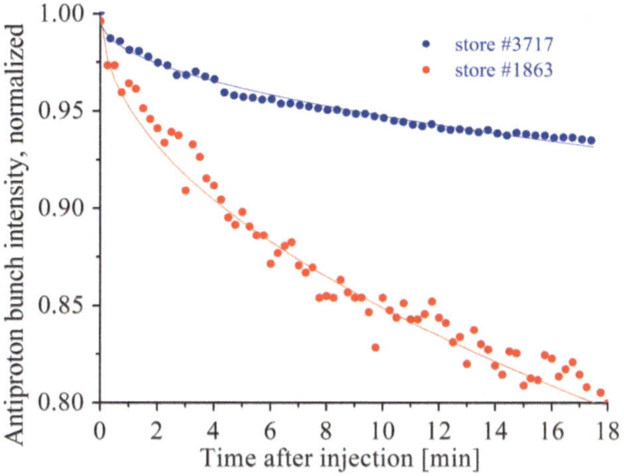

Fig. 8.4 Decay of (normalized) intensity for antiproton bunch #1 at injection. The *red dots* are for store #1863 (October 16, 2002) and the *blue dots* are store #3717 (August 8, 2004). The *blue* and *red lines* represent fits according to $N(t) = N_0 e^{-\sqrt{t/\tau}}$ with parameters $N_0 = 32.5 \times 10^9$, $\tau = 7.4$ h and $N_0 = 55.7 \times 10^9$, $\tau = 69.8$ h, respectively [7]

8.2 Beam–Beam Phenomena in Tevatron Operation

8.2.1 Long-Range Beam–Beam Interaction Effects at Injection, on the Energy Ramp and During Low-Beta-Squeeze

Before being brought to collisions, the Tevatron beams are transversely separated during the entire injection process, during energy ramp from 150 to 980 GeV and low-beta squeeze. Long-range beam–beam interaction leads to particle losses at these stages. In the absence of opposite beam, the combined losses are small, of the order of ~2–4 %, as measured during dedicated proton-only and antiproton-only stores.

Although both the proton and antiproton beams stay at 150 GeV for less than an hour, a significant particle loss occurred during that time at the beginning of the Run II. The particle losses for both beams were driven by diffusion and exacerbated by small transverse and longitudinal apertures. Figure 8.4 presents the intensity lifetimes of single antiproton bunches after injection for typical stores in 2002 and 2004. It is clearly seen for both stores that the intensity decay is not exponential. Figure 8.4 shows that the intensities are approximated well by the expression $N(t) = N_0 e^{-\sqrt{t/\tau}}$ that was used for the lifetime fits. Similar $\sqrt{t}$ dependence has been observed for the bunch length "shaving" (slow reduction of the rms bunch length), while transverse emittances do not exhibit such dependence on $\sqrt{t}$ and usually either stay flat or slightly grow [7].

During approximately 20 min needed to load antiprotons into the Tevatron, the proton lifetime degrades as more antiproton bunches are injected. Figure 8.5 shows an approximately linear dependence of the proton loss rate at 150 GeV on the number of antiprotons in the Tevatron. The proton loss rate without antiprotons is about 4 % per hour (25 h lifetime), whereas it grows to about 16 % per hour (6 h

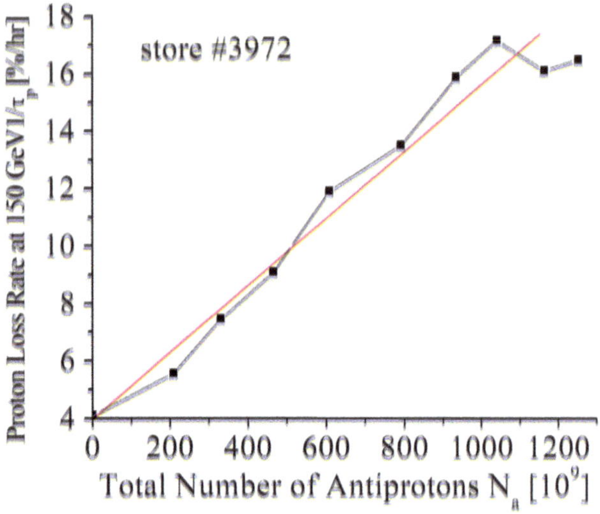

Fig. 8.5 Proton intensity loss rate at the 150 GeV helix in units of %/h vs. total number of antiprotons injected into the Tevatron N_a during shot setup #3972 (February 8, 2005). The points are the results of ordinary exponential decay fits over 2 min after each antiproton injection. The *solid red line* is the linear fit $1/\tau_p[\%/h] = 4 + 11.6(N_a/1{,}000)$ (from [1])

lifetime) when all antiproton bunches are loaded. A similar linear dependence of the antiproton loss rate on proton intensity can be seen as well.

Besides being dependent on the intensity of the opposing beam, the particle losses due to the long-range beam–beam interaction at injection, ramp, and squeeze are found to be dependent on beam emittances and chromaticities, approximately as [7, 8]

$$\frac{\Delta N_{\mathrm{a,p}}}{N_{\mathrm{a,p}}} = 1 - \frac{N(t)}{N(t=0)} \propto \sqrt{t} \cdot \varepsilon^2_{\mathrm{a,p}} \frac{N_{\mathrm{p,a}}}{\varepsilon_{\mathrm{p,a}}} Q'^2{}_{\mathrm{a,p}} \cdot F\left(\varepsilon_{\mathrm{L}}, Q_{x,y}, S_{\mathrm{a-p}}\right), \tag{8.4}$$

where the index a or p stands for antiprotons or protons, ε is transverse emittance, N is total number of particles in the opposite beam, Q' is the chromaticity on the corresponding helix, and the factor F emphasizes the fact that losses also depend on the longitudinal emittance ε_L, separation S (size of the helix and cogging stage), and tune Q. Over years of operation, the betatron tunes on both helices at injection and ramp were optimized to be close to $Q_x/Q_y = 20.584/20.576$, i.e., above seventh-order resonances at 4/7=0.5714, but close to the twelfth-order resonance 7/12=0.5833. Significant variations of the tune (in excess of ±0.002) often led to lifetime reduction, especially if the vertical tune approached the 4/7 resonance.

Equation (8.4) above emphasizes the importance of chromaticity for reducing the losses of both protons and antiprotons. Since the proton and antiproton orbits are separated using the electrostatic separators, their tunes and chromaticities can be controlled independently by using sextupole and octupole circuits, respectively. The major obstacle in attaining the desired chromaticity reduction was a weak head-tail instability in high intensity proton bunches [9]. Early in Run II, avoiding this instability required chromaticities as high as 8–12 units at 150 GeV. Reducing the proton chromaticities down to +(3–4) units became possible after removing unused high-impedance extraction Lambertson magnets, reducing the impedance of the injection Lambertson magnets by installing conductive liners, and commissioning active bunch-by-bunch instability dampers for the protons [10]. Decreasing the chromaticities to zero has become possible after reconfiguring octupole circuits to introduce Landau damping to suppress the head-tail instability. The antiproton bunches do not suffer from that instability since the intensity is much smaller than that of protons. Consequently, both Q'_x and Q'_y are set closer to 0 by using differential chromaticity octupole circuits.

The observed $\sqrt{t}$ dependence of beam intensity decay (see Fig. 8.4) and bunch length is believed to be due to particle diffusion that leads to particle loss at physical or dynamic apertures. The major diffusion mechanisms are intrabeam scattering (IBS), scattering on the residual gas, and diffusion caused by RF phase noise. For example, if the available machine aperture is smaller than the beam size of the injected beam, the beam is clipped on the first turn with an instantaneous particle loss. Such a clipping creates a step-like discontinuity at the boundary of the beam distribution that causes very fast particle loss due to diffusion. The diffusion wave propagates inward, so that the effective distance is proportional to $\sqrt{t}$. Consequently, the particle loss is also proportional to $\sqrt{t}$. To estimate such a "worst-case loss," consider an initially uniform beam distribution: $f(I) = f_0 \equiv 1/I_0$, where I_0 is the action at the boundary. For sufficiently small time, $t \ll I_0/D$, where D is diffusion coefficient, the diffusion can be considered one-dimensional in the vicinity of the beam boundary. Solving the diffusion equation

$$\frac{\partial f}{\partial t} = D\frac{\partial}{\partial I}\left(I\frac{\partial f}{\partial I}\right) \tag{8.5}$$

gives the result:

$$f(I,t) = \frac{2f_0}{\sqrt{\pi}}\int_0^{(I_0-I)/\sqrt{4I_0Dt}} e^{-\xi^2}d\xi. \tag{8.6}$$

By integrating it over I, one obtains the dependence of particle population on time:

$$\frac{N(t)}{N_0} \approx 1 - \sqrt{\frac{t}{\tau}}, \quad \tau = \frac{\pi I_0}{4D}, \quad t \ll \tau. \tag{8.7}$$

In the transverse degree of freedom, the Tevatron acceptance at 150 GeV on the helical orbit is about ${I^{tr}}_0 \approx 8$–13π mm mrad, depending on the pre-shot machine tune-up, while the emittance growth rate is about $D^{tr} \approx 0.15$–0.25π mm mrad/h chiefly from scattering on residual gas. So from (8.7), one can obtain a lifetime of $\tau \approx 30$–80 h. In addition, diffusion in the longitudinal plane with a rate $D^{long} \approx 0.03$–0.3 rad^2/h can lead to lifetimes of $\tau \approx 10$–100 h in the case where the longitudinal aperture is limited only by the RF bucket size $\sqrt{{I_0}^{long}} \approx 2$ rad. The above numbers are not well known, but we believe they are in the indicated ranges.

In reality, the machine acceptance is set by the interplay between the physical and dynamic apertures. The latter is a strong function of the synchrotron action, and beam–beam interactions drastically reduce the dynamic aperture for synchrotron oscillation amplitudes close to the bucket size. Naturally, such an aperture reduction is stronger for larger values of chromaticity.

The problem was alleviated significantly by a comprehensive realignment of many Tevatron elements in 2003–2004, as well as a reduction in the longitudinal emittances due to improvements in the Main Injector's bunch coalescing, and an increase of the Tevatron's dynamic aperture.

8.2.2 *Beam–Beam Interaction Effects During Colliding HEP Stores*

After the beams are brought into collisions at the main IPs, there are two head-on and 70 long-range collision points per bunch. Correspondingly, the beam–beam phenomena in the Tevatron collider are characterized by a complex mixture of long-range and head-on interaction effects, record high beam–beam parameters for both protons and antiprotons (the head-on tune shifts up to about $\xi = 0.03$ for both protons and antiprotons, in addition to long-range tune shifts of $\Delta Q^{\mathrm{p}} = 0.003$ and $\Delta Q^{\mathrm{a}} = 0.006$, respectively), and remarkable differences in beam dynamics of individual bunches. All that may result in the significant emittance growth and particle losses in both beams. During the running prior to the 2006 shutdown the beam–beam effects at HEP mostly affected antiprotons. The long-range collision points nearest to the main IPs were determined to be the leading cause for poor lifetime. Additional electrostatic separators were installed in order to increase the separation at these IPs from 5.4 to 6 [10]. Also, the betatron tune chromaticity was decreased from 20 to 10 units. Since then, the antiproton lifetime was dominated by losses due to luminosity and no emittance growth is observed provided that the betatron tune working point is well controlled. Electron cooling of antiprotons in the Recycler and increased antiproton intensities and brightness drastically changed the situation for protons. Figure 8.6 shows the evolution of total head-on beam–beam tune shift $\xi_{\mathrm{p.a}}$ for protons and antiprotons. Note that prior to the 2006 shutdown the proton ξ_{p} was well under 0.01 and a big boost occurred in 2007 when both beam–beam parameters became essentially equal. It was then when beam–beam-related losses and emittance blowup started to be observed in protons.

The analysis [11] showed that deterioration of the proton lifetime was caused by a decrease of the dynamical aperture for off-momentum particles due to head-on collisions. It was discovered that the Tevatron optics had large chromatic perturbations, e.g., the value of β^* for off-momentum particles could differ from that of the reference particle by as much as 20 %. Also, the high value of second-order betatron tune chromaticity $Q''=d^2Q/d(\Delta p/p)^2$ generated a tune spread of about 0.002. A rearrangement of sextupole magnet circuits in order to correct the

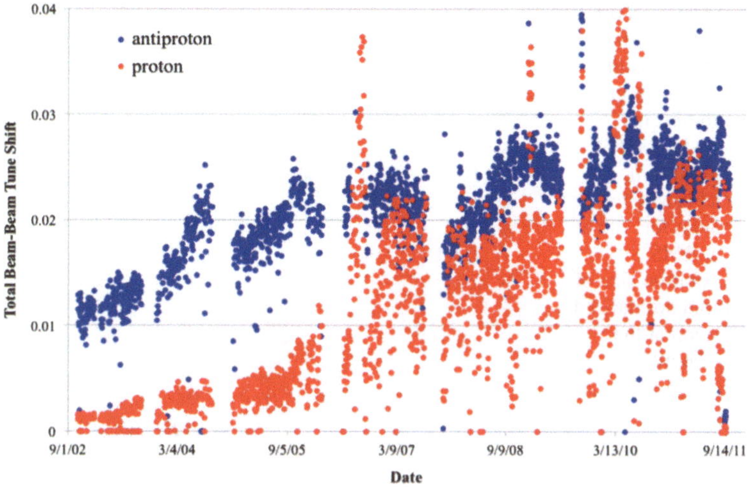

Fig. 8.6 Head-on beam–beam tune shift parameters for protons and antiprotons vs. time [8]

second-order chromaticity was planned and implemented before the 2007 shutdown [11] and led to some 10 % increase in the luminosity integral per store.

Another step up in the proton ξ_p happened after the 2007 shutdown when the transverse antiproton emittance decreased because of improvements in injection matching. The total attained head-on beam–beam tune shift for protons exceeded that of antiprotons and reached 0.028. This led to high sensitivity of the proton lifetime to small variations of the betatron tunes, and to severe background conditions for the experiments. The reason is believed to be the large betatron tune spread generated by collisions of largely different size bunches [12]. Indeed, at times the antiproton emittance was a factor of 5–6 smaller than the proton emittance. To decrease the proton to antiproton emittance ratio a system has been commissioned which increases the antiproton emittance after the top energy is reached by applying wide band noise to a directional strip line [13]. At the end of Run II, the optimal emittance ration was kept at $\varepsilon_p/\varepsilon_a \approx 3$. Below we summarize major beam–beam phenomena during HEP stores.

The beam–beam effects in the Tevatron cause nearly every measurable indicator of beam dynamics to vary as a function of position within a bunch train. As mentioned, the 36 bunches for each beam are arranged in three trains of 12 bunches each, and the variation of intensities and emittances among the proton bunches is small. Consequently, a threefold symmetry is expected [14] in the antiproton bunch dynamics. We have observed such behavior in essentially every indicator. For example, Fig. 8.7 shows that the helical orbits of antiproton bunches at the low-beta stage differ by some 40–50 μm in a systematic, ladder-like fashion (due to symmetry, the plot refers only to a single train of 12 bunches). Such variation in the closed orbits was predicted before the start of the Collider Run II [15] and

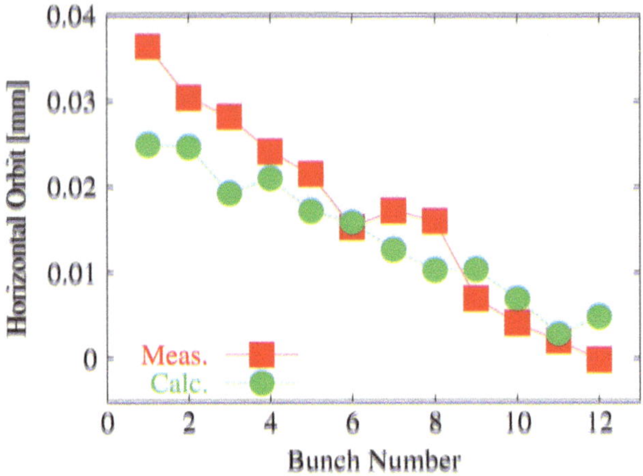

Fig. 8.7 Antiproton orbit variations along the bunch train: horizontal positions measured by synchrotron radiation monitor in collisions in store #3530 (May 23, 2004) (*red squares*) and calculated (*green circles*); for comparison, RMS horizontal betatron size of 2.5π mm mrad normalized emittance beam at the location of the monitor is equal to 0.3 mm [7]

agrees well with analytical calculations (see the comparisons in Fig. 8.7 and discussion in the next section). Vertical variation is similar and of the same order, proton orbits exhibit proportionally (to intensity) smaller bunch-by-bunch variations.

Two (vertical and horizontal) 1.7 GHz Schottky detectors [16] allow continuous, nondestructive measurements of betatron tunes and chromaticities for each proton and antiproton bunch during HEP stores. The tunes measured by the detectors represent an average over all particles in a bunch. The tune and chromaticity accuracies for single bunch measurements are better than 0.001 and 1 unit, respectively. A single measurement can be made in approximately 20 s.

Figure 8.8 presents the distribution of antiproton vertical and horizontal tunes and chromaticities along antiproton bunch train. It is remarkable that bunches #1 and #12 have vertical and horizontal tunes, respectively, much lower (by more than 0.003) than the other ten bunches. Long-range beam–beam interactions at the parasitic IPs produce such significant bunch-by-bunch tune differences ΔQ_{LR}. The data shown in Fig. 8.8 agree with analytic calculations [17, 18] if one takes into account that the measured tune is averaged over a weighted particle distribution, and, thus, the effective head-on tune shift is approximately half of the maximum beam–beam incoherent tune shift:

$$\Delta Q \approx \Delta Q_{\mathrm{LR}} + 0.5 \cdot \xi, \quad \xi = \frac{r_{\mathrm{p}} N_{\mathrm{p}}}{4\pi\varepsilon_{\mathrm{p}}} \times 2. \tag{8.8}$$

For nominal bunch parameters at the beginning of an HEP stores, the head-on tune shift is about $\xi \approx 0.020$ for antiprotons. Figure 8.9 displays the Tevatron beam tunes at the beginning of a high-luminosity HEP store on a resonance plot. Particles with up to 6σ amplitudes are presented. Small amplitude particles have tunes near the tips of the "ties" depicted for all 36 proton and 36 antiproton bunches. The most detrimental effects occur when particle tunes approach the resonances. For

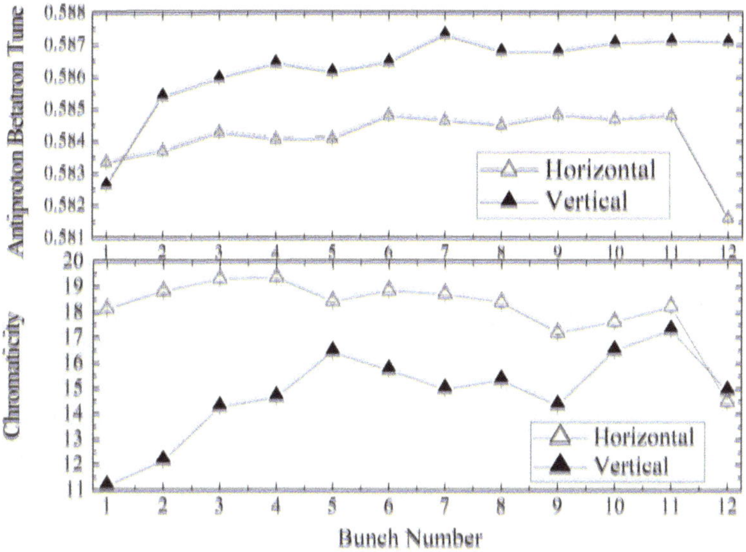

Fig. 8.8 Antiproton tunes (*top*) and chromaticities (*bottom*) measured by the 1.7 GHz Schottky monitor vs. bunch number for store #3678 (July 27–28, 2004). The tune data were taken over a period of 3 h, starting 3 h after the beginning of the store and extrapolated linearly to the time $t = 3$ h into the store. The chromaticities were assumed to be constant, and so the measurements were averaged over the entire store. The symbol size reflects the size of the statistical error bars [7]

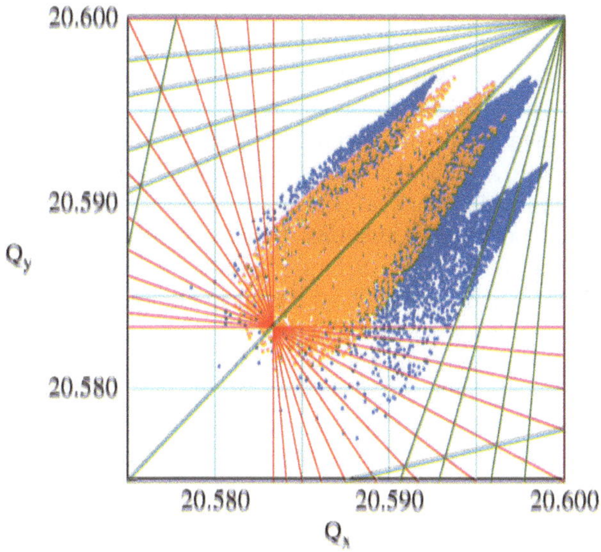

Fig. 8.9 Tevatron proton and antiproton tune distributions superimposed onto a resonance line plot. The *red* and *green lines* are various sum and difference tune resonances of up to twelfth order. The *blue dots* represent calculated the tune distributions for all 36 antiproton bunches; the *yellow* represent the protons. The tune spread for each bunch is calculated for particles up to 6σ amplitude taking into account the *measured* intensities and emittances (from [19])

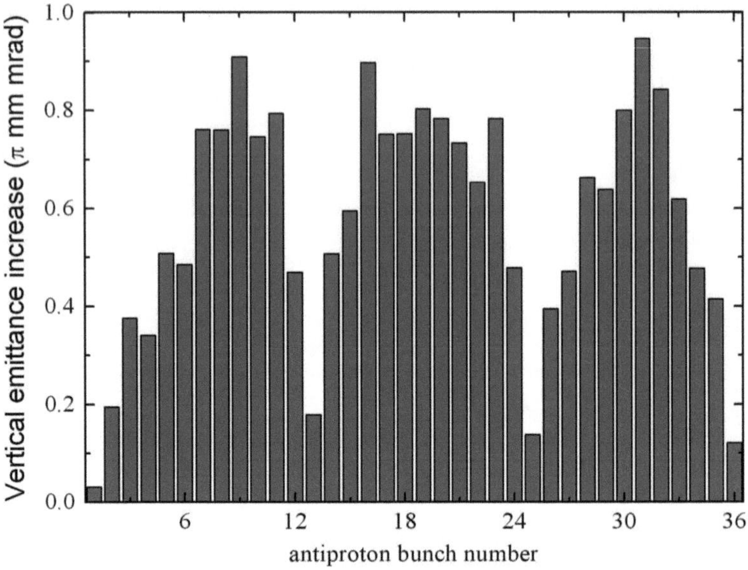

Fig. 8.10 Antiproton bunch emittance increase over the first 10 min after initiating collisions for HEP store #3231 with an initial luminosity $L = 48 \times 10^{30} \mathrm{cm}^{-2}\,\mathrm{s}^{-1}$ [19]

example, an emittance growth of the core of the beam is observed near the fifth-order resonances (defined as $nQ_x + mQ_y = 5$, such as $Q_{x,y} = 3/5 = 0.6$) or fast halo particle loss near twelfth-order resonances (for example, $Q_{x,y} = 7/12 \approx 0.583$).

The measured antiproton tunes decrease over the course of a store by some 0.005–0.007 with characteristic decay times of 12–16 h, caused by the reduction of the head-on tune shift, which itself is mostly due to the increase of proton emittances (by more than factor of 2) and the decrease of proton bunch intensities (by more than 25 %). Such excursions were found detrimental for luminosity lifetime were minimized by manual tuning as soon as 1.7 GHz Schottky monitors were made operational in 2005, resulting in increased beam lifetime. The chromaticities measured by the 1.7 GHz Schottky monitor are remarkably stable within 1 unit during the store and vary by about 6 units in both planes along a bunch train, and that is in acceptable agreement with theory.

It is not surprising that with such significant differences in orbits, tunes, and chromaticities, the antiproton bunch intensity lifetime and emittance growth rates vary considerably from bunch to bunch. As an illustration, Fig. 8.10 shows the vertical emittance blowup early in an HEP store for all three trains of antiproton bunches. One can see a remarkable distribution along the bunch train which gave rise to the term "scallops" (three "scallops" in three trains of 12 bunches) for this phenomenon—the end bunches of each train exhibit lower emittance growth than the bunches in the middle of the train. Because of the threefold symmetry of the proton loading, the antiproton emittance growth rates are the same within 5–20 % for corresponding bunches in different trains (in other words, bunches #1, #13, and

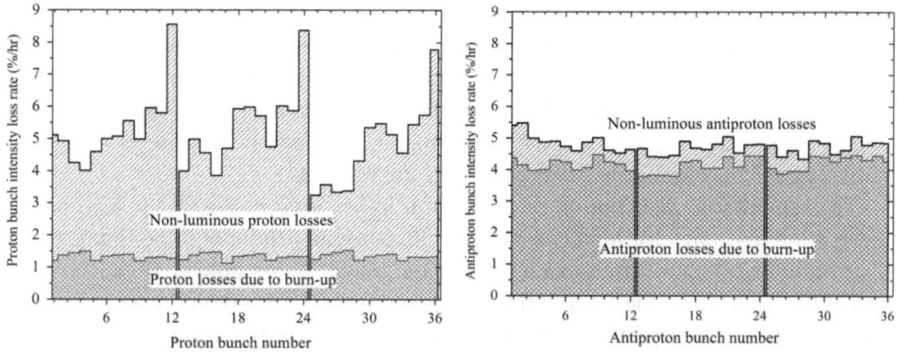

Fig. 8.11 (**a**) *Left*—proton-bunch intensity loss rates and (**b**) *right*—antiproton bunch intensity loss rates at the beginning of the Tevatron store #5155, Dec. 30, 2006, with an initial luminosity $L = 250 \times 10^{30}$ cm^{-2} s^{-1} (from [19])

#25 have similar emittance growths). The effect is dependent on the antiproton tunes, particularly on how close each bunch is to some important resonances—in case of the Tevatron working point, these are fifth-order (0.600), seventh-order (0.5714), and twelfth-order (0.583) resonances. For example, "the scallops" occur near the fifth-order resonances $nQ_x{+}mQ_y = 5$, such as $Q_{x,y} = 3/5{=}0.6$. Smaller but still definite "scallops" were also seen in protons if the proton tunes are not optimally set. After the initial 0.5–1 h of each store, the growth rate of each bunch decreased significantly. Various methods have been employed to minimize the development of scallops (including a successful attempt to compensate one bunch emittance growth with a Tevatron Electron Lens—see [19] and next section), but carefully optimizing the machine tunes was found to be the most effective—e.g., the vertical tune changes as small as −0.002 resulted in significant reduction of the amplitude of the "scallops."

As mentioned above, significant attrition rate of protons and antiprotons due to their interaction with opposite beam, both in the main IPs and in the numerous long-range interaction regions is one of the most detrimental effects of the beam–beam interaction in the Tevatron. The effect varies bunch-by-bunch and it is especially large at the beginning of the HEP stores where the total proton beam–beam tune shift parameter is peaked. Figure 8.11a shows a typical distribution of proton loss rates $(dN_p/N_p)/dt$ at the beginning of a high-luminosity HEP store. Bunches #12, 24, and 36 at the end of each bunch train typically lose about 9 % of their intensity per hour while other bunches lose only 4–6 % per hour. These losses are a very significant part of the total luminosity decay rate of about 20 % per hour (again, at the beginning of the high-luminosity HEP stores). The losses due to luminosity "burn-up" — inelastic proton–antiproton interactions $dN_p/dt = -\sigma_{int} \cdot L$ at the two main IPs ($\sigma_{int} = 0.07$ barn) are small (1–1.5 %/h) compared to the total losses. Losses due to inelastic interaction with the residual vacuum and due to leakage from the RF buckets are less than 0.3 %/h. The single largest source of proton losses is the beam–beam interaction with the antiprotons. Such conclusion is also supported by Fig. 8.11a, which shows a large bunch-to-bunch variation in the proton loss rates within each bunch train, but very similar rates for equivalent bunches, e.g., bunches #12, 24, and 36. On the contrary, antiproton intensity losses dN_a/dt are about the same for all the bunches—see Fig. 8.11b—as they are mostly due to luminosity burn-up and not determined by beam–beam effects (the latter indicated as "non-luminous" component of the loss rate).

The remarkable distribution of the proton losses seen in Fig. 8.11, e.g., particularly high loss rates for bunches #12, 24, 36, is usually thought to be linked to the distribution of betatron frequencies along the bunch trains bunch. Bunches at the end of the trains have their vertical tunes closer to the $7/12 \approx 0.583$ resonance lines, and, therefore, the higher losses. The average Tevatron proton tune Q_y of about 0.588–0.589 lies just above this resonance, and the bunches at the end of each train, whose vertical tunes are lower by $\Delta Q_y = -(0.002\text{–}0.003)$ due to the unique pattern of long-range interactions, are subject to stronger beam–beam effects. The tunes Q_y and Q_x are carefully optimized by the operation crew to minimize the overall losses of intensity and luminosity. For example, an increase of the average vertical tune by quadrupole correctors is not possible because it usually results in higher losses and "scallops" as small amplitude particle tunes move dangerously close to the $3/5 = 0.600$ resonance. The Tevatron Electron Lenses did reduce by a factor of >2 the proton losses out of the bunches #12, 24, 36 (see [19, 20] and next section).

The proton loss rate was also strongly affected by transverse size mismatch for head-on collisions of larger size proton bunches with smaller size antiproton bunches. Our studies of this phenomenon in 2003–2005 can be summarized by the following scaling formulae [7]:

$$\frac{1}{\tau_p} = \frac{1}{N_p}\frac{dN_p}{dt} \propto N_a \cdot \left(\frac{\varepsilon_p}{\varepsilon_a}\right)^2 F_2(Q_{x,y}, Q', Q'', M), \tag{8.9}$$

where M stands for bunch position in bunch train. In order to avoid large emittance ratio $\varepsilon_p/\varepsilon_a$, the antiproton emittances are routinely diffused at the beginning of HEP stores by a wide band transverse noise to a directional strip line, so the ratio is kept about 3. Factor F_2 in Eq. (8.9) shows significant dependence of the losses on the second-order betatron tune chromaticity $Q''=d^2Q/d(\Delta p/p)^2$. As mentioned at the beginning of this section, the second-order chromaticity was corrected in 2007 [11] and that resulted in significant improvement of the proton lifetime.

At the end of Run II, the antiproton intensity lifetime deterioration due to the beam–beam effects was much smaller than the proton one, and was found to scale approximately as [7]

$$\left(\frac{1}{\tau_a}\right)_{BB} = \left(\frac{dN_a}{N_a dt}\right)_{BB} \propto N_p \frac{\varepsilon_a^2}{S^3}, \tag{8.10}$$

where S stands for beam–beam separation (helix size).

Table 8.1 Tevatron luminosity and intensity loss rates averaged over the first 2 h of 2010–2011 HEP stores with initial luminosity from 300 × 10^{30} to 430 × 10^{30} cm^{-2} s^{-1} in 2010–2011

	%/h	Incl. due to beam–beam, %/h
Proton loss rate, $1/\tau_p$	2.8–3.2	1.5–2.0
Antiproton loss rate, $1/\tau_a$	5.5–6.2	1.0–1.5
Emittance growth rate, $1/\tau_\varepsilon$	9–11	Small
H-factor decay rate, $1/\tau_H$	1.2–1.4	Small
Luminosity decay rate, $1/\tau_L$	19–21	2.5–3.5

8.2.3 *Impact of Beam–Beam Effects on the Integrated Luminosity*

The collider luminosity lifetime is determined by the speed of the emittance growth, beam intensity loss rates, and bunch lengthening (that affects hourglass factor H):

$$\tau_L^{-1} = \frac{dL(t)}{L(t)dt} = \left|\tau_\varepsilon^{-1}\right| + \tau_{Na}^{-1} + \tau_{Np}^{-1} + \tau_H^{-1}. \tag{8.11}$$

At the end of Run II, the luminosity loss rates were in the range 19–21 %/h at the beginning of stores—see Table 8.1. For the 2010–2011 HEP stores in range of initial luminosities between 3.0 and 4.3 × 10^{32} cm^{-2} s^{-1}, the largest contribution to luminosity decay came from beam emittance growth with a typical time of $\tau_\varepsilon \sim$ 9–11 h. The growth is dominated by IBS in the proton bunches, with small contributions from the IBS in antiprotons and external noises. Beam–beam effects, if noticeable, usually manifest themselves in reduction of the beam emittances or their growth rates rather than in increases. The antiproton bunch intensity lifetime $\tau_a \sim$ 16–18 h is dominated by the luminosity burn rate which accounts for 80–85 % of the lifetime, while the remaining 10–15 % comes from parasitic beam–beam interactions with protons. Proton intensity loss varies in a wide range $\tau_p \sim$ 25–45 h and is driven mostly (~50 %) by the head-on beam–beam interactions with smaller size antiprotons at the main IPs. The proton lifetime caused by inelastic interactions with antiprotons in collisions and with residual gas molecules varies from 300 to 400 h. The hourglass factor decays with $\tau_H \sim$ 70–80 h due to the IBS, again, mostly in proton bunches. Beam–beam effects may lead to reduction of the proton bunch length growth (longitudinal "shaving") in a poorly tuned machine. Combining all of these loss rates together, one can estimate the hit on the luminosity lifetime τ_L due to the beam–beam effects as 12–17 % (that is equal to (2.5–3.5 %/h)/(19–21 %/h)). As concluded in [7], the luminosity integral $I = \int L dt$—the sole critical parameter for HEP experiments—depends on the product of peak luminosity and the luminosity lifetime, e.g., for a single store with initial luminosity L_0 and duration $T \sim 16$ h, the integral is $I \approx L_0 \tau_L \ln(1 + T/\tau_L)$. Therefore, the full impact of the beam–beam effects on the luminosity integral should include beam–beam-driven proton and antiproton losses at the injection energy (about 5 and 1 %,

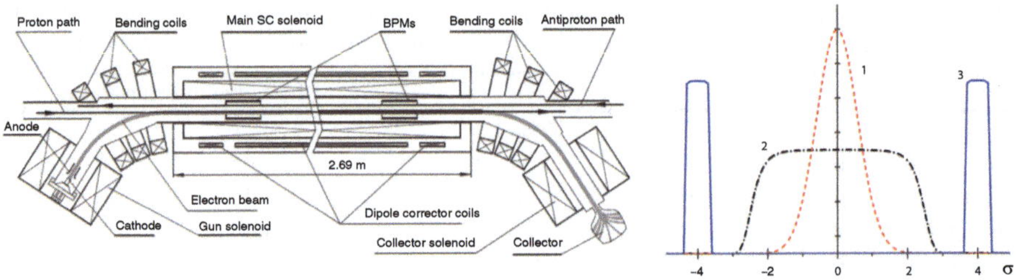

Fig. 8.12 General layout of the Tevatron electron lens; (*right*) transverse electron current profiles for (1) space-charge and head-on beam–beam compensation, (2) for bunch-by-bunch tune spread compensation, (3) halo collimation [19]

correspondingly), on the energy ramp (2 and 3 %), and in the low-beta squeeze (1–2 and 0.5 %) which proportionally reduce the initial luminosity L_0. So, altogether, at the last operational stage of the Tevatron collider present, the beam–beam effects reduce the luminosity integral by 23–33 %.

8.3 Tevatron Electron Lenses for Compensation of Beam–Beam Effects and Beam Collimation

Electron lenses employ electromagnetic fields of strongly magnetized high intensity electron beams and were originally proposed for compensation of the head-on beam–beam effects in the SSC [21] and for compensation of the long-range beam–beam effects in the Tevatron [22]. The lens employs a low energy beam of electrons which collides with the high-energy proton or antiproton bunches over an extended length. Electron space-charge forces are linear at distances smaller than the characteristic beam radius $r < a_e$ but scale as $1/r$ for $r > a_e$. Correspondingly, such a lens can be used for linear long-range beam–beam and nonlinear head-on beam–beam force compensation depending on the beam-size ratio a_e/σ and the current-density distribution $j_e(r)$. Electron lenses have also been proposed for compensation of space-charge forces in high intensity hadron accelerators [23]. Main advantages of the e-lenses are: (a) the electron beam acts on high-energy beams only through EM forces, with no nuclear interactions; (b) fresh electrons interact with the high-energy particles each turn, leaving no possibility for coherent instabilities; (c) the electron current profile (and, thus, the EM field profiles) can easily be changed for different applications—see Fig. 8.12; (d) the electron-beam current can be quickly varied, e.g., on a time scale of bunch spacing in accelerators.

Two electron lenses were built and installed in two different locations of the Tevatron p–pbar collider ring A11 and F48 [24]. They met specifications for the bunch-by-bunch tune spread compensation [17] and were used to counteract beam lifetime deterioration due to the long-range beam–beam effects [20] and for the abort gap beam removal [25] and for beam halo collimation [26]. Up to 3 A, 6–10 kV e-beam was generated at the 10–15 mm diameter thermocathode immersed in 0.3 T longitudinal magnetic field and aligned onto (anti)proton beam orbit over about 2 m length inside 6 T SC solenoid. The deviations of the magnetic field lines from a straight line are less than ±100 μm over the entire length of the SC solenoid. The electron beam, following the field lines, therefore does not deviate from the straight Tevatron beam trajectory by more than 20 % of the Tevatron beam rms size $\sigma \approx 0.5$–0.7 mm in the location of the TELs. In order to enable operation on a single bunch in the Tevatron with bunch spacing of 396 ns, the anode voltage, and consequently the beam current, is modulated with a characteristic on-off time of about 0.6 μs and a repetition rate equal to the Tevatron revolution frequency of $f_0 = 47.7$ kHz by using a HV Marx pulse generator [27] or a HV RF tube base amplifier. The electron pulse timing jitter is less than 1 ns and the peak current is stable to better than 1 %, so, the TEL operation does not incur any significant emittance growth.

The high-energy protons are focused by the TEL and experience a positive betatron tune shift:

$$dQ_{x,y} = +\frac{\beta_{x,y} L_e r_p}{2\gamma e c} \cdot j_e \cdot \left(\frac{1-\beta_e}{\beta_e}\right). \tag{8.12}$$

In the long-range beam–beam compensation (BBC) experiments, large radius electron beam was generated $a_e \approx 3\sigma$; therefore, the tune shift was about the same for most protons in the bunch. The tuneshift for the antiprotons is of about the same magnitude, but negative. Maximum measured tuneshift for 980 GeV protons was about 0.01.

In the BBC demonstration experiment [20], the electron beam of the TEL-2 installed at the A11 location with large vertical beta-function of $\beta_y = 150$ m was centered and timed onto bunch #12 without affecting any other bunches. When the TEL peak current was increased to $J_e = 0.6$ A, the lifetime $\tau = N/(dN/dt)$ of bunch #12 went up to 26.6 h from about 12 h—see Fig. 8.13. At the same time, the lifetime of bunch #36, an equivalent bunch in the third bunch train, remained low and did not change significantly (at 13.4 h lifetime). When the TEL current was turned off for 15 min, the lifetimes of both bunches were, as expected, nearly identical (16 h). The TEL was then turned on again, and once again the lifetime for bunch #12 improved significantly to 43 h while bunch #36 stayed poor at 23.5 h. This experiment demonstrates a factor of two improvement in the proton lifetime due to compensation of beam–beam effects with the TEL.

The proton lifetime, dominated by beam–beam effects, gradually improves and reaches roughly 100 h after 6–8 h of collisions; this is explained by a decrease in antiproton population and an increase in antiproton emittance, both contributing to a reduction of the proton beam–beam parameter. To study the effectiveness of BBC later in the store, the TEL was repeatedly turned on and off every half hour for 16 h, again on bunch #12. The relative improvement R, defined as the ratio of the proton

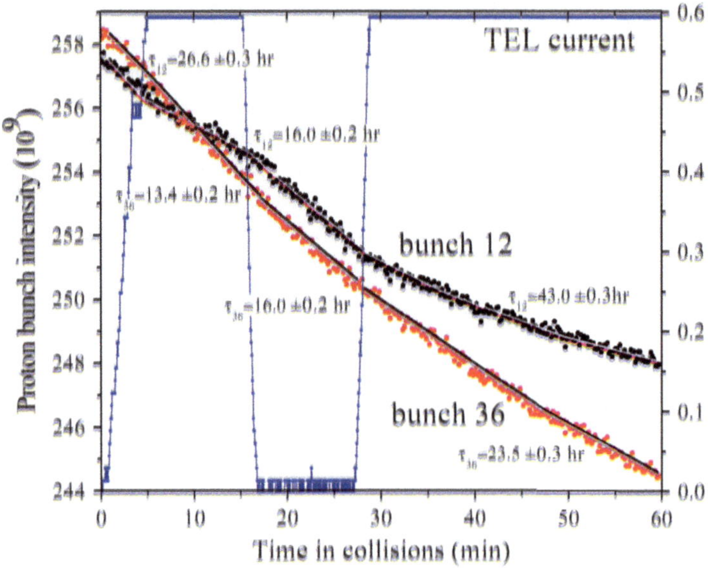

Fig. 8.13 Intensities of proton bunches #12 and #36 early in store #5119 with $L_0 = 1.6 \times 10^{32}\ \text{cm}^{-2}\ \text{s}^{-1}$ [20]

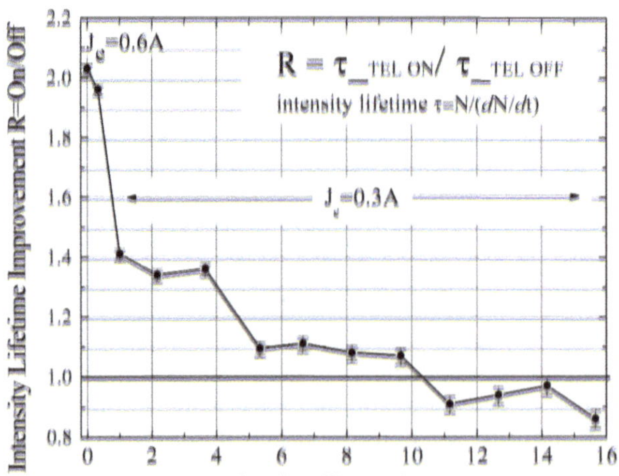

Fig. 8.14 Relative improvement of proton bunch #12 lifetime induced by TEL vs. time in store #5119 [20]

lifetime with the TEL and without, is plotted in Fig. 8.14. The first two data points correspond to $J_e = 0.6$ A (as is Fig. 8.13 and the above description), but subsequent points were taken with $J_e = 0.3$ A to observe dependence of the compensation effect on electron current. The change of the current resulted in a drop of the relative improvement from $R = 2.03$ to $R = 1.4$. A gradual decrease in the relative improvement is visible until after about 10 h, where the ratio reaches 1.0 (no gain in lifetime). At this point, the beam–beam effects have become very small, providing little to compensate. Similar experiments in several other stores with initial luminosities ranging from $L_0 = 1.5 \times 10^{32}\ \mathrm{cm^{-2}\,s^{-1}}$ to $2.5 \times 10^{32}\ \mathrm{cm^{-2}\,s^{-1}}$ repeated these results.

The lifetime improvement due to the TEL can be explained in part by the positive shift of vertical tune of protons $dQ_y \approx 0.0015$ which makes the detrimental effects of the twelfth-order resonance $Q_y = 7/12 = 0.583$ weaker. The average Tevatron proton tune $Q_y = 0.589$ (which is carefully optimized to minimize overall losses) is just above this resonance, and the bunches at the end of each train, which have vertical tunes lower by $\Delta Q_y = -(0.002\text{–}0.003)$ due to unique pattern of long-range interactions, are subject to stronger beam–beam effects (see preceding section). The TEL moves those protons away from the resonance, thus, resulting in significant reduction of the losses. It is noteworthy, that the TEL operation with $J_e = 0.6$ A resulted in bunch #12 having one of the lowest loss rates among all bunches, while its tune still remained lower $dQ_y < |\Delta Q_y|$.

Results of many experiments with TEL are reported in [19], studies of nonlinear BBC with Gaussian electron-beam current profile are presented in [28]. TELs were not used routinely for the BBC in the Tevatron because beam–beam losses were effectively controlled by other means as described in Sect. 8.2. Numerical simulations [29] predict beneficial effect of electron lenses on the ultimate intensity LHC beam lifetime.

8.4 Modeling and Simulation of Beam–Beam Effects in Tevatron

In this section we describe the models and simulation tools, which were used to study beam–beam effects in the Tevatron. Simulations correctly describe many observed features of the beam dynamics, have predictive power, and have been particularly useful for supporting and planning changes of the machine configuration. For the sake of brevity we mostly concentrate on effects occurring during high-energy physics operation.

8.4.1 Store Beam Physics Analysis

Beam–beam interaction is not the single strongest effect determining evolution of beam parameters at collisions. There are many sources of diffusion causing emittance growth and particle losses, including but not limited to intrabeam scattering, noise of accelerating RF voltage, and scattering on residual gas. Parameters of these mechanisms were measured in beam studies, and then a model was built in which

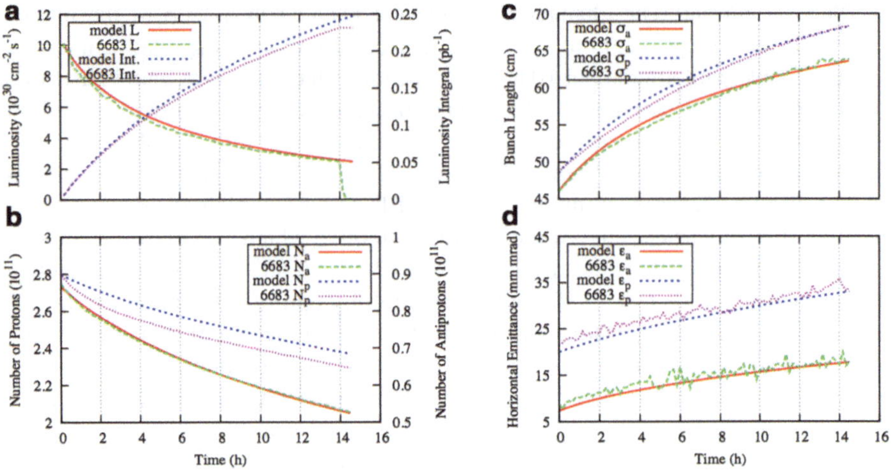

Fig. 8.15 Observed beam parameters in store 6683 compared to store analysis calculation (model). $L_0 = 3.5 \times 10^{32}$ cm^{-2} s^{-1}. (**a**) Single bunch Luminosity and Luminosity integral. (**b**) Intensity of proton bunch no. 6 and of antiproton bunch colliding with it (no. 13). (**c**) Bunch lengths. (**d**) Horizontal 95 % normalized bunch emittances [8]

the equations of diffusion and other processes are solved numerically. The model, which is described in detail in Sect. 6.4, is able to predict evolution of the beam parameters in the case of weak beam–beam effects. When these effects are not small, it provides a reference for evaluation of their strength. This approach was used on a store-by-store basis to monitor the machine performance in real time because such calculations are very fast compared to a full numerical beam–beam simulation. Figure 8.15 presents an example comparison of evolution of beam parameters in an actual high-luminosity store to calculations. Note that there is no transverse emittance blow up in both beams, and processes other than beam–beam interaction determine the emittance growth. The same is true for antiproton intensity and bunch length. The most pronounced difference between the observation and the model is seen in the proton intensity. Beam–beam effects cause proton lifetime degradation during the initial 2–3 h of the store until the proton beam–beam tune shift drops from 0.02 to 0.015. The corresponding loss of luminosity integral is about 5 %.

8.4.2 Weak-Strong Numerical Simulations

Simulations, in which the "strong" beam is considered as having constant and known distribution and is usually represented by a formula, while the other, "weak," beam is modeled as a bunch of macro-particles, are a convenient tool for predicting evolution of beam intensity and emittance caused by incoherent effects. Since such simulation does not necessitate multi-bunch treatment of beam dynamics, the tracking of 10^4 macro-particles through the Tevatron lattice with two head-on and 70 long-range collision points for 10^7 turns (which correspond to approximately 3 min of real time) takes about 20 h. One of the codes that found wide use for simulation of the Tevatron beam–beam phenomena is Lifetrac [30]. Originally, Lifetrac was developed for simulation of equilibrium distribution of particles in circular electron-positron colliders. In 1999 new features have been implemented, which allowed simulating non-equilibrium distributions, for example proton beams. In this case the goal of simulations is not to obtain the equilibrium distribution but to observe how the initial distribution is changing with time. Number of simulated particles typically varies in the range of 10^3–10^6. The tracking time is divided into "steps," typically 10^3–10^5 turns each. The statistics obtained during the tracking (1D histograms, 2D density in the space of normalized betatron amplitudes, luminosity, beam sizes, and emittances) is averaged over all particles and all turns for each step. Thus, a sequence of frames representing evolution of the initial distribution is obtained.

Another important quantity characterizing the beam dynamics is the intensity lifetime. It is calculated by placing an aperture restriction in the machine and counting particles reaching the limit. The initial and final coordinates of the lost particle are saved. This information is valuable for analysis of various beam dynamics features.

The initial 6D distribution of macro-particles can be either Gaussian (by default), or read from a separate text file. Besides, the macro-particles may have different "weights." This allows representing the beam tails more reliably with limited number of particles. Usually we simulate the Gaussian distribution with weights: particles initially located in the core region have larger weight while the "tail" particles with smaller weight are more numerous.

When performing tracking through a head-on IP, the "strong" bunch is divided into slices longitudinally. The higher are the orders of significant betatron resonances, which make effect on the distribution, the greater must be the number of slices. In our simulations 12 slices were used in the main IPs where beta-functions are approximately equal to the bunch length and only one slice in long-range collision points where beta-functions are much greater and one can neglect the betatron phase advance on the bunch length.

The transverse density distributions within "strong" slices are bi-Gaussian, allowing to apply the well-known formulae [31] for 6D symplectic beam–beam kick. However, a simple modification allowed simulating non-Gaussian strong bunches. Namely, the strong bunch is represented as a superposition of a few (up to three) Gaussian distributions with different betatron emittances. The kicks from all these "harmonics" are added. The calculation time is increased but the transformation remains 6D symplectic.

To study the dependence of beam–beam effects on various machine parameters, the following features were incorporated into the code:

- Realistic machine optics via linear 6D maps calculated from actual beam measurement data (Sect. 4.2), with full account of betatron coupling and optics differences on the proton and antiproton orbits. It was estimated that resonances generated by known Tevatron nonlinearities, such as the final focus triplets and lattice sextupoles, are much weaker than those driven by beam–beam collisions at the present betatron tune working point. Hence, inclusion of nonlinear lattice elements into the simulation was deemed unnecessary. Still, the code has the capability to include thin multipoles up to the tenth order.
- Collision point pattern individual for each bunch within the train, with beams separations obtained from beam measurements.
- First- and second-order chromaticity implemented as symplectic "chromatic drifts." In the Hamiltonian theory the chromaticity of beta-functions does not come from energy-dependent focusing strength of quads (as one would intuitively expect) but from drift spaces where the transverse momentum is large (low-beta regions). The symplectic transformations for that are

$$\begin{array}{l} x = x - L \cdot x' \cdot \delta \\ y = y - L \cdot y' \cdot \delta \\ z = z - L(x'^2 + y'^2)/2 \end{array},$$

where x, y, and z are the particle coordinates, $\delta = \Delta p/p$ is the momentum deviation, and L is the "chromatic drift" length. Then, it is necessary to adjust the betatron tune chromaticities, which are also affected by "chromatic drift." For that, an artificial element (insertion) is used with the following Hamiltonian:

$$H = I_x(2\pi Q_x + C_x\delta) + I_y(2\pi Q_y + C_y\delta), \tag{8.13}$$

where I_x and I_y are the action variables, Q_x and Q_y are the betatron tunes, C_x and C_y are the [additions to the] chromaticities of betatron tunes.

- Diffusion and noise, in the form of a random Gaussian kick applied to macroparticles once per turn. Strength of the kick on different coordinates is given by a symmetrical matrix representing correlations between Gaussian noises. In the Tevatron, the diffusion is rather slow in terms of the computer simulation—the characteristic time for the emittance change is around an hour or 10^8 turns. In simulations aimed at evaluation of the antiproton beam dynamics during the 2004–2005 run the noise was artificially increased by three orders of magnitude in order to match the diffusion and the computer capabilities [4].
- Beam–beam compensator (electron lens) element implemented as a thin nonlinear lens.

We have validated the code using available experimental data. As an example, Figs. 8.16 and 8.17 show a good reproduction of the two distinct effects in bunch-to-bunch differences caused by beam–beam effects: variation of vertical bunch centroid position due to long-range dipole kicks, and variation of transverse emittance blowup caused by difference in tunes and chromaticities.

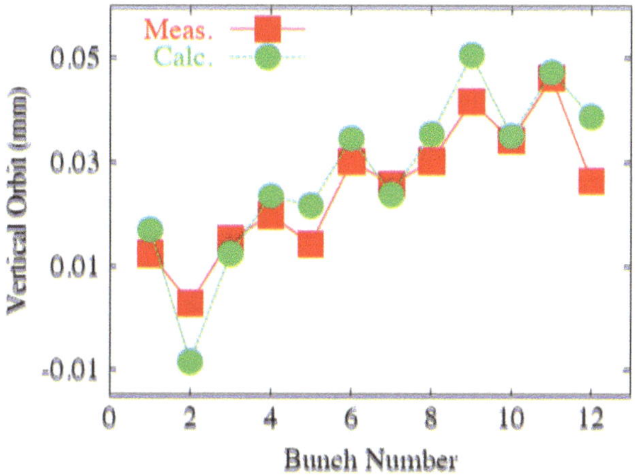

Fig. 8.16 Bunch-by-bunch antiproton vertical orbit [8]

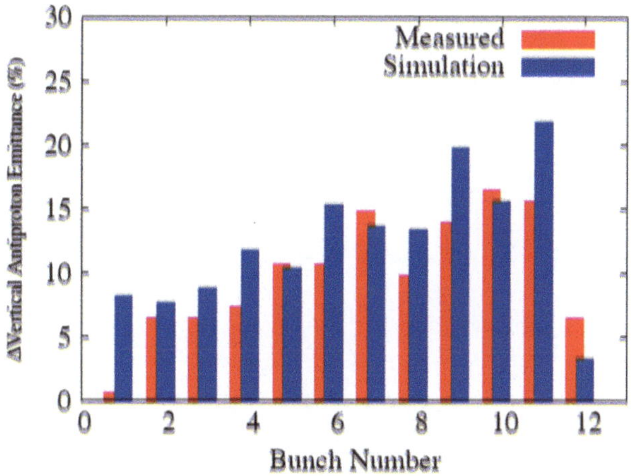

Fig. 8.17 Bunch-by-bunch antiproton emittance growth. Measured in store #3554 (*red*) and simulated with lifetrac (*blue*) [8]

Lifetrac simulations proved to be a useful tool in justification and development of machine upgrades, such as

- The decrease of antiproton betatron tune chromaticity, reduction of the β^* from 0.35 to 0.28 m (both in 2005).
- Demonstration of the importance of separation at long-range collision points nearest to the main IPs, and subsequent implementation of the new collision helix.
- Identification of the large chromaticity of β^* as a possible source of lifetime deterioration following the increase of the antiproton intensity. Simulations revealed an interesting feature in the behavior of the proton bunch length at high values of beam–beam parameter ξ—the so-called bunch shaving, when the bunch length starts to decrease after initiating head-on collisions instead of steady growth predicted by the diffusion model (Fig. 8.18). This behavior was

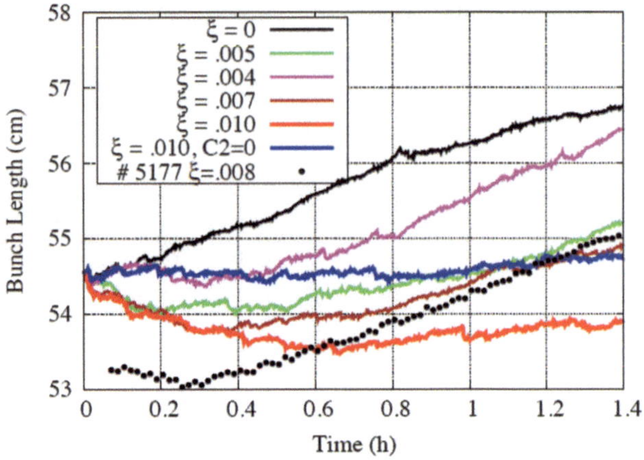

Fig. 8.18 Effect of corrected second-order chromaticity on the proton bunch length evolution. *Solid lines*—simulation for different ξ, *black dots*—experimental data [8]

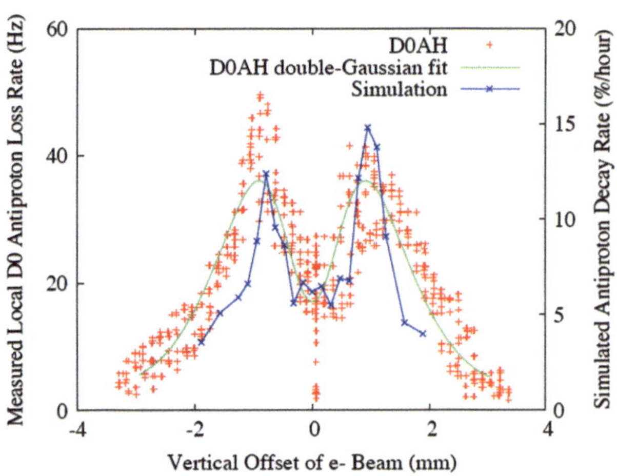

Fig. 8.19 Measured (*red crosses*, D0AH), and simulated particle loss rate during a vertical TEL-beam separation scan

observed multiple times during HEP stores in 2007, being especially pronounced when the vertical proton betatron tune was set too high.

- With the use of Lifetrac, it was shown that a change of the tune working point from 0.58 to near the half integer resonance would allow as much as 30 % increase of intensities but such upgrade required a lengthy commissioning period and was not realized during Run II.
- Lifetrac was routinely used to support beam physics studies, e.g., the experiments on BBC with electron lenses (see preceding section 8.3). For instance, Fig. 8.19 presents the measured and simulated particle loss rate during a transverse separation scan between the circulating beam and the Gaussian TEL beam.

8.4.3 Strong–Strong Numerical Simulations

Although coherent beam–beam effects did not present a limitation of the machine performance, extensive work has been done to create an accurate model of multi-bunch collective dynamics [32]. A comprehensive Tevatron simulation was created including a fully 3D strong–strong beam–beam particle-in-cell Poisson solver, interactions among multiple bunches with both head-on and long-range collisions, a linear optics model using measured coupled lattice functions, a helical trajectory consistent with beam-orbit measurements, and machine chromaticity and impedance.

The starting point for the simulation is the extended BeamBeam3d code [33, 34]. Bunches of macro-particles in two beams are generated with a random distribution in phase space with parameters that match the lattice. The accelerator ring is conceptually divided into arcs with potential interaction points at the ends of the arcs. All bunches from both beams are individually tracked. When bunches from two beams arrive at the same IP, a Poisson field solver is employed to determine the electromagnetic forces on each particle produced by the charged particles in the opposing beam bunch. Beam–beam forces in extended length bunches are computed by slicing the bunch longitudinally and moving the bunches through other in steps, applying the beam–beam forces at each step. The optics of each arc is modeled with a 6×6 linear map that transforms the phase space $\{x, x', y, y', z, \delta\}$ coordinates of each macro-particle from one end of the arc to the other. For our Tevatron simulations, the maps were calculated using the measured coupled lattice functions (see Chap. 2). The synchrotron motion is put in as a sinusoidal oscillation with the periodicity of the machine synchrotron tune. A shifted Greens function is employed in the Poisson field solver calculation to efficiently account for the mean beam transverse offset at each IP. The validity of the 3D beam–beam calculation has been verified [34] by reproducing the evolution [35] of synchrobetatron modes observed at the VEPP-2M $e+e-$ collider as a function of beam–beam parameter ξ.

The impedance model applies a momentum kick to the particles generated by the dipole component of resistive wall wakefields [36]. Each beam bunch is divided longitudinally into slices containing approximately equal numbers of particles. As each bunch is transported through an arc, particles in each slice receive a transverse kick from the wakefield induced by the dipole moment of the particles in forward slices. The impedance model has been verified to agree with analytic calculations of instability thresholds and growth rates for the two macro-particle model of strong and weak head-tail instabilities [34, 36].

During the Tevatron operation in 2009 the limit for increasing the initial luminosity was determined by particle losses in the squeeze [37]. With proton bunch intensities approaching 3.2×10^{11} particles, the chromaticity of the Tevatron had to be managed carefully to avoid the development of a head-tail instability. It was determined experimentally that, after the head-on collisions are initiated, the Landau damping introduced by beam–beam interaction is strong enough to maintain beam stability at chromaticity of +2 units. At the earlier stages of the collider cycle, when beam–beam effects are limited to long-range interactions, the chromaticity was kept as high as 15 units since the concern was that the Landau damping is insufficient to suppress the instability. At the same time, high chromaticity causes particle losses, which are often large enough to quench the superconducting magnets, and hence it is desirable to keep it at a reasonable minimum.

The strong–strong simulation was used to determine the safe lower limit for chromaticity. Simulations demonstrated that for the Tevatron parameters, long-range beam–beam interactions provide stabilization of the head-tail instability. Based on these findings, the chromaticity in the squeeze was lowered by a factor of 2, and was kept at 8–9 units. This resulted in a significant decrease of the observed particle loss rates (see, e.g., Fig. 5 in [37]).

References

1. T. Sen, arXiv:1102.2432 (2010); also in ICFA Beam Dyn. Newslett. **52**, 14–32 (2010)
2. R. Moore, A. Jansson, V. Shiltsev, JINST **4**, P12018 (2009)
3. T. Sen et al., Rev. ST Accel. Beams **7**, 041001 (2004)
4. Y. Alexahin, in *Proceedings of 2005 IEEE Particle Accelerator Conference,* Knoxville (2005), p. 544
5. S. Saritepe, G. Goderre, S. Peggs, in *Frontiers of Particle Beams: Intensity Limitations*, Springer, Lecture Notes in Physics (1991)
6. V. Bharadwaj, et al., *FNAL Preprint Fermilab-TM-1970* (1996)
7. V. Shiltsev et al., Phys. Rev. ST Accel. Beams **8**, 101001 (2005)
8. A. Valishev, et al., JINST **7**, P12002 (2012); also arXiv:0906.0386 (2009)
9. P. Ivanov, et al., in *Proceedings of the 2003 IEEE Particle Accelerator Conference*, Portland, OR (2003), p. 3062
10. Y. Alexahin, in *Proceedings of the 2007 IEEE Particle Accelerator Conference*, Albuquerque, NM (2007), pp. 3874–3876
11. A. Valishev, G. Annala, V. Lebedev, R. Moore, in *Proceedings of the Particle Accelerator Conference*, Albuquerque, NM (2007), p. 3922
12. M. Syphers, *Beam-Beam Tune Distributions with Differing Beam Sizes,* Fermilab internal document Beams-Doc-3031 (2008)
13. C.Y. Tan, J. Steimel, in *Proceedings of the 2009 IEEE Particle Accelerator Conference*, Vancouver, Canada (2009), p. 1668
14. P. Bagley, et al., *FNAL Preprint Conf-96-392* (1996)
15. L. Michelotti, *FNAL Preprint TM-1738* (1995)
16. R. Pasquinelli, A. Jansson, Phys. Rev. ST Accel. Beams **14**, 072803 (2011)
17. Y. Alexahin, *Preprint FERMILAB-PUB-00-120-T* (2000)
18. T. Sen et al., Phys. Rev. ST Accel. Beams **7**, 0410101 (2004)
19. V. Shitsev et al., New J. Phys. **10**, 043042 (2008)
20. V. Shiltsev et al., Phys. Rev. Lett. **99**, 244801 (2007)
21. E. Tsyganov, A. Taratin, A. Zinchenko, Phys. Part. Nuclei **27**, 279 (1996)
22. V. Shiltsev et al., Phys. Rev. ST Accel. Beams **2**, 071001 (1999)
23. A. Burov, G. Foster, V. Shiltsev, *Fermilab Preprint FNAL-TM-2125* (2000)
24. V. Shiltsev et al., Phys. Rev. ST Accel. Beams **11**, 103501 (2008)
25. X.-L. Zhang et al., Phys. Rev. ST Accel. Beams **11**, 051002 (2008)
26. V. Shiltsev, in *Proceedings of BEAM'06* (Valencia, Spain), Yellow Report CERN-2007-002; G. Stancari, et al., Phys. Rev. Lett. **107**, 084802 (2011)

27. H. Pfeffer, G. Saewert, JINST **6**, P11003 (2011)
28. A. Valishev, et al., in *Proceedings of 2010 International Particle Accelerator Conference*, Kyoto, Japan (2010), p. 2084
29. A. Valishev, V. Shiltsev, in *Proceedings of the 2009 IEEE Particle Accelerator Conference*, Vancouver, Canada (2009), p. 2567
30. D. Shatilov, Part. Accel. **52**, 65 (1996)
31. K. Hirata, H. Moshammer, F. Ruggiero, *KEK Report 92-117* (1992)
32. E. Stern, J.F. Amundson, P.G. Spentzouris, A.A. Valishev, Fully 3D multiple beam dynamics processes simulation for the Tevatron. Phys. Rev. ST Accel. Beams **13**, 021003 (2010)
33. J. Qiang, M.A. Furman, R.D. Ryne, J. Comp. Phys. **198**, 278–294 (2004)
34. E. Stern, et al., in *Proceedings of IEEE PAC07*, Albuquerque, NM (2007), p. 905
35. I.N. Nesterenko, E.A. Perevedentsev, A.A. Valishev, Phys. Rev. E, **65**, 056502 (2002)
36. A. Chao, *Physics of Collective Beam Instabilities in High Energy Accelerators* (Wiley, New York, 1993), pp. 56–60, 178–187, 333–360
37. A. Valishev, et al., *in Proceedings of the 2009 IEEE Particle Accelerator Conference*, Vancouver, Canada (2009), p. 4230

Chapter 9
Beam Instrumentation

A. Jansson, V. Lebedev, R. Moore, and V. Shiltsev

Operation of a superconducting magnet hadron collider, like Fermilab's Tevatron, requires a great deal of care, understanding of beam conditions, and trust in the beam diagnostics, because comparatively innocent little imperfections can lead to either beam blow-up and luminosity loss or to beam loss and quench of superconducting magnets. In the Tevatron such a quench results in 2–4 h of magnet recovery time and up to 8–16 h of no-luminosity time needed to produce the antiprotons needed for the next high energy physics (HEP) store. Over 8 years of operations we witnessed machine downtimes due to 0.5–1 % of beam intensity loss, poor beam lifetime, 0.5–1 mm orbit error, collimator malfunctioning, sequencer error, excursions of tunes or coupling of the order of few 0.001 or several units of chromaticity, instability occurrences, or malfunctioning of kickers, separators, or one of hundreds of power supplies, etc.—a detailed discussion can be found in [1]. Naturally, these peculiarities were reflected in the kinds of beam diagnostics we developed (e.g., minimization of their invasiveness) and the way they were exploited (fast data-logging, convenience for postmortem analysis, etc.).

Challenges in operation of the room-temperature accelerators in the proton injector chain machines are of a different sort and mostly related to necessity of tight control of beam losses in order to keep residual level of beam-induced radiation under certain limit, usually under 1 W/m. In addition, antiproton production, accumulation, and cooling require reliable diagnostics of low intensity beams and very slow beam cooling processes.

A. Jansson • V. Lebedev • R. Moore • V. Shiltsev (✉)
Accelerator Physics Center, Fermi National Accelerator Laboratory, MS221, PO Box 500, Batavia, IL 60510, USA
e-mail: val@fnal.gov; shiltsev@fnal.gov

V. Lebedev and V. Shiltsev (eds.), *Accelerator Physics at the Tevatron Collider*, Particle Acceleration and Detection, DOI 10.1007/978-1-4939-0885-1_9,

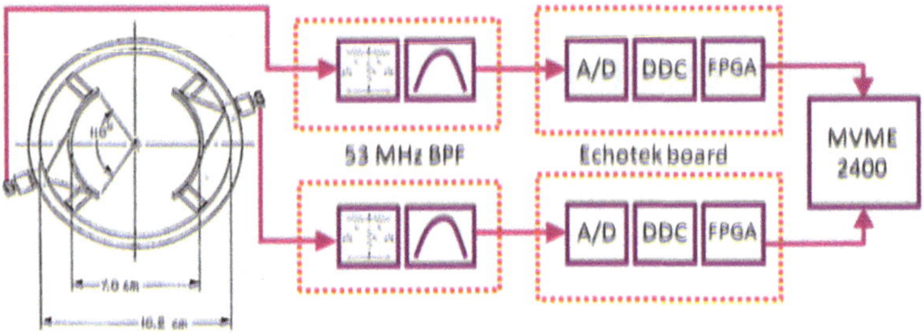

Fig. 9.1 Block diagram of the upgraded Tevatron BPM electronics. Signals from the BPM pickups go through a 53 MHz bandpass filter before being digitized and down-converted on an Echotek model ECDR-GC814-FV-A digital receiver board. A Motorola MVME-2400 processor provides the interface to ACNET for BPM readings and configuration control [1]

9.1 Beam Position Monitors

In the Tevatron, the protons and antiprotons circulate within a single beam pipe, so electrostatic separators are used to kick the beams onto distinct helical orbits to allow head-on collisions only at the desired interaction points. At 150 GeV, separation is limited to ~10–22 mm by physical aperture, while the separation above 600 GeV, ~3–6 mm, is limited by the breakdown (spark) rate of the separators at high voltage. Long-range beam effects degrade beam lifetime and machine performance, and having a good model of the optics is essential to understanding problems and how to improve operations. Reliable beam position monitors (BPMs) with good resolution are needed to measure the optics and construct the model (see chap. 2).

There were several problems with the BPM system developed for the Collider Run I [2] that hampered machine operations and diagnosis of possible problems in the Collider Run II. The orbit would deviate significantly from the desired reference orbit, 0.5 mm RMS (root mean square) differences in only 1–2 weeks, so global orbit smoothing was needed regularly. The BPM response to coalesced beam (a transfer concentrated in a single 53 MHz bucket, used for HEP stores) and uncoalesced beam (30 or so consecutive buckets, used for tune-up) differed enough so that a direct comparison between orbits recorded during HEP stores could not be compared easily to proton-only stores used to tune the machine or do orbit smoothing. The BPM position resolution was only ~150 μm and limited optics measurements to at best 20 % uncertainty. The turn-by-turn (TBT) capability was unreliable, and the system was blind to antiprotons. All these issues motivated the decision to upgrade the BPM electronics and take advantage of current technology [3]. The 240 Tevatron BPM pickups remained unchanged.

Figure 9.1 shows a block diagram of the upgraded BPM electronics system. A cross-sectional view of the pickup is shown on the left. Each detector has two 18 cm long copper electrodes in a 10.2 cm diameter stainless steel pipe. Each electrode is bent to a 3.5 cm radius of curvature, and subtends 110° of arc for a

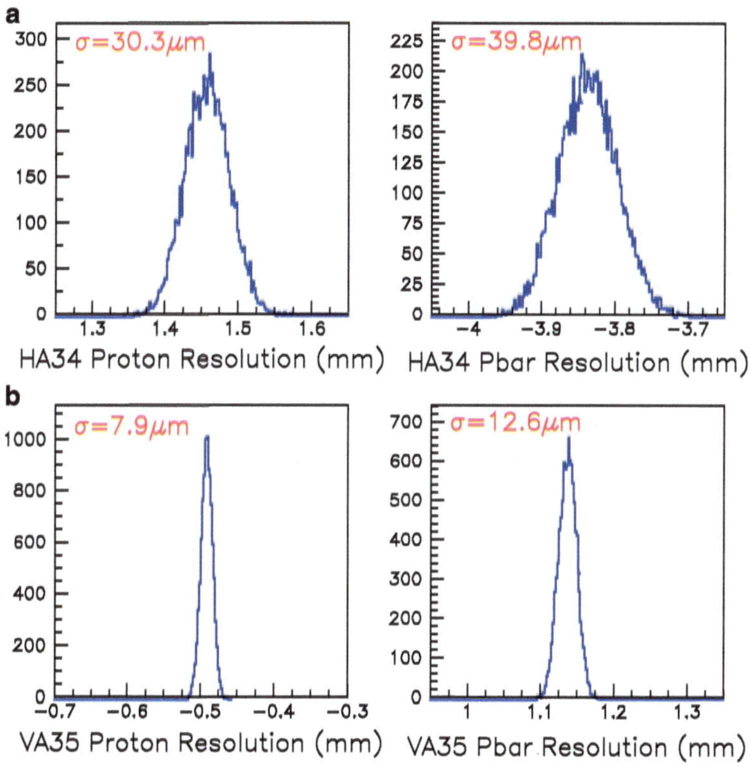

Fig. 9.2 (**a**) (*top*) Simultaneous horizontal position measurements from one BPM for protons (*left*) and antiprotons (*right*). (**b**) (*bottom*) Simultaneous vertical position measurements from one BPM for protons (*left*) and antiprotons (*right*). The given RMS values include all effects: resolution of the BPM and electronics, real beam motion (especially synchrotron oscillations for the horizontal data), and imperfect cancellation of the proton contamination into the antiproton signal [1]

centered beam. The available aperture is about 7 cm, the same as the accelerator beam pipe. The arc length and the spacing from the wall of the stainless steel pipe were selected to provide a uniform position response (including lack of sensitivity to motion in the orthogonal plane), while rigorously maintaining a 50 Ω transmission line impedance. RG-8 coaxial cables carry the signals from both ends of each pair of BPM pickups to VME racks in service buildings. In the VME crate are analog filter boards (53 MHz bandpass and attenuation), Echotek 8-channel 80 MHz digital receiver boards (ECDR-GC814-FV-A), as well as a Motorola MVME-2400 processor and a module providing clock and trigger signals. The new electronics were installed and commissioned bit by bit, usually between HEP stores, so that only a small number of BPMs would be affected at any one time. This strategy minimized the impact on operations and led to a successful implementation of the new system.

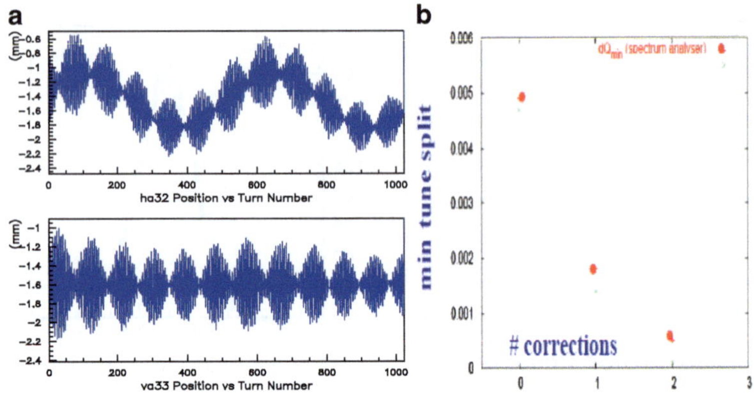

Fig. 9.3 (**a**) (*left*) Example of turn-by-turn measurements (*top*—horizontal, *bottom*—vertical) from the upgraded BPM electronics after intentionally kicking the proton beam. (**b**) (*right*) Measurements of minimum tune split during attempts to reduce coupling during machine tune-up. The *red points* are data from tune measurements made by looking at Schottky signals using a spectrum analyzer, while the *green points* are derived from turn-by-turn BPM measurements after kicking the beam. The turn-by-turn measurements achieve better results more quickly and more reliably than the spectrum analyzer method [1]

An example of the improved resolution of the new BPM electronics is shown in Fig. 9.2. The plots show distributions of the proton and antiproton closed orbit position measurements for one horizontal and one vertical BPM. The noted RMS values include the effects of true beam motion, e.g. synchrotron oscillations, and the imperfect cancellation of the proton signal onto the signal from the smaller intensity antiprotons. The intrinsic resolution from the BPMs themselves is ≈5 μm, much better than the 150 μm resolution of the old system.

The new electronics provide up to 8192 TBT position measurements at injection and on-demand. Figure 9.3 shows TBT measurements from one horizontal and one vertical BPM after intentionally kicking the beam horizontally in order to measure coupling during machine tune-up. The effect from coupling and synchrotron oscillations are clearly visible. The TBT capabilities are being exploited to develop faster and more reliable methods of measuring and correcting the beam optics.

The improved resolution of the BPMs has allowed better measurements of the machine optics which has led to lattice corrections and improvements [4, 5]. For example, the beta-functions can be measured to better than 5 % accuracy, and a new low-beta lattice with smaller $\beta^* = 28$ cm was created and increased luminosity by ≈10 %.

We have observed significant beam orbit motion during stores caused by motion of the low-beta quadrupoles and have been able to understand the source and implement an automated orbit-smoothing algorithm [6] that keeps the orbit from wandering during stores (see Fig. 9.4). In addition, the BPM response no longer

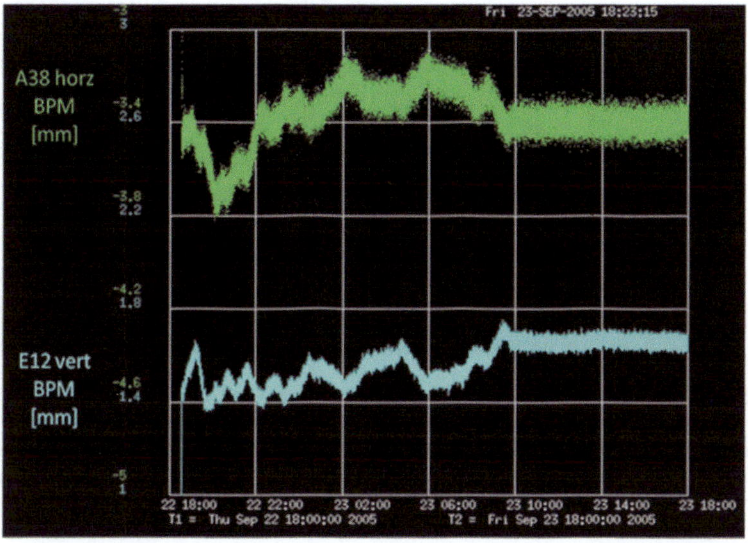

Fig. 9.4 Proton beam positions from a horizontal BPM (*top*) and a vertical BPM (*bottom*) over a 24-hour period during a high-energy physics store. The vertical scale is 400 μm per division. Before an orbit stabilization algorithm was enabled at 10:00, the orbit could wander by over 400 μm in a short period. After orbit stabilization was turned on, the orbit drift was reduced successfully to less than 50 μm. The algorithm uses several dipole correctors near the interaction regions to counteract motion of the low-beta quadrupoles caused by thermal and pressure differences between the Tevatron tunnel and the experimental halls [1]

depends on the bunch structure, so we can use orbit data from HEP stores to make global orbit corrections when needed.

9.2 Diagnostics for Low-Beta Quadrupoles and IPs

As mentioned above in Chapter 2.4, vibrations of low-beta quadrupoles are primarily responsible for orbit oscillations in the Tevatron, so we equipped each of the magnets with a fast 1 μrad resolution tiltmeter and 0.1 μm resolution hydrostatic level sensors (HLS) to detect vertical motion [7].

Some remarkable examples of orbit and magnet vibrations excited by fire trucks passing by the CDF building and remote earthquakes are shown in Fig. 9.5a, b. The HLS systems also track magnet motion due to continuous sinking of the CDF detector with rate of 0.25–0.5 mm/year. Such movements lead to a slow drift of the interaction point (IP) position inside the CDF silicon vertex detector (SVX). This and other beam-related information (like loss rates of various counters) can be monitored by Tevatron operators and physicists. For example, both CDF and D0 detectors provide data on the beta-functions at the IPs [8] (see Fig. 9.6) which is very helpful and

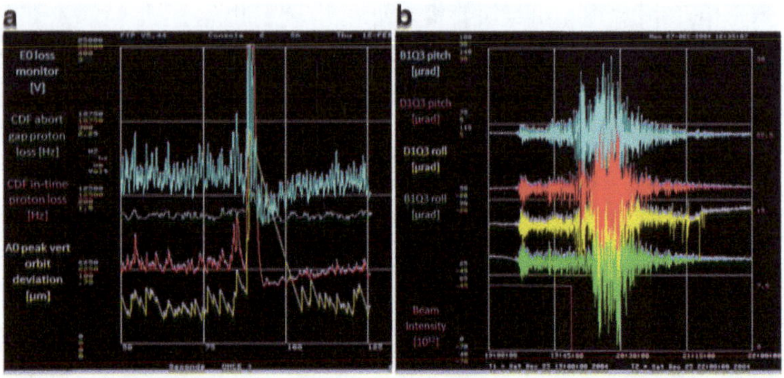

Fig. 9.5 (**a**) (*left*) 200 μm orbit oscillations, beam losses, and low-beta quadrupole vibrations excited by a 40,000 lb fire truck passing near the CDF Detector Hall. The effect was greatly reduced after installation of new quad girder supports in the 2005 shutdown; (**b**) (*right*) Disastrous M8.9 earthquake in Sumatra December 25, 2004 resulted in ±50 μrad motion as seen by the tiltmeters on CDF and D0 low-beta quadrupoles. The Tevatron beam (lower *red line* with a step down) was intentionally terminated before the arrival of the S-wave. The event lasted over 2 h [1]

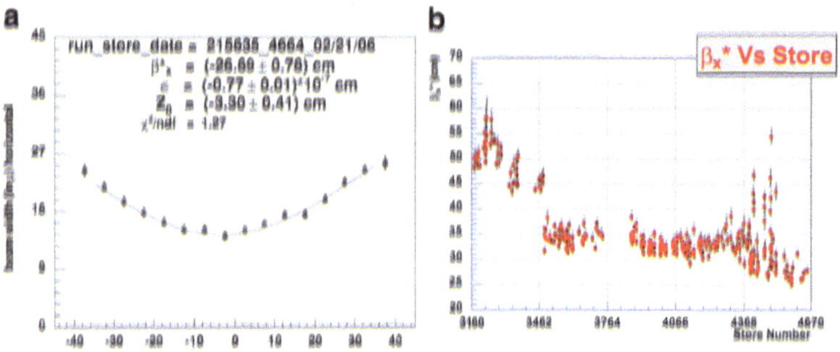

Fig. 9.6 (**a**) (*left*) RMS horizontal width of D0 luminous region vs longitudinal position. The parabolic fit is for hour-glass effect with $\beta^*_x = 27$ cm; (**b**) (*right*) January 2004–March 2006 history of the horizontal beta-function at the D0 IP measured by the D0 silicon vertex detector [1]

provides an additional insight into beam collision effects. Vertex analysis also allows separate determination of the proton and antiproton RMS bunch lengths [9].

9.3 Schottky Monitors

Measuring Schottky noise is a powerful nondestructive diagnostic for the Tevatron accelerators. It can be used for both continuous and bunched beams, in transverse and longitudinal degrees of motion, and for determination of a number of important beam parameters.

9.3.1 Schottky Noise in Continuous Beam

We start our consideration from the continuous beam. For the case of non-overlapping Schottky bands the spectral density of fluctuations of the beam current in the pickup in vicinity of n-th harmonic is [10]:

$$P_n(\delta\omega_n) = \frac{1}{\varepsilon_n(\delta\omega_n)} \frac{e^2 f_0 N \psi(\delta\omega_n/\omega_0 n\eta)}{2\pi|n\eta|}, \qquad \delta\omega_n = \omega_0 n\eta \frac{\Delta p}{p}. \tag{9.1}$$

Here N is the number of particles in the beam, $\eta = \alpha - 1/\gamma^2$ is the slip factor, $f_0 = \omega_0/2\pi$ is the revolution frequency, $\Delta p/p$ is the relative momentum deviation, and $\varepsilon_n(\delta\omega_n)$ is the beam dielectric function describing particle interaction. We also assume that the particle distribution function, $\psi(\Delta p/p)$, is normalized to 1, $\int\psi(x)dx = 1$, and that the spectral density and the correlation function are related by the following equation:

$$K_n(\tau) \equiv \langle I_n(t)^* I_n(t+\tau)\rangle_t = \int_{-\infty}^{\infty} P_n(\omega) e^{i\omega\tau} d\omega. \tag{9.2}$$

Here $I_n(t)$ is the complex amplitude of the beam current fluctuations, and * denotes a complex conjugate. The magnitude of beam current fluctuations has a Gaussian distribution:

$$f_I(|I|) = \frac{|I|}{K_n(0)} \exp\left(-\frac{|I|^2}{K_n(0)}\right), \tag{9.3}$$

while the rate at which the amplitude and phase of fluctuations are changing is determined by the correlation function of Eq. (9.2). It is assumed in Eq. (9.3) that the harmonic number takes both positive and negative values, $n = -\infty, \ldots \infty$. If one considers only positive n, the result of Eq. (9.3) has to be multiplied by factor of 2.

If the particle interaction is sufficiently small at the revolution harmonic where measurements are performed, the beam dielectric function is close to 1 and it can be neglected in Eq. (9.1). In this case the beam longitudinal distribution is proportional to the spectral density of beam current fluctuations. If, however, the particle interaction cannot be neglected, the analysis of the beam spectra becomes much more complicated task. For a low energy machines their longitudinal impedance is dominated by the beam space charge impedance. In this case $|Z_n/n|$ stays constant up to very high frequency, $\omega \approx \pi c\gamma/a$, and beam interaction is equally strong for all harmonics. Here a is the vacuum chamber radius, and γ is the relativistic factor. However for high energy machines $|Z_n/n|$ is usually decreasing with frequency and therefore measurements at higher frequencies can reduce an effect of particle interaction. The first work devoted to the extraction of beam parameters from the longitudinal Schottky noise of deeply cooled beam, where Schottky spectrum is completely dominated by particle interaction, was reported in [11]. Below we assume that the particle interaction is sufficiently small and can be neglected.

A choice of data acquisition technique used for measurement of beam spectrum depends on time available for measurements. If time is large enough, a general purpose scanning spectrum analyzer addresses the problem. If time allocated for a measurement is small, the beam signal is usually down-converted to a smaller frequency where a vector signal analyzer (VSA) can be used. If the beam distribution changes fast, the process is not repeatable, and a detailed evolution of beam distribution is desired, a direct digitization of down-converted beam noise by fast ADC with subsequent digital Fourier transform can be a better choice. In this case the entire process is recorded to a single data array resulting in no loss of information. After the measurements the data are split into smaller arrays, and the Fourier transform is performed for each of them. After averaging of nearby spectra one finally obtains a sequence of spectra representing an evolution of particle distribution.

Let us estimate the time required for a single measurement of the distribution function. Let the noise be digitized with sampling rate f_s. After the measurement the data are split into M arrays of length N; each array is subjected to a digital Fourier transform and corresponding spectral densities of the beam noise are computed for each array, $S_n = |a_n|^2$; and finally the obtained M spectral densities are summed to find an average. Such procedure determines that the maximum frequency in the spectrum is $f_s/2$, and the frequency resolution is f_s/N. After Fourier transform the complex amplitude of each harmonic, a_n, has a random values of phase and amplitude. The distribution over phase is uniform, and the distribution for modules of squared amplitudes, $|a_n|^2$, is described by Eq. (9.3). Typically the frequency resolution, $\delta f = f_s/N$, is chosen to be much smaller than a width of the spectrum $\Delta\omega$. In this case the decoherence time, $\sim 1/\Delta\omega$, is smaller than the time required to sample one array, $1/\delta f$, and, consequently, amplitudes of each next spectrum are statistically independent. Therefore the final spectral density obtained by averaging of M initial spectral densities has a relative value of statistical fluctuations equal to $1/\sqrt{M}$. Thus we finally obtain that the time required to obtain the distribution function with relative accuracy $\sqrt{\langle\delta\psi^2\rangle}/\psi$ is:

$$T = \frac{1}{\Delta\omega}\frac{\Delta\omega}{\delta f}\frac{1}{\langle\delta\psi^2\rangle/\psi^2}. \tag{9.4}$$

For a typical measurement the accuracy is ~10 % ($M = 100$) and a relative frequency resolution $\delta f/\Delta\omega \approx 0.06/(2\pi) \approx 0.01$. That results in that the required time is $\approx 10^4/\Delta\omega$. That determines that the measurement time is inversely proportional to the harmonic number.

In difference to the longitudinal Schottky noise taking place at the revolution harmonics the transverse Schottky noise comes at sidebands of the betatron frequencies, $\omega_{\perp n} = \omega_0(n + Q)$. For the case of non-overlapping Schottky bands the effective spectral density of beam current transverse noise is:

$$P_{\perp n}(\delta\omega_{\perp n}) = \frac{1}{\varepsilon_{\perp n}(\delta\omega_{\perp n})} \frac{e^2 f_0 N \psi(\delta\omega_{\perp n}/\omega_0(n\eta+\xi))}{2\pi|n\eta+\xi|} \frac{\overline{x^2}}{A^2}, \quad \delta\omega_{\perp n}$$
$$= \omega_0(n+Q)\frac{\Delta p}{p}, \tag{9.5}$$

where we assume that the transverse distribution does not depend on the momentum, $\sqrt{\overline{x^2}}$ is the rms value of particle transverse motion (in x- or y-plane), A is the effective aperture of pickup, Q is the betatron tune, and ξ is the tune chromaticity. The transverse Schottky noise is completely determined by the rms particle amplitude in the pickup and is not coupled to the details of transverse particle distribution. For high frequencies, $f/f_0 \gg \xi/\eta$, where the Schottky noise is less affected by particle interaction the distribution width is determined rather by the revolution frequency spread than by the betatron frequency spread. The noise integral over one transverse Schottky band is directly related to the rms beam size and, consequently, to the beam emittance; and the difference of widths for the positive and negative sidebands, $(n \pm Q) f_0$, (also called lower and upper sidebands) is directly related to the tune chromaticity:

$$\xi = n\eta \frac{\Delta f_+ - \Delta f_-}{\Delta f_+ + \Delta f_-}. \tag{9.6}$$

Here Δf_+ and Δf_- are the Schottky band widths for positive and negative betatron bands. Thus, measurements of transverse Schottky noise yield simple and noninvasive measurements of the beam emittance and the tune chromaticity.

9.3.2 Longitudinal Schottky Noise in Bunched Beam

In difference to the continuous beam the electromagnetic signal of bunched beam consists of two parts: the bunch coherent signal (observed on the revolution frequency harmonics) and the incoherent signal of particles (observed in vicinity of these harmonics).

Firstly, let us consider the spectrum of a single particle. We assume that the particle performs a synchrotron motion so that its position relative to the bunch center is described by a periodic function $z(t)$ with period $T_s = 2\pi/\omega_s$. Then, the spectrum of particle current is:

$$S(\omega) = \frac{e}{2\pi}\int_{-\infty}^{\infty} dt e^{-i\omega t} \sum_{n=-\infty}^{\infty} \delta\left(t - nT_0 - \frac{z(nT_0)}{\beta c}\right)$$
$$= \frac{e}{2\pi}\sum_{n=-\infty}^{\infty} \exp\left(-i\omega\left(nT_0 + \frac{z(nT_0)}{\beta c}\right)\right). \tag{9.7}$$

Here to find the time of pickup crossing by the particle we assumed that the synchrotron tune is small, $\nu_s \equiv \omega_s/\omega_0 \ll 1$. To compute the sum in the right-hand part of Eq. (9.7) we expand the exponential term in the Fourier series:

$$\exp\left(-i\frac{\omega z(t)}{\beta c}\right) = \sum_{m=-\infty}^{\infty} z_m \mathrm{e}^{im\omega_s t}, \quad z_m = \frac{1}{T_s}\int_{-T_s/2}^{T_s/2} \exp\left(-i\frac{\omega z(t)}{\beta c}\right)\mathrm{e}^{-im\omega_s t}dt. \tag{9.8}$$

Substituting Eq. (9.8) into Eq. (9.7) and using the following identities,

$$\sum_{n=-\infty}^{\infty} \exp(ixn) = 2\pi \sum_{n=-\infty}^{\infty} \delta(x - 2\pi n), \quad \int \delta(ax)dx = \frac{1}{a}\int \delta(x)dx, \tag{9.9}$$

we finally obtain:

$$S(\omega) = \frac{e}{T_0}\sum_{n=-\infty}^{\infty}\sum_{m=-\infty}^{\infty} z_m \delta(\omega - n\omega_0 - m\omega_s). \tag{9.10}$$

The beam current in the pickup can be presented as a sum of currents of individual particles:

$$I(t) = e\sum_k \sum_{n=-\infty}^{\infty} \delta\left(t - nT_{0_k} - \frac{z_k(nT_{0_k})}{\beta c}\right),$$

where the index k numerates particles (from 1 to N_{p}), and we assume that different particles have different synchrotron amplitudes, phases, and periods. Performing the inverse Fourier transform of each particle spectrum [see Eq. (9.10)] one obtains:

$$I(t) = e\sum_k \sum_{n=-\infty}^{\infty}\sum_{m=-\infty}^{\infty} \frac{z_{m_k}}{T_{0_k}} \exp(i(n\omega_{0_k} + m\omega_{s_k})t). \tag{9.11}$$

In the further consideration we will neglect the difference in particle periods in the denominator in the right-hand side of Eq. (9.11) ($T_{0k} \to T_0$). Taking into account the relationship between the spectral density of beam current and its correlation function:

$$P(\omega) = \frac{1}{2\pi}\int_{-\infty}^{\infty} K(\tau)\mathrm{e}^{-i\omega\tau}d\tau. \tag{9.12}$$

We obtain for the beam spectral density:

$$P(\omega) = \frac{e^2}{2\pi T_0{}^2} \int_{-\infty}^{\infty} \mathrm{e}^{-i\omega\tau} d\tau \sum_{k,\,l=-\infty}^{\infty} \sum_{n,\,p=-\infty}^{\infty} \sum_{m,\,q=-\infty}^{\infty}$$
$$\left\langle z_{m_k} z_{q_l}{}^* \exp(i(n\omega_0 + m\omega_{s_k})t) \exp(-i(p\omega_0 + q\omega_{s_l})(t+\tau)) \right\rangle_t$$
$$= \frac{e^2}{2\pi T_0{}^2} \int_{-\infty}^{\infty} \mathrm{e}^{-i\omega\tau} d\tau \sum_{k,\,l=-\infty}^{\infty} \sum_{n,\,p=-\infty}^{\infty} \sum_{m,\,q=-\infty}^{\infty}$$
$$\exp(-i(p\omega_0 + q\omega_{s_l})\tau) \left\langle z_{m_k} z_{q_l}{}^* \exp(i((n-p)\omega_0 + (m\omega_{s_k} - q\omega_{s_l}))t) \right\rangle_t, \tag{9.13}$$

where * denotes the complex conjugate, and $\langle \ldots \rangle_t$ denotes the averaging over time and initial distributions. The averaging over t leaves in the sum only terms with $k = l$ and $n = p$ if both m and q are not equal to zero. For $m = q = 0$ and $n = p$ all terms with $k \neq 1$ also contributes. That yields:

$$P(\omega) = \frac{e^2}{2\pi T_0{}^2} \int_{-\infty}^{\infty} \mathrm{e}^{-i\omega\tau} d\tau \left[\sum_{k,\,l=-\infty}^{\infty} \sum_{n=-\infty}^{\infty} \langle z_{0_k} z_{0_l}{}^* \rangle \mathrm{e}^{-in\omega_0\tau} \right.$$
$$\left. + \sum_{k=-\infty}^{\infty} \sum_{n=-\infty}^{\infty} \sum_{\substack{m=-\infty \\ m\neq 0}}^{\infty} \langle z_{m_k} z_{m_k}{}^* \rangle \mathrm{e}^{-i\left(n\omega_0 + m\omega_{s_k}\right)\tau} \right]. \tag{9.14}$$

After integration one obtains:

$$P(\omega) = \frac{e^2}{T_0{}^2} \left[\sum_{k,l} \sum_{n=-\infty}^{\infty} \langle z_{0_k} z_{0_l}{}^* \rangle \delta(\omega - n\omega_0) \right.$$
$$\left. + \sum_{k} \sum_{n=-\infty}^{\infty} \sum_{\substack{m=-\infty \\ m\neq 0}}^{\infty} \left\langle \left| z_{m_k} \right|^2 \right\rangle \delta(\omega - n\omega_0 - m\omega_{s_k}) \right]. \tag{9.15}$$

Note that Eq. (9.8) directly results in that for $m = 0$ the amplitude z_m is a positive real number which does not depend on the initial particle phase, while for $m \neq 0$ the amplitude is a complex number which argument depends on the initial phase of particle synchrotron motion. That yields that $z_{0_k} z_{0_l}{}^*$ is a real and positive number even for different particles ($k \neq l$). Averaging over initial distribution sets and replacing the summing in Eq. (9.15) by integration over the distribution finally results in:

$$P(\omega) = \frac{e^2}{T_0{}^2} \left[N_\mathrm{p}(N_\mathrm{p} - 1) \sum_{n=-\infty}^{\infty} \delta(\omega - \omega_{n0}) \left| \int dJ z_0(J,\omega) f(J) \right|^2 \right.$$
$$\left. + N_\mathrm{p} \sum_{n=-\infty}^{\infty} \sum_{m=-\infty}^{\infty} \int \delta(\omega - \omega_{nm}) \left| z_m(J,\omega) \right|^2 f(J) dJ \right], \tag{9.16}$$

where $\omega_{nm} = n\omega_0 + m\omega_s(J)$, J is the action of longitudinal motion so that $\int f(J)\,dJ = 1$, N_P is the number of particles in the bunch,

$$z_m(J,\omega) = \frac{1}{T_s(J)}\int_{-T_s(J)/2}^{T_s(J)/2} \exp\left(-i\frac{\omega z(J,t)}{\beta c}\right) \mathrm{e}^{-im\omega_s(J)t}\,dt, \tag{9.17}$$

and the function $z(J, t)$ describes a dependence of longitudinal particle position on time. The first addend in Eq. (9.16) represents the coherent signal of the bunch, and the second one represents the incoherent signal or the Schottky noise. Taking into account large number of particles in the beam we omit 1 in the product $N_p\ (N_p - 1)$ in further consideration

Using Eqs. (9.9) and (9.17) one can easily prove that in the general case

$$\sum_{m=-\infty}^{\infty} |z_m(J,\omega)|^2 = 1. \tag{9.18}$$

Integrating the spectral density of Schottky noise [second addend of Eq. (9.16)] over one revolution harmonic one obtains:

$$\int_{(n-1/2)\omega_0}^{(n+1/2)\omega_0} P_{ic}(\omega)d\omega = \frac{e^2}{{T_0}^2}N_p \int_{(n-1/2)\omega_0}^{(n+1/2)\omega_0} d\omega \sum_{m=-\infty}^{\infty} \int \delta(\omega - \omega_{nm})|z_m(J)|^2 f(J)dJ = \frac{e^2}{{T_0}^2}N, \tag{9.19}$$

where Eq. (9.18) was used to perform the summation over m. One can see that the Schottky noise integral over one revolution harmonic does not depend on particle motion and is equal to the integral of Schottky noise of continuous beam with the same number of particles.

The coherent part of beam signal presented by the first addend in Eq. (9.16) can be expressed as sum of revolution frequency harmonics

$$P_c(\omega) = \frac{e^2}{{T_0}^2}{N_p}^2 \sum_{n=-\infty}^{\infty} \delta(\omega - \omega_{n0})|A_n|^2, \tag{9.20}$$

where coefficients A_n can be expressed as corresponding Fourier harmonics of bunch longitudinal density as follows from the string of equations below.

$$A_n = \int dJ z_0(J, n\omega_0) f(J) = \int dJ f(J) \frac{\omega_s(J)}{2\pi} \int_{-T_s(J)/2}^{T_s(J)/2} \exp\left(-i\frac{n\omega_0 z(J,t)}{\beta c}\right) dt$$
$$= \int dJ f(J) \frac{\omega_s(J)}{2\pi} 2 \int_{-z_{\max}(J)}^{z_{\max}(J)} \exp\left(-i\frac{n\omega_0 z}{\beta c}\right) \frac{dz}{(dz/dt)_I} = \int dz f_z(z) \exp\left(-i\frac{n\omega_0 z}{\beta c}\right). \tag{9.21}$$

Here the factor of 2 appearing at the transition from integration over t to z is related to the integration over half of synchrotron motion period for the integration over z, and we used the following equation relating distribution functions over action, $f(J)$, and over longitudinal coordinate, $f_z(z)$,

$$f_z(z) = \frac{1}{\pi} \int dJ \frac{f(J)\omega_s(J)}{(dz/dt)_I}, \tag{9.22}$$

where $\int f_z(z) \mathrm{d}z = 1$.

The second addend in Eq. (9.16) represents the incoherent part of beam signal. For sufficiently high frequencies it can be reduced to the form which directly relates the spectral density to the distribution function over momentum. In this case the major contribution to the integral of Eq. (9.17) comes from the area where

$$m\omega_s = -\frac{\omega}{\beta c} \frac{dz(J, t_r)}{dt}. \tag{9.23}$$

There are two points in one revolution period where this condition is satisfied. That yields that Eq. (9.17) can be approximated as following:

$$z_m(J,\omega) = \frac{1}{T_s(J)} \sum_{r=1,2} \exp\left(-i\left(\frac{\omega z(J,t_r)}{\beta c} + m\omega_s(J)\right)\right) \int_{-\infty}^{\infty} \exp\left(-i\frac{\omega}{2\beta c} \frac{d^2 z(J,t_r)}{dt^2} \tau^2\right) d\tau, \tag{9.24}$$

where time t_r is determined by Eq. (9.23). Here and below we do not imply the symmetry of potential well. For a symmetric potential well the modulus of $d^2z(J, t_r)/dt^2$ would be equal for both resonance points (r=1, 2). Performing integration and squaring the result one obtains:

$$|z_m(J, \omega)|^2 = \frac{2\pi}{T_s(J)^2} \frac{\beta c}{\omega} \sum_{r=1,2} \left| \frac{d^2 z(J, t_r)}{dt^2} \right|^{-1}, \tag{9.25}$$

where we omitted the interference term between two addends. This interference term oscillates fast with m and J and makes negligible contribution after averaging over distribution. Substituting Eq. (9.25) to Eq. (9.16) one obtains:

$$P_{ic}(\omega) = \frac{e^2}{T_0^2} N_p \sum_{n=-\infty}^{\infty} \sum_{m=-\infty}^{\infty} \int \delta(\omega - \omega_0 n - \omega_s m) \frac{2\pi}{T_s(J)^2} \frac{\beta c}{\omega} \sum_{r=1,2} \left| \frac{d^2 z(J, t_r)}{dt^2} \right|^{-1} f(J) dJ. \tag{9.26}$$

At high frequencies the synchrotron number m is large and summing over m can be replaced by integration.[1] Introducing the momentum as $\widetilde{p} = dz(J,t)/dt$, replacing the integration variable m by $\widetilde{p}$ and using Eq. (9.23) one obtains:

$$P_{ic}(\omega) = \frac{2\pi e^2}{T_0^2} N_p \sum_{n=-\infty}^{\infty} \int \frac{\omega}{\omega_s(J)} \frac{d\widetilde{p}}{\beta c} \int \delta\left(\omega\left(1 + \frac{\widetilde{p}}{\beta c}\right) - \omega_0 n\right)$$
$$\left(\frac{\omega_s(J)}{2\pi}\right)^2 \frac{\beta c}{\omega} \sum_{r=1,2} \left| \frac{1}{d\widetilde{p}(J, p_r)/dt} \right| f(J) dJ. \tag{9.27}$$

Taking into account that the distribution function over momentum is equal to

$$f_p(\widetilde{p}) = \frac{1}{2\pi} \int \sum_{r=1,2} \left| \frac{1}{d\widetilde{p}(J, \widetilde{p})/dt} \right| \omega_s(J) f(J) dJ, \tag{9.28}$$

and performing integration one finally obtains:

$$P_{ic}(\omega) = \frac{e^2}{T_0^2} N_p \sum_{n=-\infty}^{\infty} f_p\left(\frac{n\omega_0 - \omega}{\omega_0 n} c\beta\right). \tag{9.29}$$

where $\int f_p(\widetilde{p}) d\widetilde{p} = 1$. Note that by definition the momentum $\widetilde{p}$ is related to the particle momentum deviation as $\widetilde{p} = R_0 \eta \omega_0 \Delta p/p$, where R_0 is the average machine radius.

Although the functional form of Eq. (9.16) is quite different from Eq. (9.1) the corresponding spectral densities are numerically close if frequency is sufficiently large as it follows from Eq. (9.29). To demonstrate the considered above formalism we consider Schottky spectra for cases of two quite different potential wells: (1) a barrier bucket with the rigid walls and (2) a parabolic well where particle motion is linear and the synchrotron frequency does not depend on frequency.

In the case of barrier bucket a dependence of particle position on time during one synchrotron period can be expressed in the following form:

[1] It implies that the resolution of the measurements is larger than the synchrotron frequency, ω_s, or sufficiently large dependence of synchrotron frequency on the action, J, so that an averaging over particle distribution makes a smooth spectral density.

$$z_m(\Delta p, t) = L_b\left(\left|\frac{2t}{T_s(\Delta p)}\right| - \frac{1}{2}\right), \quad |t| < \frac{T_s(\Delta p)}{2}, \tag{9.30}$$

where L_b is the span of longitudinal particle motion, $T_s(\Delta p) = 4\pi\kappa_b/(\omega_0\eta\Delta p/p)$ is the synchrotron period, $\kappa_b = L_b/2\pi R_0$ is the fraction of the orbit taken by the particle motion. Substituting Eq. (9.30) to Eq. (9.17) and performing integration at frequencies $\omega_{nm} = n\omega_0 + m\omega_s(\Delta p)$ one obtains the spectrum of a single particle motion:

$$z_{nm} = \mathrm{e}^{-i\pi m/2}\frac{\sin\left(\pi(m - 2\kappa_b n)/2\right)}{\pi(m - 2\kappa_b n)} + \mathrm{e}^{i\pi m/2}\frac{\sin\left(\pi(m + 2\kappa_b n)/2\right)}{\pi(m + 2\kappa_b n)}. \tag{9.31}$$

One can see that the amplitudes z_{nm} do not depend on $\Delta p/p$ but the spectrum width is proportional to $\omega_s(\Delta p)$ and, consequently, to $\Delta p/p$. In the case of continuous beam the single particle spectrum in vicinity of n-th harmonic is presented by a single line shifted by $n\omega_0\eta\Delta p/p$. In the case of particle in the barrier bucket the particle is moving back and forward with momentum deviation of $\pm\Delta p/p$ but instead of being presented by only two lines with frequency shifts $\pm n\omega_0\eta\Delta p/p$ its spectrum is presented by large number of lines. However for sufficiently large frequency, $2\kappa_b n \gg 1$, the relative width of this spectrum is narrow and the spectrum is peaked in vicinity of these two lines. For harmonic numbers above $n \gg 1/\kappa_b \equiv 2\pi R_0/L_b$ the spectrum of bunched beam is close to its momentum distribution.

In the case of parabolic potential well a dependence of particle position on time during one synchrotron period is:

$$z_m(\Delta p, t) = \frac{L_b}{2}\cos\left(\omega_s t\right). \tag{9.32}$$

Substituting it to Eq. (9.17) and performing integration one obtains the spectrum of a single particle motion:

$$z_{nm} = \mathrm{e}^{-i\pi m/2} J_m(\pi\kappa_b n), \tag{9.33}$$

where the same as for a beam in the barrier bucket, $\kappa_b = L_b/2\pi R_0$. A larger spectrum width for the case of parabolic well is related to the $\pi/2$ larger maximum momentum deviation for the same values of bunch length and synchrotron period. One can see that $|z_{nm}|$ is a fast oscillating function of m. However in practical applications the observed spectrum is created by signals of many particles having different amplitudes which effectively smears these fast oscillations in the spectrum. Applying Eq. (9.25) to the parabolic potential well one obtains the high frequency asymptotic

$$\left|z_{nm}\right|_{\mathrm{HF}}^2 = \frac{1}{\pi^2\kappa_b n\sqrt{1 - (m/\pi\kappa_b n)^2}}, \quad |m| < \pi\kappa_b n. \tag{9.34}$$

which describes the averaged distribution sufficiently well.

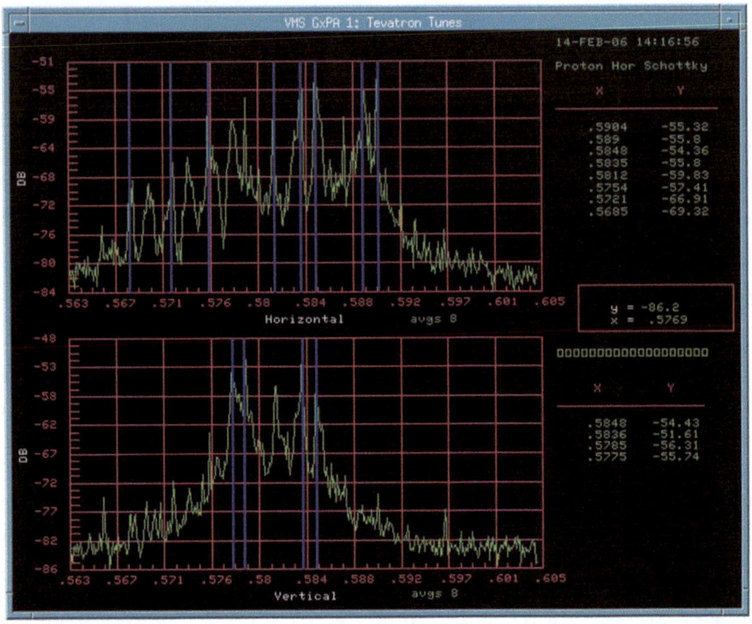

Fig. 9.7 Schottky spectra from the 21.4 MHz pickups. The spectrum typically contains many peaks, and it may be hard to determine which one is the tune. The system is also unable to resolve antiprotons, due to the much stronger proton signal [1]

Thus, in most practical applications the Schottky noise properties of the bunched beam are not much different from the case of continuous beam. If frequency is not sufficiently high ($f < \ln(N_p)/(2\pi\sigma_\tau)$), the coherent peak at the revolution frequency harmonics is well visible but does not distort the spectrum if electronics has sufficient dynamic range. Here σ_τ is the rms bunch duration. However, in the case of intense beams a small amplitude coherent longitudinal motion can persist for very long time (many hours in the case of normal colliding bunches in Tevatron at 980 GeV). In this case an observation of "true" Schottky noise requires significantly higher frequencies than is determined by the above estimate.

9.3.3 21 MHz Tevatron Schottky Detectors

There are several systems that measure tunes in the Tevatron. The 21 MHz Schottky monitor [12] is the workhorse for tune measurements during shot setup and studies. The tune is determined by the operator, looking at the Schottky spectrum on a signal analyzer in the Control Room—see Fig. 9.7. The result may be somewhat subjective, since the spectra typically contain numerous coherent peaks, and it may not be

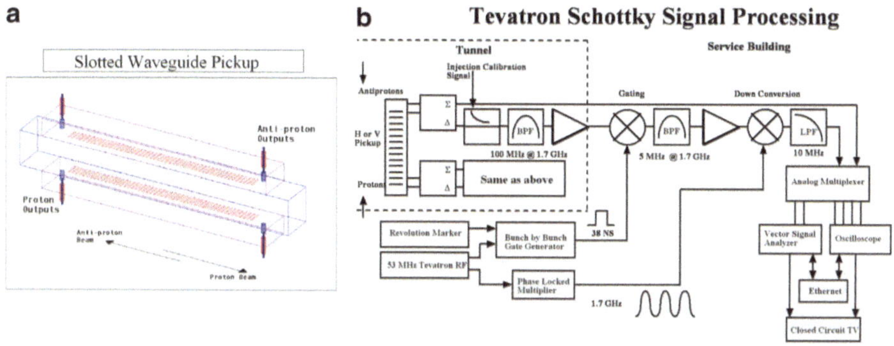

Fig. 9.8 (**a**) (*left*) Schematic of a 1.7 GHz slotted waveguide pickup. (**b**) (*right*) Schematic of the electronics and readout system [14]

immediately clear which one (if any) represent the real tune. To enhance the signal and make the tunes visible, noise can be injected into the beam through the transverse damper system.

The 21 MHz Schottky system was originally designed with movable pickup plates to maximize sensitivity. The original incarnation also had two pickups which could be added with a variable phase to suppress the proton tune in favor of the antiproton tune, although the practical usefulness of this feature in operation was very limited. The pickups are resonant with a tunable resonance frequency. This was intended mainly to compensate for the change in capacitance when the plates were moved. In Run II, the plates are left in a fixed position, and hence tuning is only done occasionally. The 21 MHz Schottky system is used to measure the horizontal and vertical tunes of the proton beam (antiproton signal is attenuated by 20 dB) without the possibility of gating on individual bunches. It has high resolution of about 0.0001.

9.3.4 1.7 GHz Tevatron Schottky Detectors

The 1.7 GHz Schottky pickups [13] are slotted waveguide structures (see Fig. 9.8). The high operating frequency was chosen to be above the coherent spectrum of the beam, thus measuring "true" Schottky signals. Since the devices are not resonant, it is possible to gate on select bunches, making it possible to measure the antiproton tune in the presence of protons. Chromaticity, momentum spread, and emittance can also be extracted from the signals, making the 1.7 GHz Schottky a very versatile tool [14].

An advantage of these pickups is that they can be used to measure tunes during normal operation without additional beam excitation. In order to maximize the usefulness of these devices, open access client (OAC) software was developed to

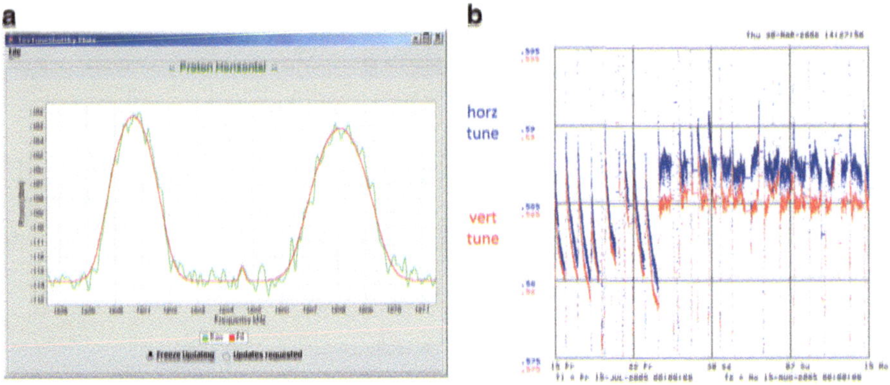

Fig. 9.9 (**a**) (*left*) Schottky spectra from the 1.7 GHz Schottky, showing both upper and lower betatron sidebands; (**b**) (*right*) Logged antiproton tune values over several weeks showing the effect of tune compensation during stores. Before the operators started using the 1.7 GHz Schottky tune readback to compensate for tune drifts, the tune would change significantly during a store as a result of the decreasing beam–beam parameter (*top line*—horizontal tune, *bottom*—vertical tune) [1]

run continuously, analyze the data, and publish the resulting tune, chromaticity, momentum spread, and emittance on ACNET. Among other things, this allows the tunes to be logged. The 1.7 GHz tune readings are also used in everyday operation to adjust the antiproton tunes as the beam–beam tune shift changes over the course of a store (see Fig. 9.9).

A peculiarity with the system is that due to the high frequency, the Schottky bands are very wide, and therefore it is not possible resolve the normal modes by frequency. The effect causes an underestimation of the tune separation in the presence of coupling (it can be shown that it approximately measures the uncoupled tunes). Though the 1.7 GHz Schottky monitors are capable of measuring the horizontal and vertical tunes of individual proton and antiproton bunch but to get to the precision of 0.0001, a few minutes of averaging time is required.

One of the original reasons for developing the system was to be able to extract emittance from the Schottky spectrum during stores. However, it has been observed that even in the microwave range, the Schottky spectrum still has a significant coherent contribution. The reason for this has yet to be fully understood. In the meantime, new thinner carbon filaments have enabled the use of the "flying wires" (see below) during stores, reducing the need for Schottky pickups for this particular task.

Identical 1.7 GHz Schottky detectors are used to monitor the betatron band power of the 8 GeV antiproton beam in the Recycler ring. The computed betatron power is scaled to reflect the transverse emittance of the beam based on the constants obtained from calibration with the mechanical scrapers.

9.4 Other Tune Diagnostics

9.4.1 *Digital Tune Monitor and Tunetracker*

It has been estimated that the beam in the Tevatron oscillates with ≈0.1 μm amplitude at betatron frequencies; at lower frequencies, the oscillation amplitudes can be larger ~1–10 μm due to various noise sources, not all well known, including ground motion, jitter in magnet and separator power supplies, and vibrations from the cryogenic system (see Sect. 2.4). In an attempt to use this effect to measure the tune without excitation, a very sensitive system called Digital Tune Meter (DTM) has been developed. This system is quite similar to the 3D-BBQ (Direct Diode Detection Base Band Tune) system developed at CERN in that it uses a diode-based sample and hold circuit, but it includes some novel features. Rather than measuring only the positive or negative peak from a stripline doublet, it measures both and takes the difference. It also employs slow feedback to remove baseline variations that can be quite large, thus enhancing the dynamic range. Fourier transformation of the position signal allows to determine the tune. The RMS position resolution of the DTM in the FFT averaging mode was estimated to be of the order of 100 nm. The averaged spectra for all 36 individual bunches are available from the DTM in about 80 s (2.2 s are required to record 25 sets of 4,096 turns [15]). The DTM has been successfully tested with proton beam, showed very good tune resolution for individual bunches of about 0.0002 at a low level of beam excitation by external noise, and has been used during dedicated beam studies [16].

The Tunetracker system uses a phase-locked loop (PLL) around the beam response. The beam is excited at a given frequency using a stripline pickup as a kicker, and the response is measured on another stripline. The PLL locks to a given frequency in the tune spectra, defined by a pre-selected phase response value, and tracks any changes in the tune. A typical tune tracking rms error is of about 0.0003–0.0006. A novelty in the Tevatron system compared to previous tune tracker implementations is the capability of pulsed excitation. When measured with high resolution, the beam phase response exhibits large excursions from the synchrotron sidebands, which can cause the PLL to jump from one synchrotron band to another. By pulsing the excitation, the measured phase response is smoothed out to follow the slow underlying phase response more closely, resulting in a more reliable measurement [17].

9.4.2 *Tune Measurements from Turn-by-Turn Data*

Tune determination from turn-by-turn beam position measurements after a transverse kick still remains the most reliable diagnostics in many Fermilab accelerators. In fast ramping synchrotrons like the Fermilab Booster the conventional method of betatron tune evaluation does not work due to rapid changes of the tunes (sometimes in a course of a few dozens of turns) and a high level of noise. Standard method for tune evaluation using fast Fourier transform (FFT) has resolution ~1/N, N being the number of turns, which is insufficient in the case of rapid changes of the tunes and/or fast decoherence of the betatron oscillations. Much better precision can be achieved with the so-called Continuous Fourier Transform (CFT) method [18] which consists in evaluation of the sum:

$$X(\nu) = \frac{1}{N}\sum_{n=1}^{N} x_n e^{-2\pi\nu(n-1)} \tag{9.35}$$

as a function of continuous variable ν and finding the maximum of $|X(\nu)|$. In absence of random noise CFT provides precision ~1/N^2. There are methods—i.e. the Hanning windowing technique—which further improve precision, up to 1/N^4, but they fail in the presence of noise. In the case of white noise the CFT provides tune with the r.m.s. error of $\sigma_\nu \approx \frac{\sqrt{6}\sigma}{\pi N^{3/2} a}$ where σ is the r.m.s. value of BPM errors and a is the betatron oscillations amplitude, which is better than the FFT error even in absence of noise. But it may be still not enough in a situation when the noise level is high and only a small number of turns are available.

A new technique based on phasing of signals from a large number of BPMs which significantly increases the signal to noise ratio has been proposed in [19] and implemented for routine operational use in the Booster. The turn-by-turn position data from many (all) BPMs are used to construct a phased sum (phasor)

$$\widetilde{x}_n = \sum_{k=1}^{N_{\mathrm{BPM}}} x_n^{(k)} \mathrm{e}^{-i\varphi(k)} \tag{9.36}$$

where k being the BPM index and $\phi(k)$ is the betatron phase advance between. It was shown that the phase advances from a theoretical optics model can be used as effectively as real measured ones. Performing the CFT analysis of the phased sum gives the tune value. One can easily see that the proper part of the signal propagating with expected phase advance is amplified by a factor of N_{BPM} whereas the alien modes and random noise are amplified only as $\sqrt{N_{\mathrm{BPM}}}$ so that the signal to noise ratio is improved by a factor of $\sqrt{N_{\mathrm{BPM}}}$.

In the Booster, a kicker—horizontal or vertical—is set up to kick the beam every 500 turns. On completion of the ramp the application reads out the turn by turn BPM data for all turns and all BPM's. The horizontal or vertical BPM's readings are then combined for each turn according to Eq. (9.36) and the moments of pings on the beam are identified by the oscillation. CFTs are then performed for each ping. Each spectrum is normalized so that the peak value is the same for all pings. Figure 9.10 compares contour plots of the CFT spectra obtained from a single BPM and from the phased sum. The achieved clarification of the spectra allowed

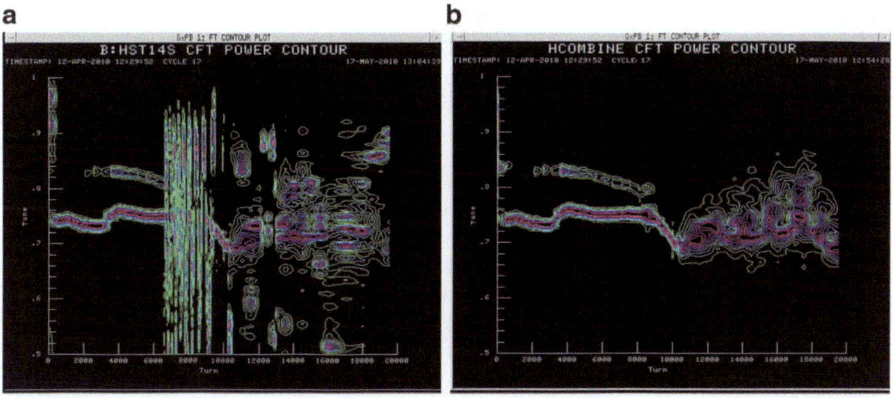

Fig. 9.10 (**a**) (*left*) Contour plot of the CFT spectra of the horizontally pinged beam obtained from one BPM; (**b**) (*right*) Contour plot of the CFT spectra of the phase-combined sum from 24 BPMs [19]

better tuning of the Booster which resulted in noticeably improved performance. The remaining fuzziness in the second part of the ramp is a result of systematic noise probably produced by the BPM electronics.

9.5 Data-Logging and On/Offline Presentation

Online and offline access to the vast amount of accelerator data is crucial to evaluating and improving machine performance and diagnosing failures. Retaining bunch-by-bunch values is especially useful since the beam dynamics vary over the bunch positions within a train. In the Tevatron collider complex, the readings and settings of accelerator devices are obtained via Fermilab's own ACNET control system. Device data can be plotted live at up to 1,440 Hz. Device data can be logged at various fixed rates or periods, e.g. 15 Hz or 1 min, or on a specific event, e.g. when the energy ramp is complete. Logged data is stored in circular buffers on ~70 nodes hosting a MySQL database and ~80 GB of storage for compressed data. The data in the circular buffers wrap-around in a time that depends upon the number of devices and their logged rate for a given logger. Logged data up to a 1 Hz maximum rate is also copied to a "backup" logger for long-term storage.

There are several means of accessing and plotting accelerator data: standard C-language based console applications used in operations, Java applications via a web-based interface, exports to Excel spreadsheets and Java Analysis Studio files, as well as programmatic APIs. Each method has its own advantages and disadvantages, but the flexibility allows users, both on-site and off-site, to access the data

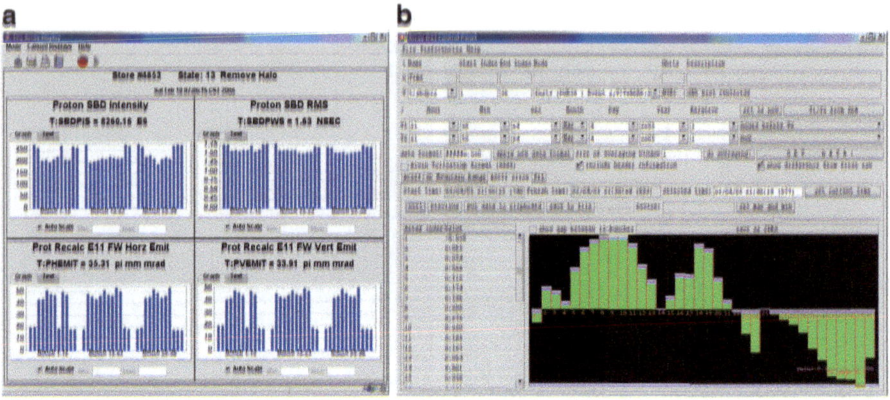

Fig. 9.11 (**a**) (*left*) The window of a Java applications showing live, bunch-by-bunch data for Tevatron proton bunches including intensity, RMS bunch length, and transverse emittances; an instability during antiproton injections had caused emittance growth and beam loss for particular proton bunches. (**b**) (*right*) A Java application showing a snapshot of logged proton bunch centroid positions within their RF buckets; a longitudinal, coupled-bunch instability with $\pm 4^{\circ}$ of RF phase oscillation amplitude was occurring at the time [1]

how they want or need. Figure 9.11 shows two examples of accessing Tevatron data.

In addition to the above data-logging scheme, data for all Tevatron shots is automatically collected and stored via a package called SDA, for Sequence Data Acquisition. The desired data and plots for all stages of a shot (injection, low-beta squeeze, etc.) can be easily configured. SDA also stands for Shot Data Analysis; SDA software automatically generates summary reports and tables for each store. These data are readily accessible by various means and allow for convenient analysis of the accelerator complex on a shot-by-shot basis [20].

9.6 Longitudinal Beam Diagnostics

The Tevatron resistive wall current monitors (WCM) [21] consists of a short ceramic vacuum pipe with 80 resistors across it. A copper casing around the ceramic break filled with ferrite provides a low impedance bypass for DC currents while forcing AC currents to flow through the resistors. There are also ferrite cores inside the vacuum to improve the signal quality. With the larger number of bunches and higher intensities, the resistive wall monitors developed vacuum problems early in Run II. The cause was beam-related heating of the ferrite cores. This problem was solved by replacing the ferrite with a different type. Signals from four locations around the ceramic pipe are summed to provide an intensity measurement. There are two resistive wall monitors in the Tevatron. One is dedicated to the Fast Bunch Integrator (FBI) and

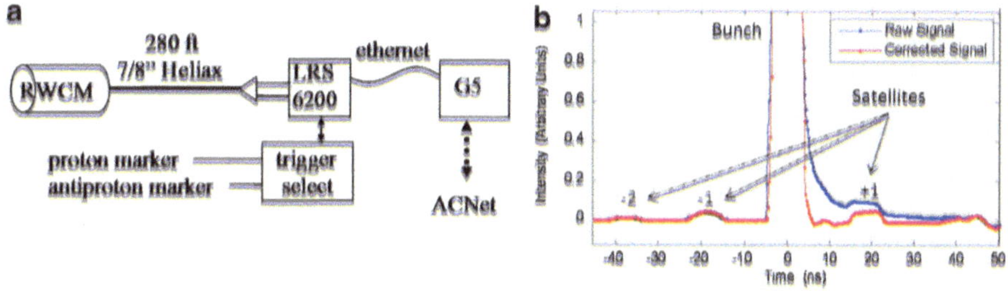

Fig. 9.12 (**a**) (*left*) Schematics of the signal acquisition from resistive wall current monitor. (**b**) (*right*) Raw and corrected signals for a Tevatron proton bunch and satellites. One can easily see the effect of removing the dispersion [1]

Sampled Bunch Display (SBD), and the other is for general use. This is to avoid errors in intensity readings due to improper terminations.

The schematic of the Tevatron SBD [22] is shown in Fig. 9.12a. The source of the signal is a wide-band resistive WCM installed in the F0 straight section of the Tevatron which also houses the RF cavities. In the Tevatron, since protons and antiprotons circulate oppositely within the same beampipe, the WCM is located at F11 where the proton and antiproton bunches are maximally separated (~30 m from the center of the straight section). The WCM has a >4 GHz bandwidth with a ~1 Ω gap resistance. The WCM signal travels from the tunnel to an above ground service building through a 7/8 in. Heliax cable. There it is split and digitized by two separate channels of LeCroy Corporation Wave Runner 6200 series digital scope having a 2 GHz bandwidth and 5 GSamples/s sampling rate. The splitter produces two identical signals which are captured by high-gain and low-gain channels in the oscilloscope that allows to effectively increase the dynamic range. In the Tevatron, where the coexisting protons and antiprotons vary in intensity by more than a factor of 3, the disparate gains allow simultaneous measurements of both particle species with equal sensitivity. The oscilloscope trigger generation is controlled by a separate board which is programmed to send 64 triggers to the oscilloscope, once every other revolution. The first 32 are triggered by the proton beam marker and the second 32 are triggered by the antiproton beam marker to obtain the relative antiproton timing. After digitizing the total number of beam revolutions, the digitized data is sent to an Apple Mac Pro computer for signal processing (combining of the 2 different gains to form one signal, correction of the dispersion in the cable from the WCM to the oscilloscope, and subtraction of the baseline) and extraction of longitudinal beam parameters (individual bunch intensity, the mean phase and rms width, the skew and kurtosis, and the emittance and momentum spread). All the processing is done within the National Instruments LabVIEW framework and the results are made available to the Accelerator Controls Network (ACNet) once a second continuously.

The longitudinal phase monitor (LPM) [23] is using the signal from a stripline pickup. The original idea was to use the antisymmetric shape of the bunch signal

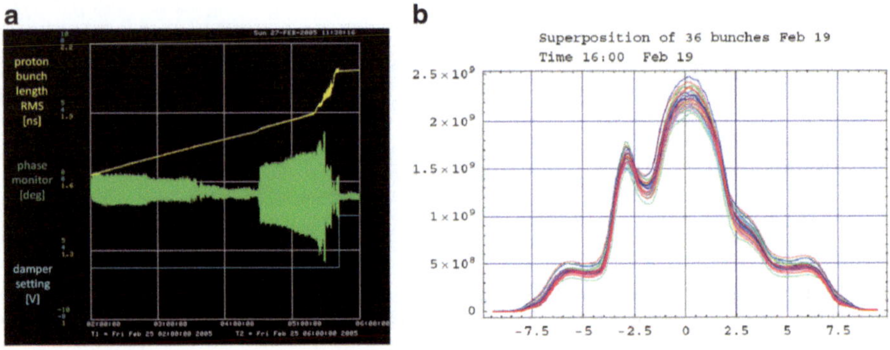

Fig. 9.13 (**a**) (*left*) Longitudinal phase monitor readings during an instability. (**b**) (*right*) Shapes of all 36 proton bunches as detected by the Tevatron SBD after longitudinal instability had developed [1]

from a stripline pickup, by multiplying it with a sine and cosine function locked to the RF. The two signals are then integrated over the bunch, and the phase can be extracted from the ratio of the two integrals. This was implemented in analog electronics using mixers and gated integrators, and the result was digitized and processed in an FPGA. The FPGA calculated the average over all bunches and output it as an analog voltage through a DAC. Turn-by-turn values were also saved in buffer memory and could be retrieved via an Ethernet interface (Fig. 9.13).

9.7 Abort Gap Monitors

Longitudinal instabilities, RF noise, and intra-beam scattering can cause particles to leak out of RF buckets and into satellites or into the abort gaps [24, 25]—see Fig. 9.14. There are three 2.6 μs gaps between 3 trains of 12 bunches each separated by 396 ns. The presence of even a small fraction (few 10^9 or 0.0001 of the total) of the beam in the abort gaps can induce quenches of the superconducting magnets, as these particles are sprayed onto the magnets when an abort kicker fires, and inflict severe radiation damage on the silicon detectors of the CDF and D0 experiments. Synchrotron radiation (SR) from these unwanted 980 GeV protons is collected for monitoring their intensity.

A very sensitive gated monitor of the SR from the beam in the gap was developed on the base of Hamamatsu R5916U-50 micro-channel plate (MCP) PMT with a minimum gating time of 5 ns. This tube can be used to measure DC beam intensity immediately following a bunch of protons. The DAQ system consists of a fast integrator, to which the anode of the PMT is connected, and a VME digitizer that is read by an application on a processor board residing in the

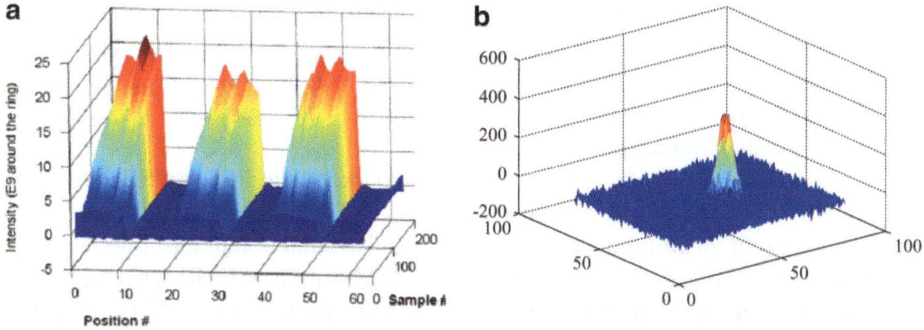

Fig. 9.14 (**a**) (*left*) Time evolution of beam in the gaps and between main bunches during a store (sample #200 corresponds to about 20 h, position # reflects 340 ns gap synchronization with respect to the revolution marker. Intensity is given in the units of equivalent number of particles if uniformly distributed around the circumference). (**b**) (*right*) Profile of DC beam in the gap imaged by a CID camera [1]

VME crate. Data is collected for 1,000 revolutions and averaged in the processor board. This cycle is repeated every 3 or 4 s. The application controls the timing of both the PMT and integration boards. Figure 9.14a shows how the intensity of the beam outside of the main 36 bunches is growing over a course of a HEP store. A standard synchrotron light monitor equipped with an image intensifier can see the DC beam profile, but only if enough camera frames can be summed together. A LabVIEW-controlled Windows PC system does such integration of CID camera RS-170 video images captured by a frame-grabber card. Figure 9.14b presents an example of the proton DC beam profile in the Tevatron. Details of calibrating and measuring the intensity of beam in the abort gap using synchrotron light and a gated photomultiplier tube can be found in [26].

9.8 Intensity Measurements

In the Tevatron, a DC Current Transformer (DCCT) and a Resistive Wall Current Monitor (RWCM or RWM) are the pieces of instrumentation that allow beam intensity measurements [27, 28]. The DCCT can only provide a measurement of the total beam intensity (sum of proton and antiproton currents). The RWM does distinguish between protons and antiprotons because of its high bandwidth and location where the protons and antiprotons are well separated in time.

The DCCT front-end contains an Interactive Circuits and Systems (ICS) ICS-110BL-8B 24-bit, 8-channel ADC to digitize the DCCT signal and a Motorola MVME-2401 processor. The ADC samples at 6.9 MHz and outputs a 128-sample average measurement at 54 kHz. The crate CPU performs additional averaging and provides the interface to ACNET. There is also a circular buffer that can be stopped upon a beam abort in order to help diagnose the cause of beam loss. The DCCT provides the most precise intensity measurements with a resolution of $\approx 0.5 \times 10^9$ for typical Tevatron total beam intensities of 10^{11}–10^{13} particles. The DCCT is calibrated via an external pulser.

Bunched-beam intensity measurements are made by the FBI and SBD systems, both of which use the RWM as their signal source. The FBI uses ADCs to integrate the RWM output gated on the individual RF buckets and obtain baseline measurements taken in the gaps between each train. A Motorola MVME-2401 processor performs the baseline correction and acts as the interface to ACNET. The FBI system provides narrow-gate (single bucket) and wide-gate (five buckets) intensity measurements for all proton and antiproton bunches at a rate of up to a few hundred Hz. Comparing the narrow and wide-gate values provides a measure of the intensity of satellite bunches, typically a few percent of the main bunch intensity.

The SBD configuration was described previously. The resolution of the bunch intensity measurements is $\approx 0.5 \times 10^9$ for present typical intensities of $20–80 \times 10^9$ for antiprotons and $240–300 \times 10^9$ for protons. The SBD can update measurements at approximately 1 Hz rate.

Both the FBI and SBD intensities can be calibrated via the very well-known measurement provided by the DCCT via the equation:

$$I_{\mathrm{DCCT}} = I_{\mathrm{P,true}} + I_{\mathrm{A,true}} = I_{\mathrm{P,RWM}} \times \left(1 + \frac{I_{\mathrm{A,RWM}}}{I_{\mathrm{P,RWM}}}\right) \times A_{\mathrm{calib}}, \tag{9.37}$$

where the DCCT intensity should be equal to the sum of the measured bunch intensities. A few percent correction needs to be made for satellites and other beam observed by the DCCT but not the FBI or SBD. This method requires no knowledge of the RWM, but only the relative gains of the proton and antiproton channels of the system being calibrated.

9.9 Beam Loss Monitoring System

Careful diagnostics and control of beam losses is crucial for the operation of the collider. Beam-Loss Monitors (BLM) are installed at each quadrupole magnet and at a number of critical components. The physical particle detector is an Ar-filled glass cylindrical ionization chamber with nickel electrodes and was chosen to be extremely radiation hard [29]. The gas is 1 atm pure Ar with an active volume about 110 cm^3. No CO_2 was used, because CO_2 could eventually dissociate under ionizing radiation and alter the response. The detector calibration is 70 nC/Rad and is extremely stable. The ionization chamber design was a compromise among high gain, fast response, size, and low cost. The design included placing the anode (center electrode, positive high voltage) and cathode (outer electrode, signal) feedthroughs at opposite ends of the glass bottle with a guard ring on the glass envelope to minimize the dark current leakage (roughly 100 pA at 2,500 V). Pulsed-beam measurements at Argonne National Lab showed that the chamber could detect an instantaneous radiation dose (in 1 or 2 μs) of about 10 Rads with less than 20 % charge loss. This large amount of space charge (about 700 nC) severely modifies the electric fields in the gas as the ions and electrons drift toward the electrodes, leading to both recombination losses and gas multiplication. For comparison, the inter-electrode capacitance is about 2 pF, leading to a displacement charge on the electrodes of about 5 nC at 2,500 V. The signal from the cathode is brought out of the tunnel enclosure on RG-58 cable, which is not particularly noise resistant. Complicating the noise situation is the legacy high voltage distribution which is supplied via a single RG-58 cable daisy chained through typically 12 ionization chambers producing myriad ground loops with the signal cables. Additionally, noise sources vary depending on the location within the accelerator. To combat the noise conditions, a variety of schemes are utilized to reduce the noise pickup to acceptable levels. An inductive choke is used on each signal cable just before the acquisition electronics to reduce large common-mode noise pickup on the cable. The size of the choke was chosen to give the proper frequency response. In parallel with the hardware efforts at noise reduction, a software algorithm was developed to subtract coherent noise from the signal in cases where the analog noise rejection techniques were insufficient [30].

The original BLM system ca 1980s was not adequate to the challenges of high luminosity Collider Run II operation, the BLM system was disabled during collider operation and protection of the Tevatron magnets relied on the quench protection system. In 2008, it was replaced by a sophisticated VME-based data acquisition system [30] which allowed appropriate abort logic and abort thresholds on losses to be set over the full set of collider operating conditions. The upgrade also provided better post abort diagnostics through its turn-by-turn buffer and the three buffers of integrated loss types. The upgrade has proven to be robust resulting in less down-time and maintenance requirements.

9.10 Chromaticity Diagnostics

High intensity proton and antiproton beam stability and lifetime depend strongly on machine chromaticity [31]. So, one of the operational challenges in the Tevatron is to measure accurately and control both vertical and horizontal chromaticities so that they are high enough to keep beam stable, yet low enough to avoid high beam losses. Several methods are employed.

The standard one—observation of the tune change while changing RF frequency—works well and is accurate to ~0.5 unit of Q' with ± 40 Hz change of the $f_{RF} = 53.1$ MHz if measured by 21 MHz Schottky tune detector. A much faster,

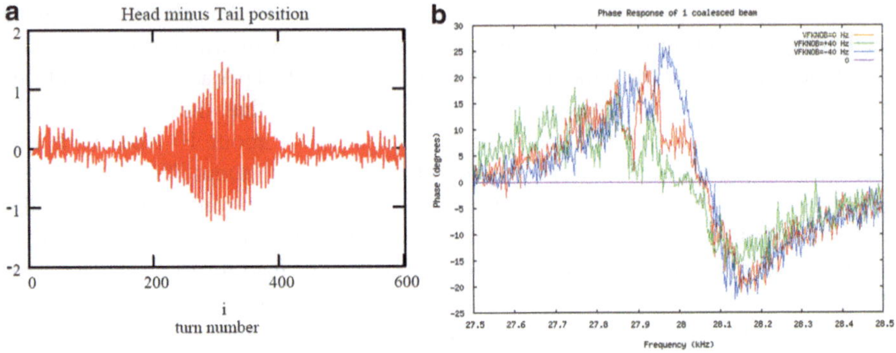

Fig. 9.15 (**a**) (*left*) Differential motion (in mm) between the head (+4 ns off bunch center) and tail (−4 ns) of a high intensity proton bunch in the Tevatron at 150 GeV vs turn number after a 1 mm vertical kick. (**b**) (*right*) Tune Tracker Beam Response Function spectra measured with $\approx 200 \times 10^9$ per bunch at 150 GeV, the zero-crossing frequency (betatron frequency) varies with the RF frequency change of ±40 Hz

but just as accurate head-tail method has been developed [32]. In that method, beam is kicked (causing a slight ~5 % emittance growth) and the differential motion of bunch head and bunch tail, as measured by a stripline pickup, is recorded by a fast digital scope (Tektronix TDS7000, 1.5 GHz analog bandwidth, 5 GS/s). An example is shown in Fig. 9.15a. The amplitude of the motion has a maximum at half of the synchrotron period (about 300 turns in Fig. 9.15a), and is proportional to Q'. Chromaticity found by that method agrees with the RF method to within ~±0.5 unit. Another fast and even less destructive technique is to take advantage of the superb accuracy and precision of the Tune Tracker for the RF method. Figure 9.15b shows BTF function measurements with the TT for different RF frequencies—again, chromaticity can be found from the tune shift as $Q' = -dQ/(df_{RF}/f_{RF})/\eta$, where $\eta = 0.00283$ is the Tevatron lattice momentum compaction factor. The tune measurement accuracy is better than 0.0001 with 3 Hz bandwidth, resulting in Q' accuracy of about 0.2 units. There are systematic Q' differences of about 0.5 unit between the three methods which are due to the second-order chromaticity induced by octupoles.

9.11 Beam Profile Monitors

9.11.1 Flying Wires

The so-called Flying Wires [33] have been the main source of determining transverse emittances and profiles of the protons and antiprotons. There are three flying wire cans in the Tevatron: one horizontal and one vertical at a low dispersion area,

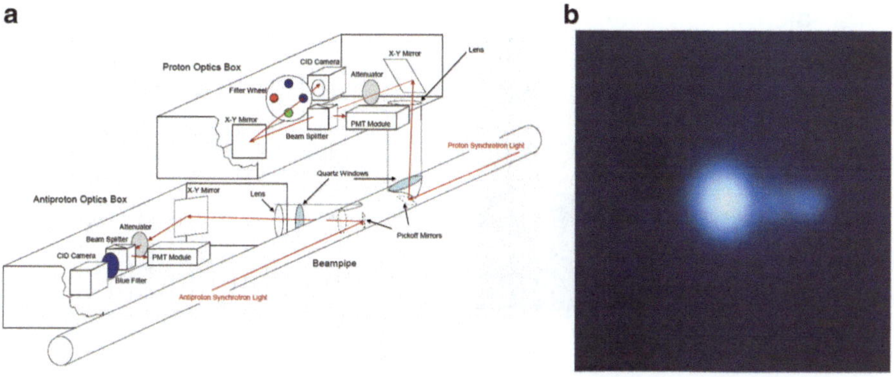

Fig. 9.16 (**a**) (*left*) 3D view of the SyncLite systems; (**b**) (*right*) Image of the synchrotron light emitted by 980 GeV antiprotons; the peak on the left is the dipole edge radiation which is used for beam size calculations, the tail on the right is radiation from the body of the magnet which is omitted in the image analysis [34]

and one horizontal at a high dispersion location. The original flying wire cans had 33 μm diameter carbon fibers, but those thick wires caused high loss spikes in the experiments when they were used during HEP stores and therefore were of limited use. Since the original fibers were changed to 5 μm, the wires are used routinely in every stage of operation. Wires are flown following most proton injections, after each pbar injection, several times during ramp, squeeze and halo removal (scraping), and every hour during HEP stores. The flying wires provide the rms normalized emittance measurements with about 0.2 π mm mrad uncertainty. Uncertainties in the lattice parameters at those locations were a major systematic error for the emittance measurement. Similar Flying Wires systems are also in daily operation in the Main Injector (one vertical and one horizontal), and in the Recycler ring (one vertical and one horizontal).

9.11.2 Synchrotron Light Monitor

The synchrotron radiation emitted by protons and antiprotons from the edges of the Tevatron dipoles is enhanced in the blue optical region, one can point a telescope at it and produce a transverse image of the beams. The Synclite system [34] does this and thus offers a nondestructive method for measuring the transverse beam emittances. For the Tevatron, synchrotron radiation is non-negligible only when the beam energy gets above 600–700 GeV. The physical layout of the Synclite system is shown in Fig. 9.16 and comprises a vacuum insertion mechanism to which is attached a pickoff mirror, and a light tight box containing the necessary optics for forming an image. The optical path contains a single lens for focusing, motorized

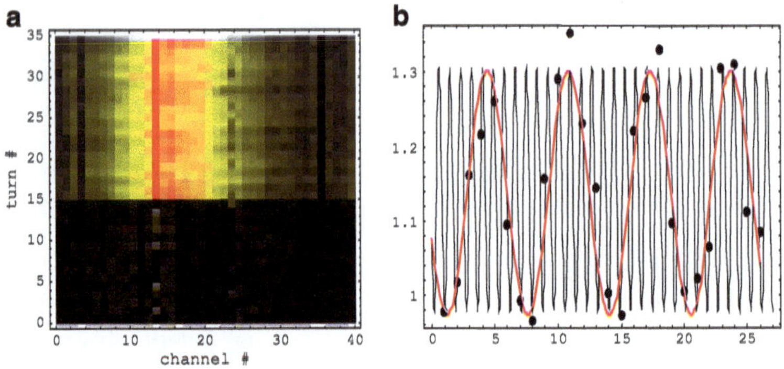

Fig. 9.17 (**a**) (*left*) TBT vertical profile of a coalesced proton bunch at injection using the IPM. The bunch was injected at turn #15 on this plot. This horizontal coordinate corresponds to ≈1 cm. (**b**) (*right*) Measured profile widths calculated from the turn-by-turn data. *Black dots* are IPM data, *red line* is a fit to $\cos((2\nu_y - 41)n)$, *black line* is a fit to $\cos(2\nu_y n)$ [1]

mirrors for positioning the image, a 440/10 nm band pass filter (400/40 nm for pbars), and a gated, Image Intensified CID (Charge Injection Device) camera for obtaining the image. The image is retrieved from a frame grabber in a PC running Windows XP and LabVIEW. The image has its background subtracted before the beam sizes are extracted from fits to the horizontal and vertical profiles in a window around the peak. The fits are simple Gaussians with linear baselines. The values of the fits are then corrected for nonlinearities in the intensified camera response and for theoretical distortions in the image (most notably, the broadening of the peak due to diffraction—the impact is about 100 μm on a beam size which for pbars is in the range of 200–400 μm—and the impact from the longitudinal extension of the source). Comparisons have been made with the profile measurements from the flying wires and reported emittances agree within ±10 % [35].

9.11.3 Ionization Profile Monitors (IPM)

The Tevatron IPM's [36] were designed to be single turn, single bunch devices, in order to resolve any turn-by-turn beam size oscillation caused by injection mismatch—see Fig. 9.17—as well as to allow noninvasive, nearly continuous measurement of the beam profiles at all stages of operation. The IPM data from multiple turns can be combined in software to create an average beam profile from which the emittance can be calculated. The IPM measures ionization electrons collected using a parallel electric and magnetic field. A localized pressure bump obtained by controlled injection of pure N2 is required to obtain single turn sensitivity. The magnets and gas injection is left on continuously, while the high voltage sweep field is pulsed only for acquisitions. The detector granularity is

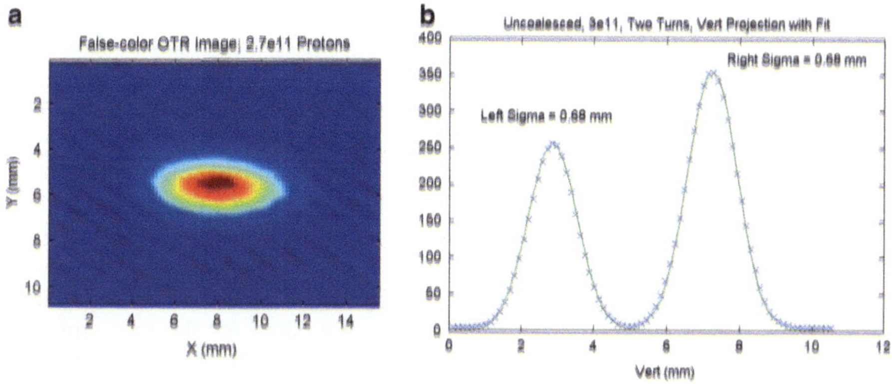

Fig. 9.18 (**a**) (*left*) Transverse 2D bunch profile as measured by the OTR. (**b**) (*right*) Vertical profiles of a single proton bunch from the OTR on two consecutive turns. The second turn profile is offset from the first, and the images are summed together by the slow camera. Note that over the two first turns, the OTR does not show evidence of the quadrupole oscillations seen in the IPM. However, from the IPM data only a 5 % effect is expected between these two turns [1]

1/4 mm, and the readout uses electronics borrowed from HEP experiments, enabling close to single electron sensitivity. Similar IPM systems but without bunch-by-bunch diagnostics capability are used for the emittance measurements in the Main Injector and Booster.

9.11.4 Optical Transition Radiation Detector

The Optical Transition Radiation (OTR) instrument [37] was installed close to the Tevatron IPM and used as a cross-check of the IPM at injection; we cannot leave the OTR foils inserted for routine operation as they tolerate only a few turns of beam. The OTR could also in principle be used for multi-turn acquisition, but the camera that is currently used does not have enough time resolution. However, by injecting a mis-steered beam, non-overlapping profiles from the two first turns can be obtained (see Fig. 9.18).

9.12 Special BPMs

9.12.1 Injection Beam-Line Tuners

To reduce emittance dilution caused by mis-steering at injection, dipole corrector magnets are adjusted by a beam-line tuner system, based on measurements of turn-by-turn orbit positions from directional stripline pickups. The 1 m long striplines are separated by an 83 mm gap, have ≈30 dB directionality and 0.65 dB/mm sensitivity. Measurements from one injection are used to make corrections for the subsequent shot, and usually injection offsets can be reduced from 1 mm to less than ¼ mm. Measurements of synchrotron oscillations can be used to correct energy and RF phase differences between the Main Injector and Tevatron. In addition, the tunes and coupling at injection can also be extracted from the stripline signals. Two different beam-line tuners have been used. In the first, a Tektronix TDS7104 oscilloscope digitizes the sum and difference signals from the striplines, and the embedded PC performs the signal processing. That system has been used only for closure of antiproton injections; it was too slow for the more frequent proton injections. A faster system, based on the Struck SIS3300 digital receiver module, provides 20–40 μm position measurements for 1,000 turns [38]. The digitized data is transferred to a PC which performs digital down-conversion at 30 MHz and calculates the positions and time-of-arrival for the transferred bunches. This system can be used for both proton and antiproton injections. During stores, it can also continuously store position data into a circular buffer that is stopped on a beam abort. The buffered data can be used in the postmortem diagnosis of a lost store.

9.12.2 AC Dipole

An AC dipole is a magnet with a field oscillating close to the betatron frequency. By exiting the beam close to a resonance, a relatively small field can drive a large amplitude oscillation, and keep the oscillation amplitude constant over many turns—as demonstrated in Fig. 9.19—allowing precise measurements of the beam response without any emittance blow-up. Dedicated AC dipole magnets have been built before, e.g. for RHIC. In the Tevatron, an existing vertical "pinger" magnet was converted to a resonant circuit using an external capacitor bank, and driven by a commercial high-power audio amplifier. While very successful R&D project, both demonstrating the use of commercial audio equipment for this purpose and furthering the understanding of the analysis of driven oscillation data to extract the beam optics [39, 40], much like the kicked beam method, the AC dipole in the Tevatron has not been used much in operation. This is partially due to the fact that the system only measures one plane, but to a larger degree it is due to the success of the orbit response method. However, the tests performed in the Tevatron were carried over to the LHC, where a very similar system (based on audio amplifiers and resonant kicker magnets) is the prime method for measuring optics [41].

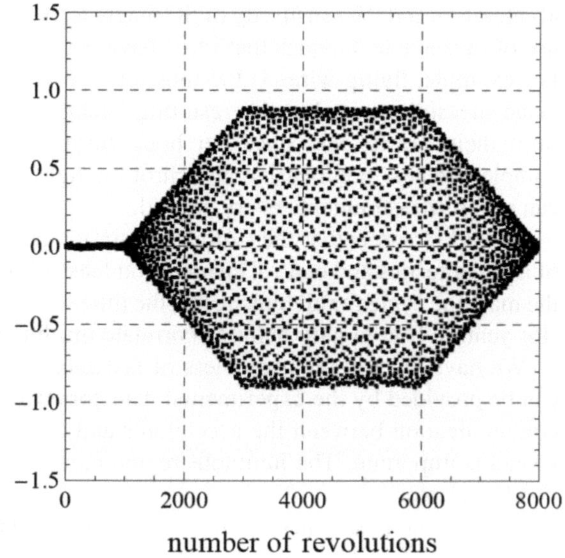

Fig. 9.19 Measured turn-by-turn oscillations of the beam driven by an AC dipole in the Tevatron. In contrast to a kicked beam, the driven particles do not lose the coherence. By adiabatically ramping up and down the amplitude of the AC dipole's field, an excitation is produced without increasing the beam size [39]

9.13 Discussion on the Beam Diagnostics for Tevatron Collider Accelerators

Development of many new diagnostic tools for the Tevatron Collider accelerators was needed to provide insights into serious issues of coherent instabilities, beam losses, and beam–beam interactions. As a result, almost two dozen various instruments were either developed or significantly improved, and that eventually paid off in the integrated luminosity delivered to the CDF and D0 detectors.

There are several lessons learned during this campaign. First, we realized the importance of multiple instruments for cross-checking and cross-calibrating one another. For example, there are several instruments to measure beam intensity: DC Current Transformer (DCCT), FBI, and SBD. The DCCT is the most precise but it has limited application range, e.g. it cannot report individual bunch intensities. The FBI and SBD are not as precise but they are really multi-functional, operating on a bunch-by-bunch basis, and calibrating them within 1 % of the DCCT made them trustworthy and very useful in operations. In addition, the fast LPM was cross-checked with the SBD. Three tune monitors—21 MHz Schottky (used for injection tune-up), 1.7 GHz Schottky (most versatile) detectors, and Tune Tracker (the fastest and most precise of the three)—are employed in operations for different tasks after being carefully cross-calibrated. A lot of effort over many years was needed to bring the three emittance measurement tools—Flying Wires (FWs), Synchrotron Light Monitor (SyncLite), and 1.7 GHz Schottky detector—into satisfactory agreement; currently, they agree within ±5 %.

Another lesson is the need for noninvasive beam diagnostics for nearly continuous monitoring of beam parameters. The lack of any natural damping in proton accelerators and the sensitivity of SC magnets to beam losses (quenches) restrict the use of invasive techniques that often have better resolution than noninvasive ones. For example, flying wires is the most precise and understood technique for emittance measurements, but the resulting background spikes and emittance growth limit their use to only once per hour during high-energy collision stores. The complementary, noninvasive Synchrotron light monitor and 1.7 GHz Schottky can report measurements every second.

A third lesson is that the Collider operation team needs fast data collection rate of all diagnostics and control channels (at least 1 Hz) for all channels at all stages of the machine cycle for all bunches all the time—and the data should be saved forever (for years)! That greatly helps to correlate machine behavior now with the past.

We have learned the usefulness of fast access to beam-related information that can be provided by the experimental detectors (CDF and D0, in our case), so good communication between the accelerator and experiment control systems and personnel is important. The luminous region parameters information noted above is a good example.

We also have benefited from help and ideas from other groups and laboratories that have expertise in a number of specific areas: for example, Fermilab's Computing Division experts took a leading role in development of DAQ for the Tevatron BPM upgrade; FNAL Particle Physics Division leads Tevatron BLM upgrade and provides luminous region analysis data (β^* monitors); Berkeley Lab contributed in the development of the MCP-PMT based Abort Gap Monitor, etc.

And finally, we realized that constructing a new instrument is fast compared to the time needed to make it "fully operational," i.e. satisfactory to operators and physicists. A lot of effort went into the debugging, tune-up, cross-calibration, and "polishing" of beam diagnostics. So, we teamed up diagnostics developers and users (physicists and engineers) from the very start of instrument development until the end of its commissioning. Such teams of two to four were very efficient in developing or overhauling about two dozen beam diagnostics instruments for the Tevatron Run II.

References

1. R.S. Moore, A. Jansson, V. Shiltsev, JINST **4**, P12018 (2009)
2. R.E. Shafer, R.C. Webber, T.H. Nicol, IEEE NS-28 **3**, 3390 (1981)
3. G. Annala et al., JINST **6**, T11005 (2011)
4. Yu. Alexahin, E. Gianfelice-Wendt, JINST **6**, P10006 (2011)
5. A. Petrenko, A. Valishev, V. Lebedev, Rev. ST Accel. Beams **14**, 092801 (2011)
6. V. Ranjbar, in *Proceedings of the 2005 IEEE Particle Accelerator Conference*, Knoxville, 2005, p. 1353
7. J.T. Volk et al., JINST **7**, P01004 (2012)
8. J. Slaughter et al., in *Proceedings of the 2003 IEEE Particle Accelerator Conference*, Portland, 2003, p. 1763
9. M. Jones et al., Preprint FNAL-TM-2172 (2002)
10. D. Boussard, CERN Accelerator School, CERN-Report 87-03 (1987), pp. 417–445
11. V. Parkhomchuk, D. Pestrikov, Sov. Phys. Tech. Phys. **25**(7), 818 (1980)

12. D. Martin et al., in *Proceedings of the 1989 IEEE Particle Accelerator Conference*, Chicago, 1989, pp. 1483 and 1486
13. R. Pasquinelli et al., in *Proceedings of the 2003 IEEE Particle Accelerator Conference*, Portland, 2003, p. 3068
14. R. Pasquinelli, A. Jansson, Phys. Rev. ST Accel. Beams **14**, 072803 (2011)
15. A. Semenov et al., *in Proceedings of the 2007 IEEE Particle Accelerator Conference*, Albuquerque, 2007, p. 3877; V. Kamerdzhiev et al., Preprint FERMILAB-CONF-08-139-AD-APC (2008)
16. V. Shiltsev, G. Stancari, A. Valishev, JINST **6** P08002 (2011)
17. C.Y. Tan, NIM A **557**, 615 (2006)
18. R. Bartolini et al., Part. Accel. **56**, 167–199 (1996)
19. Yu. Alexahin, E. Gianfelice-Wendt, W.L. Marsh, in *Proceedings of the International Particle Accelerator Conference*, Kyoto, 2010, p. 1179
20. T. Bolshakov et al., in *Proceedings of the 2005 IEEE Particle Accelerator Conference*, Knoxville, 2005, p. 1099; also T. Bolshakov, E.S. McCrory, in *Proceedings of the European Particle Accelerator Conference*, Edinburgh, 2006, p. 3101
21. J. Crisp, B. Fellenz, JINST **6**, T11001 (2011)
22. R. Thurman-Keup et al., JINST **6**, T10004 (2011)
23. J.P. Carneiro et al., in *Proceedings of the 2005 Particle Accelerator Conference*, Knoxville, 2005, p. 1428
24. X.L. Zhang et al., Phys. Rev. ST Accel. Beams **11**, 051002 (2008)
25. J. Steimel et al., in *Proceedings of the 2003 IEEE Particle Accelerator Conference*, Portland, 2003, p. 48
26. R. Thurman-Keup, in *Proceedings of the IEEE Particle Accelerator Conference*, Knoxville, 2005, p. 2440
27. S. Pordes et al., in *Proceedings of the 2003 IEEE Particle Accelerator Conference*, Portland, 2003, p. 2491
28. S. Pordes et al., in *Proceedings of the 2005 IEEE Particle Accelerator Conference*, Knoxville, 2005, p. 1362
29. R.E. Shafer et al., in *Proceedings of the XII International Conference on High Energy Accelerators*, Batavia, 1983, p. 609
30. A. Baumbaugh et al., JINST **6**, T11006 (2011)
31. P. Ivanov et al., in *Proceedings of the 2003 IEEE Particle Accelerator Conference*, Portland, 2003, p. 3062
32. V. Ranjbar, P. Ivanov, Phys. Rev. ST Accel. Beams **11**, 084401 (2008)
33. J. Gannon et al., in *Proceedings of the 1989 IEEE Particle Accelerator Conference*, Chicago, 1989, pp. 68–70; also in J. Zagel et al., in *Proceedings of the 1991 IEEE Particle Accelerator Conference*, San Francisco, 1991, pp. 1174–1176
34. R. Thurman-Keup et al., JINST **6**, T09003 (2011)
35. J. Zagel et al., in *Proceedings of the Beam Instrumentation Workshop BIW08*, Tahoe City, 2008, p. 282
36. A. Jansson et al., in *Proceedings of the Beam Instrumentation Workshop BIW06* (Fermilab), AIP Conference Proceedings 868, 2006, ed. by T. Meyer, R. Webberp, p. 159
37. V. Scarpine et al., in *Proceedings of the Beam Instrumentation Workshop*, AIP Conference Proceedings 868, 2006, ed. by T. Meyer, R. Webber, p. 473
38. W. Schappert et al., in *Proceedings of the 2003 IEEE Particle Accelerator Conference*, Portland, 2003, p. 2494
39. R. Miyamoto et al., in *Proceedings of the 2007 IEEE Particle Accelerator Conference*, Albuquerque, 2007, p. 3868
40. R. Miyamoto et al., Phys. Rev. ST Accel. Beams **11**, 084002 (2008)
41. R. Thomas et al., Phys. Rev. ST Accel. Beams **13**, 121004 (2010)

Appendix: Ph.D. Theses on Accelerator Physics at the Tevatron

1. M. Syphers (1987), University of Illinois, Chicago
   An improved 8-GeV beam transport system for the Fermi National Accelerator Laboratory.
2. Leonid Sagalovsky (1989), University of Illinois in Urbana, Champagne
   Third-order charged particle beam optics.
3. Nikilitsa Merminga (1989), University of Michigan
   A study of nonlinear dynamics in the Tevatron.
4. Xiao-Qing Wang (1991), Illinois Institute of Technology
   A study of longitudinal coherent effects of unbunched beams near transition in the Fermilab accumulator.
5. Peilei Zhang (1991), Houston University
   A study of tunes near integer values in hadron colliders.
6. Steven Stahl (1991), Northwestern University
   Beam dynamics in the Fermilab booster in the presence of space charge.
7. John Palkovic (1991), University of Wisconsin
   Gabor lens focusing and emittance growth in a low-energy proton beam.
8. Katherine Harkay (1993), Purdue University
   A study of longitudinal instabilities and emittance growth in the Fermilab booster synchrotron.
9. Ping Zhou (1993), Northwestern University
   A study of ion trapping and instability in the Fermilab anti-proton accumulator.
10. Todd Satogata (1993), Northwestern University
    Nonlinear resonance islands and modulational effects in a proton synchrotron.
11. William Graves (1994), University of Wisconsin
    Measurement of transverse emittance in the Fermilab booster.
12. Xian-Ping Lu (1994), University of Colorado
    Study of a longitudinal coupled bunch instability in the Fermilab main ring.
13. Ping-Jung Chou (1995), Northwestern University
    The nature of transverse beam instabilities at injection in the Fermilab main ring.

V. Lebedev and V. Shiltsev (eds.), *Accelerator Physics at the Tevatron Collider*,
Particle Acceleration and Detection, DOI 10.1007/978-1-4939-0885-1,

14. Donna Siergiej (1995), University of New Mexico
Beam-beam interaction effects in the Fermilab collider.
15. Linda Spentzouris (1996), Northwestern University
Direct measurement of diffusion rates in high energy synchrotrons using longitudinal beam echoes.
16. David Olivieri (1996), University of Massachusetts
A dynamic momentum compaction factor lattice for improvements to stochastic cooling in storage rings.
17. Oleg Krivosheev (1998), Tomsk Polytechnic University, Russia
Object oriented integrated system for beam induced energy deposition simulations for Tevatron and upgrades.
18. Christina Dimopoulou (2001), CERN
Experimental tests of a high perveance gun at Fermilab.
19. Kip Bishofberger (2005), UCLA
Tevatron beam-beam compensation.
20. Ludovic Nicolas (2005), University of Glasgow, UK
Radiation environment simulations at the Tevatron, studies of the beam profile and measurement of the Bc meson mass.
21. Sergei Seletskiy (2005), Rochester University
Attainment of electron beam suitable for medium energy electron cooling.
22. Robert Zwaska (2005), University of Texas, Austin
Accelerator systems and instrumentation for the NuMI neutrino beam.
23. Xiaobiao Huang (2005), Indiana University
Beam diagnosis and lattice modeling of the Fermilab booster.
24. Pavel Snopok (2007), Michigan State University
Capture of a large phase space beam.
25. Phil Yoon (2007), University of Rochester
Error-induced beam degradation in Fermilab's accelerators.
26. Alexei Poklonsky (2008), Michigan State University
Optimization and control of Tevatron parameters.
27. Ryoichi Miyamoto (2008), University of Texas, Austin
AC dipole diagnostics of Fermilab's Tevatron.
28. Valentina Previtali (2010), University of Milano, Italy
Performance evaluation of a crystal enhanced collimation system for the LHC.
29. Dan McCarron (2010), Illinois Institute of Technology
Measurement and simulations of intensity dependent effects in the Fermilab booster synchrotron.

Index

V. Lebedev and V. Shiltsev (eds.), *Accelerator Physics at the Tevatron Collider*, Particle Acceleration and Detection, DOI 10.1007/978-1-4939-0885-1,

C

MIX
Papier aus verantwortungsvollen Quellen
Paper from responsible sources
FSC® C105338

If you have any concerns about our products,
you can contact us on
ProductSafety@springernature.com

In case Publisher is established outside the EU,
the EU authorized representative is:
Springer Nature Customer Service Center GmbH
Europaplatz 3, 69115 Heidelberg, Germany

Printed by Libri Plureos GmbH
in Hamburg, Germany